Biotechnology & Genetic Engineering Reviews

Volume 11

BIOTECHNOLOGY AND GENETIC ENGINEERING REVIEWS

Biotechnology & Genetic Engineering Reviews

Volume 11

Editor:
MICHAEL P. TOMBS
Special Professor, University of Nottingham

Intercept

Andover

British Library Cataloguing in Publication Data
Biotechnology & genetic engineering reviews.—
Vol.11
1. Biotechnology—Periodicals

A CIP catalogue record for this book is available from the British Library.
ISBN 0 946707 77 4
ISSN 0264–8725

Published in December 1993 by Intercept Limited,
PO Box 716, Andover, Hants SP10 1YG, England.

Filmset in 'Linotron' Times by
Ann Buchan (Typesetters), Shepperton, Middlesex.
Printed by Athenaeum Press, Ltd, Newcastle-upon-Tyne.

Contributors

C.W. BAMFORTH, *BRF International, Lyttel Hall, Coopers Hill Road, Nutfield, Surrey, RH1 4HY, UK*

N.B. CAROZZI, *Ciba Agricultural Biotechnology Research Unit, 3054 Cornwallis Road, Research Triangle Park, NC 27709–2257, USA*

W. COCKBURN, *Plant Antibody Group, Department of Botany, University of Leicester, Leicester LE1 7RH, UK*

C. COLACO, *Quadrant, Maris Lane, Trumpington, Cambridge CB2 2SY, UK*

R.S. CRESPI, *Blackbird Cottage, Walton Avenue, Bognor Regis, Sussex PO21 1NP, UK*

T.C. CURRIER, *Ciba Agricultural Biotechnology Research Unit, 3054 Cornwallis Road, Research Triangle Park, NC 27709–2257, USA*

S.V. EVOLA, *Ciba Agricultural Biotechnology Research Unit, 3054 Cornwallis Road, Research Triangle Park, NC 27709–2257, USA*

G.K. FARBER, *Departments of Chemistry and Molecular Biology and Center for Biomolecular Structure and Function, The Pennsylvania State University, 152 Davey Laboratory, University Park, PA 16802, USA*

A.R. GANDECHA, *Plant Antibody Group, Department of Botany, University of Leicester, Leicester LE1 7RH, UK*

J.R.M. HAMMOND, *BRF International, Lyttel Hall, Coopers Hill Road, Nutfield, Surrey RH1 4HY, UK*

S.E. HARDING, *National Centre for Macromolecular Hydrodynamics, University of Nottingham, School of Agriculture, Sutton Bonington LE12 5RD, UK*

M.G. KOZIEL, *Ciba Agricultural Biotechnology Research Unit, 3054 Cornwallis Road, Research Triangle Park, NC 27709–2257, USA*

P.A. LAZZERI, *AFRC Institute of Arable Crops Research, Rothamsted Experimental Station, Harpenden, Hertfordshire AL5 2JQ, UK*

J.B. LORENS, *Department of Biochemistry and Molecular Biology, University of Bergen, Årstadveien 19, N-5009 Bergen, Norway*

R. MALE, *Center of Biotechnology, University of Bergen, HIB, N-5020 Bergen, Norway*

A.H. NERLAND, *Norbio AS, HIB, Thormåhlensgt. 55, N-5008 Bergen, Norway*

Y.M. NEWMAN, *AFRC Institute of Food Research, Norwich Research Park, Colney Lane, Norwich, NR4 7UA, UK*

M.R.L. OWEN, *Plant Antibody Group, Department of Botany, University of Leicester, Leicester LE1 7RH, UK*

M. RADMAN, *Institut Jacques Monod, 2 Place Jussieu, Tour 43, 75251 Paris Cedex 05, France*

S.G. RING, *AFRC Institute of Food Research, Norwich Laboratory, Norwich Research Park, Norwich NR4 7UA, UK*

E.Y. SHAMI, *Hybrisens Ltd and Department of Biology, York University, Keele St Campus, Farq Building 104, Toronto, Ontario M3J 1P3, Canada*

P.R. SHEWRY, *Department of Agricultural Sciences, University of Bristol, AFRC Institute of Arable Crops Research, Long Ashton Research Station, Long Ashton, Bristol BS18 9AF, UK*

E. SLINDE, *MATFORSK, Norwegian Food Research Institute, Osloveien 1 N-1430 As, Norway*

B.L. STODDARD, *Fred Hutchinson Cancer Research Center, Division of Basic Sciences, Program in Structural Biology, 1124 Columbia St, Seattle, WA 98104, USA*

R.K. STRONG, *Fred Hutchinson Cancer Research Center, Division of Basic Sciences, Program in Structural Biology, 1124 Columbia St, Seattle, WA 98104, USA*

R. WAGNER, *Institut Jacques Monod, 2 Place Jussieu, Tour 43, 75251 Paris Cedex 05, France*

G.W. WARREN, *Ciba Agricultural Biotechnology Research Unit, 3054 Cornwallis Road, Research Triangle Park, NC 27709–2257, USA*

G.C. WHITELAM, *Plant Antibody Group, Department of Botany, University of Leicester, Leicester LE1 7RH, UK*

Contents

1

Heterologous Protein Production in Transgenic Plants

GARRY C. WHITELAM, BILL COCKBURN, ATUL R. GANDECHA AND MERAN R.L. OWEN

Plant Antibody Group, Department of Botany, University of Leicester, Leicester, UK

Introduction

For the purposes of this review, a heterologous protein is defined as being the product of a gene isolated from a species different from the receiving organism and is further restricted to proteins which do not exist as closely similar analogues in the donor and receiver species. Justifications for the production of such heterologous proteins in plants fall into two basic categories: (1) 'factory farming', in which the unique properties of plants can be exploited in the production of desirable proteins, and (2) 'phenotype modification', in which the heterologous protein may be used to alter the characteristics of the plant itself.

Factory farming

It is difficult to envisage a protein production system, more economical and more amenable to large-scale production than plant agriculture. All that is required is a suitable plant, sunlight, mineral salts from the soil (or fertilizers) and water. Furthermore, modern agricultural machinery allows efficient harvesting of large quantities of plant material. Traditional agriculture exploits these characteristics of plants in the production of immense quantities of materials including, for example, the proteins of wheat and soybean.

All of these advantages apply equally to the production of heterologous proteins and, particularly where very large-scale production is contemplated, the plant system represents a realistic alternative to microbial or animal

Abbreviations: BiP, binding protein; CaMV, cauliflower mosaic virus; cDNA, complementary DNA; CDR, complementarity determining region; C_H, constant domain of the heavy chain; CL, constant domain of the light chain; EPSP synthase, enolpyruvylshikimate-3-phosphate synthetase; ER, endoplasmic reticulum; HAMA, human anti-mouse antibody; HSA, human serum albumin; IgG, immunoglobulin G; IgM, immunoglobulin M; mRNA, messenger RNA; PCB, polychlorinated biphenyl; PCR, polymerase chain reaction; TMV, tobacco mosaic virus; V_H, heavy-chain variable region; V_L, light-chain variable region.

Biotechnology and Genetic Engineering Reviews — Vol. 11, December 1993
0264–8725/93/11/1–29 $20.00 + $0.00 © Intercept Ltd, P.O. Box 716, Andover, Hampshire SP10 1YG, UK

cell-based systems for heterologous protein production. Furthermore, as eukaryotes, plants are capable of the post-translational processing of proteins of eukaryotic origin (plant or animal) which may be essential to their proper functioning and which may present problems in bacterial and other prokaryotic systems. The ultimate determinant of the utility of an expression system is, of course, the overall cost of production. In this context, the existence of an established efficient infrastructure and facilities for the growth, harvesting and processing of plants will represent a significant component in the cost equation.

Although it may well turn out to be of transient concern, any programme aiming to exploit transgenic technology must ultimately come to terms with consumer acceptability. Here again plants offer advantages in that the general public appears to be less concerned about the production and exploitation of transgenic plants than transgenic animals or micro-organisms. Presumably the ease with which the layman can identify with experimental animals and the association of micro-organisms with disease, along with the benign though alien image of plants, are significant factors here.

High-value proteins for which there is a sustained demand and limited supply by conventional means represent the most attractive prospects for factory farming. Examples of such proteins which plants have already been shown to be capable of synthesizing are human serum albumin and leu-enkaphalin (Krebbers and Vanderkerckhove, 1990; Sijmons *et al.*, 1990). Plants have also been shown to be capable of expressing functional antibodies and their derivatives (discussed in detail below), and these important therapeutic and diagnostic reagents represent another of the many potential candidates for factory farming.

Additional to the ability to synthesize and post-translationally modify such proteins is, of course, the requirement for yields of products which are both sufficient to be economical and are amenable to extraction and downstream processing. These two crucial requirements are linked to the site of accumulation of the product and are discussed in detail below.

An intriguing suggestion which illustrates the potential sophistication of the heterologous protein farming approach is the suggestion that a microbial α-amylase, an enzyme involved in processing potato starch, could be secreted to the potato apoplast where it has no access to starch in the living tissue. After-harvest processing, however, would result in substrate and enzyme being brought together, thus obviating the step in conventional processing in which exogenous amylase is added (Pen *et al.*, 1993).

An attractive possibility which may be viewed as a variant of factory farming is to use plant expression as a means of distributing and delivering oral vaccines. This currently requires specialist knowledge and often specialized storage including refrigeration, often creating difficulties in the third world. It should be possible to express the appropriate vaccine proteins in plants and take advantage of the plant's ability to conveniently package proteins in seeds as a means of simplifying storage and distribution. Expression of vaccine proteins in fruits or vegetables may also be envisaged as perhaps the ultimate arrangement for distribution and administration of the vaccine. Only seeds (usually very robust and long-lasting) need to be transported and distributed and even in the most hard-pressed countries, provided that the seeds themselves are inedible, the most likely outcome is that plants grown from them will be eaten.

It is hoped that this brief survey of but a few of the potential applications of factory farming will serve to show that where large-scale production is required, plants are worthy of consideration for the expression of any high-value protein.

Phenotype modification

The range of strategies and possibilities for the modification of the plant phenotype through the expression of heterologous proteins is very broad indeed, and encompasses virtually every activity of the plant. A conceptually straightforward application is in the correction of the well-documented deficiencies (from the point of view of human nutrition) of existing crop storage proteins, legumes in sulphur amino acids and tryptophan, and cereals in lysine, threonine and tryptophan. These deficiencies are amenable to correction by transferring genes coding for proteins containing balancing quantities of these amino acids and by the use of appropriate promoters and targeting signals to ensuring that they accumulate in sufficient amounts in the seed storage tissue (Altenbach and Simpson, 1990).

Reversible male sterility is a desirable agronomic trait which is valuable in the production of defined hybrids in outbreeding crops. The expression of the gene encoding the *Bacillus amyloliquefaciens* RNase, Barnase, in the tapetum of developing anthers, which prevents the anther from producing functional pollen but which can be inactivated by the expression of a second heterologous protein, barstar, when required, achieves this aim (Mariani *et al.*, 1990, 1992).

The expression of heterologous enzymes feature in numerous strategies aimed to improve crop value through simple increase in yield of desired product and also through modification of the nature of the product. Changes in assimilate distribution may be brought about by the expression of heterologous proteins which catalyse key steps in a pathway. The accumulation of glucose and fructose rather than sucrose as a result of the expression of yeast invertase in the apoplast of potato indicates that such approaches may be realistic (Sonnewald, von Schaewen and Willmitzer, 1993).

The accumulation of desirable natural products in plants which do not normally produce them is also feasible. The expression of bacterial fructosyl transferases in tobacco results in the accumulation of polyfructans in tobacco, a plant which does not normally accumulate these commercially useful polymers (Smeekens *et al.*, 1991). This example serves to illustrate the potential complexity which may attend execution of an apparently simple strategy. In fructan accumulation, the site to which the transgenic enzyme is targeted plays a vital part in the outcome of expressing the heterologous gene. When targeted to the cytosol or apoplast, the transgenic enzyme (presumably through accumulation of fructans) is damaging to the plant. Targeting the enzyme to the vacuole, the normal site of fructan metabolism and accumulation, is much less damaging.

Thus, an understanding of the spatial dimension and targeting to appropriate locations is likely to be very significant in the successful implementation of strategies involving heterologous protein production. The characteristics of fructan accumulation in transgenic tobacco also demonstrates that attention should be paid to temporal and developmental aspects of heterologous protein expression. Sink leaves which receive carbon as sucrose store it as fructan, whereas source leaves which accumulate carbon through photosynthesis rather than transport, accumulate starch (in the chloroplasts) as well as fructans.

Bioconversions resulting in the accumulation of compounds not naturally accumulating in any plant species are also possible. Through the expression of genes encoding two bacterial enzymes, transgenic *Arabidopsis* plants have been produced which convert acetoacyl-coenzyme A (a normal metabolite) to polyhydroxybutyrate, a compound not naturally accumulating in plants and which forms the basis for a biodegradable thermoplastic (Poirier *et al.*, 1992). Thus, there is the possibility of the development of entirely novel crops producing a range of materials not even limited by the plants' already remarkable capacity for chemical gymnastics.

In addition to the expression of heterologous enzymes which participate directly in bioconversions, there is also the possibility of modulating the activities of existing metabolic pathways through expression of proteins which influence the activity of pathway enzymes. Enzymes which synthesize or degrade effector molecules such as fructose 2,6-bisphosphate or antibody derivatives which bind to an active site would fall into this category. The activities of regulatory proteins are also accessible to modulation through the expression of antibody derivatives (immunomodulation), and plants expressing an antibody derivative directed against the photoregulatory protein phytochrome show modified phytochrome-mediated photomorphogenic responses (Owen *et al.*, 1992a).

An important area where the expression of heterologous proteins will certainly find application is in plant protection. The expression of proteins which interfere in some way with pathogenicity can extend potential defence mechanisms beyond the plant's normal genetic repertoire, and therefore potentially outside the range of defence mechanisms to which particular pathogens have evolved responses. From the agronomic point of view, heterologous protein-mediated protection (or indeed any modification of plant activities) is particularly attractive because it involves only minor changes to the plant, often the addition of a single heterologous protein gene. Thus the important characteristics of cultivars for which commercial demand and growth, harvesting and processing parameters are already established are likely to remain unchanged.

It is possible to envisage heterologous protein-mediated protection against all manner of pathogens and in a few cases the strategies have already been shown to be successful. For example, resistance to viruses can result from the expression of viral coat proteins and resistance to insects by expression of the

Bacillus thuringiensis toxins and proteinase inhibitors (Sturtevant and Beachy, 1993; Perlak *et al.*, 1991; Hilder *et al.*, 1987). Recently, potato plants expressing a gene encoding T4 lysozyme have been shown to possess increased resistance to a bacterial pathogen (Düring *et al.*, 1993).

Protection of crop plants against weeds has been achieved by the introduction of herbicide resistance through the expression of heterologous proteins. For instance, resistance to glyphosate, which inhibits the enzyme enolpyruvylshikimate-3-phosphate synthetase (EPSP synthase) and so blocks aromatic amino acid synthesis, has been obtained by transformation with a *Salmonella typhimurium* gene that encodes a mutated, glyphosate-insensitive EPSP synthase (Shah *et al.*, 1986). Such strategies are attractive to agrochemical companies, since they tie the customer to both specific herbicide and resistant plants. They are not so attractive to environmentalists, since they are perceived as being likely to encourage increased use of herbicides.

It seems certain that there will be many successful applications of the largely speculative heterologous protein-mediated strategies outlined above. Realization of the full potential of the approach will require further information on the factors that control heterologous protein production in plants. We have elected to centre our discussion of these factors around antibodies and their derivatives in order to provide a coherent exploration of the subject which is nevertheless relevant to the expression of any heterologous protein in plants. Antibodies, and their derivatives, are prime candidates for exploitation in heterologous protein farming and are also involved in a wide range of strategies for phenotypic modification. In addition, antibodies are complex multimeric glycosylated proteins with measurable *in vitro* function. Thus the expression of antibodies in plants provides a paradigm covering all aspects of heterologous protein expression.

Factors controlling heterologous protein production in plants

WHOLE PLANT FACTORS

Economics of protein farming

One of the key factors in determining whether crop plants can represent a viable heterologous system for protein farming is the overall cost of foreign protein production in plants, relative to the economics of protein production in alternative systems. The economic viability of protein farming in crop plants will depend upon the demand for, and commercial value of, the recombinant protein being produced. Where a high-value protein is needed in large quantities (hundreds of kilograms), then the high capacity and flexibility of agricultural production offers advantages over other production systems, such as microbial fermenters. In terms of biomass production, the growing of crops in the field is extremely cost-effective and can compete with any other system (Pen *et al.*, 1993). Crops that are already harvested and processed for some other, non-proteinaceous, primary product are particularly attractive as candidate sources of biomass for the production of heterologous proteins. Thus, crops such as starch potatoes and oilseed rape, which are grown for starch and oil, have been the focus of attention. Clearly, the use of a non-dedicated crop for protein production requires that the extraction of the primary product does not interfere with the extraction and downstream processing of the heterologous protein. The wet milling processes used in the extraction of starch from potatoes are considered to be compatible with the separation of proteins from the 'waste' liquid fraction. However, oilseed rape is less attractive, since the extraction of oil is performed at high temperatures that are only likely to be compatible with thermostable proteins.

The economic potential of developing plants as heterologous systems for protein production can be clearly demonstrated when, for example, the production of a heterologous protein in soybeans is considered. If soybeans could accumulate the protein to a level of approximately 1% of their total protein, then its production cost, prior to extraction from soybean meal, will be approximately $0.1 per gram. However, in the case of, say, pharmaceutical proteins, this low cost of production might be offset by the high cost of meeting the requirement of a very high level of purity. At present, the lack of established techniques and procedures for the downstream processing of heterologous proteins produced in crops makes plants less attractive than competitive systems, such as bacterial fermentation systems and mammalian cell cultures, for the production of high-value immunotherapeutic and pharmaceutical proteins.

Transformability of major crop plants

In order to produce recombinant proteins on a commercial scale, a number of factors have to be taken into consideration, one of which is the ability to transform a suitable host crop species. Considerable effort has been directed to developing reliable transformation systems for a number of different plant species, and it is now possible to routinely transform many of the major crop species such as potato, tomato, oilseed rape and sugarbeet. Consequently, it is likely that it will be one of these crops that will be the first to be used as a protein factory. However, it seems certain that when a routine and reliable method becomes available for transforming monocots, the major cereal crops will also be used for this purpose.

A number of different techniques are available for transforming plants based either on a vector or a vector-free method. The most frequently used vector is *Agrobacterium tumefaciens* (reviewed in Klee, Horsch and Rogers, 1987). However, *A. tumefaciens* does have a rather narrow host range, limiting its use to a small number of crop species. As there has only been a limited amount of success with the *Agrobacterium*-mediated transformation of monocot species, vector-free systems or direct gene transfer methods which are not restricted to a particular host range, have attracted a considerable degree of attention. There are a number of different vector-free techniques, such as the chemical stimulation of protoplasts (Hain *et al.*, 1985; Krens *et al.*, 1982), microinjection (Griesbach, 1983; Reich, Iyer and Miki, 1986), electroporation (Lindsey and Jones, 1987) and microprojectile bombardment (Klein *et al.*, 1987), which have all shown some degree of success.

It is envisaged that since a number of major crop species can now be routinely transformed, and the routine transformation of the others is a realistic expectation, the choice of crop for recombinant protein production will largely depend on the ability to extract and process the recombinant protein economically from a suitable harvestable tissue.

Tissue specificity of protein accumulation

The site of protein accumulation will be an additional important factor in determining which plant species will be used as protein factories. Specific promoters can be used to drive transgene expression only in the harvested tissues and not in tissues that are by-products of the harvesting process. For example, in potatoes, the promoter from the gene encoding the major storage protein of the tuber, patatin, can be used to drive expression specifically in the tuber. Storage proteins, such as patatin and seed proteins, are normally accumulated to very high levels in a relatively pure form within the vacuole or protein body of specific harvestable tissues. Consequently, these sites of protein accumulation are attractive for targeting heterologous proteins. The production of bioactive peptides in the form of fusions to seed storage proteins represents an example of the exploitation of the seed protein body as a site that has evolved to accumulate and store protein (see below).

Containability, storage and harvesting

The containability of plant genetic material presents a unique problem in the development of plants as a heterologous system for recombinant protein production. Nevertheless, the problems of both physical containability and genetic containability are not insurmountable. Physical containability will rely upon secure systems for growing the crop, whereas genetic containability will probably necessitate the growing of male sterile lines and/or F1 hybrids.

One of the main advantages of developing plants as a heterologous system for recombinant protein production is that the methods of harvesting and storing plant material are already well established and consequently will require very little refinement. It is envisaged that the crops synthesizing recombinant proteins will be harvested in conventional manners and either removed for the immediate extraction of the recombinant protein or placed in storage. Methods for the long-term storage of many crops are well established, such as for cereal grains and potato tubers. The ability to store plant material uner conditions that require very little maintenance and supervision is an additional advantage in developing plants as a heterologous protein expression system. Additionally, the transportation of harvested plant material is also well established and in general is relatively simple and inexpensive. Consequently, crop production and processing facilities need not be located at a single site.

Downstream processing

The problems associated with the large-scale downstream processing of plant material for protein extraction have not previously been addressed. Consequently, the economic factors have only been investigated in hypothetical cases. The costs of manufacturing recombinant proteins in a few transgenic crops have been estimated and the calculations would indicate that, at the currently obtainable levels of transgene product accumulation, plants would make an ideal heterologous system for the large-scale production of recombinant proteins that do not require a high level of purity (Pen *et al.*, 1993). As techniques for increasing the level of transgenic protein accumulation and for protein purification improve, the manufacturing costs of high-grade pharmaceutical proteins will become less of an economic barrier.

Probably the most significant economic outlay in the development of plants as a source of recombinant protein will be in the development of downstream processing. It seems likely that the development of fusion proteins will be of importance in the improvement of methods for the downstream processing of recombinant proteins from plants. Protein fusions can increase protein stability and protect the recombinant protein from degradation. Purification tags (proteins, ligands, etc) for the affinity-purification of recombinant proteins are the most frequently used method in the downstream processing of proteins (Lowe, 1984; Narayanan and Cranen, 1990). Ideally, the fusion partner should be small and the fusion tag should be capable of being cleaved from the target protein following purification. A number of expression vectors which possess these characteristics are available for producing recombinant proteins in *Escherichia coli*, and similar systems could easily be developed for plants.

MOLECULAR AND CELLULAR FACTORS

Promoters and promoter activity

A key factor determining the level of heterologous protein in transgenic plants is the level of mRNA transcription. Attempts to increase transgene transcription have formed the basis for most strategies aimed at increasing heterologous protein yield. In particular, considerable effort has been invested in the development of strong promoters to drive the expression of heterologous genes in transgenic plants. The promoter most commonly used for this purpose is the cauliflower mosaic virus (CaMV) 35S promoter (Harpster *et al.*, 1988; Kay *et al.*, 1987; Odell *et al.*, 1988; Sanger, Daubert and Goodman, 1990; Teeri *et al.*, 1989). The CaMV 35S promoter acts as a strong, constitutive promoter in most organs of transgenic plants (Jensen *et al.*, 1986). Numerous attempts have been made to modify and improve the CaMV 35S promoter. For instance, it has been reported that duplication of some elements of the CaMV 35S promoter has a strong (up to 100-fold) enhancing effect on the expression of a number of reporter transgenes in tobacco (Kay *et al.*, 1987). However, multiplication of elements of the CaMV

35S promoter has been shown to lead to no significant enhancement of the expression of an ovalbumin gene in the tissues of transgenic plants (Wandelt *et al.*, 1991). Similarly, Comai, Moran and Maslyar (1990) found that reiteration of a CaMV 35S upstream region (the 35S enhancer) did not have a major efect on the expression of the *gusA* reporter gene in either tobacco or tomato transformants. However, a hybrid promoter called Mac, in which a region of the mannopine synthase (*mas*) gene promoter, derived from the TR-DNA of an *Agrobacterium tumefaciens* octopine Ti plasmid, was joined to the enhancer region of the CaMV 35S promoter, was observed to drive higher *gus* expression than the double CaMV 35S promoter. The expression of *gus* driven by the Mac promoter was 3–5 times higher in leaves, and 10–15 times higher in roots and hypocotyls than expression driven by the double CaMV 35S promoter (Comai, Moran and Maslyar, 1990). One feature of the Mac promoter that is thought to contribute to the higher expression driven by this promoter is the 5′ transcribed untranslated region of *mas*. It is possible that this region may be an mRNA leader that favours translation initiation, as is the case for several viral RNA 5′ leaders (Gallie *et al.*, 1987).

In some situations, inducible expression of a transgene may be necessary or desirable. For instance, a gene whose product confers resistance to a pathogen need only be expressed when the plant is challenged with the pathogen. The constitutive expression of the transgene may be detrimental to the normal growth and development of the plant. There are a number of inducible plant promoters that could be useful in inducible expression vectors. These include the promoters of genes normally induced by wounding and/or pathogen attack (e.g. Williams *et al.*, 1992), the promoter of the anaerobically inducible alcohol dehydrogenase gene (*Adh*), as well as the promoters from a range of light-inducible genes and genes induced by plant growth regulators. A tightly regulated, inducible promoter has been described by Ainley and Key (1990), who constructed a heat-inducible expression cassette containing the promoter and all but five bases of the mRNA leader from soybean heat shock protein gene, *Gmhsp17.5E*, linked to the GUS coding sequence. In a transient expression system, at heat shock temperatures, this promoter–GUS construct was expressed at roughly 80-fold higher levels, per unit time, than a CaMV 35S–GUS construct.

There are many situations in which a heterologous protein needs to be synthesized in a particular organ or tissue. Thus, for protein farming, it is likely that the heterologous protein will need to be synthesized in a readily harvestable and storable tissue such as seed, tuber or fruit (see above). While promoters that direct transgene expression in a wide range of organs, tissues and cell types have been identified and analysed, relatively few have been assessed for their use in protein farming. The promoter regions of seed storage proteins, together with the amino acid signals that direct seed storage proteins to storage organelles, have been employed in the production of some bioactive peptides (see below).

DNA methylation and transgene silencing

DNA methylation has been implicated in the control of endogenous gene expression in animal cells, where an inverse correlation exists between gene methylation and expression. Methylation also appears to be involved in the regulation of plant gene expression and similar inverse correlations between methylation and expression of endogenous genes have been reported (e.g. Bianchi and Viotti, 1988). In addition, methylation seems to be a mechanism for the inactivation of chimaeric transgenes in plants (e.g. Hobbs, Kpodar and Delong, 1990; Matzke *et al.*, 1989), clearly undesirable in a heterologous protein expression system. It has been suggested that, at least in some cases, transgene methylation and hence inactivation occurs in *Agrobacterium tumefaciens* prior to transfer to the cell (Palmgren, Mattson and Okkels, 1993). It is known that *in vitro* methylation of transgenes is maintained in plant cells with attendant reduced expression (e.g. Weber, Ziechmann and Graessmann, 1990), and that *Agrobacterium* DNA is methylated. In support of their view, Palmgren, Mattson and Okkels (1993) have shown that treatment of *Agrobacterium* with the demethylating agent 5-azacytidine leads to increased transient and stable expression of a reporter gene. Not all transgenes are subject to silencing and appropriate selection of suitable transformants will circumvent problems from this source.

Transcript processing

Post-transcriptional modification processes are extremely important to gene expression, and can therefore influence the yield of heterologous protein in transgenic plants. The presence of introns in both plant and animal genes can affect the level of transcription of those genes. The methanism by which this occurs is not fully understood. Enhancer sequences in the introns are thought to increase expression levels; however, the efficiency of splicing and other processing aspects of mRNA stability may also be involved.

In plants, a number of examples exist where the presence of introns has been shown to increase expression of a gene. Normal expression of the maize alcohol dehydrogenase 1 (*Adh 1*) gene is dependent upon the presence of at least one of the nine introns normally present in the gene (Callis, Fromm and Walbot, 1987). Intron 2 or 6 of the *Adh 1* gene have been shown to enhance chloramphenicol acetyl transferase gene expression in electroporated maize protoplasts by up to 20-fold (Mascarenhas *et al.*, 1990). McElroy *et al.* (1990) have demonstrated that the intron from the rice actin gene can enhance expression of the *gus* reporter gene in transformed rice. Similarly, Tanaka *et al.* (1990) have reported that the first intron of the castor bean catalase gene (*cat-1*) also enhances the expression of the *gusA* gene in transformed rice, but not in transgenic tobacco.

The mechanism by which these introns increase expression levels in plants is unknown, but it appears to be correlated to splicing efficiency. Where splicing is efficient, gene expression is increased, whereas limited splicing results in little or no enhancement (Simpson *et al.*, 1992).

Translation initiation and codon usage

It is well established that not all eukaryotic RNAs are translated with equal efficiency *in vivo* or *in vitro*. The RNAs of several plant viruses are known to be well translated and highly competitive and the 5′-untranslated leader sequences of these are known to act in *cis* to enhance the translation of foreign mRNAs both *in vitro* and *in vivo* (Gallie *et al.*, 1987; Sleat *et al.*, 1987). Numerous plant transformation vectors have been designed that incorporate viral translational enhancers. Guerineau, Lucy and Mullineaux (1992) have investigated the influence of sequences around the translation initiation codon on the expression of the *gus* reporter gene in tobacco. Sequences corresponding to the vertebrate consensus sequence, CCACCATGG (Kozak, 1987) and the plant consensus sequence AACAATGG (Lutcke *et al.*, 1987), were fused with the GUS-coding region and transient expression, driven by a duplicated CaMV 35S promoter, was determined in tobacco mesophyll protoplasts. The constructs containing translational fusions between the consensus translation initiation sequences gave over three times more GUS activity than a control construct containing a transcriptional fusion. There was no significant difference between the GUS activity seen for the two consensus sequences.

There is evidence that efficiency of translation of transgenic mRNAs in plants can be increased by modification of the coding sequence of the transgene. Genes that encode the insecticidal δ-endotoxins of *Bacillus thuringiensis*, such as *crylA*, have been expressed in transgenic plants and insect resistance has been reported (e.g. Perlak *et al.*, 1990). However, the level of expression of these insect control proteins in plants are typically very low, although expression of a truncated gene encoding only the *N*-terminal half of the protein does result in a modest increase in transgenic protein levels (Vaeck *et al.*, 1987). Substantial increases, up to 100-fold, in insect control protein accumulation have been observed following modification of the coding regions of both the *crylA(b)* and *crylA(c)* genes (Perlak *et al.*, 1991). Partial modification of the coding sequences, involving removal of a subset of potential polyadenylation signals and ATTTA sequences, increased the number of transformants displaying insect control, as well as increasing protein accumulation. Full modification of the genes, involving elimination of all potential polyadenylation sequences, all ATTTA sequences and increasing the G + G content led to even greater accumulation of the insect control proteins (Perlak *et al.*, 1991). Since more protein was produced for a given amount of the fully modified mRNA, compared with the partially modified mRNA, the presence of predominantly plant-preferred codons in the former may have increased translational efficiency.

Protein stability

In general, yields of transgene products in plants are low, although a very few heterologous proteins have been reported to accumulate to levels of 1% of total soluble plant protein. Considerable effort has been invested in devising strategies to increase transgene transcription and transcript processing and translation (see above), yet relatively little is known about the factors that determine the stability of proteins in plants. Clearly, if the half-life of a transgenic protein could be increased, then an increase in yield of that protein could be achieved without the need for additional energetic expenditure by the plant. The molecular determinants of protein stability in eukaryotes are beginning to be identified, with particular progress being made in defining the steps that lead to degradation of short-lived cytoplasmic proteins (Jentsch and Bachmair, 1992). Plants are only rarely employed in these types of studies (Bachmair, Becher and Schell, 1993).

There are indications from several studies on transgenic protein accumulation that cellular factors, specifically cellular location, can have a profound effect on protein stability and hence accumulation. Starch potatoes have been assessed as a target crop for the agricultural production of human serum albumin (HSA). A number of modified forms of the HSA gene were transformed into potato and tobacco plants using an *Agrobacterium* binary vector system, in which transgene expression was driven by a modified CaMV 35S promoter and which included an alfalfa mosaic virus leader sequence. In plants transformed with the coding region of mature HSA fused to the coding region of a signal sequence derived from a secreted tobacco protein (PR-S, a pathogenesis-related protein), correctly processed HSA protein was found to accumulate up to 0.02% of the total protein (Sijmons *et al.*, 1990; Pen *et al.*, 1993). Similarly, potato and tobacco plants transformed with the gene sequence encoding the native pre-pro-HSA also accumulated a processed HSA protein. In both cases, the transgenic HSA was localized in the extracellular fluid, indicating that both the native and the tobacco signal sequences had directed secretion of the HSA via the endoplasmic reticulum. However, in tobacco plants transformed with gene sequences encoding the mature HSA without a signal or pro-sequence, no accumulation of the protein could be detected. The HSA mRNA was readily detectable in these transformed plants. Thus, the presence of the signal sequence was necessary for detectable transgene product accumulation. While it is possible that the presence of the signal sequence coding region influences translatability of the HSA mRNA, these observations suggest that non-secreted HSA is not stable in the cytoplasm of tobacco cells. Similar increased transgene product accumulation as a result of secretion has been reported for transgenic antibody and antibody fragments from tobacco. In the case of multimeric, intact immunoglobulins, however, secretion via the endoplasmic reticulum is probably also required for correct assembly of the protein (see below). Also, in a recent study on the production of the *Trichosanthes kirilowii* ribosome inactivating protein, α-trichosanthin, in tobacco, it was speculated that the high level accumulation of the transgene product (at least 2% of total soluble protein) was partly due to the native α-trichosanthin signal peptide targeting the protein to the extracellular space (Kumagai *et al.*, 1993).

The observation that transgene products that are secreted to the extracellular space show increased accumulation is not general. As part of a study aimed at evaluating a model system based on the transient synthesis of heterologous proteins in tobacco protoplasts, it has been reported that the presence of signal sequences correlates with a reduction in the level of the encoded gene products (Denecke, Bottermann and Deblaere, 1990). Thus, whereas the levels of neomycin phosphotransferase II mRNA were essentially unaffected by fusion of the neomycin phosphotransferase II gene to the coding regions of either the tobacco PR1 signal sequence or the *Hyalophora cecropia* cecropin B signal sequence, the levels of neomycin phosphotransferase II were always 5- to 10-fold lower for the secretory constructs. This observation is similar to that reported for the synthesis of wheat α-amylase in transgenic yeast, in which deletion of the α-amylase signal sequence coding region was found to increase protein level by about 10-fold (Rothstein *et al.*, 1984). A decrease in protein stability or translation efficiency are the most plausible explanations for these observations. That the protein : mRNA ratio of some transgene products is increased by signal sequence-mediated secretion of the transgene protein, while in other cases it is decreased, suggests that the effect may be gene-dependent.

Recently, a method for identifying transgenic plants with a high level of expression of the introduced gene, that relies upon secretion of the gene product, has been described (Pen *et al.*, 1992). In this case, tobacco protoplasts were transformed with an expression construct containing a translational fusion between mature *Bacillus licheniformis* α-amylase and the signal sequence of the tobacco PR-S protein. The transformed protoplasts were cultured to microcalli, then transplanted to solid media containing starch. Following a suitable period of incubation, the starch plates were stained with iodine solution and the individual calli scored for halo size. The levels of α-amylase protein in plants regenerated from the calli were found to be related to halo size. This method, which could be applied to a wide range of extracellular proteins by, say, scoring the formation of antibody–antigen complexes in solid media, could provide a very convenient early selection for high-level transgene expression.

Further evidence that transgene product accumulation can be significantly increased by targeting the protein to a different cellular compartment, thereby increasing its stability, has been provided by a recent study of the expression of the pea seed storage protein, vicilin, in transgenic tobacco and alfalfa plants (Wandelt *et al.*, 1992). Seed storage proteins normally accumulate to high levels in protein storage vacuoles, protein bodies, and are stable over long periods of time. The genes encoding some seed storage proteins, such as the cowpea trypsin inhibitor, can be modified to replace the upstream regions that confer seed specificity of expression, allowing the transgenic storage proteins to accumulate in the leaves of transgenic host plants (Hilder *et al.*, 1987). However, for some storage protein genes, expression in the leaves of transgenic host plants leads to very low accumulation of protein (Higgins and Spencer, 1991). It has been shown that the protein body targeting information contained within seed storage proteins leads to transport of transgenic storage proteins to the protease-rich vacuoles of host plant leaves (Chrispeels, 1991). The failure to detect the soybean storage protein, conglycinin, in the soluble (vacuolar) fraction isolated from transgenic tobacco leaves is consistent with the notion that this storage protein is unstable in the leaf vacuole (Lawton *et al.*, 1987). To overcome the problem of storage protein instability in leaf vacuoles, Wandelt *et al.* (1992) have described the redirection of transgenic vicilin to the endoplasmic reticulum of host plants using the tetrapeptide, KDEL, endoplasmic reticulum lumen retention signal. The presence of the KDEL sequence at the C-terminus of transgenic vicilin was found to increase accumulation of the protein by up to 20- to 100-fold in the leaves of transgenic alfalfa and tobacco plants. The increased accumulation of vicilin-KDEL protein compared with native vicilin was not accompanied by a detectable change in vicilin mRNA level or translatability. Rather, the increased levels of vicilin-KDEL reflected a significant reduction in the degradation of newly synthesized vicilin-KDEL (half-life 48 h) compared with vicilin (half-life 4.5 h). Tobacco leaves expressing the vicilin-KDEL gene under control of the CaMV 35S promoter were found to accumulate vicilin at up to 3.7% of total leaf protein.

From the available evidence it appears that, for some proteins at least, there may be scope for increasing heterologous protein stability and, hence, accumulation through targeting of the protein to a more suitable cellular compartment. At present, this is supported by studies on a small number of heterologous proteins in relation to only three cellular compartments: the cytoplasm, the apoplast (extracellular space) and the lumen of the endoplasmic reticulum. It has been demonstrated that heterologous proteins can be targeted to other plant cell organelles, for instance the chloroplast (Schreier *et al.*, 1985). Indeed, many crop improvement based applications of heterologous protein production require that the foreign protein be resident in a cellular compartment other than the cytoplasm. However, the consequences, in terms of protein stability and yield, of targeting heterologous proteins to these locations have yet to be determined.

In bacterial expression systems, the expression of a foreign protein as fusion protein has been found to increase its stability both *in vivo* and *in vitro* (e.g. Hellebust, Veide and Enfors, 1988). Similar strategies may be effective for the production of some heterologous proteins in plants. For example, Hightower *et al.* (1991) produced transgenic tobacco and tomato plants which express the *afa3* gene, a synthetic gene encoding an analogue of the winter flounder anti-freeze protein, under control of the CaMV 35S promoter. Although the *afa3* gene was expressed to high steady-state mRNA levels in leaves of transformed plants, anti-freeze protein activity could not be detected. On the other hand, transformation of tomato with a chimaeric gene encoding a fusion protein between truncated *Staphylococcal* protein A and the *afa5* anti-freeze protein gene resulted in anti-freeze protein activity in tissue extracts. It is possible that the size and nature of the protein A portion of the fusion protein protects the small anti-freeze protein from proteolysis in plant cells.

A fusion protein strategy has also been adopted for the production of bioactive peptides in transgenic plants (Krebbers and Vanderkerckhove, 1990; Krebbers, van Rompaey and Vanderkerckhove, 1993). Sequences encoding leu-enkephalin (a neuropeptide) or magainin (an antibacterial peptide) have been introduced into the coding regions of the 2S albumin seed storage protein from both *Arabidopsis* and *Brassica napus*. Expression of the fusion protein-encoding genes, driven by the *Arabidopsis* 2S1 promoter, in either transgenic *Arabidopsis* or transgenic *B. napus*, led to accumulation of the peptides at levels up to 100–200 nmol of peptide per gram of seed. The use of the seed-specific 2S1 promoter, together with exploitation of the protein body targeting signals present on the 2S albumin fusion partner, ensures that the peptides are produced as stable, stored fusions in the seeds of the transformed plants. At present, little is known about the downstream processing of these types of fusion proteins, but it is anticipated that incorporation of sites recognized by restriction proteases will allow specific cleavage of fusion proteins and release of the peptide.

Production of antibodies and antibody fragments in plants

As a case study to illustrate the complexity and problems associated with heterologous proteins in plants, whether from the point of view of factory-scale production or for crop improvement, the expression of antibody-encoding genes, or fragments of such genes, can be used to illustrate most of the points raised above.

USES OF PLANT-PRODUCED ANTIBODIES

Broadly speaking, there are three reasons for producing antibodies in transgenic plants: (1) the large-scale production of commercial antibodies, i.e. antibody farming; (2) the antibody-mediated modification of antigen activity *in planta*; and (3) the study of the plant cell protein targeting and assembly machinery.

Antibody farming

Monoclonal antibodies are produced commercially on a very large scale for a wide range of uses. For instance, therapeutic agents that consist of a cell type-specific monoclonal antibody, which may or may not be conjugated to a drug, isotope or toxin, are used in the treatment of cancer and in the diagnosis of cardiovascular diseases. In addition, monoclonal antibodies are used as vaccines in passive immunization. It has been estimated that the value of therapeutic antibodies will be about $1000 million in 1994. Outside of therapeutics, the exquisite binding specificities of monoclonal antibodies are also employed in a range of affinity separations and in a plethora of diagnostic kits.

The production of reagent monoclonal antibodies usually involves either complex and expensive procedures for hybridoma cell growth and antibody purification or the isolation of antibody from ascites fluid produced in animals. There has been considerable interest in the development of more convenient and cheaper methods for the production of antibodies. Crop plant-produced antibodies may prove to be significantly less expensive than antibodies from other sources. This may be especially true where existing agricultural infrastructure can be employed for protein farming. Also, it has been speculated that transgenic crops may provide a convenient means of producing and delivering antibodies for use in passive immunization (Hiatt and Ma, 1992). In particular, it has been suggested that therapeutic monoclonal antibodies that are effective following oral delivery, such as those used to prevent the establishment of *Streptococcus mutans*, the agent that causes dental caries, could be administered directly in edible plant tissues.

Modification of antigen activity in planta

Antibody gene expression in heterologous cells provides a means of stably modifying the activity of antigen molecules present within the expressing cells. The feasibility of this approach was first demonstrated by Carlson (1988), who showed that transgenic yeast expressing gene sequences encoding truncated immunoglobulin heavy and light chains (signal sequences removed), derived from an anti-alcohol dehydrogenase antibody, displayed reduced activity of the target enzyme. Also, a gene encoding a recombinant antibody fragment displaying chorismate mutase activity has been shown to complement a chorismate mutase-deficient mutant of yeast (Tang, Hicks and Hilvert, 1991). More recently, Owen *et al.* (1992a) have shown that transgenic tobacco plants expressing a synthetic gene encoding an antibody fragment reactive with the plant regulatory photoreceptor, phytochrome, display aberrant phytochrome physiology. This latter demonstration opens the way to numerous possible applications of antibody gene expression for crop improvement, as well as for the study of plant metabolism (Hiatt, 1990; Owen *et al.*, 1992a, b). It has been speculated that the *in planta* synthesis of antibodies that bind to antigens necessary for pathogenesis may provide a means to enhance plant resistance to pathogens. For example, Saunal, Witz and van Regenmortel (1993) have identified a monoclonal antibody that interferes with the co-translational disassembly of tobacco mosaic virus (TMV) particles, and have suggested that expression of genes encoding this antibody in tobacco may provide resistance to the virus. Similarly, a strategy for engineering resistance to the potato cyst nematode, *Globodera rostochiensis*, has been described, which involves *in planta* expression of antibodies that bind to, and inactivate, the saliva proteins of the nematode (Schots *et al.*, 1992).

In addition to plant protection, the antibody-mediated modulation of plant antigen activity could be applied to the manipulation of plant growth and development through, for example, the *in situ* binding and inactivation of plant hormones or other regulatory molecules. It is envisaged that antibody synthesis may be complementary to antisense approaches for the creation of loss-of-function mutations. Also, plants synthesizing antibodies that bind to or detoxify environmental pollutants could be used in environmental clean-up and could be particularly appropriate for use with large bodies of water containing persistent pollutants such as polychlorinated biphenyls (PCBs) at low concentrations. A key determinant of the success of these approaches will be establishing conditions that allow the synthesis of functional antibodies in the appropriate subcellular compartment of the plant cell.

The study of protein targeting, assembly and modification

Intact immunoglobulins are multimeric proteins, the product of two different genes, that undergo post-translational modification and require a specialized environment for assembly. In this regard, they represent an ideal model for studying the synthesis of complex heterologous proteins in plant cells. Also, the ease with which two plants expressing different genes encoding individual monomers of a multimeric protein can be sexually crossed to produce progeny that express both genes is very convenient in the study of complex protein assembly.

ANTIBODY ENGINEERING

The advent of recombinant DNA technology has led to the development of methods for the expression of immunoglobulin-encoding genes in a wide range of heterologous organisms, including non-lymphoid mammalian cells, insect cells, yeast, bacteria and plants (Wright, Shin and Morrison, 1992). The isolation of antibody-encoding genes from hybridoma cells and their expression in these heterologous hosts led directly to a revolution in monoclonal antibody technology in which recombinant DNA techniques have been used to alter the structure, properties, specificity, function and immunogenicity of antibodies. These developments in antibody engineering have extended so far that now it is possible to express the entire immune repertoire such that the antibodies are displayed on the surface of bacteriophage (Chiswell and McCafferty, 1992). Once displayed on phage surfaces, antibodies can be subjected to molecular evolution, thus extending the natural immune repertoire (Marks *et al.*, 1992).

Although there are several different classes of antibodies that differ in some important characteristics, the antibodies most commonly employed as reagents and as the starting point for antibody engineering are members of the immunoglobulin G (IgG) family. These proteins are Y-shaped and are formed by four polypeptide chains – two identical glycosylated heavy chains and two identical non-glycosylated light chains. The heavy chains are joined to each other by disulphide bonds, and each of the light chains is joined to one of the heavy chains by a disulphide bond (*Figure 1*). The domains that form the tips of the arms of the Y, the Fv domains, are solely responsible for binding to the antigen and are formed by the variable region of the heavy chain (V_H) and the variable region of the light chain (V_L). Within each of the variable regions lie three hypervariable segments, the complementarity determining regions (CDRs), which form the antigen-binding site. Proteolytically derived fragments of immunoglobulins, such as the $F(ab)_2$ fragment, that possess the Fv region, show antigen-binding activity that is essentially the same as that of the intact antibody molecule. The constant, Fc, region of the immunoglobulin molecule is involved in activation of the complement system leading to elimination of antibody–antigen complexes.

The domain structure of the immunoglobulin molecule makes antibodies particularly suitable for protein engineering (Winter and Milstein, 1991). Thus, gene sequences that encode fragments carrying the Fv regions of immunoglobulins, for instance Fab and Fv fragments, have been expressed in heterologous hosts to yield active antigen-binding immunoglobulin fragments (see Wright, Shin and Morrison, 1992). These small, recombinant antibody fragments have attracted much interest in applications where the Fc region of an intact antibody is unnecessary or undesirable. For example, a major drawback of using murine antibodies for human therapy is their slow clearance times and their immunogenicity leading to the induction of the HAMA (human anti-mouse antibody) response. These problems are reduced by reducing antibody size. Further improvements in the design of recom-

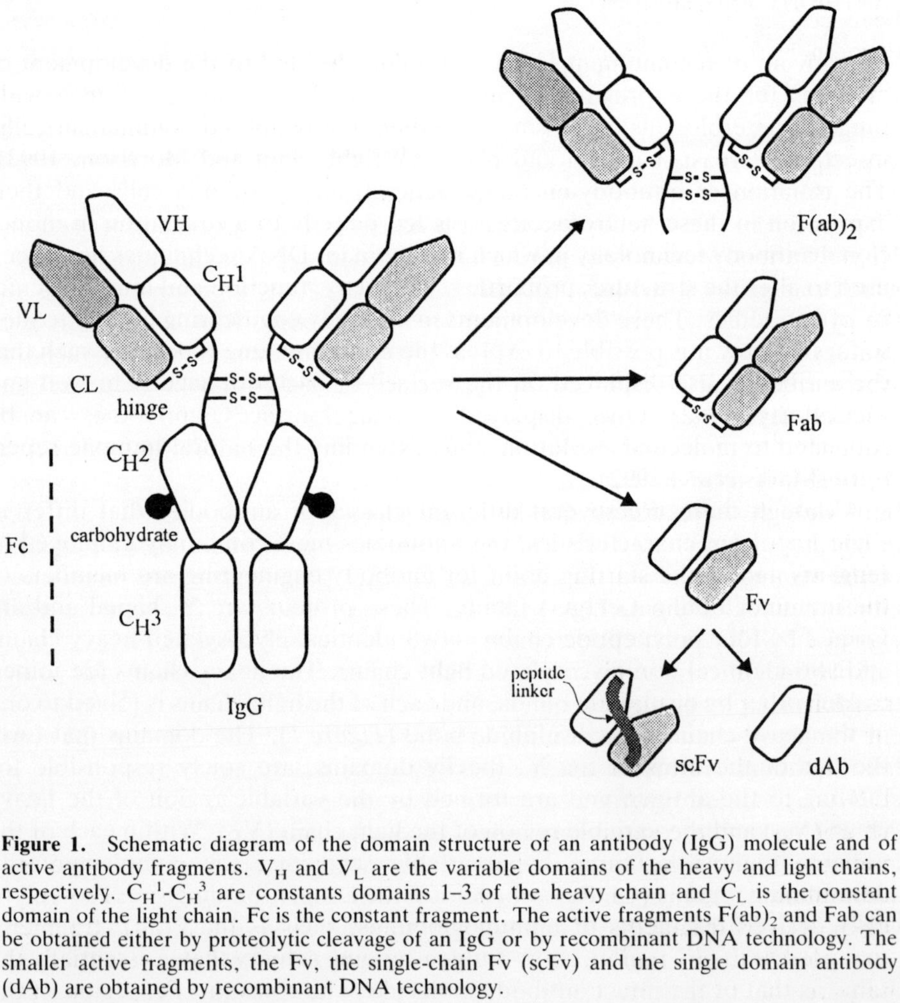

Figure 1. Schematic diagram of the domain structure of an antibody (IgG) molecule and of active antibody fragments. V_H and V_L are the variable domains of the heavy and light chains, respectively. $C_H{}^1$-$C_H{}^3$ are constants domains 1–3 of the heavy chain and C_L is the constant domain of the light chain. Fc is the constant fragment. The active fragments F(ab)$_2$ and Fab can be obtained either by proteolytic cleavage of an IgG or by recombinant DNA technology. The smaller active fragments, the Fv, the single-chain Fv (scFv) and the single domain antibody (dAb) are obtained by recombinant DNA technology.

binant immunoreagents for human therapy have been achieved by humanization of murine antibodies by CDR-grafting. In this technique, the hypervariable antigen-binding loops of a murine antibody with the desired specificity are transplanted to the variable region framework of a human antibody or antibody fragment (Winter and Milstein, 1991). The Fv, a small recombinant antibody fragment that retains a complete antigen-binding site, is formed by the association of two separate polypeptides and is not especially stable. The single-chain Fv (scFv), in which the V_H and V_L domains of an immunoglobulin are linked together by a peptide (*Figure 1*), is a more stable reagent and is the product of a single synthetic gene. The scFv is the focus of much activity in the design of new therapeutic, diagnostic and affinity reagents.

Recombinant, antigen-binding fragments, such as the scFv, can be endowed with new properties by creating fusions between the scFv gene and sequences encoding proteins with other functions. Thus, recombinant immunotoxins designed to target and destroy specific types of cancer cells, have been constructed by fusing an anti-cancer cell scFv gene to a cytotoxin-encoding gene (e.g. Kreitman *et al.*, 1990). Active immunotoxin has been recovered following expression of these gene fusions in *Escherichia coli*.

SYNTHESIS AND ASSEMBLY OF INTACT IMMUNOGLOBULINS IN PLANTS

The production of a recombinant antibody in plants was first described by Hiatt, Cafferkey and Bowdish (1989). The catalytic IgG1 antibody 6D4, which recognizes a synthetic phosphonate ester, P3, and which can catalyse the hydrolysis of certain carboxylic esters, was chosen for study. A two-step strategy was adopted in order to produce transgenic tobacco plants capable of synthesizing intact immunoglobulin. In the first step, cDNAs encoding the γ-heavy chain and its signal sequence or the κ-light chain and its signal sequence, were placed behind the CaMV 35S promoter and used for *Agrobacterium*-mediated tobacco transformation. In addition, transformations were performed with truncated heavy- or light-chain encoding cDNAs in which the signal sequence coding regions were removed. The second step of the strategy involved the sexual crossing of individual plants expressing either heavy- or light-chain cDNA, in order to produce progeny that expressed cDNAs encoding both chains.

Analysis of transgenic plants from the first step revealed that only transformed plants expressing the intact, full-length heavy- or light-chain encoding cDNAs accumulated detectable immunoglobulin protein. In contrast, although transformants expressing the truncated heavy- or light-chain cDNAs contained readily detectable levels of appropriate transcripts, they did not accumulate significant amounts of the proteins. Thus, the native immunoglobulin signal sequences were necessary in order to observe protein accumulation. The F1 plants produced by crossing plants expressing individual, full-length cDNAs were found to contain significant amounts of assembled antibody that, upon extraction, bound P3 with the same affinity as the parental, hybridoma-derived antibody (Hiatt, Cafferkey and Bowdish,

1989; Hiatt and Mostov, 1993). The levels of intact immunoglobulin protein present in F1 plants expressing both heavy- and light-chain cDNAs was found to be significantly higher than the levels of individual chains produced in the parent plants. Furthermore, crossing of plants producing heavy and light chain at vastly different levels did not lead to the creation of F1 plants that accumulated an excess of one chain; only assembled antibodies were detected. Because co-expression of both chains increases the yield of each chain, it appears that assembly of the chains enhances their stability. The yield of 6D4 from F1 plants was in the range of 1% of the total protein, representing one of the highest reported yields for a heterologous protein produced in transgenic plants. It appears possible that with appropriate modifications to the immunoglobulin cDNA codons and with optimization of promoter and untranslated leader sequences, that even higher yields could be obtained.

Assembled antibodies were not detected in F1 plants expressing truncated cDNAs, suggesting that co-translational insertion into the endoplastic reticulum (ER) lumen is necessary for chain assembly and stability. In antibody-producing B cells, the assembly of heavy- and light-chain complexes occurs in the ER and is thought to be facilitated by (a) protein disulphide isomerase, which catalyses the formation of disulphide bonds both in and between nascent polypeptide chains, and (b) a heavy-chain binding protein (BiP), a chaperonin related in structure to hsp70 (Haas and Wabl, 1983). The ER lumen of plant cells contains protein disulphide isomerase and a BiP homologue (Fontes *et al.*, 1991). The absence of these proteins from the cytoplasm, together with the reducing nature of this compartment that would prevent disulphide bond formation, may lead to the conclusion that intact immunoglobulins could not be produced in the cytoplasm of plant, or other heterologous, cells. However, transient expression of truncated cDNAs encoding the μ-heavy chain and λ-light chain of an IgM antibody, following their microinjection into the nucleus of *Acetabularia*, has been shown to result in synthesis and assembly of antibody in the cytoplasm (Stieger *et al.*, 1991). Similarly, truncated cDNAs encoding the heavy and light chains of this, and other antibodies, have been expressed in cultured mammalian or yeast cells, and evidence for the synthesis of low levels of functional antibody in the cytoplasm has been presented (Carlson, 1988; Piccoli, *et al.*, 1991). Thus, for some antibodies at least, targeting to the Er is not an absolute requirement for synthesis, assembly and accumulation.

In F1 plants expressing both intact cDNAs, the assembled antibody protein is located in the intercellular spaces (Hein *et al.*, 1991), indicating that following targeting to the ER the polypeptides enter the default pathway and are secreted. In callus cell suspension cultures established from the F1 plants, functional, whole immunoglobulins are actively secreted to, and can be isolated from, the culture medium at concentrations up to 20 mg l^{-1} (Hein *et al.*, 1991). The native mouse immunoglobulin signal sequence coding regions of the 6D4 heavy- and light-chain cDNAs have been replaced with the coding region of the yeast α-mating factor pre-pro sequence, resulting in successful synthesis, assembly and secretion of functional immunoglobulin in transgenic

F1 plants (Hein *et al.*, 1991). Levels of antibody were similar to those observed when secretion was facilitated by the native signal sequences and assembled, secreted immunoglobulin was produced in F1 plants expressing heavy and light chains possessing different signal sequences (Hein *et al.*, 1991). In contrast to these observations, During *et al.* (1990) have reported that the expression in tobacco plants of chimaeric genes composed of the coding region of the barley α-amylase signal sequence fused to cDNAs encoding mature heavy and light chains of a monoclonal IgM antibody leads to assembly of a functional immunoglobulin that is not secreted from the cell. In this case, tobacco plants were transformed using an *Agrobacterium* vector containing both the μ-heavy chain and λ-light chain under the control of the pNOS and pT_R promoters. Assembled, functional antibody was detected in transformed plant tissues, with immuno-gold electron microscopy showing that this immunoglobulin protein was present in the ER and, surprisingly, the chloroplasts, but not the intercellular spaces. No feature of the engineered antibody protein was expected to lead to chloroplast targeting. The reason for the different targeting of this immunoglobulin, in which plant signal sequences are used to target the polypeptides to the ER, compared with the 6D4 antibody, in which mouse or yeast signal sequences have been used, is not known.

In the ER of B-cells, a pre-formed dolichol phosphate donor glycosylates the nascent immunoglobulin heavy chain. The added glycan moiety is of the endoglycosidase H-sensitive, high mannose type, containing *N*-acetylglucosamine and mannose, but lacking galactose and sialic acid. Following immunoglobulin assembly and passage to the golgi, the glycan undergoes modification, involving the addition of galactose and sialic acid producing a modified glycan that is resistant to endoglycosidase H (Tartakoff and Vasalli, 1979; Goding and Herzenberg, 1980). The γ-heavy chain of the 6D4 IgG antibody produced in tobacco plants was found to bind to the lectin concanavalin A (specific for mannose and glucose), indicating that it is glycosylated (Hein *et al.*, 1991). However, in contrast to the heavy chain from parental, hybridoma-produced 6D4, the plant-produced 6D4 heavy chain was not reactive with *Ricinus communis* agglutinin and wheat germ agglutinin, indicating an absence of galactose and sialic acid residues. Nevertheless, the glycan of the plant-produced γ-chain was found to be resistant to endoglycosidase H digestion, suggesting that it was of the modified type. Furthermore, the glycan of the plant-produced γ-chain was observed to bind to concanavalin A with the same affinity as that of the parental antibody, suggesting that both antibodies had undergone similar processing.

Although the difference in targeting of plant-produced IgG and IgM antibody awaits explanation, it does seem that plants are capable of producing significant amounts of processed, assembled, functional immunoglobulin virtually indistinguishable from antibodies produced in animal cells. At present, intact immunoglobulins have only been stably produced in tobacco plants and no attempts have been made to express of the heavy- and light-chain cDNAs in storage tissues. Also, little is known about whether the isolation of antibodies is compatible with current crop-processing practices.

Nevertheless, plants do have the potential to be exploited for the large-scale production of immunoglobulins.

SYNTHESIS OF ANTIBODY FRAGMENTS IN PLANTS

Tobacco plants have been shown to be capable of producing a functional scFv protein in both the cytoplasm and the extracellular space (Owen *et al.*, 1992a; Firek *et al.*, in press). A non-antigen-binding, smaller fragment of a rat IgG monoclonal antibody directed against the tachykinin neuropeptide, substance P, has been synthesized in transgenic tobacco (Benvenuto *et al.*, 1991). Gene sequences encoding a fusion between the leader sequence of the *Erwinia* pectate lyase, *pelB*, protein and the V_H domain of the anti-substance P antibody, under the control of the CaMV 35S promoter, were used to transform tobacco leaf discs. Although this isolated V_H domain displayed no antigen-binding activity, similar single domains from other monoclonal antibodies have been shown to bind antigen (Ward *et al.*, 1989). The V_H domain polypeptide was reported to accumulate in the transgenic tobacco plants to concentrations up to 1% of the soluble protein fraction, a similar level to that reported for intact, assembled immunoglobulins. It is not clear why the construct employed by Benvenuto *et al.* (1991) included the *pelB* leader sequence. This sequence is extensively employed in bacterial expression systems, where it directs transport to the periplasmic space (e.g. Ward *et al.*, 1989). The leader was reported to have been cleaved from the plant-produced V_H domain polypeptide, but the antibody fragment appeared to be predominantly located in the intracellular fraction of tobacco cells. It is possible that, as was reported for the assembled IgM antibody (Düring *et al.*, 1990), the V_H polypeptide was retained in the ER.

For functional scFv protein expression in transgenic tobacco, Owen *et al.* (1992a) used PCR amplification to isolate the V_H and V_L coding regions derived from cDNA prepared from a mouse hybridoma. The hybridoma line, AS32, secretes an IgG1 antibody directed against a conserved epitope on the plant regulatory photoreceptor, phytochrome A. The synthetic scFv gene, shown to encode an antigen-binding protein following its expression in *Escherichia coli*, was placed under the control of the CaMV 35S promoter and was used for *Agrobacterium*-mediated leaf disc transformation. More than 100 individual transformants were screened by Northern blot analysis and the plant displaying the highest expression levels selected. The accumulation of scFv protein in the homozygous progeny of this individual was low, amounting to a maximum of only about 0.06% of total soluble leaf protein. Many transformants that were clearly positive for scFv mRNA were found to accumulate undetectable amounts of scFv protein. The anti-phytochrome scFv protein produced in the high-expressing tobacco line was found to be functional following extraction from plant tissues. Thus, the scFv could be purified to near homogeneity by passage of crude protein extracts through a phytochrome A-Sepharose column (Owen *et al.*, 1992a). Evidence that the scFv protein was functional *in vivo* was obtained from analysis of phytochrome-mediated developmental responses of the transformed plants.

Phytochrome A has been implicated in the control of the de-etiolation responses of higher plant seedlings, including germination, growth and greening. Homozygous seeds expressing the anti-phytochrome A scFv gene were found to show poor phytochrome-dependent, light-mediated promotion of germination (Owen *et al.*, 1992a). In addition, other aspects of the photomorphogenesis of the transformed plants were perturbed, including the photocontrol of hypocotyl elongation, the photocontrol of cotyledon expansion and the photocontrol of greening (Owen *et al.*, unpublished observations). These observations provide validation of the use of *in planta* synthesized antibodies, or antibody fragments, for the modulation of plant antigens. Progress is being made on the *in planta* expression of gene sequences that encode antibodies reactive with a range of other plant cell antigens, including regulatory molecules and plant pathogens.

Since phytochrome A is cytoplasmic, the anti-phytochrome scFv gene was designed to lead to cytoplasmic synthesis of the encoded protein. The perturbation of phytochrome responses seen in the transformed seedlings is strongly suggestive that at least some of the scFv protein was present in the cytoplasm. In order to determine the consequences, in terms of yield and activity, of secretion of an scFv in plants, a translational fusion between the anti-phytochrome scFv and the signal sequence of the tobacco PR1 a protein was constructed and used for tobacco transformation (Firek *et al.*, in press). A processed, functional scFv protein was found to accumulate in the extracellular space of the transformed plants. Cultured callus cells derived from transformed plants were observed to secrete functional scFv protein into the culture medium, such that it accumulated to a concentration of about 50% of total medium protein. In plants expressing the secretion construct, scFv protein was readily detectable and was found to accumulate to levels up to 0.5–1.0% of total soluble leaf protein. The elevated levels of scFv protein, in plants expressing the secretion construct compared with those expressing the 'native' scFv gene, were observed against a background of lower scFv mRNA accumulation. Thus, the protein : mRNA ratio was very significantly higher for plants expressing the secretion construct. This suggests that the presence of the signal sequence coding region leads to increased stability of the scFv protein and/or to increased translatability of the scFv mRNA.

The behaviour of this scFv is therefore similar to the behaviour of the whole 6D4 immunoglobulin and of HSA, in which the addition of a signal sequence was found to be necessary for significant accumulation of the transgene product. In all three cases, the most likely explanation for this observation is an instability of the transgenic protein in the cytoplasm of plant cells. Recently, we have transformed tobacco plants with gene sequences that encode cytoplasmically located fusions between the anti-phytochrome scFv and the GUS reporter enzyme. Preliminary analyses of the plants suggests that fusion of the scFv to a stable enzyme have increased its level of accumulation in tobacco cell cytoplasm. This work also shows that plants may be appropriate hosts for the large-scale production of a range of immunoreagents that are based around the scFv.

Prospects

At the present time, the number of agronomically important species which can be reliably, routinely transformed with foreign genes is relatively limited and the major monocot crops rice, maize and wheat are not yet routinely transformable. However, there can be no doubt that this is a temporary problem, which is currently being solved through the more sophisticated use of *Agrobacterium*-based gene transfer methods and the introduction and optimization of new direct transfer technology. Vector systems based on viral genomes may also play a part in this process.

Transformation technology relies heavily on the ability to grow cells and tissues of the crop plant in question and to reliably regenerate whole plants from these cultures and the technology for this, too, is advancing rapidly.

It seems virtually certain that in the near future, essentially all major crops including the monocots will become fully accessible to transgenic technology and, therefore, that the expression of heterologous proteins will represent an important and many-faceted tool for the improvement of current crops, the development of novel crops and for the evolution of entirely novel strategies for the exploitation of the unique characteristics of plants.

References

AINLEY, W.M. AND KEY, J.L. (1990). Development of a heat shock inducible expression cassette for plants: Characterization of parameters for its use in transient expression assays. *Plant Molecular Biology* **14**, 949–967.

ALTENBACH, S.B. AND SIMPSON, R.B. (1990). Manipulation of methionine-rich protein genes in plant seeds. *Trends in Biotechnology* **8**, 156–160.

BACHMAIR, A., BECHER, F. AND SCHELL, J. (1993). Use of a reporter transgene to generate *Arabidopsis* mutants in ubiquitin-dependent protein degradation. *Proceedings of the National Academy of Sciences, USA* **90**, 418–421.

BENVENUTO, E., ORDAS, R.J., TAVAZZA, R., ANCORA, G., BIOCCA, S., CATTANEO, A. AND GALEFFI, P. (1991). 'Phytoantibodies': A general vector for the expression of immunoglobulin domains in transgenic plants. *Plant Molecular Biology* **17**, 865–874.

BIANCHI, M.W. AND VIOTTI, A. (1988). DNA methylation and tissue-specific transcription of the storage protein genes of maize. *Plant Molecular Biology* **11**, 203–214.

CALLIS, J., FROMM, M. AND WALBOT, V. (1987). Introns increase gene expression in cultured maize cell. *Genes and Development* **1**, 1183–1200.

CARLSON, J.D. (1988). A new means of inducibly inactivating a cellular protein. *Molecular and Cellular Biology* **8**, 2638–2646.

CHISWELL, D.J. AND MCMAFFERTY, J. (1992). Phage antibodies: Will new 'coliclonal' antibodies replace monoclonal antibodies? *Trends in Biotechnology* **10**, 80–84.

CHRISPEELS, M.J. (1991). Sorting of proteins in the secretory system. *Annual Review of Plant Physiology and Plant Molecular Biology* **42**, 21–53.

COMAI, L., MORAN, P. AND MASLYAR, D. (1990). Novel and useful properties of a chimeric plant promoter combining CaMV 35S and MAS elements. *Plant Molecular Biology* **15**, 373–381.

DENECKE, J., BOTTERMANN, J. AND DEBLAERE, R. (1990). Protein secretion in plant cells can occur via the default pathway. *The Plant Cell* **2**, 51–59.

DÜRING, K., HIPPE, S., KREUZALER, F. AND SCHELL, J. (1990). Synthesis and self-assembly of a functional monoclonal antibody in transgenic *Nicotiana tabacum*. *Plant Molecular Biology* **15**, 281–293.

DÜRING, K., PORSCH, P., FLADUNG, M. AND LORZ, H. (1993). Transgenic potato plants resistant to the phytopathogenic bacterium *Erwinia carotovora. The Plant Journal* **3**, 587–598.

FIREK, S., DRAPER, J., OWEN, M.R.L., GANDECHA, A., COCKBURN, W. AND WHITELAM, G.C. (in press). Secretion of a functional single-chain Fv protein in transgenic tobacco plants and cell suspension cultures. *Plant Molecular Biology*.

FONTES, E.B.P., SHANK, B.B., WROBEL, R.L., MOOSE, S.P., O'BRIAN, G.R., WURTZEL, E.T. AND BOSTON, R.S. (1991). Characterization of an immunoglobulin binding protein homolog in the maize *floury-2* endosperm mutant. *The Plant Cell* **3**, 483–496.

GALLIE, D.R., SLEAT, D.E., WATTS, J.W., TURNER, P.C. AND WILSON, T.M.A. (1987). A comparison of eukaryotic viral 5′ leader sequences as enhancers of mRNA expression *in vivo. Nucleic Acids Research* **15**, 8693–8711.

GODING, J.W. AND HERZENBERG, L.A. (1980). Biosynthesis of lymphocyte surface IgD in the mouse. *Journal of Immunology* **124**, 2540–2547.

GRIESBECH, R.J. (1983). Protoplast microinjection. *Plant Molecular Biology Reporter* **1**, 32–37.

GUERINEAU, F., LUCY, A. AND MULLINEAUX, P. (1992). Effect of two consensus sequences preceding the translation initiator codon on gene expression in plant protoplasts. *Plant Molecular Biology* **18**, 815–818.

HAAS, I.G. AND WABL, M. (1983). Immunoglobulin heavy chain binding protein. *Nature* **306**, 387–389.

HAIN, R., STABEL, P., CZERNILOFSKY, A.P., STEINBIS, H.H., HERRERA-ESTRELLA, L. AND SCHELL, J. (1985). Uptake, integration, expression and genetic transmission of a selectable chimaeric gene by plant protoplasts. *Molecular and General Genetics* **199**, 161–168.

HARPSTER, M.H., TOWNSEND, J.A., JONES, J.D.G., BEDBROOK, J. AND DUNSMUIR, P. (1988). Relative strengths of the 35S cauliflower mosaic virus, 1′, 2′, and nopaline synthase promoters in transformed tobacco, sugarbeet, and oilseed rape callus tissue. *Molecular and General Genetics* **14**, 433–443.

HEIN, M., TANG, Y., MCLEOD, D.A., JANDA, K.D. AND HIATT, A. (1991). Evaluation of immunoglobulins from plant cells. *Biotechnology Progress* **7**, 455–461.

HELLEBUST, H., VEIDE, A. AND ENFORS, S.O. (1988). Proteolysis degradation of fused protein A-β-galactosidase in *Escherichia coli. Journal of Biotechnology* **7**, 185–198.

HIATT, A. (1990). Antibodies produced in plants. *Nature* **344**, 469–470.

HIATT, A. AND MA, J.K.-C. (1992). Monoclonal antibody engineering in plants. *FEBS Letters* **307**, 71–75.

HIATT, A. AND MOSTOV, K. (1993). Assembly of multimeric proteins in plant cells: Characteristics and uses of plant-derived antibodies. In *Transgenic Plants: Fundamentals and Applications* (A. Hiatt, Ed.), pp. 221–238. Marcel Dekker, New York.

HIATT, A., CAFFERKEY, R. AND BOWDISH, K. (1989). Production of antibodies in transgenic plants. *Nature* **342**, 76–78.

HIGGINS, T.J.V. AND SPENCER, D. (1991). The expression of a chimeric cauliflower mosaic virus (CaMV-35S)-pea vicilin gene in tobacco. *Plant Science* **20**, 89–98.

HIGHTOWER, R., BADEN, C., PENZES, E., LUND, P. AND DUNSMUIR, P. (1991). Expression of anti-freeze proteins in transgenic plants. *Plant Molecular Biology* **17**, 1013–1021.

HILDER, V.A., GATEHOUSE, A.M.R., SHEERMAN, S.E., BARKER, R.F. AND BOULTER, D. (1987). A novel mechanism of insect resistance engineered into tobacco. *Nature* **330**, 160–163.

HOBBS, S.L.A., KPODAR, P. AND DELONG, C.M.O. (1990). The effect of T-DNA copy number, position and methylation on reporter gene expression in tobacco transformants. *Plant Molecular Biology* **15**, 851–846.

JENSEN, J.S., MARCKER, K.A., OFFEN, L. AND SCHELL, J. (1986). Nodule-specific expression of a chimaeric soybean leghaemogloblin gene in transgenic *Lotus corniculatus*. *Nature* **321**, 669–674.

JENTSCH, S. AND BACHMAIR, A. (1992). Principles of protein turnover: Possible manipulations. In *Protein Engineering: A Practical Approach* (A.R. Rees, M.J.E. Sternberg and R. Wetzel, Eds), pp. 221–228. IRL Press, Oxford.

KAY, R., CHAN, A., DALY, M. AND MCPHERSON, J. (1987). Duplication of CaMV 35S promoter sequences creates a strong enhancer for plant genes. *Science* **230**, 1299–1302.

KLEE, H., HORSCH, R. AND ROGERS, S. (1987). *Agrobacterium* mediated plant transformation and its further applications to plant biology. *Annual Review of Plant Physiology* **38**, 467–486.

KLEIN, T.M., WOLF, E.D., WU, R. AND SANDFORD, J.C. (1987). High-velocity microprojectiles for delivering nucleic acids into living cells. *Nature* **327**, 70–73.

KOZAK, M. (1987). An analysis of 5′-noncoding sequences from 699 vertebrate messengers RNAs. *Nucleic Acids Research* **15**, 8125–8148.

KREBBERS, E. AND VANDERKERCKHOVE, J. (1990). Production of peptides in seeds. *Trends in Biotechnology* **8**, 1–3.

KREBBERS, E., VAN ROMPAEY, J. AND VANDERKERCKHOVE, J. (1993). Expression of modified seed storage proteins in transgenic plants. In *Transgenic Plants: Fundamentals and Applications* (A. Hiatt, Ed.), pp. 37–60. Marcel Dekker, New York.

KREITMAN, R.J., CHAUDHARY, V.K., WALDMANN, T., WILLINGHAM, M.C., FITZGERALD, D.J. AND PACASTAN, I. (1990). The recombinant immunotoxin anti-Tac(Fv)-*Pseudomonas* exotoxin 40 is cytotoxic toward peripheral blood malignant cells from patients with adult T-cell laukemia. *Proceedings of the National Academy of Sciences, USA* **87**, 8291–8297.

KRENS, F.A., MOLENDIJK, L., WULLEMS, G.J. AND SCHILPEROORT, R.A. (1982). *In vitro* transformation of plant protoplasts with Ti-plasmid DNA. *Nature* **296**, 72–74.

KUMAGAI, M.H., TURPEN, T.H., WEINZETTL, N., DELLA-CIOPPA, G., TURPEN, A.M., DONSON, J., HILF, M.E., GRANTHAM, G.L., DAWSON, W.O., CHOW, T.P., PIATAK, M. AND GRILL, L.K. (1993). Rapid, high-level expression of a biologically active α-trichosanthin in transfected plants by an RNA viral vector. *Proceedings of the National Academy of Sciences, USA* **90**, 427–430.

LAWTON, M.A., TIERNEY, M.A., NAKAMURA, I., ANDERSON, E., KOMEDA, Y., DUBE, P., HOFFMAN, N., FRALEY, R.T. AND BEACHY, R.N. (1987). Expression of a soybean β-conglycinin gene under the control of the cauliflower mosaic virus 35S and 19S promoters in transformed petunia tissues. *Plant Molecular Biology* **9**, 315–324.

LINDSEY, K. AND JONES, M.G.K. (1987). The permeability of electroporated cells and protoplasts of sugarbeet. *Planta* **172**, 346–355.

LOWE, C.R. (1984). New developments in downstream processing. *Journal of Biotechnology* **1**, 3–12.

LUTCKE, H.A., CHOW, K.C., MICKEL, F.S., MOSS, K.A., KERN, H.F. AND SCHEELE, G.A. (1987). Selection of AUG initiation codon differs in plants and animals. *EMBO Journal* **6**, 43–48.

MARIANI, C., DEBEUCKELEER, M., TRUETTNER, J., LEEMANS, J. AND GOLDBERG, R.B. (1990). Induction of male sterility in plants by a chimeric ribonuclease gene. *Nature* **347**, 737–741.

MARIANI, C., GOSSELE, V., DEBEUCKELEER, M., DEBLOCK, M., GOLDBERG, R.B., DEGREEF, W. AND LEEMANS, J. (1992). A chimeric ribonuclease-inhibitor gene restores fertility to male sterile plants. *Nature* **357**, 384–387.

MARKS, J.D., HOOGENBOOM, H.R., GRIFFITHS, A.D. AND WINTER, G. (1992). Molecular evolution of proteins on filamentous phage. Mimicking the strategy of the immune system. *Journal of Biological Chemistry* **267**, 16007–16010.

MASCARENHAS, D., METTLER, I.J., PIERCE, D.A. AND LOWE, H.W. (1990). Intron-mediated enhancement of heterologous gene expression in maize. *Plant Molecular Biology* **15**, 913–920.

MATZKE, M.A., PRIMIG, M., TRNOVSKY, J. AND MATZKE, A.J.M. (1989). Reversible methylation and inactivation of marker genes in sequentially transformed tobacco plants. *EMBO Journal* **8**, 643–649.

MCELROY, D., ZHANG, W., CAO, J. AND WU, R. (1990). Isolation of an efficient actin promoter for use in rice transformation. *Plant Cell* **2**, 163–171.

NARAYANAN, S.R. AND CRANEN, L.J. (1990). Affinity chromatography supports: A look at performance requirements. *Trends in Biotechnology* **8**, 12–16.

ODELL, J.T., KNOWLTON, S., LIN, W. AND MACAUVAIS, C.J. (1988). Properties of an isolated transcription-stimulating sequence derived from the cauliflower mosaic virus 35S promoter. *Plant Molecular Biology* **10**, 263–272.

OWEN, M., GANDECHA, A., COCKBURN, W. AND WHITELAM, G. (1992a). Synthesis of a functional anti-phytochrome single-chain Fv protein in transgenic tobacco. *Bio/Technology* **10**, 790–794.

OWEN, M.R.L., GANDECHA, A., COCKBURN, W. AND WHITELAM, G.C. (1992b). The expression of antibodies in plants. *Chemistry and Industry* **11**, 406–408.

PALMGREN, G., MATTSON, O. AND OKKELS, F.T. (1993). Treatment of *Agrobacterium* or leaf disks with 5-azacytidine increases transgene expression in tobacco. *Plant Molecular Biology* **21**, 429–435.

PEN, J., VAN OOYEN, A.J.J., VAN DEN ELZEN, P.J.M., RIETVELD, K. AND HOEKEMA, A. (1992). Direct screening for high-level expression of an introduced α-amylase gene in plants. *Plant Molecular Biology* **18**, 1133–1139.

PEN, J., SIJMONS, P.C., VAN OOIJEN, A.J.J. AND HOEKEMA, A. (1993). Protein production in transgenic crops: Analysis of plant molecular farming. In *Transgenic Plants: Fundamentals and Applications* (A. Hiatt, Ed.), pp. 239–251. Marcel Dekker, New York.

PERLAK, F.J., DEATON, R.W., ARMSTRONG, T.A., FUCHS, R.L., SIMS, S.R., GREENPLATE, J.T. AND FISCHOFF, D.A. (1990). Insect resistant cotton plants. *Bio/Technology* **8**, 939–943.

PERLAK, F.J., FUCHS, R.L., DEAN, D.F., MCPHERSON, S.L. AND FISCHOFF, D.A. (1991). Modification of the coding sequences enhances plant expression of insect control protein genes. *Proceedings of the National Academy of Sciences, USA* **88**, 3324–3328.

PICCOLI, P., RUBERTI, F., BIOCCA, S., DI LUZIO, A., WERGE, T.M., BRADBURY, A. AND CATTANEO, A. (1991). Neuroantibodies: Molecular cloning of a monoclonal antibody against substance P for expression in the central nervous system. *Proceedings of the National Academy of Sciences, USA* **88**, 5611–5615.

POIRIER, Y., DENNIS, D.E., KLOMPARENS, K. AND SOMERVILLE, C. (1992). Polyhydroxybutyrate, a biodegradable thermoplastic, produced in transgenic plants. *Science* **256**, 520–523.

REICH, T.J., IYER, V.J. AND MIKI, B.L. (1986). Efficient transformation of alfalfa protoplasts by the intranuclear microinjection of Ti plasmids. *Bio/Technology* **8**, 1001–1003.

ROTHSTEIN, S.J., LAZARUS, C.M., SMITH, W.E., BAULCOMBE, D.C. AND GATENBY, A.A. (1984). Secretion of wheat α-amylase expressed in yeast. *Nature* **308**, 662–665.

SANGER, M., DAUBERT, S. AND GOODMAN, R.M. (1990). Characterization of a strong promoter from Figwort Mosaic Virus: Comparison with the analogous 35S promoter from Cauliflower Mosaic Virus and the regulated mannopine synthase promoter. *Plant Molecular Biology* **14**, 433–443.

SAUNAL, H., WITZ, J. AND VAN REGENMORTEL, M.H.V. (1993). Inhibition of *in vitro* cotranslational disassembly of tobacco mosaic virus by monoclonal antibodies to the viral coat protein. *Journal of General Virology* **74**, 897–900.

SCHOTS, A., DE BOER, J., SCHOUTEN, A., ROOSIEN, J., ZILVERENTANT, J.F., POMP,

H., BOUWEN-SMITS, L., OVERMARS, H., GOMMERS, F.J., VISSER, B., STIEKEMA, W.J. AND BAKKER, J. (1992). 'Plantibodies': A flexible approach to design resistance against pathogens. *Netherlands Journal of Plant Pathology* **98**, 183–191.

SCHREIER, P.H., SEFTOR, E.A., SCHELL, J. AND BONHERT, H.J. (1985). The use of nuclear-encoded sequences to direct the light-regulated synthesis and transport of a foreign protein into plant chloroplasts. *EMBO Journal* **4**, 25–32.

SHAH, D.M., HORSCH, R.B., KLEE, H.J., KISHMORE, G.M., WINTER, J.A., TUMER, N.E., HIRONAKA, C.M., SANDERS, P.R., GASSER, C.S., AYKENT, S., SIGEL, N.R., ROGERS, S.G. AND FRALEY, R.T. (1986). Engineering herbicide tolerance in transgenic plants. *Science* **233**, 478–481.

SIJMONS, P.C., DEKKER, B.M.M., SCHRAMMEIJER, B., VERWOERD, T.C., VAN DEN ELZEN, P.J.M. AND HOEKEMA, A. (1990). Production of correctly processed human serum albumin in transgenic plants. *Bio/Technology* **8**, 217–221.

SIMPSON, C.G., SIMPSON, G.G., CLARK, G., LEADER, D.J., VAUX, P., GUERINEAU, F., WAUGH, R. AND BROWN, J.W.S. (1992). Splicing of plant pre-mRNAs. *Proceedings of the Royal Society of Edinburgh* **99B**, 31–50.

SLEAT, D.E., GALLIE, D.R., JEFFERSON, R.A., BEVAN, M.W., TURNER, P.C. AND WILSON, T.M.A. (1987). Characterisation of the 5′-leader sequence of tobacco mosaic virus RNA as a general enhancer of translation *in vitro*. *Gene* **217**, 217–225.

SMEEKENS, S., ANGENANT, G., EBSKAMP, M. AND WEISBEEK, P. (1991). Molecular biology of fructan accumulation in plants. *Biochemical Society Transactions* **19**, 565–569.

SONNEWALD, U., VON SCHAEWEN, A. AND WILLMITZER, L. (1993). Subcellular manipulation of sucrose metabolism in transgenic plants. *Journal of Experimental Botany* **44**, 293–296.

STIEGER, M., NEUHAUS, G., MOMMA, T., SCHELL, J. AND KREUZALER, F. (1991). Self-assembly of immunoglobulins in the cytoplasm of the alga *Acetabularia mediterranea*. *Plant Science* **73**, 181–190.

STURTEVANT, A.N. AND BEACHY, R.N. (1993). Virus resistance in transgenic plants: Coat protein-mediated resistance. In *Transgenic Plants: Fundamentals and Applications* (A. Hiatt, Ed.), pp. 93–112. Marcel Dekker, New York.

TANAKA, A., MITA, S., OHTA, S., KYAZUKA, J., SHIMAMOTO, K. AND NAKAMURA, K. (1990). Enhancement of foreign gene expression by a dicot intron in rice but not in tobacco is correlated with an increased level of mRNA and an efficient splicing of the intron. *Nucleic Acids Research* **18**, 6767–6770.

TANG, Y., HICKS, J.B. AND HILVERT, D. (1991). *In vivo* catalysis of a metabolically essential reaction by an antibody. *Proceedings of the National Academy of Sciences, USA* **88**, 8784–8786.

TARTAKOFF, A. AND VASALLI, P. (1979). Plasma cell immunoglobulin M molecules: Their biosynthesis, assembly and intracellular transport. *Journal of Cell Biology* **83**, 284–299.

TEERI, T.H., LEHVASLAIHO, H., FRANK, M., UOTILA, J., HEINO, P., PALVA, E.T., VAN MONTAGU, M. AND HERERRA-ESTRELLA, L. (1989). Gene fusions to lacZ reveal new expression patterns of chimeric genes in transgenic plants. *EMBO Journal* **8**, 343–350.

VAECK, M., REYNAERTS, A., HOFTE, H., JANSENS, S., DE BEUKELEER, M., DEAN, C., ZABEAU, M., VAN MONTAGU, M. AND LEEMANS, J. (1987). Transgenic plants protected from insect attack. *Nature* **328**, 33–37.

WANDELT, C.I., KNIBB, W., SCHROEDER, H.E., KHAN, M.R.I., SPENCER, D., CRAIG, S. AND HIGGINS, T.J.V. (1991). The expression of an ovalbumin and a seed protein gene in the leaves of transgenic plants. In *Plant Molecular Biology 1990* (R.G. Hermann and B.A. Larkins, Eds), pp. 153–159. Plenum Press, New York.

WANDELT, C.I., KHAN, M.R.I., CRAIG, S., SCHROEDER, H.E., SPENCER, D. AND

HIGGINS, T.J.V. (1992). Vicilin with carboxy-terminal KDEL is retained in the endoplasmic reticulum and accumulates to high levels in the leaves of transgenic plants. *Plant Journal* **2**, 181–192.

WARD, E.S., GUSSOW, D., GRIFFITHS, A.D., JONES, P.T. AND WINTER, G. (1989). Binding activities of a repertoire of single Ig variable domains secreted from *Escherichia coli*. *Nature* **341**, 544–546.

WEBER, H., ZIECHMANN, C. AND GRAESSMANN, A. (1990). *In vitro* DNA methylation inhibits gene expression in transgenic tobacco. *EMBO Journal* **9**, 4409–4415.

WILLIAMS, S., FRIEDRICH, L., DINCHER, S., CAROZZI, N., KESSMANN, H., WARD, E. AND RYALS, J. (1992). Chemical regulation of *Bacillus thuringiensis* δ-endotoxin expression in transgenic plants. *Bio/Technology* **10**, 540–543.

WINTER, G. AND MILSTEIN, C. (1991). Man-made antibodies. *Nature* **349**, 293–299.

WRIGHT, A., SHIN, S.-U. AND MORRISON, S.L. (1992). Genetically engineered antibodies: Progress and prospects. *Critical Reviews in Immunology* **12**, 125–169.

2

Biotechnology in Aquaculture, with Special Reference to Transgenic Salmon

RUNE MALE,[1] JAMES B. LORENS,[2] AUDUN H. NERLAND[3] AND ERIK SLINDE[4*]

[1] Center of Biotechnology, [2] Department of Biochemistry and Molecular Biology, University of Bergen, [3] Norbio A/S, Bergen and [4] MATFORSK, Norwegian Food Research Institute, Ås, Norway

Introduction

Almost 10% of the fish consumed in the world are produced by aquaculture. This regular supply of cultured fish is important to the food industry (Slinde, 1992). The aquaculture industry is expanding at a rapid pace, much of this growth coming from the increasing culture of salmonid species. Since its inception in the 1970s, the production of cultured salmon has grown to over 250 000 tons. The red colour, high quality, uniform size (approximately 4 kg) and year-round availability give cultured salmon its marketing appeal. Salmon represent a valuable source of ω-3 and ω-6 fatty acids, which may prevent 'lifestyle diseases' (Skjervold, 1992), and increased consumption of fish is recommended.

AQUACULTURE OF SALMON

As an anadromous species, Atlantic salmon (*Salmo salar*) spends the juvenile part of its lifecycle in fresh water, migrates to the sea where the major growth phase occurs, and returns to spawn in the freshwater stream where it was born. Spawning occurs in the brisk currents of shallow cold waters. The female deposits eggs into an indentation in the river bed, and the male releases sperm over the eggs. In aquaculture, a broodstock is kept for production of eggs and sperm. A few months after fertilization, the fry hatches and 'start-feeding' commences (*Figure 1*). The fingerlings grow until

---

Abbreviations: ACTH, adrenocorticotropin; AFP, anti-freeze protein; ATPase, adenosine 5′-triphosphate hydrolase; cAMP, adenosine 3′5′-monophosphate; cDNA, complementary DNA; CNS, central nervous system; CREB, cAMP response element binding protein; ES cells, embryonic stem cells; GH, growth hormone; GnRH, gonadotropin releasing hormone; GRF, growth hormone releasing factor; IGF-I, insulin-like growth factor I; PCR, polymerase chain reaction; Pit-1/GHF-1, pituitary-specific transcription factor I; SS, somatostatin.

---

* To whom all correspondence should be addressed.

Biotechnology and Genetic Engineering Reviews — Vol. 11, December 1993
0264–8725/93/11/31–56 $20.00 + $0.00

the parr marks disappear and smoltification occurs. This change from fry to smolt requires an enormous physiological adaptation to the impending rise in osmotic pressure. Smolts develop a silvery colour caused by subcutaneous deposits of guanine. The minimum size at which a salmon will smolt is 15 g, depending on genetic and environmental factors, although seawater survival is better at 30 g. The seawater grow-out phase is carried out in enclosed inlets, submerged cages or tanks.

Salmon growth is influenced by aquaculture production conditions. Salmon along the western Norwegian coast require over 4 years in freshwater streams before smolting. By raising the water temperature during egg incubation and the post-hatching period, 1-year-old smolts (i.e. fry that smolt in the spring of their second year) can be produced at commercial hatcheries. Totland *et al.* (1987) compared the growth rate of Atlantic salmon (2 kg, 56 cm, 17 kg fish m^{-3}) in a commercial scale swimming raceway (length 20 m, depth 4.5 m, width 4 m; streamsetters giving water speed of 28 cm se^{-1}) with salmon raised in normal sea-cages. Raceway-cultured fish gained 40% more weight than siblings raised in the ordinary sea pens. The difference was primarily due to the size distribution of the muscle fibres: white muscle fibres were hypertrophied, while the inner red fibres were smaller. Commercially it was found that the raceway contained 9.2% more 'superior quality' fish, a visual classifica-

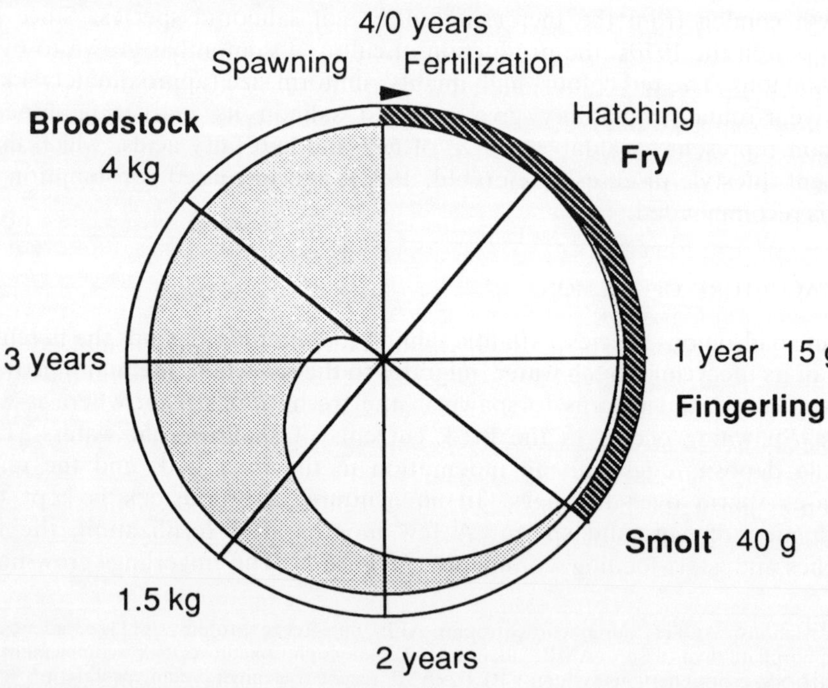

Figure 1. Schematic of Atlantic salmon life-cycle. In aquaculture of Atlantic salmon in the northern hemisphere, fertilization of eggs occurs in November–December and the eggs hatch in the following spring (April–May). The fry are kept in fresh water until the smolt stage is reached (1 year later for a large proportion of the fry). The smolt (30–50 g) are transferred to sea water and grow rapidly and may reach 6–8 kg after 2 years. Hatched area indicates freshwater phase. Usual growth rate is indicated by the shaded area.

tion (based, for example, on fin damage, lacerations) relating to market appeal. Optimal, controlled culturing conditions are necessary both to conduct reproducible experiments and produce high-quality fish for market. Continued improvement of culturing conditions, including refined feed, better caging and cautious selection of suitable aquaculture facility locations, are important future endeavours.

Aquaculture requires large capital investments in buildings, cages, equipment, broodstock, labour and feed. In Norway, smolt culture and the seawater grow-out phase are generally two separate operations. Expenditures related to mortality, inferior quality and disease are highly variable; however, feed quality is paramount to the economic survival of an aquaculture installation. Fish feed must contain all the necessary nutritients for growth, and maintain taste, size and shape in an aquatic environment. This is determined by water absorption, texture and buoyancy factors. The feed must also be acceptable to the fish at different stages of their development. During the seawater grow-out phase, feed comprises a third to a half of the total operating cost. Hence, improving feed utilization represents a major goal for aquaculture biotechnology.

A unique feature of salmon is the 'salmon-red' muscle colour. This is due to the presence of astaxanthin (3,3′-dihydroxy-β-β-carotene-4,4′-dione). Obtained in the wild from shrimps, it is today chemically synthesized and added to fish feed (Bernhard, 1990). Synthetic astaxanthin is expensive, adding 10–15% to the total feed cost. Astaxanthin produced by an algae (Burbrick, 1991) or a yeast such as *Phaffia rhodozyma* (Johnson and An, 1991) represent alternative natural sources.

Diseases caused by different fungi, algae, protozoa, parasites and crustacea create significant problems in aquaculture. Vibriosis, caused by *Vibrio anguillarum* (Knappskog *et al.*, 1993), has inflicted substantial losses on a variety of fish species. In salmon aquaculture, furunculosis (*Aeromonas salmonicida*; McCarthy and Roberts, 1980), bacterial kidney disease (*Renibacterium salmoniarum*; Sanders and Freyer, 1980) and viral diseases (e.g. from infectious pancreatic necrosis virus) can devastate an aquaculture farm. Adequate feed, high hygienic standards, reliable diagnostic methods and efficient pharmaceutical and chemical treatments minimize disease-associated problems. Efficient vaccines are available against different bacterial diseases, but currently no commercial vaccines exist to pre-empt viral diseases. Expenses incurred by fish diseases are substantial and further vaccine development is required.

BREEDING OF ATLANTIC SALMON

Traditional breeding of domesticated animals for phenotypic qualities like growth and disease resistance has been quite successful. Atlantic salmon have been in captivity for only 20 years and traditional breeding has only recently addressed specific traits. The traits targeted by contemporary breeding programmes include: a rapid growth rate, low feed conversion ratio (i.e. dry feed mass used to produce 1 kg of fish), tolerance of high stock densities (i.e.

kg fish per m^3 tank volume), disease resistance, late sexual maturation, large egg size, high fecundity and high market quality of the meat.

Atlantic salmon siblings demonstrate a tremendous (20-fold) growth variation, partly abolished in commercial production by frequent sorting. However, this reveals the potential for enhanced growth. Photoperiod manipulation may improve growth rate and delay sexual maturation of cultured salmon (Stefansson *et al.*, 1991; Hansen, Stefansson and Taranger, 1992). Breeding using specific genetic markers associated with a desirable phenotype is also possible, as illustrated by the specific trypsin isoenzyme marker recently linked to enhanced growth (Torrissen, 1987, 1991). Modern biotechnological techniques for the manipulation and transfer of single genes has engendered the possibility of controlled production of domesticated animals, including fish, with altered or new phenotypes.

Development of transgenic salmon in aquaculture

Transgenic animals have proved important in the study of gene structure and function. A range of species have been employed including several fish species. Various aspects of the production of transgenic fish were recently reviewed (Hew and Fletcher, 1992). Fish are especially suitable for transgenic study, as most have external fertilization and generate a large number of eggs. The rearing of the eggs is similarly simple, and virtually thousands of potentially transgenic embryos may be produced at relatively low cost. Most experiments with transgenic fish have been conducted in the interest of basic research; however, several applied aspects important to aquaculture, including disease resistance, control of sexual maturation, increased growth rate and feed utilization, may benefit from transgenic study. The development of transgenic animals for food production dictates several limitations regarding the choice of methods and their application. It is not our intention to evaluate the marketing of food produced from transgenic animals, but to consider the scientific basis and criteria for the commercial success of transgenic fish (*Table 1*). Further discussion is therefore limited to the use of genes, promoters and methods for DNA transfer which may meet these criteria.

TECHNICAL REQUIREMENTS AND CHOICE OF METHOD FOR INJECTION

Eggs from most fish must survive outside the mother and are therefore equipped with a robust outer egg shell (chorion or zona radiata), which must be penetrated or removed before microinjection. Fish eggs are usually relatively large in size and filled with a yolk mass that often obscures the pronucleus or zygote (*Figure 2*). Some researchers have succeeded in nuclear microinjection of medaka (*Oryzias latipes*) pre-ovulated oocytes followed by artificial fertilization (Ozato, Inoue and Wakamatsu, 1992). Generally, fertilized eggs are microinjected in the cytoplasm. After fertilization, the chorion hardens and may be difficult to penetrate with a microinjection needle. This problem has been circumvented by various methods such as mild trypsinisation (Zhu, 1992), drilling or cutting of a small hole (Rokkones *et al.*,

Table 1. Criteria for production of transgenic fish in commerical aquaculture

	General considerations	Requirements: Ecological	Requirements: Physiological
Gene construct	Available gene sequence from same or closely related species Employ suitable promoter/enhancer Well characterized *in vitro* Stable when incorporated into genome Possible to detect *in vitro* and *in vivo* Avoid retroviral sequences Remove all cloning vector sequences prior to gene transfer Fulfils legislative criteria	Non-virulent Non-transposable Preferably encoding phenotypes which might arise by natural mutation	Use optimal promoter/enhancer to obtain expression in desired organ/tissue
Gene product	Non-toxic Effects of protein must be fully understood	Understanding of repercussions if passed to natural stocks	Non-toxic to animal Regulation and consequences must be fully understood
Transgenic fish	Fulfil legislative criteria Possible to detect transgenic individuals Commercially desirable traits	Comprehensive evaluation of the effects on natural stocks following inter-breeding Development of biological containment strategies (e.g. sterility)	Ethically acceptable changes Produces healthy fish in good condition Genetically stable and inheritable phenotype

1989; Chourrout, Guyomard and Houdebine, 1986) or by physically removing the egg shell with a pair of forceps (Ozato *et al.*, 1986). Salmon eggs are large, typically 5–6 mm in diameter. The pronuclei are not visible due to large amounts of fat droplets (*Figure 2*). The micropyle, an orifice for sperm docking, is visible and readily penetrated with a fine glass needle (Shears *et al.*, 1992; Lossius *et al.*, in press). This technique also ensures that the DNA solution is injected in close proximity to the nuclear area or blastodisc, located just beneath the micropyle after fertilization (*Figure 2a*). It is our experience that a skilled person can inject 100–150 eggs with this technique in a few hours. Normally, 0.1–0.5 μl is injected containing up to 10^6 copies of the DNA construct (2 to 3-fold more DNA than in the zygote), without influencing the survival rate. Several alternative methods for introducing foreign DNA into cells have been developed for use in cell cultures and their potential for production of transgenic fish is discussed in other reviews (Chen and Powers, 1990; Houdebine and Chourrout, 1991; Zhu, 1992). Gene transfer using electroporation has proven successful with eggs from medaka (*Oryzias latipes*: Inoue *et al.*, 1990; Ozato, Inoue and Wakawatsu, 1992) and loach (*Misgurnus anguillicaudatua*: Zhu, 1992), both of which produce relatively small eggs (1 mm). In salmon, however, this technique is limited by the relatively large egg size. An exciting technique which may prove valuable in the future is the culture of totipotent fish embryonic stem (ES) cells, derived from blastocysts. Genetically altered ES cells may be reintroduced into blastocysts to produce viable chimeric animals (Evans and Kaufman, 1981). Another alternative is nuclear transplantation and replacement of the egg pronucleus as discussed by Zhu (1992). These encouraging methods could facilitate rapid, large-scale testing for the recombinant products in cell cultures, a prerequisite for the detection of homologous recombination events in insertional mutagenesis (Thomas and Capecchi, 1986). The technique of gene transfer in ES cells is only reported in cells of mammalian origin; however, blastocysts from zebrafish (*Brachydanio rerio*: S. Lin *et al.*, 1992) and rainbow trout (Nilsson and Cloud, 1992) may be transferred to recipient eggs and contribute to the germ line.

Evidence from several experiments with transgenic fish have revealed that injected DNA can survive, replicate and be expressed extrachromosomaly (see Vielkind, 1992 and references therein). The DNA is apparently incorporated into the genome at a later developmental stage, resulting in a mosaic animal (Guyomard *et al.*, 1989; Fletcher and Davies, 1991). Success in incorporating injected DNA varies between species and researchers from 1 to 75% (Chen and Powers, 1990; Houdebine and Chourrout, 1991). The injected DNA seems to form large concatenates of end-to-end linked molecules (Maclean, Penman and Zhu, 1987; Stuart, McMurray and Westerfield, 1988; Chong and Vielkind, 1989; Guyomard *et al.*, 1989; Winkler, Vielkind and Schartl, 1991). The production of large molecules of non-integrated DNA may have resulted in an overestimation of the number of transgenic fish in some cases. An interesting observation in our laboratory is that protamine mixed with DNA at a 2 : 1 w/w ratio prior to microinjection may increase the rate of transgenesis (Lossius *et al.*, 1993). Injected super-

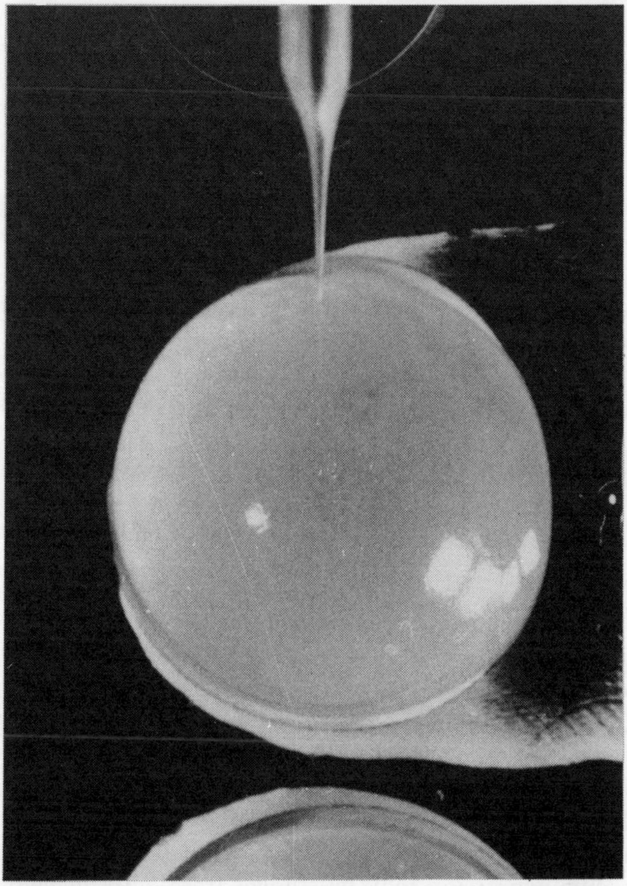

Figure 2. Microinjection in Atlantic salmon eggs. The eggs were fertilized 1–4 h prior to microinjection and kept in tap water on ice. Microinjection was performed through the micropyle using an ~10 μm glass needle connected to a micromanipulator. Approximately 10 nl containing 10^6 copies of Atlantic salmon GH-I gene was injected.

coiled plasmids are transiently expressed but less efficiently incorporated in the genome (Chong and Vielkind, 1989; Penman *et al.*, 1990; Winkler, Vielkind and Schartl, 1991).

The promoter/enhancer utilized in the gene construct dictates under what conditions and in which cell type(s) the gene will be expressed. It is critical to use promoter constructs recognized by the transcriptional machinery of the host organism, either by isolation of the promoter sequences from the same or a closely related species, or by using a strong promoter of viral origin competent to direct transcription in a broad variety of species and cell types. The supply of such sequences for fish systems is presently limited. Fortunately, this will be overcome in the future as more fish genes are characterized. A recent search in GenBank (May 1993) for teleost sequences revealed 1040 records, of which 116 had been added in the last 6 weeks.

To ensure expression from genes transferred into fish embryos, different promoter/enhancer constructs have been tested (reviewed by Chen and

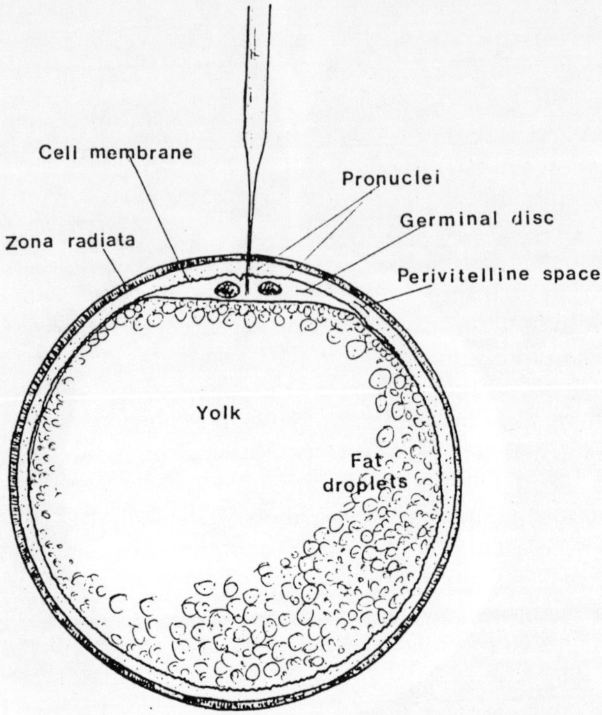

Figure 2a.

Powers, 1990; Moav *et al.*, 1992). Suitable fish cell culture models have recently been developed for analysis of promoters and enhancers (Friedenreich and Schartl, 1990; Collodi *et al.*, 1992; Sharps *et al.*, 1992). Furthermore, cell type-specific expression may be studied transiently *in vivo* in order to identify requisite promoter elements (Rinder *et al.*, 1992; Westerfield *et al.*, 1992; Gong, Hew and Vielkind, 1991). Most transgenic studies have been performed with non-fish promoters/enhancers, usually from animal viruses, which drive efficient gene expression in fish cells. Some recent reports describe transgenic fish which express injected gene constructs controlled by promoters from fish genes, including carp (*Cyprinus carpio* β-actin (Liu *et al.*, 1990; Moav *et al.*, 1992), Winter flounder (*Pseudopleuronectus americanus*) anti-freeze protein (Shears *et al.*, 1991; Fletcher, Davies and Hew, 1992) and ocean pout (*Macrozoarces americanus*) anti-freeze protein (Du *et al.*, 1992). Typically, the expression of injected gene constructs has varied between transgenic individuals, conceivably the result of integration within active or inactive regions of the genome. Preliminary assessment of gene constructs in cell culture models and rapidly growing teleosts (Zebrafish/medaka) is especially important for salmon, which have a long generation time (*Figure 1*).

Introduction of novel salmon phenotypes by genetic engineering

Several fish species are protected from freezing in ice-laden sea water by high levels of systemic anti-freeze protein (AFP: DeVries, Komatsu and Feeney, 1970; Hew and Fletcher, 1985; Davies, Hew and Fletcher, 1988). Salmon lack these proteins and may freeze if kept in sea water below –0.5 to –0.8°C. To improve the winter resistance of Atlantic salmon, the AFP gene from Winter flounder was microinjected into fertilized eggs (Shears *et al.*, 1991; Fletcher, Davies and Hew, 1992). The researchers used approximately 10^6 copies of an AFP construct in a plasmid (pUC9) as linear DNA and screened with a polymerase chain reaction (PCR) assay for transgenic salmon. Approximately 3% were found to carry the gene construct in full blood with 40% of these expressing the gene. Crossing of transgenic males with wild-type females generated F1 progeny carrying the gene construct below the expected Mendelian frequency, indicating that the transgenic male was mosaic for the AFP gene (Fletcher, Davies and Hew, 1992). Expression of the AFP gene was detected, albeit 100-fold less than that considered necessary for efficient protection at low temperatures. Furthermore, the AFP protein was not processed to remove the *N*-terminal precursor, as in the flounder. These observations indicated that Atlantic salmon lack the necessary proteolytic processing mechanisms for the AFR protein and possibly specific transcription factor(s) needed for efficient transcription of the flounder AFP gene.

MANIPULATING EXISTING GENES: REGULATION OF SEXUAL MATURATION

Early sexual maturation of cultured salmon incurs substantial losses in commercial aquaculture. It has been suggested that manipulation of gonadotropin releasing hormone (GnRH) may postpone or even totally inhibit sexual maturation in fish (Alestrøm *et al.*, 1992). GnRH is delivered by the hypothalamic portal system to the pituitary and stimulates the release of the pituitary gonadotropins which regulate gametogenesis and reproductive capability. Both cDNA and genomic clones of Atlantic salmon GnRH have been characterized (Klungland *et al.*, 1992). This facilitates the development of GnRH antagonists and gene constructs directing expression of these peptides to the hypothalamus to compete with native GnRH for receptor (Alestrøm *et al.*, 1992). Alternatively, sterile fish may be produced by hormonal sex reversal or by inducing tetraploidity using shock treatment of fertilized eggs with elevated temperature or pressure (Lincoln and Bye, 1989; Pepper, 1990 and references therein). These methods are relatively simple and applicable to eggs from cultured transgenic salmon to avert breeding of escaped farmed fish with natural salmon populations.

ENHANCEMENT OF SOMATIC GROWTH

The majority of transgenic fish experiments have been conducted with the growth hormone gene to enhance the growth rate, since it is both a convenient marker and a commercially important phenotype (Chen and

Powers, 1990; Houdebine and Chourrout, 1991; Moav *et al.*, 1992). The physiological and molecular regulation of growth have been studied intensely in mammals, but also in fish. Thus, manipulation of the growth rate of fish on an aquaculture scale may be accomplished in the near future. However, very few (if any) of the studies reported meet the limitations for use of transgenic fish in aquaculture outlined in *Table 1*. To fulfil these requirements, given that somatic growth is a highly regulated process, ramifying to other physiological processes, it is necessary to evaluate the current understanding of growth regulation and evaluate possible strategies for growth enhancement through gene transfer and the potential consequences of such manipulation.

The regulation of growth hormone

Growth hormone (GH), a pituitary polypeptide hormone, plays a key role in the somatic growth of vertebrate organisms. Growth hormone is synthesized by somatotrophic cells located in the anterior lobe (adenohypophysis) of the pituitary gland. Growth hormone secretion is under the dual control of opposing hypothalamic hormones: growth hormone releasing factor (GRF) and the GH-release inhibitor, somatostatin (SS: *Figure 3*). In mammals, the adenohypophysis is vascularized with a hypophyseal portal system, which delivers neuropeptides directly from the hypothalamic neurosecretory neurons. Although several fish species may control pituitary function by direct innervation (Ball, 1981), a study of the Atlantic salmon pituitary suggest that the brain utilizes neurovascular links to convey neuropeptides (Friedberg and Ekengren, 1977). Recent studies have revealed mammalian-like hypophyseal portal systems in other teleost species (Baskaran and Sathyanesan, 1992). Teleost GRF- and SS-hypothalamic neurosecretory pathways have been identified by immunohistochemistry (Pan, Liu and Bancroft, 1985; Marivoet, Moons and Vandesande, 1988; Olivereau *et al.*, 1984). Both human and a recently purified carp GRF stimulate GH release *in vitro* (Luo *et al.*, 1990; Le Bail *et al.*, 1991; Vaughan *et al.*, 1992) and *in vivo* (Peter *et al.*, 1984). Somatostatin has been characterized from a number of teleost species, including salmon (Conlon, 1990). Somatostatin inhibits GH secretion *in vitro* and *in vivo*, and auto-immunization against SS increases GH levels and somatic growth rates (Fryer, Nishioka and Bern, 1979; Luo *et al.*, 1990; Hall, Harvey and Scanes, 1986 and references therein; Le Bail *et al.*, 1991).

Mammalian GH secretion profiles are characterized by episodic bursts (Tannenbaum and Ling, 1984; Plotsky and Vale, 1985; Jansson, Edén and Isaksson, 1985), and pulsatile GH secretion was recently described in trout (Le Bail *et al.*, 1991). Episodic GH regulation is generated by the alternating secretion of GRF and SS: SS withdrawal sets the timing and GRF determines the magnitude of GH release (Plotsky and Vale, 1985; Kraicer *et al.*, 1988). This ultradian secretory pattern is critical to effective growth (Robinson and Clark, 1987). Pulsatile infusion of GRF, in contrast to continuous treatment, is required to achieve effective growth stimulation in rats (Clark and Robinson, 1985).

Control of the GH-regulating neuropeptides by the central nervous system

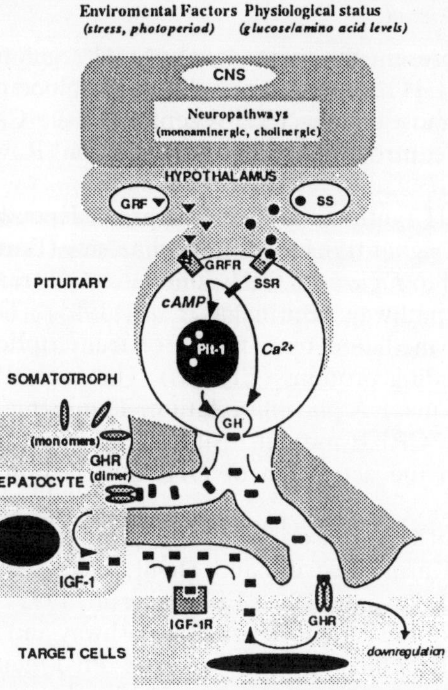

Figure 3. Schematic of growth hormone regulation and its action. Physiological stimuli, both external and internal, which influence GH levels in the body (top) act via neuropathways in the central nervous system (CNS). These neuropathways modulate hypothalamic neurosecretory neurons, which synthesize growth hormone releasing factor (GRF) and somatostatin (SS) (Rawlings and Mason, 1989). For example, cholinergic control of SS release is connected to peripheral feedback pathways mediated by glucose and GH (Page, Dieguez and Scanlon, 1989). The central adrenaline system exerts dual GH control: α_2-adrenoreceptor activation stimulates GRF secretion, while α_1-receptors exclusively stimulate SS release (Cella *et al.*, 1987). GRF and SS secreted into the hypothalamic–pituitary portal system act on GH-producing cells (somatotrophs) in the pituitary. GRF binds to its receptor, a member of the classical G-protein-coupled, seven-helix transmembrane domain protein superfamily (C. Lin *et al.*, 1992; Mayo, 1992). GRF cell-signalling operates by raising cAMP levels through the G_s-protein-linked receptor/adenylate cyclase pathway (Barinaga *et al.*, 1985). The rise in intracellular cAMP serves to stimulate GH secretion and synthesis. In addition, GRF cues proliferation of the somatotrophic cell lineage (Billestrup, Swanson and Vale, 1986). GRF stimulates GH synthesis independent of GH secretion, via the cAMP-dependent activation of the pituitary-specific transcription factor, Pit-1 (McCormick *et al.*, 1990; Chen *et al.*, 1990). Pit-1 interacts at the GH promoter with other ubiquitous transcription factors to enhance GH transcription. Somatostatin binding to its G_i-coupled receptor inhibits Ca^{2+}-influx through voltage-operated calcium channels (Rosenthal *et al.*, 1988), blocking exocytosis of storage vesicles containing Zn^{2+}-coordinated GH dimers (Cunningham, Mulkerrin and Wells, 1991). Once in the systemic circulation, GH is borne as a monomer by serum binding proteins. GH binds to its cell-surface receptor, located on several cell types. GH binds in a sequential manner to two GH receptor monomers (Cunningham *et al.*, 1991). Dimerization of the GH receptor generates the signalling mechanism, resulting in the synthesis and secretion of IGF-I in hepatocytes. Hepatic IGF-I will in turn circulate inducing proliferation at several target tissues. In addition, GH-induced IGF-I secretion may act in an autocrine fashion inducing proliferation and differentiation of different cell types. Feedback pathways exist at all levels: for example, GRF and SS stimulate each other's secretion while down-regulating their own release; both GH and IGF-1 decrease somatotroph GRF responsiveness while stimulating SS release; and there is down-regulation of the GH receptor on target tissues (Hall, Harvey and Scanes, 1986; Page, Dieguez and Scanlon, 1989; Gluckman and Breier, 1989). In addition, each of these different modes of GH regulation are subject to modulation by other factors, exemplified by the GH synthesis-enhancing effects of glucocorticoids and thyroid hormones and estradiol induction of hepatic high-affinity GH binding sites (Gluckman and Breier, 1989).

(CNS) appears to represent the major level of GH regulation. Physiological stimuli that influence GH release, such as changes in blood metabolites, stress and photoperiod, in most cases act via receptors in the CNS which activate brain neuropathways controlling hypothalmic function (Rawlings and Mason, 1989).

In the pituitary, GH synthesis and release are independent events, mediated through distinct signal transduction mechanisms (Barinaga *et al.*, 1985; outlined in the legend to *Figure 3*). GRF stimulates GH transcription through a cAMP-dependent pathway (Barinaga *et al.*, 1985). The transcriptional effects of cAMP are mediated by a family of transcription factors, cAMP response element binding proteins (CREBs) (Habener, 1990; Karin, 1989) activated by protein kinase A phosphorylation. Transgenic mice expressing a non-phosphorylatable CREB mutant resulted in a dwarf phenotype, clearly implicating CREB in the activation of GH transcription (Struthers *et al.*, 1991).

The key regulator of GH transcription is the pituitary-specific transcription factor Pit-1/GHF-1 (Karin, Castrillo and Theill, 1990), recently described in salmonids (Ono and Takayama, 1992; Elsholtz *et al.*, 1992). Pit-1 represents a major nuclear target of the cAMP-dependent pathway and Pit-1 transcription is induced by CREB. Once 'turned on', the Pit-1 gene remains active throughout the cell life-span and cooperates with other transcription factors at the GH promoter to induce GH transcription (Karin, Castrillo and Theill, 1990).

The effects of growth hormone

The immediate metabolic effects of GH administration in mammals resemble those of insulin (Press, 1988). Following this acute response, GH stimulates adipocyte lipolysis and hepatic carbohydrate catabolism. This insulin-antagonistic effect apparently balances the increased insulin secretion. Growth hormone is also an important protein anabolic hormone, increasing amino acid transport, nitrogen retention and stimulating protein synthesis. The metabolic effects of GH in fish are less well understood, although where investigated they are congruent with the mammalian studies. For example, GH treatment stimulated lipid mobilization and liver lipolytic enzyme activity in coho salmon parr (Sheridan, 1986).

The growth-inducing (mitogenic) effects of GH are largely mediated by insulin-like growth factor I (IGF-I). Originally, the liver was considered the target organ of GH action leading to the production of systemic IGF-I (Daughaday *et al.*, 1972). This 'somatomedin hypothesis' designated IGF-I as the effector molecule for GH, a model in concert with other pituitary hormone pathways. Indeed, IGF-I treatment leads to enhanced growth (Schoenle *et al.*, 1982; Guler *et al.*, 1989) and transgenic mice strains over-expressing IGF-I display a 1.3-fold increase in weight (Mathews *et al.*, 1988). However, it became clear that GH receptors are present on multiple cell types, also capable of IGF-I synthesis stimulated by GH binding (see Waters *et al.*, 1989; Humbel, 1990). Howard Green and colleagues have recently proposed the concept of a dual effector system to describe the growth-inducing effects of GH (Green, Morikawa and Nixon, 1985). The theory predicts that GH creates the preferred target cells of IGF-I action through differentiation of precursor cells into mature, IGF-I responsive cells: GH was shown to promote the differentiation of precursor pre-adipocyte cells (Zezulak and Green, 1986). The young, differentiated adipocytes become sensitive to the mitogenic effects of IGF-I. Other cell types have also been implicated in this scheme: chondrocytes (Isaksson, Jansson and Gause, 1982), granulocytes (Hochberg *et al.*, 1989) and T-lymphocytes (Geffner *et al.*, 1990). It therefore would appear that GH regulates IGF-I production at two levels: (1) by generation of endocrine IGF-I from the liver, creating a large reservoir of serum IGF-I, bound to binding proteins; and (2) locally produced IGF-I, acting in a autocrine/paracrine fashion. The relative contributions of these two sources of IGF-I are still debated (see Humbel, 1990).

The role of IGF-I was recently addressed in salmonids. Recombinant mammalian IGF-I was shown to have a growth-enhancing effect in coho salmon only when constantly infused into animals exhibiting submaximal growth rates (McCormick *et al.*, 1992). Skyrud *et al.* (1989) did not observe enhanced growth following weekly (bolus) injections of mammalian IGF-I. This study revealed a dramatic IGF-I-induced hypoglycaemic and hypoaminoacidaemic effect, previously encountered in mammalian studies, possibly due to saturation of insulin receptors. Recently, several salmonid IGF-Is have been cloned, displaying 80% identity with the mammalian protein (Cao *et al.*, 1989; Shamblott and Chen, 1992; Duguay *et al.*, 1992). Cao *et al.* (1989) also demonstrated a transcriptional enhancement of hepatic salmon IGF-I following GH treatment, consistent with data from mammalian systems (Mathews, Norstedt and Palmiter, 1986).

The growth-promoting effects of GH in teleosts are well studied (Donaldson *et al.*, 1979; Weatherley and Gill, 1987). Pickford showed 40 years ago that administration of mammalian GH restores the effect of hypophysectomy and stimulates body weight and length in normal teleosts (Pickford, 1954). Several studies with partially purified fish hormones have documented a general growth-promoting effect (see Donaldson *et al.*, 1979; Weatherley and Gill, 1987). The availability of recombinant mammalian and later fish GH facilitated more extensive studies (Gill *et al.*, 1985; Sekine *et al.*, 1985; Agellon *et al.*, 1988). Gill *et al.* (1985) demonstrated that intraperitoneal injections of natural and recombinant chicken or bovine GH resulted in dramatic dose-dependent increases in growth. The GH-treated fish were longer, a consequence of accelerated bone growth (Takagi *et al.*, 1992), and demonstrated a greater feed conversion efficiency. Similar results were obtained by Agellon *et al.* (1988) using recombinant rainbow trout GH. Interestingly, the GH-treated fish retained their growth increase several weeks after the final GH treatment, growing to twice the size of control fish.

Several transgenic fish studies have assessed the effects of GH overproduction in fish (see Chen and Powers, 1990). In a recent study, Du *et al.* (1992) fused chinook salmon GH cDNA to an anti-freeze protein gene promoter. Salmon shown to carry the microinjected gene construct displayed between 2- and 6-fold enhanced growth rate. Hence, our current understanding of GH action in fish indicates that many of the effects found in mammals are present. In addition, there are interesting fish-specific actions of GH. Growth hormone was recently shown to regulate the production of antifreeze protein in Winter flounder (Idler *et al.*, 1989), and to participate in the seawater adaptation of several salmonid species (Barron, 1985). Growth hormone also enhances the activity of gill Na^+/K^+-ATPase and the steroidogenic capacity and ACTH responsiveness of the inter-renal gland, indicating both a direct and indirect role for GH during smoltification (Young, 1988; Richman and Zaugg, 1987).

The cellular effects of GH are mediated through interaction with membrane-associated receptors. The GH receptor was recently cloned from several mammalian species (see Mathews, 1991). The GH receptor belongs to the haematopoietic receptor superfamily, a group of receptors generally involved in cell growth and differentiation (Bazan, 1990). Recently, the three-dimensional structure of the extracellular domain of the human GH receptor bound to GH was determined (de Vos, Ultsch and Kossiakoff, 1992). A single GH molecule binds two receptor molecules, in a sequential manner, at two different regions of the GH molecule (Cunningham *et al.*, 1991). The nature of the intracellular signal from the GH receptor is still unknown, but requires the formation of a receptor dimer (Fuh *et al.*, 1992). At high GH levels, receptor binding is predominatly monomeric, attenuating the biological response (Fuh *et al.*, 1992). This could explain the lack of correlation between plasma GH levels and growth observed by Du *et al.* (1992) – individuals possessing the highest levels of GH showed no increase in growth compared with non-transgenic controls.

In general, the physiology and molecular biology of GH appear evolutionarily conserved between teleosts and mammals. Hence, the recent surge of understanding of GH actions and regulation will greatly aid aquaculture applications.

EVALUATION OF CANDIDATE GENES AND PROMOTERS FOR MANIPULATION OF GROWTH

The generation of transgenic salmon and their utilization in commerical aquaculture entail conflicting interests. The use of transgenic animals in food production is highly regulated and subject to public concern (Hallerman and Kapuscinski, 1992). This argues for a conservative approach in the design of gene constructs. The use of salmonid genes and promoters/enhancers for the generation of transgenic salmon would represent a *de facto* gene copy number increase and/or modest adjustment of existing physiological processes and would, presumably, be acceptable.

Injection of genes under control of their native promoters would ensure expression exclusively in the natural organ/cells and conceivably maintain a near normal physiological regulation. Such a 'natural' construct, as with all gene constructs, may be unduly influenced by the integration vicinity in the chromosomal DNA. Considerable proportions of chromosomal DNA in somatic cells are not actively transcribed and the chance of integration in a silent area are substantial. Indeed, integrated test constructs driven by strong viral promoters are often not expressed in the F1 generation (Culp, Nüsslein-Volhard and Hopkins, 1991). Hence a large number of transgenic fish would have to be produced and tested for at least two generations to obtain the desired phenotypic effect. The effectiveness of a promoter may be manipulated once detailed knowledge of the transcription factor recognition sites (short DNA sequences required for specific interactions) is obtained. For instance, addition of multiple Pit-1 recognition sites to reporter constructs enhances transcription several fold in Pit-1-expressing cells (Ding *et al.*, 1991; Yan and Bancroft, 1991). The removal of repressor sequences from the rat GH promoter resulted in a 3- to 4-fold increase in activity (Pan, Liu and Bancroft, 1990). However, most experiments have utilized promoters that direct expression of the gene to an organ/cell-type which does not normally synthesize the protein (see Chen and Powers, 1990).

Salmon parr (pre-smolts) treated with GH for short periods maintained their growth advantage relative to untreated siblings (Agellon *et al.*, 1988). This could reduce the relatively long freshwater phase, as size is a major determinant of smoltification. Transgenic fish which express higher GH levels during this critical life-stage may later benefit. Thus a gene promoter primarily active at this stage would be useful. One candidate is the prolactin promoter. In salmon, prolactin has been associated with freshwater osmoregulation prior to smoltification (Barron, 1985). This may imply that the prolactin promoter is mainly active prior to seawater entry, but this must be confirmed experimentally.

Many of the factors involved in the regulation of somatic growth are potential targets for transgenic manipulation of growth. GRF, a key regulator of GH, is an obvious candidate. Transgenic mice over-expressing GRF exhibit enhanced somatic growth (Hammer *et al.*, 1985). Vaughan *et al.* (1992) recently reported the carp GRF peptide sequence. Even so, targeted expression to the cognate hypothalamic neurons will require characterization of the corresponding gene sequence. Pancreas is capable of expressing and processing the active GRF peptide; in fact, GRF was originally isolated from pancreatic neoplasias of acromegalic patients (Rivier *et al.*, 1982; Guillemin *et al.*, 1982). Hence, promoters from genes expressed in the endocrine pancreas could be employed. Supraphysiological GRF levels induce somatotrophic cell hyperplasia and could potentially damage adjacent hormone-secreting cells (Asa *et al.*, 1984; Hammer *et al.*, 1985; Billestrup, Swanson and Vale, 1986). The mammalian GRF receptor was recently cloned and characterized (C. Lin *et al.*, 1992; Mayo, 1992). Over-expression of the receptor in a transgenic strain, if not strictly limited to somatotrophic cells, could induce undesirable side-effects, as the cAMP signalling pathway mediates a variety of effects in different cells. However, mice carrying a cholera toxin gene construct under control of the GH promoter, generating somatotroph-specific cAMP elevation, resulted in gigantism (Burton *et al.*, 1991). The GRF receptor and GH are both regulated by the transcription factor Pit-1 (C. Lin *et al.*, 1992). But as

Pit-1 also regulates lactotroph and thyrotroph development (Li *et al.*, 1990), it may not be easily manipulated in transgenic animals.

Growth hormone has been the favourite reporter gene in transgenic fish studies; GH is a well-characterized system (see above), inducing an easily scorable phenotype, and many teleost genes are available. Several laboratories have generated transgenic fish exhibiting spectacular growth enhancement (Chen and Powers, 1990; Du *et al.*, 1992; Zhu, 1992). Growth hormone expressed in non-somatotrophic cells escapes hypothalamic regulation and feedback down-regulation (with the exception of the GH receptor; *Figure 3*). Constitutively high GH levels do not necessarily result in enhanced growth (Du *et al.*, 1992). Indeed, high GH levels could antagonize growth by sequestering the GH receptor in non-productive monomeric interactions (Fuh *et al.*, 1992).

The gene encoding the GH receptor could potentially represent a viable target for transgene manipulation. The gene is only characterized in mammals (Leung *et al.*, 1987; Mathews, 1991) and would require cloning from salmon. The GH receptor may be a limiting step in somatic growth regulation; however, the signal transduction mechanism is still unknown. As with the GRF receptor, over-expression may cause unacceptable effects in heterologous cells. The use of the GH receptor gene promoter or possibly the IGF-I gene promoter could avert this problem. The final effector of somatic growth, IGF-I, is probably produced in all GH-responsive cells. Thus IGF-I remains a promising target for gene transfer, but as pointed out above, expression would have to be fine-tuned to avoid cross-reaction with the insulin receptor. The conflicting results of IGF-I as a growth enhancer necessitate further study of the physiological effects of high levels of IGF-I before it may be useful in aquaculture.

MONITORING TRANSGENIC ANIMALS

Rapid and reliable methods for the identification of fish harbouring injected DNA are a prerequisite for large-scale transgenic studies. The polymerase chain reaction (PCR) (Saiki *et al.*, 1985) is ideal: it is specific and sensitive, requiring only minute tissue samples; a large number of samples can be assayed at relatively low cost; and it is amenable to automation. Various applications of PCR in screening transgenic fish have been reviewed by Davies and Gauthier (1992). The traditional PCR technique will not discriminate between integrated and extrachromosomal DNA and may therefore only be employed in preliminary screenings. Progeny from putative transgenic fish must be shown to carry the construct. Furthermore, this F1 generation should manifest the desired phenotype. The injection of native gene constructs presents a special problem; namely, how to differentiate between the endogenous and transferred genes. We solved this problem for Atlantic salmon GH-I gene transfers by modifying the gene sequence slightly to provide construct-specific PCR primers, without disturbing the encoded amino acid sequence (Lossius *et al.*, 1993). This strategy is outlined in *Figure 4*. With this technique, only a minute blood sample was required for detection

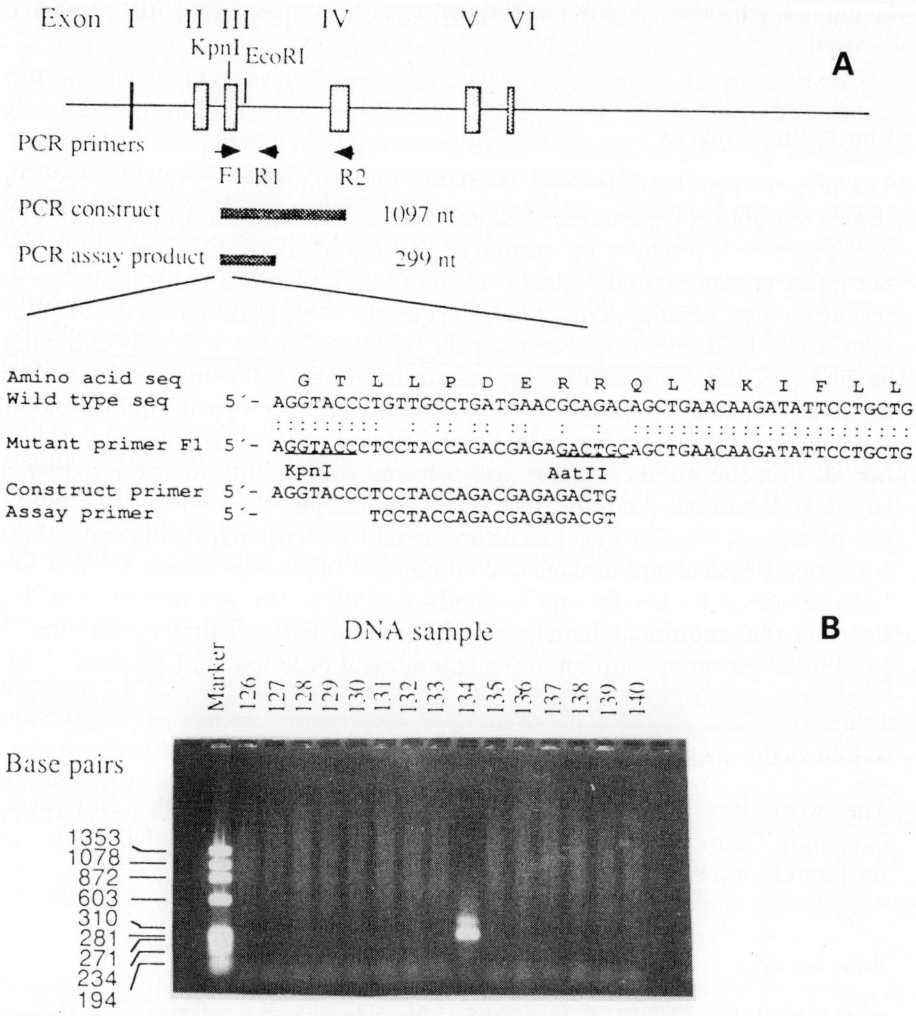

Figure 4. Mutagenesis and detection of Atlantic salmon GH gene in transgenic fish.
(A) The Atlantic salmon GH-I gene (Male *et al.*, 1992) was modified using PCR. The reaction was initiated with a 52 nucleotides (nt) primer from exon II containing a unique Kpn I restriction enzyme site followed by 20 nt selected to obtain maximum divergence from the wild type sequence, while conserving the encoded amino acids sequence, and flanked by 24 nt identical to the wild type GH-I sequence. The PCR reaction was supplemented with the reverse primer R2 from exon IV and a primer composed of the first 28 nt of F1 (construct primer). The resulting PCR fragment was cleaved with Kpn I and Eco RI and exchanged with the cognate wild type sequence. The resulting modified GH-I gene was injected into fertilized Atlantic salmon eggs as described (Lossius *et al.*, 1993).
(B) Analysis of 15 PCR reactions from individual fish. Two-year-old fish (n=140) injected with the GH-I construct shown in (A) were analysed by PCR using blood samples. One positive individual was identified. The assay primer and a reverse primer in intron III were employed to generate a construct-specific 299 bp DNA fragment. *Methods*: Blood (3 μl) from individual fish was mixed with 100 μl of 100/mM NaOH, boiled for 3 min, and centrifuged briefly (3 min, 12 000 × g); 1.5 μl from these samples were assayed in standard 50 μl PCR reactions (Innis *et al.*, 1991), incubated for 5 min at 94°C and run for 35 cycles; 94°C, 49.6°C, 72°C each for 1 min. The PCR reaction products were resolved in a 1.2% agarose gel and visualized by ethidium bromide staining.

(salmon erythrocytes are nucleated), although any tissue (e.g. fin clips) may be used.

Concluding remarks

The technical prerequisites are available for establishing transgenic salmon strains suitable for commercial aquaculture. The acquculture of transgenic species not only requires a thorough evaluation of applicable gene constructs, but raises ecological and ethical considerations, discussed in detail in several recent reviews (Tiedje *et al.*, 1989; Ferguson, 1990; Kapuscinski and Hallerman, 1990; Hallerman and Kapuscinski, 1992). Salmon are usually cultivated in submerged sea-pens, and escapees are inevitable, allowing the interaction of transgenic salmon with wild populations. The ecological implications of this interaction depend foremost on the fitness of the particular phenotype, but also on the number of escaped fish and their ability to enter spawning streams. A biological control to reduce or eliminate the reproductive potential of transgenic fish in aquaculture would at least partly alleviate these concerns. Physical and biological containment of transgenic fish has recently been reviewed by Devlin and Donaldson (1992). The production of sterile fish has the additional benefit of eliminating early sexual maturation, a substantial economic burden in the commercial production of salmon.

Acknowledgements

This work was supported by Marine Genetics A/S and the Norwegian Research Council for Sciences and Humanities (NAVF). J.B.L. is the recipient of a postdoctoral fellowship from NAVF.

References

AGELLON, L.B., EMERY, C.J., JONES, J.M., DAVIES, S.L., DINGLE, A.D. AND CHEN, T.T. (1988). Promotion of rapid growth of rainbow trout (*Salmo gairdneri*) by a recombinant fish growth hormone. *Canadian Journal of Fisheries and Aquatic Science* **45**, 146–151.

ALESTRØM, P., KISEN, G., KLUNGLAND, H. AND ANDERSEN, Ø. (1992). Fish gonadotropin-releasing hormone gene and molecular approaches for control of sexual maturation: Development of a transgenic fish model. *Molecular and Marine Biology and Biotechnology* **1**, 376–379.

ASA, S.L., SCHEITHAUER, W., BILBAO, J.M., HORVATH, E., RYAN, N., KOVACS, K., RANDALL, R.V., LAWS, E.R., SINGER, W., LINFOOT, J.A., THORNER, M.O. AND VALE, W. (1984). A case for hypothalamic acromegaly: A clinicopathological study of six patients with hypothalamic gangliocytomas producing growth hormone-releasing factor. *Journal of Clinical Endocrinology and Metabolism* **58**, 796–803.

BALL, J.N. (1981). Hypothalamic control of the pars distalis in fishes, amphibians, and reptiles. *General and Comparative Endocrinology* **44**, 135–170.

BARINAGA, M., BILIZIKJIAN, L.M., VALE, W.W., ROSENFELD, M.G. AND EVAN, R.M. (1985). Independent effects of growth hormone releasing factor on growth hormone release and gene transcription. *Nature* **314**, 279–281.

Barron, M.G. (1985). Endocrine control of smoltification in anadromous salmonids. *Journal of Endocrinology* **108**, 313–319.

Baskaran, G. and Sathyanesan, A.G. (1992). Tetrapod-like hypothalamo-hypophysial portal system in the teleost *Megalops cyprinoides* (Broussonet). *General and Comparative Endocrinology* **86**, 211–219.

Bazan, J.F. (1990). Structural design and molecular evolution of a cytokine receptor superfamily. *Proceedings of the National Academy of Sciences, USA* **87**, 6934–6938.

Bernhard, K. (1990). Synthetic astaxanthine: The route of a carotenoid from research to commercialization. In *Carotenoids: Chemistry and Biology* (N.I. Krinsky, M.M. Mathews-Roth and R.F. Taylor, Eds), pp. 337–363. Eighth International Symposium on Carotenoids. Plenum Press, New York.

Billestrup, N., Swanson, L.W. and Vale, W. (1986). Growth hormone-releasing factor stimulates proliferation of somatotrophs *in vitro*. *Proceedings of the National Academy of Sciences, USA* **83**, 6854–6857.

Burbrick, P. (1991). Production of astaxanthin from Haematococcus. *Bioresource and Technology* **38**, 237–239.

Burton, F.H., Hasel, K.W., Bloom, F.E. and Sutcliffe, J.G. (1991). Pituitary hyperplasia and gigantism in mice caused by a cholera toxin transgene. *Nature* **350**, 74–77.

Cao, Q.-P., Duguay, S.J., Plisetskaya, E., Steiner, D.F. and Chan, S.J. (1989). Nucleotide sequence and growth hormone-regulated expression of salmon insulin-like growth factor I mRNA. *Molecular Endocrinology* **3**, 2005–2010.

Cella, S.G., Locatelli, V., de Gennaro, V., Wehrenberg, W.B. and Müller, E.E. (1987). Pharmacological manipulations of α-adrenoreceptors in the infant rat and effects on growth hormone secretion. *Endocrinology* **120**, 1639–1643.

Chen, S., Ingraham, H.A., Treacy, M.N., Albert, V.R., Wilson, L. and Rosenfeld, M.G. (1990). Autoregulation of pit-1 gene expression mediated by two *cis*-active promoter elements. *Nature* **346**, 583–586.

Chen, T.T. and Powers, D.A. (1990). Transgenic fish. *Trends in Biotechnology* **8**, 209–215.

Chen, T.T., Lin, C.M., Zhu, Z., Gonzales-Villasenor, L.I., Dunham, R.A. and Powers, D.A. (1990). Gene transfer, expression and inheritance of rainbow trout and human growth hormone genes in carp and loach. In *Transgenic Models in Medicine and Agriculture* (R. Church, Ed.), pp. 127–139. Wiley-Liss, New York.

Chong, S.S.C. and Vielkind, J.R. (1989). Expression and fate of CAT reporter gene microinjected into fertilized medaka (*Oryzias latipes*) eggs in the form of plasmid DNA, recombinant phage particles and its DNA. *Theoretical Applications of Genetics* **78**, 369–380.

Chourrout, D., Guyomard, R. and Houdebine, L.M. (1986). High efficiency gene transfer in rainbow trout (*Salmo gairdneri* Rich.) by microinjection into egg cytoplasm. *Aquaculture* **5**, 143–150.

Clark, R.G. and Robinson, I.C.A.F. (1985). Growth induced by pulsatile infusion of amidated fragment of human growth hormone releasing factor in normal and GHRF-deficient rats. *Nature* **314**, 281–283.

Collodi, P., Kamei, Y., Ernst, T., Miranda, C., Buhler, D.R. and Barnes, D.W. (1992). Culture of cells from zebrafish (*Brachydanio rerio*) embryo and adult tissues. *Cell Biology and Toxicology* **8**, 43–61.

Conlon, J.M. (1990). Somatostatin: Aspects of molecular evolution. In *Progress in Comparative Endocrinology* (A. Epple, C.G. Scanes and M.H. Stetson, Eds). *Progress in Clinical and Biological Research* **342**, 10–15.

Culp, P., Nüsslein-Volhard, C. and Hopkins, N. (1991). High-frequency germ-line transmission of plasmid DNA sequences injected into fertilized zebrafish eggs. *Proceedings of the National Academy of Sciences, USA* **88**, 7953–7957.

CUNNINGHAM, B.C., MULKERRIN, M.G. AND WELLS, J.A. (1991). Dimerization of human growth hormone by zinc. *Science* **253**, 545–548.

CUNNINGHAM, B.C., ULTSCH, M., DE VOS, A.M., MULKERRIN, M.G., KLUASER, K.R. AND WELLS, J.A. (1991). Dimerization of the extracellular domain of the human growth hormone receptor by a single hormone molecule. *Science* **254**, 821–825.

DAUGHADAY, W.H., HALL, K., RABEN, M.S., SALMON, W.D., VAN DEN BRANDE, J.L. AND VAN WYK, J.J. (1972). Somatomedin: Proposed designation for sulphation factor. *Nature* **235**, 107.

DAVIES, P.L. AND GAUTHIER, S.Y. (1992). The application of PCR to aquaculture. In *Transgenic Fish* (C.L. Hew and G.L. Fletcher, Eds), pp. 61–71. World Scientific Publishing Co., Singapore.

DAVIES, P.L., HEW, C.L. AND FLETCHER, G.L. (1988). Fish antifreeze proteins: Physiology and evolutionary biology. *Canadian Journal of Zoology* **66**, 2611–2617.

DEVLIN, R.H. AND DONALDSON, E.M. (1992). Containment of genetically altered fish with emphasis on salmonids. In *Transgenic Fish* (C.L. Hew and G.L. Fletcher, Eds), pp. 229–265. World Scientific Publishing Co., Singapore.

DE VOS, a., uLTSCH, M. AND KOSSIAKOFF, A.A. (1992). Human growth hormone and extracellular domain of its receptor: Crystal structure of the complex. *Science* **255**, 306–312.

DEVRIES, A.L., KOMATSU, S.K. AND FEENEY, R.E. (1970). Chemical and physical properties of freezing point depression glycoproteins from Antarctic fishes. *Journal of Biological Chemistry* **245**, 2901–2913.

DING, Y., LU, W., ROBERSON, M.S., MOYE-ROWLEY, W.S. AND MAURER, R.A. (1991). The tissue-specific mammalian transcription factor, Pit-1, activates transcription in *Saccharomyces cerevisia. Molecular Endocrinology* **5**, 1239–1245.

DONALDSON, E.M., FAGERLUND, U.H.M., HIGGS, D.A. AND MCBRIDE, J.R. (1979). Hormonal enhancement of growth. In *Fish Physiology* (W.S. Hoar, D.J. Randall and J.R. Brett, Eds), Vol. VIII, pp. 455–597. Academic Press, New York.

DU, S.J., GONG, Z., FLETCHER, G.L., SHEARS, M.A., KING, M.J., IDLER, D.R. AND HEW, C.L. (1992). Growth enhancement in transgenic Atlantic salmon by use of an 'all fish' chimeric growth hormone gene construct. *Bio/Technology* **10**, 176–181.

DUGUAY, S.J., PARK, L.K., SAMADPOUR, M. AND DICKHOFF, W.W. (1992). Nucleotide sequence and tissue distribution of three insulin-like growth factor I prohormones in salmon. *Molecular Endocrinology* **6**, 1202–1210.

ELSHOLTZ, H.P., MAJUMDAR-SONNYLAL, S., XIONG, F., GONG, Z. AND HEW, C.L. (1992). Phylogenetic specificity of prolactin gene expression with conservation of Pit-1 function. *Molecular Endocrinology* **6**, 515–522.

EVANS, M.J. AND KAUFMAN, M.H. (1981). Establishment in culture of pluripotential cells from mouse embryos. *Nature* **292**, 154–156.

FERGUSON, M.M. (1990). The genetic impact of introduced fishes on native species. *Canadian Journal of Zoology* **68**, 1053–1057.

FLETCHER, G.L. AND DAVIES, P.L. (1991). Transgenic fish for aquaculture. In *Genetic Engineering Principles and Methods* (J.K. Setlow, Ed.), Vol. 13, pp. 331–370. Plenum Press, New York.

FLETCHER, G.L., DAVIES, P.L. AND HEW, C.L. (1992). Genetic engineering of freeze-resistant Atlantic salmon. In *Transgenic Fish* (C.L. Hew and G.L. Fletcher, Eds), pp. 190–208. World Scientific Publishing Co., Singapore.

FRIEDBERG, G. AND EKENGREN, B. (1977). The vascularization and neuroendocrine pathways of the pituitary gland in the Atlantic salmon, *Salmo salar. Canadian Journal of Zoology* **55**, 1284–1298.

FRIEDENREICH, H. AND SCHARTL, M. (1990). Transient expression directed by homologous and heterologous promoter and enhancer sequences in fish cells. *Nucleic Acids Research* **18**, 3299–3305.

FRYER, J.N., NISHIOKA, R.S. AND BERN, H.A. (1979). Somatostatin inhibition of teleost growth hormone secretion. *General and Comparative Endocrinology* **39**, 244–246.

FUH, G., CUNNINGHAM, B.C., FUKUNAGA, R., NAGATA, S., GOEDDEL, D.V. AND WELLS, J.A. (1992). Rational design of potent antagonists to the human growth hormone receptor. *Science* **256**, 1677–1680.

GEFFNER, M.E., BERSCH, N., LIPPE, B.M., ROSENFELD, R.G., HINTZ, R.L. AND GOLDE, D.W. (1990). Growth hormone mediates the growth of T-lymphoblast cells lines via locally generated insulin-like growth factor-I. *Journal of Clinical Endocrinology and Metabolism* **71**, 464–469.

GILL, J.A., SUMPTER, J.P., DONALDSON, E.M., DYE, H.M., SOUZA, L., BERG, T., WYPYCH, J. AND LANGLEY, K. (1985). Recombinant chicken and bovine growth hormones accelerate growth in aquacultured juvenile Pacific salmon *Oncorhynchus kisutch*. *Bio/Technology* **3**, 643–646.

GLUCKMAN, P.D. AND BREIER, B.H. (1989) The regulation of the growth hormone receptor. In *Biotechnology in Growth Regulation* (R.B. Heap, C.G. Prosser and G.E. Lamming, Eds), pp. 27–33. Butterworth, London.

GONG, Z., HEW, C.L. AND VIELKIND, J.R. (1991). Functional analysis and temporal expression of promoter regions from fish antifreeze protein genes in transgenic Japanese medaka embryos. *Molecular and Marine Biology and Biotechnology* **1**, 64–72.

GREEN, H., MORIKAWA, M. AND NIXON, T. (1985). A dual effector theory of growth hormone action. *Differentiation* **29**, 195–198.

GUILLEMIN, R., BRAZEAU, R., BÖHLEN, P., ESCH, F., LING, N. AND WEHRENBERG, W.B. (1982). Growth hormone-releasing factor from a human pancreatic tumor that caused acromegaly. *Science* **218**, 585–587.

GULER, H.P., ZAPF, J., BINZ, K. AND FROESCH, E.R. (1989). Growth promotion using recombinant insulin-like growth factor-I. In *Biotechnology in Growth Regulation* (R.B. Heap, C.G. Prosser and G.E. Lamming, Eds), pp. 119–122. Butterworth, London.

GUYOMARD, R., CHOURROUT, D., LEROUX, C., HOUDEBINE, L.M. AND POURRAIN, F. (1989). Integration and germ line transmission of foreign genes microinjected into fertilized trout eggs. *Biochimie* **71**, 857–863.

HABENER, J.F. (1990). Cyclic AMP response element binding proteins: A cornucopia of transcription factors. *Molecular Endocrinology* **4**, 1087–1094.

HALL, T.R., HARVEY, S. AND SCANES, C.G. (1986). Control of growth hormone secretion in the vertebrates: A comparative survey. *Comparative Biochemistry and Physiology* **84A**, 231–253.

HALLERMAN, E.M. AND KAPUSCINSKI, A.R. (1992). Ecological and regulatory uncertainties associated with transgenic fish. In *Transgenic Fish* (C.L. Hew and G.L. Fletcher, Eds), pp. 209–228. World Scientific Publishing Co., Singapore.

HAMMER, R.E., BRINSTER, R.L., ROSENFELD, M.G., EVANS, R.M. AND MAYO, K.E. (1985). Expression of human growth hormone-releasing factor in transgenic mice results in increased somatic growth. *Nature* **315**, 413–416.

HANSEN, T., STEFANSSON, S.O. AND TARANGER, G.L. (1992). Growth and sexual maturation in Atlantic salmon, *Salmo salar* L., reared in sea cages at two different light regimes. *Aquatics and Fisheries Management* **23**, 135–140.

HEW, C.L. AND FLETCHER, G.L. (1985). Biochemical adaptation to the freezing environment-structure, biosynthesis and regulation of fish antifreeze polypeptides. In *Circulation, Respiration and Metabolism* (R. Gilles, Ed.), pp. 553–563. Springer-Verlag, Berlin.

HEW, C.L. AND FLETCHER, G.L. (Eds) (1992). *Transgenic Fish*. World Scientific Publishing Co., Singapore.

HOCHBERG, Z., MAOR, G., LEWINSON, D. AND SILBERMANN, M. (1989). The direct effects of growth hormone on chondrogenesis and osteogenesis. In *Biotechnology in Growth Regulation* (R.B. Heap, C.G. Prosser and G.E. Lamming, Eds), pp. 123–128. Butterworth, London.

HOUDEBINE, L.M. AND CHOURROUT, D. (1991). Transgenesis in fish. *Experientia* **47**, 891–897.

HUMBEL, R.E. (1990). Insulin-like growth factors I and II. *European Journal of Biochemistry* **190**, 445–462.

IDLER, D.R., FLETCHER, G.L., BELKHODE, S., KING, M.J. AND HWANG, S.J. (1989). Regulation of antifreeze protein production in Winter flounder: A unique function for growth hormone. *General and Comparative Endocrinology* **74**, 327–334.

INNIS, M.A., GELFAND, D.H., SNINSKY, J.J. AND WHITE, T.J. (1990). *PCR Protocols: A Guide to Methods and Applications*. Academic Press, San Diego.

INOUE, K., YAMASHITA, S., HATA, J.-I., KABENO, S., ASADA, S., NAGAHISA, E. AND FUJITA, T. (1990). Electroporation as a new technique for producing transgenic fish. *Cell Differences and Development* **29**, 123–128.

ISAKSSON, O.G.P., JANSSON, J.O. AND GAUSE, I.A.M. (1982). Growth hormone stimulates longitudinal bone growth directly. *Science* **216**, 1237–1239.

JANSSON, J.-O., EDÉN, S. AND ISAKSSON, O. (1985). Sexual dimorphism in the control of growth hormone secretion. *Endocrinology Review* **6**, 128–150.

JOHNSON, E.A. AND AN, G.-H. (1991). Astaxanthin from microbial sources. *Critical Reviews in Biotechnology* **11**, 297–326.

KAPUSCINSKI, A.R. AND HALLERMAN, E.M. (1990). Transgenic fish and public policy: Anticipating environmental impacts of transgenic fish. *Fisheries* **15**, 2–11.

KARIN, M., CASTRILLO, J.-L. AND THEILL, L.E. (1990). Growth hormone gene regulation: A paradigm for cell-type-specific gene activation. *Trends in Genetics* **6**, 92–96.

KLUNGLAND, H., LORENS, J., ANDERSEN, Ø., KISEN, G. AND ALESTRÖM, P. (1992). The Atlantic salmon prepro-gonadotropin-releasing hormone gene and mRNA. *Molecular and Cellular Endocrinology* **84**, 167–174.

KNAPPSKOG, D.H., RØDSETH, OM., SLINDE, E. AND ENDRESEN, C. (1993). Immunochemical analysis of *Vibrio anguillarum* strains isolated from cod (*Gadus morhua* L.) suffering from vibriosis. *Journal of Fish Disease* **16**, 327–338.

KRAICER, J., SHEPPARD, M.S., LUKE, J., LUSSIER, B., MOOR, B.C. AND COWAN, J.S. (1988). Effect of withdrawal of somatostatin and growth hormone (GH)-releasing factor on GH release *in vitro*. *Endocrinology* **122**, 1810–1815.

LE BAIL, P.-Y., SUMPTER, J.P., CARRAGHER, J., MOUROT, B., NIU, P.D. AND WEIL, C. (1991). Development and validation of a highly sensitive radioimmunoassay for chinook salmon (*Oncorhynchus tshawytscha*) growth hormone. *General and Comparative Endocrinology* **83**, 75–85.

LEUNG, D.W., SPENCER, S.A., CACHIANES, G., HAMMONDS, R.G., COLLINS, C., HENZEL, W.J., BARNARD, R., WATERS, M.J. AND WOOD, W.I. (1987). Growth hormone receptor and serum binding protein: Purification, cloning and expression. *Nature* **330**, 537–543.

LI, S., CRENSHAW, E.B., RAWSON, E.J., SIMMONS, D.M., SWANSON, L.W. AND ROSENFELD, M.G. (1990). Dwarf locus mutants lacking three pituitary cell types result from mutations in the POU-domain gene pit-1. *Nature* **347**, 528–533.

LIN, C., LIN, S.-C., CHANG, C.-P. AND ROSENFELD, M.G. (1992). Pit-1-dependent expression of the receptor for growth hormone releasing factor mediates pituitary cell growth. *Nature* **360**, 765–768.

LIN, S., LONG, W., CHEN, J. AND HOPKINS, N. (1992). Production of germ-line chimeras in zebrafish by cell transplants from genetically pigmented to albino embryos. *Proceedings of the National Academy of Sciences, USA* **89**, 4519–4523.

LINCOLN, R.F. AND BYE, V.J. (1989). Genetic techniques for the commercial control of sexual maturation in rainbow trout (*Salmo gairdneri* R.). In *Aquaculture: A*

Biotechnology in Progress (N. de Pauw, E. Jaspers and N. Wilkins, Eds), pp. 551–559. European Aquaculture Society, Breden, Belgium.

LIU, Z., MOAV, B., FARAS, A.J., GUISE, K.S., KAPUSCINSKI, A.R. AND HACKETT, P.B. (1990). Development of expression vectors for transgenic fish. *Bio/Technology* **8**, 1268–1272.

LOSSIUS, I., KRYVI, H., LORENS, J.B., MALE, R., NERLAND, A.H., OLSEN, L., TELLE, W., TOTLAND, G.K. AND AASLAND, R. (1993). Cloning and characterization of growth regulating genes from Atlantic salmon: A strategy for increasing growth rate by microinjection of a modified growth hormone gene into fertilized salmon eggs. In *Physiology and Biochemistry of Marine Fish Larvae* (B.T. Walther and H.J. Fyhn, Eds), pp. 40–46. University of Bergen, Bergen.

LUO, D., MCKEOWN, B.A., RIVIER, J. AND VALE, W. (1990). *In vitro* responses of rainbow trout (*Oncorhynchus mykiss*) somatotrophs to carp growth hormone-releasing factor (GRF) and somatostatin. *General and Comparative Endocrinology* **80**, 288–298.

MACLEAN, N., PENMAN, D. AND ZHU, Z. (1987). Introduction of novel genes into fish. *Bio/Technology* **5**, 257–261.

MALE, R., NERLAND, A.H., LORENS, J.B., TELLE, W., LOSSIUS, I. AND TOTLAND, G.K. (1992). The complete nucleotide sequence of the Atlantic slamon growth hormone I gene. *Biochimica Biophysica Acta* **1130**, 345–348.

MARIVOET, S., MOONS, L., VANDESANDE, F. (1988). Localization of growth hormone releasing factor-like immunoreactivity in the hypothalamo-hypophyseal system of the frog (*Rana temporaria*) and sea bass (*Dicentrarchus labrax*). *General and Comparative Endocrinology* **72**, 72–79.

MATHEWS, L.S. (1991). Molecular biology of growth hormone receptors. *Trends in Endocrinology and Metabolism* **2**, 176–180.

MATHEWS, L.S., NORSTEDT, G. AND PALMITER, R.D. (1986). Regulation of insulin-like growth factor I gene expression by growth hormone. *Proceedings of the National Academy of Sciences, USA* **83**, 9343–9347.

MATHEWS, L.S., HAMMER, R.E., BEHRINGER, R.R., D'ERCOLE, A.J., BELL, G.I., BRINSTER, R.L. AND PALMITER, R.R. (1988). Growth enhancement of transgenic mice expressing human insulin-like growth factor I. *Endocrinology* **123**, 2827–2833.

MAYO, K.E. (1992). Molecular cloning and expression of a pituitary-specific receptor for growth hormone-releasing hormone. *Molecular Endocrinology* **6**, 1734–1744.

MCCARTHY, D.H. AND ROBERTS, R.J. (1980). Furunculosis of fish: The present state of our knowledge. In *Advances in Aquatic Microbiology* (M.A. Droop and H.W. Jannasch, Eds), pp. 293–341. Academic Press, London.

MCCORMICK, A., BRADY, H., THEILL, L.E. AND KARIN, M. (1990). Regulation of the pituitary-specific homeobox gene GHF1 by cell-autonomous and environmental cues. *Nature* **345**, 829–832.

MCCORMICK, S.D., KELLEY, K.M., YOUNG, G., NISHIOKA, R.S. AND BERN, H.A. (1992). Stimulation of coho salmon growth by insulin-like growth factor I. *General and Comparative Endocrinology* **86**, 398–406.

MOAV, B., LIN, Z., MOAV, N.L., GROSS, M.L., KAPUSCINSKI, A.R., FARAS, A.J., GUISE, K. AND HACKETT, P.B. (1992) Expression of heterologous genes in transgenic fish. In *Transgenic Fish* (C.L. Hew and G.L. Fletcher, Eds), pp. 120–141. World Scientific Publishing Co., Singapore.

NILSSON, E.E. AND CLOUD, J.G. (1992). Rainbow trout chimeras produced by injection of blastomeres into recipient blastulae. *Proceedings of the National Academy of Sciences, USA* **89**, 9425–9428.

OLIVEREAU, M., OLIVEREAU, F., VANDESANDE, F. AND OLIVEREAU, J. (1984). Somatostatin in the brain and the pituitary of some teleosts: Immunocytochemical identification and the effect of starvation. *Cell Tissue Research* **238**, 289–296.

ONO, M. AND TAKAYAMA, Y. (1992). Structure of cDNA encoding chum salmon pituitary-specific transcription factor, Pit-1/GHF-1. *Gene* **226**, 275–279.

OZATO, K., KONDOH, H., INOHARA, H., IWAMATSU, T., WAKAMATSU, Y. AND OKADA, T.S. (1986). Production of transgenic fish: Introduction and expression of chicken δ-crystallin gene in medaka embryos. *Cell Differences and Development* **19**, 237–244.

OZATO, K., INOUE, K. AND WAKAMATSU, Y. (1992). Gene transfer and expression in medaka embryos. In *Transgenic Fish* (C.L. Hew and G.L. Fletcher, Eds), pp. 27–43. World Scientific Publishing Co., Singapore.

PAGE, M.D., DIEGUEZ, C. AND SCANLON, M.F. (1989). Neuroregulation of growth hormone secretion. In *Biotechnology in Growth Regulation* (R.B. Heap, C.G. Prosser and G.E. Lamming, Eds), pp. 47–55. Butterworth, London.

PAN, J.X., LECHAN, R.M., LIN, H.D. AND JACKSON, I.M.D. (1985). Immunoreactive neuronal pathways of growth hormone-releasing hormone (GRH) in the brain and pituitary of the teleost *Gadus morhua. Cell Tissue Research* **241**, 487–493.

PAN, W.T., LIU, Q. AND BANCROFT, C. (1990). Identification of a growth hormone gene promoter repressor element and its cognate double- and single-stranded DNA-binding proteins. *Journal of Biological Chemistry* **265**, 7022–7028.

PENMAN, D.J., BEECHING, A.J., PENN S. AND MACLEAN, N. (1990). Factors affecting survival and integration following microinjection of novel DNA into rainbow trout eggs. *Aquaculture* **85**, 35–50.

PEPPER, D.P. (1991). Proceedings of the Atlantic Canada Workshop on methods for the production of non-maturing salmonids. *Canadian Technical Report on Fisheries and Aquatic Science,* No. 1789.

PETER, R.E., NAHORNIAK, C.S., VALE, W.W. AND RIVIER, J.E. (1984). Human pancreatic growth hormone-releasing factor (hpGRF) stimulates growth hormone release in goldfish. *Journal of Experimental Zoology* **231**, 161–163.

PICKFORD, G.E. (1954). The response of hypophysectomized male killifish to purified fish growth hormone, as compared with the response to purified beef growth hormone. *Endocrinology* **55**, 274–287.

PLOTSKY, P.M. AND VALE, W. (1985). Patterns of growth hormone-releasing factor and somatostatin secretion into the hypophysial-portal circulation of the rat. *Science* **230**, 461–463.

PRESS, M. (1988). Growth hormone and metabolism. *Endocrinology Review* **4**, 391–414.

RAWLINGS, S.R. AND MASON, W.T. (1989). Modulation of growth hormone release: From CNS to the secretory event. In *Biotechnology in Growth Regulation* (R.B. Heap, C.G. Prosser and G.E. Lamming, Eds), pp. 36–46. Butterworth, London.

RICHMAN, N.H. AND ZAUGG, W.S. (1987). Effects of cortisol and growth hormone on osmoregulation in pre- and desmoltified coho salmon (*Oncorhynchus kisutch*). *General and Comparative Endocrinology* **65**, 189–198.

RINDER, H., BAYER, T.A., GERTZEN, E.-M. AND HOFFMANN, W. (1992). Molecular analysis of the ependymic gene and functional test of its promoter region by transient expression in *Brachydanio rerio. DNA Cell Biology* **11**, 425–432.

RIVIER, J., SPIESS, J., THORNER, M. AND VALE, W. (1982). Characterization of a growth hormone releasing factor from a human pancreatic islet tumor. *Nature* **300**, 276–278.

ROBINSON, I.C.A.F. AND CLARK, R.G. (1987). The secretory pattern of GH and its significance for growth in the rat. In *Growth Hormone: Basic and Clinical Aspects* (O. Isaksson, C. Binder, K. Hall and B. Hökfeldt, Eds), pp. 109–127. Elsevier, Amsterdam.

ROKKONES, E., ALESTRÖM, P., SKJERVOLD, H. AND GAUTVIK, K.M. (1989). Microinjection and expression of a mouse metallothionine human growth hormone fusion gene in fertilized salmonid eggs. *Journal of Comparative Physiology* **B158**, 751–758.

ROSENTHAL, W., HESCHELER, J., HINSCH, K.-D., SPICHER, K., TRAUTWEIN, W. AND SCHULTZ, G. (1988). Cyclic AMP-independent, dual regulation of voltage-dependent Ca^{2+} currents by LHRH and somatostatin in a pituitary cell line. *EMBO Journal* **7**, 1627–1633.

SAIKI, R.K., SCHARF, S., FALOONA, F., MULLIS, K.B., HORN, G.T., ERLICH, H.A. AND ARNHEIM, N. (1985). Enzymatic amplification of β-globin genomic sequences and restriction site analysis for diagnosis of sickle cell anemia. *Science* **230**, 1350–1354.

SANDERS, J.E. AND FRYER, J.L. (1980). *Renibacterium Salmoinarum* gen. nov., the causative agent of bacterial kidney disease in salmonid fishes. *International Journal of Systematic Bacteriology* **30**, 496–502.

SCHOENLE, E., ZAPF, J., HUMBEL, R.E. AND FROESCH, E.R. (1982). Insulin-like growth factor I stimulates growth in hypophysectomized rats. *Nature* **296**, 252–253.

SEKINE, S., MIZUKAMI, T., NISHI, T., KUWANA, Y., SAITO, A., SATO, M., ITOH, S. AND KAWAUCHI, H. (1985). Cloning and expression of cDNA for salmon growth hormone in *Escherichia coli. Proceedings of the National Academy of Sciences, USA* **82**, 4306–4310.

SHAMBLOTT, M.J. AND CHEN, T.T. (1992). Identification of a second insulin-like growth factor in a fish species. *Proceedings of the National Academy of Sciences, USA* **89**, 8913–8917.

SHARPS, A., NISHIYAMA, K., COLLODI, P. AND BARNES, D. (1992). Comparison of activities of mammalian viral promoters directing gene expression *in vitro* in zebrafish and other fish cell lines. *Molecular and Marine Biology and Biotechnology* **1**, 426–431.

SHEARS, M.A., FLETCHER, G.L., HEW, C.L., GAUTHIER, S. AND DAVIES, P.L. (1991). Transfer, expression, and stable inheritance of antifreeze protein genes in Atlantic salmon (*Salmo salar*). *Molecular and Marine Biology and Biotechnology* **1**, 58–63.

SHEARS, M.A., KING, M.J., GODDARD, S.V. AND FLETCHER, G.L. (1992). Gene transfer in salmonids by injection through the micropyle. In *Transgenic Fish* (C.L. Hew and G.L. Fletcher, Eds), pp. 44–60. World Scientific Publishing Co., Singapore.

SHERIDAN, M.A. (1986). Effects of thyroxin, cortisol, growth hormone, and prolactin on lipid metabolism of coho salmon, *Oncorhynchus kisutch*, during smoltification. *General and Comparative Endocrinology* **64**, 220–238.

SKJERVOLD, H. (1992). *Lifestyle Diseases and the Human Diet: How Should the New Discoveries Influence Future Food Production?* Ås Trykk, Norway.

SKYRUD, T., ANDERSEN, Å., ALESTRØM, P. AND GAUTVIK, K.M. (1989). Effects of recombinant human growth hormone and insulin-like growth factor 1 on body growth and blood metabolites in brook trout (*Salvelinus fontinalis*). *General and Comparative Endocrinology* **75**, 247–255.

SLINDE, E. (1992). Salmon aquaculture: The importance of biotechnology. *Meat Focus International* **1**, 144–147.

STEFANSSON, S.O., BJÖRNSSON, B.T., HANSEN, T., HAUX, C., TARANGER, G.L. AND SAUNDERS, R.L. (1991). Growth, parr-smolt transformation, and changes in growth hormone of Atlantic salmon (*Salmo salar*) reared under different photoperiods. *Canadian Journal of Fisheries and Aquatic Science* **48**, 2100–2108.

STRUTHERS, R.S., VALE, W.W., ARIAS, C., SAWCHENKO, P.E. AND MONTMINY, M.R. (1991). Somatotroph hypoplasia and dwarfism in transgenic mice expressing a non-phosphorylatable CREB mutant. *Nature* **350**, 622–624.

STUART, G.W., MCMURRAY, J.V. AND WESTERFIELD, M. (1988). Replication, integration and stable germ line transmission of foreign sequences injected into early zebrafish embryos. *Development* **103**, 403–412.

TAKAGI, Y., MORIYAMA, S., HIRANO, T. AND YAMADA, J. (1992). Effects of growth hormones on bone formation and resorption in rainbow trout (*Oncorhynchus mykiss*), as examined by histomorphometry of the pharyngeal bone. *General and Comparative Endocrinology* **86**, 90–95.

TANNENBAUM, G.S. AND LING, N. (1984). The interrelationship of growth hormone (GH)-releasing factor and somatostatin in generatin of the ultradian rhythm of GH secretion. *Endocrinology* **115**, 1952–1957.

THOMAS, K.R. AND CAPECCHI, M.R. (1986). Introduction of homologous DNA sequences into mammalian cells induces mutations in the cognate gene. *Nature* **324**, 34–38.

TIEDJE, J.M., COLWELL, R.K., GROSSMAN, Y.L., HODSON, R.E., LENSKI, R.E., MACK, R.N. AND REGAL, P.J. (1989). The planned introduction of genetically engineered organisms: Ecological considerations and recommendations. *Ecology* **70**, 298–315.

TILSETH, S., HANSEN, T. AND MØLLER, D. (1991). Historical development of salmon aquaculture. *Aquaculture* **99**, 1–9.

TORRISSEN, K.R. (1987). Genetic variation of trypsin-like isozymes correlated to fish size of Atlantic salmon (*Salmo salar*). *Aquaculture* **62**, 1–10.

TORRISSEN, K.R. (1991). Genetic variation in growth rate of Atlantic salmon (*Salmo salar*) with different trypsin-like isozyme patterns. *Aquaculture* **93**, 299–312.

TOTLAND, G.K., KRYVI, H., JØDESTØL, K.A., CHRISTIANSEN, E.N., TANGERÅS, A. AND SLINDE, E. (1987). Growth and composition of the swimming muscle of adult Atlantic salmon (*Salmo salar* L.) during long-term sustained swimming. *Aquaculture* **66**, 299–313.

VAUGHAN, J.M., RIVIER, J., SPIESS, J., PENG, C., CHANG, J.P., PETER, R.E. AND VALE, W. (1992). Isolation and characterization of hypothalamic growth-hormone releasing factor from common carp, *Cyprinus carpio*. *Neuroendocrinology* **56**, 539–549.

VIELKIND, J.R. (1992). Medaka and zebrafish: Ideal as transient and stable transgenic systems. In *Transgenic Fish* (C.L. Hew and G.L. Fletcher, Eds), pp. 72–91. World Scientific Publishing Co., Singapore.

WATERS, M.J., SPENCER, S.A., LEUNG, D., HAMMONDS, R.G., CACHIANES, G., HENZEL, W.J., WOOD, W.I., BARNARD, R., QUIRK, P. AND HAMLIN, G. (1989). Growth hormone receptors and binding proteins. In *Biotechnology in Growth Regulation* (R.B. Heap, C.G. Prosser and G.E. Lamming, Eds), pp. 15–26. Butterworth, London.

WEATHERLEY, A.H. AND GILL, H.S. (1987). *The Biology of Fish Growth*. Academic Press, London.

WESTERFIELD, M., WEGNER, J., JEGALIAN, B.G., DEROBERTIS, E.M. AND PÜSCHEL, A.W. (1992). Specific activation of mammalian Hox promoters in mosaic transgenic zebrafish. *Genes Development* **6**, 591–598.

WINKLER, C., VIELKIND, J.R. AND SCHARTL, M. (1991). Transient expression of foreign DNA during embryonic and larval development of the medaka fish (*Oryzias latipes*). *Molecular and General Genetics* **226**, 129–140.

YAN, G. AND BANCROFT, C. (1991). Mediation by calcium of thyrotropin-releasing hormone action on the prolactin promoter via transcription factor Pit-1. *Molecular Endocrinology* **5**, 1488–1497.

YOUNG, G. (1988). Enhanced response of the interrenal of coho salmon (*Oncorhynchus kisutch*) to ACTH after growth hormone treatment *in vivo* and *in vitro*. *General and Comparative Endocrinology* **71**, 85–92.

ZEZULAK, K.M. AND GREEN, H. (1986). The generation of insulin-like growth factor I-sensitive cells by growth hormone action. *Science* **233**, 551–553.

ZHU, Z. (1992). Generation of fast growing transgenic fish: Methods and mechanisms. In *Transgenic Fish* (C.L. Hew and G.L. Fletcher, Eds), pp. 92–119. World Scientific Publishing Co., Singapore.

3
The Facts and Fancy of Microgravity Protein Crystallization

BARRY L. STODDARD,[1*] GREGORY K. FARBER,[2] AND ROLAND K. STRONG[1]

[1] *Division of Basic Sciences, Program in Structural Biology, Fred Hutchinson Cancer Research Center, Seattle, WA and* [2] *Departments of Chemistry and Molecular Biology and Center for Biomolecular Structure, Pennsylvania State University, Philadelphia, PA, USA*

Introduction

The first step in the elucidation of the three-dimensional structure of a molecule by X-ray crystallography is the growth of diffraction quality crystals of that molecule. Many diverse techniques have been developed to obtain crystals of the requisite quality. Growth of crystals in the absence of gravity has been hypothesized to be one such method that might be of general utility for the improvement of crystal quality. Since 1983, over a dozen experiments have been conducted in low Earth orbit on a variety of platforms by several consortiums of American and European scientists in an effort to take advantage of this method. In this review, we explore the scientific basis for this hypothesis and evaluate the published results from these experiments, a particularly timely undertaking since the US Congress is currently considering spending tens of *billions* of dollars to build a space station which, among other tasks, would be used as a microgravity crystallization laboratory.

X-ray crystallography is a powerful technique for the elucidation of the three-dimensional structures of biological macromolecules, but depends upon the availability of suitable crystals. Crystallography has yielded the structures of hundreds of proteins, nucleic acids and viruses, as can be seen by a casual perusal of the literature since the first crystal structure of a protein was determined in 1960 by Kendrew *et al.* The availability of these structures has profoundly altered our understanding of the structural basis of enzymology, immunology and molecular biology. Visualization of the structures of proteins involved in disease has extended our understanding of the mechanisms

Abbreviations: BLD, boundary layer diffusion device; HEWL, hen egg white lysozyme; PMP, polymethylpentene; RNase S, ribonuclease S; 10-THFS, 10-formyl tetrahydrofolate synthetase; VD, vapour-diffusion device.

* To whom all correspondence should be addressed.
Biotechnology and Genetic Engineering Reviews — Vol. 11, December 1993
0264–8725/93/11/57–77 $20.00 + $0.00

of illness to the molecular level, and rational drug design holds the promise of creating new treatments for many of these disorders. X-ray crystallography remains a difficult and demanding technique, though several advances have made structures much more amenable to solution in the past decade. These advances include an enormous increase in the computer power affordable by the average crystallographer as well as the development of equally powerful computer software and algorithms, the development of X-ray sources and detector systems in individual laboratories and at synchrotron light source facilities, and advances in molecular biology which have allowed for the production of large quantities of pure proteins through recombinant expression systems.

However, the process of X-ray crystallography remains dependent on the availability of diffraction-quality crystals of the molecule to be studied, hence the term 'crystallography'. The physical processes underlying the crystallization of biological macromolecules remain to be determined. Consequently, the methods used by crystallographers to crystallize proteins have been determined by trial and error for the most part; rational algorithms for determining the crystallization conditions for a new protein do not exist. Often the crystals obtained of a given sample prove to be unsuitable for diffraction experiments. In these cases, it is not always apparent what to try next, or even if there is any set of conditions that will yield suitable crystals. Because of these reasons, crystallization is often described as the 'bottle neck' of X-ray crystallography, and explains the intense interest in developing techniques that might yield diffraction-quality crystals from situations where there are only poor-quality crystals or none at all.

Protein crystallization and the microgravity environment

Biological macromolecules are crystallized by creating the necessary degree of supersaturation in an appropriate chemical solution. The traditional methods of protein crystallization (see McPherson, 1982, for a thorough review) involve mixing a protein solution in an appropriate buffer (to maintain a suitable pH and ionic strength) with a precipitating agent. For biological macromolecules, these agents include salts (such as ammonium sulphate), organic solvents (such as 2-methyl-2, 4-pentanediol) and organic polymers (such as polyethylene glycol). If the conditions have been chosen correctly, the formation of suitable nuclei occurs, inducing the growth of crystals. Otherwise, the result is usually an amorphous precipitant. Nuclei form in protein and precipitant solutions at characteristic degrees of supersaturation, which can be quite high for proteins. Nucleation can occur on dust particles or on irregularities in the surface of the container. In the simplest crystallization method, known as 'batch' crystallization, the initial solutions are set up at the necessary degree of supersaturation at the beginning of the experiment. The crystallographer simply waits for the formation of nuclei and the growth of crystals once the solutions have been mixed.

The most commonly used method of crystallization, the vapour dialysis technique, involves mixing a volume of protein solution with some initial concentration of the chosen precipitating agent. A drop of this mixture is then suspended in a sealed container together with a much larger volume, or reservoir, of a solution of the precipitating agent at the desired final concentration for the experiment. The drop then equilibrates with the reservoir through the vapour phase, resulting in a reduction in the size of the drop with a concurrent increase in the concentrations of the protein and the precipitant. If the conditions have been chosen correctly, the necessary degree of supersaturation will be reached and crystals will be obtained. The advantages of vapour diffusion experiments include the ability of the vapour diffusion experiment to be conducted on smaller quantities of protein than are usually used in batch experiments, an important point when one considers the arduous task of obtaining proteins which are difficult to purify. The vapour dialysis experiment also gives the crystallographer some control over the approach to nucleation, since unlike the batch methods, the experiment does not involve static concentrations.

In the microdialysis or free-interface diffusion experiment, a variation on the vapour dialysis method, the drop of protein or precipitant solution is separated from the precipitant solution reservoir not by an air space but either by a dialysis membrane (microdialysis) or by layering the two solutions on top of each other and depending on the difference in density of the two solutions to keep them separate (free-interface diffusion). In this technique, the protein concentration does not change, as the volumes remain constant; the diffusion of the protein is either restrained by the dialysis membrane or by the greater mass of the protein relative to the precipitating agent. Only the precipitant concentration varies, giving the crystallographer yet another means to approach nucleation. (*see Figure 1* for a graphical description of these techniques).

The elimination of gravity from crystallization experiments may reduce the detrimental effects of several processes and create conditions highly conducive to the growth of high-quality crystals. In the Earth's gravity well, concentration gradients in solutions and air spaces lead to the formation of convection currents through differences in density; everyday examples of this phenomenon include hot air balloons and oil rising to the top of salad dressing. Many consequences of this process occur in crystallization experiments. (1) Protein crystals are usually denser than the solutions they grow from, and therefore crystals not attached to a surface tend to collect in the bottom of hanging drops or crystallization vials. They bump into each other and clearly interact during their growth (Koszelak and McPherson, 1988). Crystals often grow fused together, limiting their utility in diffraction experiments, and the close proximity of growing crystals may hinder or limit growth. (2) Convection currents in sealed crystallization chambers lead to the formation of convection cells in the air spaces of vapour diffusion experiments, which hastens the approach to equilibrium, possibly to the detriment of the resultant crystals. (3) It is often impossible to set up free-interface diffusion experiments in the laboratory because the density differences between the different solutions are insufficient to establish stable solution layers. (4) Perhaps the most detrimental effect occurs at the surface of the

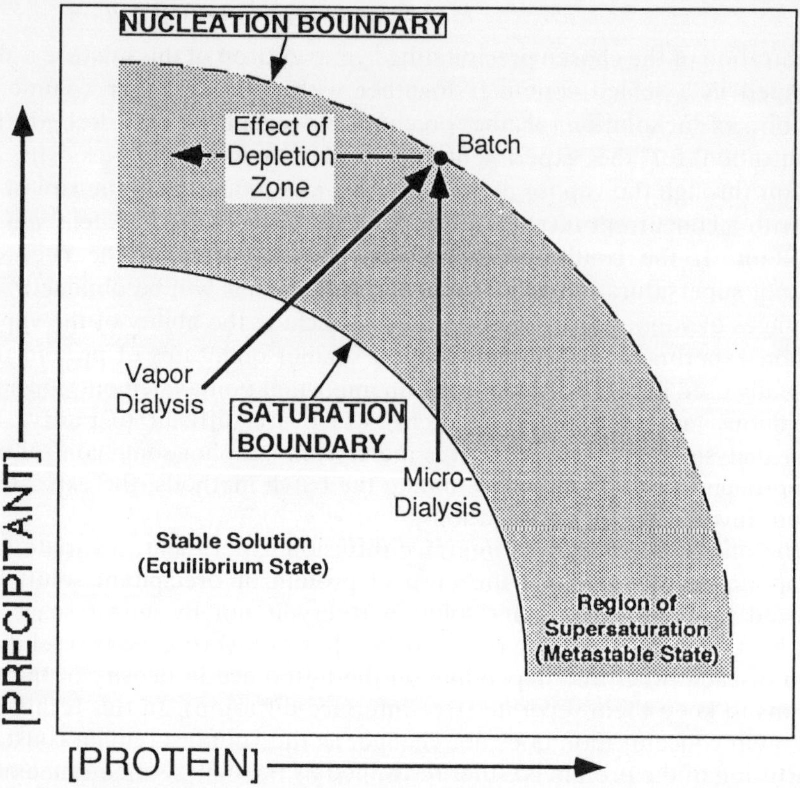

Figure 1. Phase diagram for protein crystallization. The growth of crystals involves the creation of a supersaturated solution of protein, usually at a high concentration, and a chemical precipitating agent. This concentration region is shown as the grey shaded area of protein *vs* precipitant concentration. Batch methods of crystallization are initially mixed at a concentration condition where supersaturation exists at time = 0, and nucleation then occurs without any subsequent equilibration effects. This method often works when nucleation is a very slow, kinetically rate-limited process. Vapour dialysis provides a method of concentration equilibration in which the concentrations of both protein and precipitant increase over time; often, pH gradients can also be created through the use of volatile acidic or basic agents in the protein drop. Concentration equilibration also occurs in microdialysis (liquid diffusion) experiments, but in this case only the concentration of precipitant increases. The initial concentration of protein and precipitant can represent a sub- or supersaturated solution.

growing crystal. As the crystal grows, protein is abstracted from the solution adjacent to the surface of the crystal resulting in a 'depletion zone' around an actively growing crystal. This zone is less dense than the bulk growth solution and rises in a stream from the crystal, which can be visualized photographically (DeLucas and Bugg, 1991; Shlichta, 1986). This results in turbulence near the growing surface, potentially to its detriment by encouraging the introduction of dislocations or other imperfections into the growing crystal.

None of these drawbacks would exist in a crystallization experiment conducted in the microgravity environment found in low Earth orbit. In addition, there may be some unique advantages to such an experiment. In particular, the 'depletion zone' now stably formed around a growing crystal, hanging suspended in the middle of the growth solution, undisturbed by close neighbours, would act to lower the degree of supersaturation locally around the crystal. This results in slower, potentially more controlled, growth of a crystal after nucleation (*see Figure 1*). In ground-based experiments, the crystal grows at the degree of supersaturation that is necessary for the formation of nuclei, which may be considerably higher than that necessary to sustain growth. The result is rapid growth that may limit the order, and therefore the quality of the resultant crystals. The combination of these potential benefits has been the driving force behind 10 years of microgravity crystallization experiments.

Microgravity-based macromolecular crystallization: History and results

In the past 10 years, over a dozen experiments have been conducted by several consortiums of American and European scientists on the US Space Shuttle, the Russian *Mir* Space Station and various unmanned platforms using a variety of crystallization hardware. These experiments are listed in *Table 1*, and cover a wide range of crystallization hardware, methodology and macromolecules. The majority of the experiments have been conducted by three consortiums: the University of Alabama at Birmingham group (the Center for Microgravity Research), headed by Lawrence DeLucas and Charlie Bugg, which conducts experiments on the US Shuttle; a group of primarily Western European crystallographers, which conducts experiments on unmanned, retrievable orbiters, organized by the company INTOSPACE (the COSIMA missions); and the experiments organized by Payload Systems, Inc, which fly on the Russian space station *Mir*, for which the authors of this review act as the Science Advisory Board (the PSI missions). Table 1 does not include recently performed microgravity experiments (such as USML-1 on the US Shuttle), planned experiments (such as Spacelab D-2, the PSI-3 mission or the EURECA free-flyer experiments), or failed experiments (such as the COSIMA-3 and COSIMA-4 missions), for which there are no published results. In addition, the Russian crystallographic community and a small number of collaborators have been performing microgravity crystallization experiments for many years, both on *Mir* and on the *Photon* orbiter, but few of the results of these experiments have been published. The results of these experiments must be interpreted in ways that are mindful of the difficulties inherent in conducting any experiment on an orbital platform, while acknowledging the subjective nature of the concept of crystal quality.

A number of different crystallization geometries and designs have been utilized over the past 10 years in microgravity. As discussed above, there are four basic methods for producing equilibration of protein solutions to a supersaturated state: bulk, batch, liquid-interface diffusion (which is usually performed by the use of a dialysis membrane barrier between protein and precipitant in the Earth-bound laboratory) and vapour-diffusion (using hanging or sitting drop geometries). In microgravity, virtually all reported experiments have used slow-diffusion equilibration methods, using either free liquid-interface diffusion or vapour diffusion. Due to the lack of gravity-dependent density forces, different solutions of varying density can equilibrate directly in microgravity without the use of dialysis membranes. Also, vapour dialysis can be performed without the consideration of hanging versus

Table 1. Previously reported microgravity-based crystallizations

	Platform	Date	Experiments	Duration	Comments
PSI-2	*Mir* (Russia)	Jan-March 1992	19 proteins	57 days	Vapour-diffusion experiments
IML-1	US Shuttle/ Spacelab	Jan 1992	>2 proteins	8 days	Free-interface diffusion experiments
MASER3	Sounding rocket	1991	RNase S	7 min	Short-duration microgravity
STS-31	US Shuttle	April 1990	?	?	Vapour-diffusion experiments
PSI-1	*Mir* (USSR)	Dec 1989 to Feb 1990	HEWL and Datase	56 days	Free-interface and vapour-diffusion experiments
STS-32	US Shuttle	Jan 1990	?	?	Vapour-diffusion experiments
COSIMA-2	*Photon* (USSR)	Sept 1989	>12 proteins	10 days	Vapour-diffusion experiments
STS-29	US Shuttle	March 1989	15 proteins	?	Vapour-diffusion experiments
STS-26	US Shuttle	Nov 1988	8 proteins	3 days	Vapour-diffusion experiments
COSIMA-1	*Long March* (China)	1988	>12 proteins	8 days	Vapour-diffusion experiments
Russian	*Photon* (USSR)	April 1988	5 proteins	14 days	Free-interface diffusion experiments
STS-61C	US Shuttle	Jan 1986			While many crystallization
STS-61B	US Shuttle	Nov 1985	12 proteins/	4–10	experiments were
STS-51F	US Shuttle	July 1985	3 nucleic acids	days	conducted, these flights
STS-51D	US Shuttle	April 1985			were primarily for hardware development
STS-61A Spacelab D-1	US Shuttle/ Spacelab	Oct 1985	β-gal and HEWL	?	Failure of automated hardware
STS-9 Spacelab-1	US Shuttle/ Spacelab	Nov 1983		β-gal and HEWL	Free-interface diffusion experiments

References: Day and McPherson (1992), DeLucas and Bugg (1991), DeLucas *et al*. (1986, 1991), Littke and John (1984), Stoddard *et al*. (1992), Strong *et al*. (1992), Trakhanov *et al*. (1991).

sitting drop geometries. Below, we summarize the hardware design considerations used during protein crystallization experiments on *Mir* (Strong *et al.*, 1992; Stoddard *et al.*, 1992) as being representative of one particular hardware design used in these experiments. For a good overview of the hardware design used on the US Space Shuttle, see DeLucas and Bugg (1990, 1991) and DeLucas *et al.* (1991).

VAPOUR DIFFUSION CRYSTALLIZATIONS

The vapour-diffusion device, or VD (*Figure 2*), was designed to reproduce hanging or sitting drop techniques in a microgravity environment. Equilibration in this experiment occurs through differences in osmotic pressure between the protein solution (initially at low ionic strength) and a reservoir solution (initially at high ionic strength), which leads to bulk transfer of water through the vapour phase and an increase in protein and precipitant concentration to supersaturation. The device is designed to maintain physical separation of the two solutions during flight and recovery, and to allow activation and creation of a vapour pathway between protein and reservoir on orbit. The bodies of the devices were originally constructed of polymethylpentene (PMP), a light (0.83 g cm^{-3}), strong (430 kg cm^{-2} bending strength) and inert plastic with a high melting point (235°C), which is hydrophobic and non-interactive with macromolecules. The devices are now constructed from a variety of other materials, such as polycarbonate and cel-cast acrylic plastic, depending upon the preferences of the investigator. Protein solutions, mixed with a low initial concentration of precipitant are placed in cups of 10, 25 or 50 μl (*Figure 2a*), which are surrounded by moats of 500–750 μl of precipitant at the final concentration (*Figure 2B*). The precipitant in each moat is immobilized by a coil of filter paper. A screw-plunger in the top plate of the device is closed down on the protein cup. When screwed down, the plunger presses a silicone rubber septum against the edges of the protein-filled cup, thus sealing the protein solution away from the precipitant. To allow vapour-phase equilibration (activation), the septum is raised, establishing an air path between the solutions. To deactivate the experiment, the cup is resealed with the septum. The device has a template attached to the bottom plate for photographing the protein wells. The surfaces of the optical pathway are polished to photography-grade flatness and quality.

BOUNDARY LAYER CRYSTALLIZATIONS

The basic principle of equilibration to supersaturation using boundary layer crystallization methods is similar to vapour-phase diffusion, although in this case the process of diffusion occurs across a direct solution interface. A stable interface can be created in microgravity due to the lack of density differences. The boundary layer diffusion device, or BLD (*Figure 3*), was derived from a proposal of Littke and John. Protein and precipitant solutions are held in chambers separated by a rotating shaft with connecting chambers filled with 'neutral' buffer (no salt or protein).

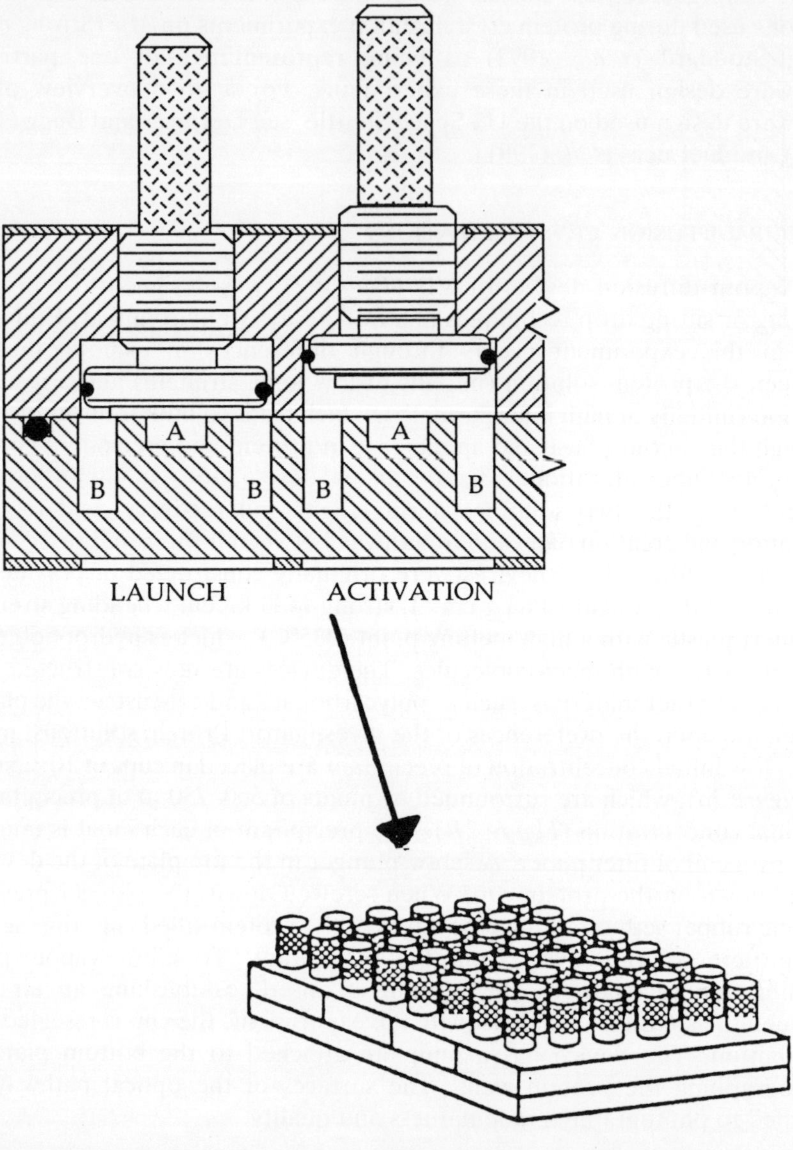

Figure 2. Vapour-diffusion crystallization device (Strong *et al.*, 1992). The device contains a protein well of variable volume (A) and a reservoir containing a high initial concentration of precipitating agent (B). A silicon rubber septa seals the protein well during launch, and is raised through the use of a manually operated thumbscrew. Variations in this design allow on-orbit mixing of protein and precipitant. Each device currently contains 48 individual crystallization wells.

The BLD device is constructed of the same materials as the vapour-diffusion device. The teflon screws sealing the individual chambers incorporate a teflon-coated silicone rubber septum. As each device is sealed, trapped air is removed by inserting needles through the septa and into the chambers. For the first flight, each chamber accommodated 500 μl of protein solution,

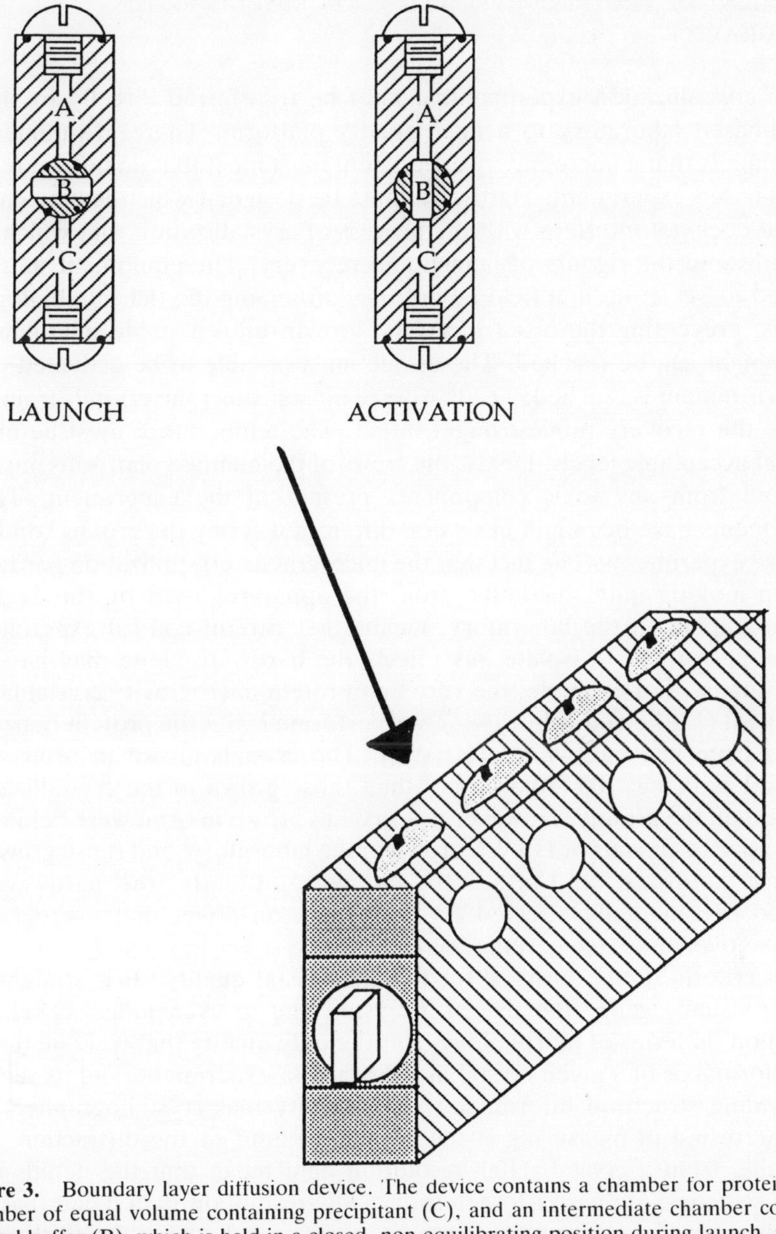

Figure 3. Boundary layer diffusion device. The device contains a chamber for protein (A), a chamber of equal volume containing precipitant (C), and an intermediate chamber containing neutral buffer (B), which is held in a closed, non-equilibrating position during launch, and then rotated to establish a pathway of liquid–liquid diffusion upon activation. Each device contains four crystallization chambers, and 16 devices are used in each flight complement, allowing 64 crystallization experiments to be performed.

500 μl of precipitant and 75 μl of neutral buffer. All photographic path surfaces were polished to the same optical clarity as the vapour-diffusion device.

ASSESSMENT OF PROTEIN CRYSTALLIZATION RESULTS FROM MICROGRAVITY

Protein crystallization experiments cannot be transferred directly from the ground-based laboratory to a microgravity platform. There are numerous requirements that a microgravity experiment must meet that do not apply to a ground-based experiment. Hardware must be designed which is biocompatible and does not interfere with the process of crystallization, but which can also withstand the rigours of launch and recovery. The equipment must be designed to be set up in a field laboratory, protecting the delicate biological samples, preventing the onset of crystal growth until a stable microgravity environment can be reached. The device must be able to be activated once this environment is reached. The hardware must protect the crystals from any damage the recovery process might inflict. The temperature must be maintained at acceptable levels. Lastly, the crews of the manned platforms must be protected from any toxic components present in the experiment. These requirements have been met in several different ways by the groups conducting these experiments. The fact that the microgravity crystallization hardware ends up looking quite dissimilar from the apparatus used by the average crystallographer in the laboratory, means that careful control experiments must be performed to isolate any effects the hardware alone may have on crystal quality. For example, the very first protein microgravity crystalliation experiment (Littke and John, 1984) was performed with the protein lysozyme isolated from hen egg whites (HEWL). The crystals grown in orbit were described as being '1000-fold larger' than those grown in the crystallization hardware on the ground. However, the crystals grown in orbit were below the average size of crystals of HEWL grown in the laboratory, and those grown in the same hardware on Earth were far worse. Clearly, the hardware as designed detrimentally affected crystallization, an effect that microgravity may have overcome to a certain degree.

What criteria *should* be used for judging crystal quality? It is straightforward to visually gauge the size of a crystal, and to even judge its relative 'perfection' in terms of morphology. But the only quality that truly matters is the performance of a given crystal in a diffraction experiment, and its success in providing structural information. The quality that crystallographers are most interested in measuring is the resolution limit of the diffraction data collectable from a crystal. High-resolution data mean that the details of a protein structure are easier to determine; low-resolution data mean that the determination of an unambiguous structure may not be possible. In the case of lysozyme crystals, despite the use of this protein in more microgravity experiments than any other, crystals of HEWL grown in microgravity do not yield higher resolution diffraction data than HEWL crystals grown on the ground. Another example, the protein canavalin (Day and McPherson, 1992), yields beautiful, flawless crystals in microgravity experiments, which nevertheless do not provide diffraction data that are significantly better than crystals grown on the ground.

The results of 10 years of microgravity crystallization experiments from a number of different groups have consistently shown that 20–25% of the macromolecular crystals grown in orbit yield modestly improved diffraction data, as a function of resolution limits and signal-to-noise ratios throughout all observable resolution shells. For those fractions of the crystals which are improved by growth in the absence of gravity, the resolution limit of the diffraction data is extended by 0.1 Å to 0.5 Å. This level of improvement is not dramatic; it would not usually make a significant difference in determining the structure of a protein from unsuitable crystals. Microgravity crystallization has never yielded crystals that were suitable for diffraction experiments from a crystallization problem where the laboratory-grown crystals were not also usable. However, a small number of space-grown crystals have provided diffraction data, which have at least been utilized during refinement and high-resolution refinement. (1) Microgravity-derived diffraction data from one protein crystallization problem, γ-interferon (Ealick *et al.*, 1991), has been used in a published structure; the determination of that structure did not depend on the improved microgravity data. (2) Crystals of a second protein, satellite tobacco mosaic virus, show a 0.3 Å improvement in diffraction resolution when grown in microgravity (Day and McPherson, 1992); these data will be used in the refinement of the structure of this virus, which had been determined from laboratory-grown crystals (Koszelak, Dodds and McPherson, 1989). While not dramatic, this level of improvement is real and can make the determination of a three-dimensional structure considerably easier or the interpretation of a structure more significant. Other protein crystals which have displayed small improvements in size, internal perfection and scattering power when grown in microgravity include porcine elastase, *Lathyrus ochrus* lectin I protein, canavalin, anti-HPr Fab, anti-Brucella Fab, isocitrate dehydrogenase, and the 'Z' crystal form of ribonuclease S (DeLucas and Bugg, 1990, 1991; McPherson, Greenwood and Day, 1991; DeLucas *et al.*, 1991; Stoddard *et al.*, 1992; Asano *et al.*, 1992).

In addition to the modest improvements in resolution limits summarized above, a small number of the proteins which have been flown and crystallized in microgravity have initially been reported as *exceptional* successes, including lysozyme (Littke and John, 1984), isocitrate lyase (DeLucas and Bugg, 1991), 10-formyl tetrahydrofolate synthetase (10-THFS) and aspartate ammonia lyase (Stoddard *et al.*, 1992). As described above, the conclusions published after the first crystallization of lysozyme in microgravity have been criticized as being misdirected and plagued by a lack of adequate control experiments. After careful ground control studies and further investigations in the laboratory, the other three 'successes' have been reproduced or dramatically exceeded on Earth, as described below. In addition, another protein (albumin) which produced improved or slightly improved crystals in microgravity, subsequently was shown to improve to an equivalent degree through the use of a gel matrix during crystallization, and another protein which was flown *unsuccessfully* on *Mir* (anti-chymotrypsin, a serpin inhibitor) was subsequently successfully crystallized after incorporation of a disulphide cross-link at the C-terminus of the protein (Stoddard *et al.*, 1992; Katz *et al.*, 1993), as described below.

Isocitrate lyase

This protein was originally crystallized as dendritic clusters, subsequently characterized as belonging to space group $P2_12_12_1$. An improved crystal morphology belonging to the same space group was grown in the microgravity experiments conducted on the shuttle missions STS-26 and STS-32. These crystals grew as well-formed prisms and diffracted to considerable higher resolution (DeLucas and Bugg, 1991). However, subsequent careful ground control experiments indicate that this improved crystal morphology was due solely to the altered evaporation and equilibration rate in the crystallization hardware, and could be duplicated on the ground under several different conditions (DeMattei, Feigelson and Weber, 1992).

10-THFS

Initial crystallization trials of tetrahydrofolate synthetase from *Clostridia acidici-urici*, spanning a 2 year period, yielded only long needle-like specimens which did not diffract measurably. Crystallization on *Mir* yielded rhombs which diffracted to approximately 30 Å resolution (Stoddard *et al.*, 1992). However, subsequent intense ground controls in the same hardware on Earth reproduced this result, which appeared to be caused by the temperature profile which existed on *Mir* during crystallization rather than by the microgravity environment. More importantly, two independent investigators have succeeded in growing large, single crystals of this enzyme which diffract to high resolution. The enzyme described above from *C. acidici-urici* produced large orthrhombic crystals, space group $P2_12_12$, which diffract to approximately 2.5 Å resolution after simply reducing the temperature from 20°C to 18°C (D'Ari *et al.*, submitted). Simultaneously, a second group has also produced large, well-diffracting crystals of the same enzyme by using an alternate biological source, *Clostridium thermoaceticum* (Lewinski *et al.*, 1993). Structure determination is proceeding in both laboratories.

Aspartate ammonia lyase

The same mission flown on the *Mir* Space Station also produced crystals of aspartate ammonia lyase (aspartase) from the vapour diffusion set-ups which diffracted to significantly higher resolution (3.2 Å) than their ground-based counterparts, which diffracted to 4.0 Å resolution (Stoddard *et al.*, 1992). However, recent work in the laboratory has yielded crystals of much greater size and volume, simply by utilizing the dialysis equilibration method rather than vapour diffusion and by adding significant amounts of the reducing agent dithiothreitol to the mother liquor. These crystals diffract to better than 2.8 Å resolution and have been characterized as belonging to space group $P2_12_12$. Structure determination is proceeding.

Human serum albumin

Crystals of human serum albumin were grown on the STS-31 shuttle mission and subsequently shown to be larger than Earth-grown specimens and to diffract slightly more strongly (DeLucas and Bugg, 1991). However, crystals grown in gels on the ground were comparable to the microgravity-grown crystals (Miller, He and Carter, 1992).

It is worth noting that the above examples of protein crystallization experiments, in which the results from microgravity were subsequently matched or vastly exceeded by further ground-based crystallization attempts, probably represents a very small number of the total number of crystallizations flown in orbit for which this result has occurred. This is due primarily to the fact that almost all the published experiments from microgravity platforms present detailed analyses only of successful crystallizations, so that accurate book-keeping and comparison of all space-based crystallization attempts with the further utilization of standard laboratory crystallization methods is extremely difficult. Only in the case of the recent crystal growth experiments on *Mir* (Strong *et al.*, 1992; Stoddard *et al.*, 1992) has the detailed publication of all attempted crystallization experiments led to the identification of at least one species (anti-chymotrypsin) which completely failed to crystallize in microgravity and which later produced excellent specimens through the use of targetted disulphide cross-linking, as discussed above.

Earth-bound crystallization: More crystals for the money, more structures for the effort

Microgravity-based crystallization has often, perhaps always, been justified to a large extent because crystallization is a 'bottle neck' in the elucidation of macromolecular three-dimensional structures. Historically, this is true, and continues to be true today. However, currently, this is due not only to the inherent difficulties of the crystallization process, but also because of the incredible advances in data collection hardware, computer power, memory and speed, and molecular modelling and refinement algorithms which have appeared in the past 5–10 years. At the time that the first several dozen X-ray structures of proteins were deduced, there were no area detectors, synchrotron radiation sources, low-temperature cryostats, powerful unix-based workstations or CRAY supercomputers, or simulated annealing algorithms for molecular dynamics and refinement. Therefore, even though today crystallization still represents the most time-consuming portion of structure determination, the rate of new structures being solved continues to accelerate, and the amount of time necessary for any crystallization project to succeed drops dramatically each year: the number of abstracted new structures in *Macromolecular Structures* (Hendrickson and Wuttrich, 1991, 1992) jumped from 134 in 1991 to 165 in 1992, a rate of increase which Hendrickson and Wuttrich take to indicate that the rate of new structure determinations will climb into the thousands each year by the end of the decade.

In addition to enormous improvements in data collection efficiency and computational power, the success rate of attempted macromolecular crystallizations has itself increased dramatically over the past 10 years. The rapid pace of novel crystallizations is due to a few technologies being exploited by structural laboratories which are increasingly adept in their use: (1) intense, thorough purification of proteins and nucleic acids; (2) genetic manipulation and over-expression of the targets of crystallization; (3) determination of aggregation state, binding stoichoimetry and solution behaviour by electrophoretic, chromatographic and dynamic light scattering techniques; and (4) accurate environmental control and variation during crystallization. In addition, improvements in scattering power (diffraction resolution limits) of data quality crystals are currently possible through four unrelated technologies: (1) optimization of molecular constructs through genetic engineering or chemical modification to eliminate disorder effects; (2) use of low temperature to reduce or eliminate radiation-dependent crystal decay and therefore improve counting statistics, data redundancy and accuracy; (3) the use of extremely sensitive area detectors, particularly phosphor-plate systems; and (4) the availability of synchrotron radiation. In this section of our review, we summarize the importance of various methodologies on macromolecular crystallization and provide an abbreviated sampling of structures solved in the past 2 years through their use. The question we then ask is simple: Do we really need microgravity-based protein crystallization, or would the field of structural biology be far better served by continued effort and investment in these 'standard' methodologies?

In contrast to microgravity-based crystalliation, which in over 10 years has yet to produce a single macromolecular crystal that has been shown to be necessary to produce a structure determination, it is clear from *Table 2* that a number of techniques which have become widely used in the past 10 years have been instrumental in producing an impressive number of important new structures. Of these, two main areas stand out: complete purification of the target macromolecular population to macro- and micro-homogeneity, and the use of DNA recombinant techniques to vary the primary structure of the protein to the advantage of the crystallographer.

CRYSTALLIZABILITY IS DIRECTLY CORRELATED TO PURITY

Studies of virtually every protein under crystallographic study shows that impurities, often undetected by simple denaturing electrophoretic analysis, are almost always deleterious to crystallization. Differences throughout the molecular population with respect to variable isoforms, conformers and post-translational modifications of the same protein are often the culprit in poor crystallization results, and are often inadequately characterized. A prime example is the enzyme alanine racemase, which has produced twinned crystals for the past 10 years (Neidhart *et al.*, 1987). Recently, a single additional purification step, which eliminated a minor subpopulation of the enzyme, has produced improved crystals, and a structure determination is now forthcoming (D. Ringe and G. Petsko, pers. comm.). Similarly, the periplasmic domain of the Aspartate receptor and the MAX oncogene product could not be crystallized until adequate purification (Milburn *et al.*, 1991; Ferre-D'Amare *et al.*, 1993). Any experienced crystallographer can produce similar cases from their own experiences. As McPherson (1982) accurately states: 'There are numerous sources of microheterogeneity in an otherwise pure preparation, and an awareness of some of the more common of these is essential in diagnosing the source of a particular [crystallization] problem . . . though there are undoubtedly other causes as well, this list can serve as a starting point for scrutiny at the molecular level.' It is not unreasonable, therefore, to state that for those molecular species which do not crystallize readily, the best course to pursue is analysis of purity and solution behaviour, followed by further purification and possibly genetic manipulation as described below.

SUCCESSFUL CRYSTALLIZATION STRATEGIES

Variation in the overall expressed protein construct, *N*- and *C*-termini locations, subunit and lattice contacts by genetic manipulation, combined with accurate and thorough characterization and purification, is the most likely strategy for crystallization. As can be seen from *Table 2*, the most successful technique by which X-ray crystallographers have improved the success rate of protein crystallization is by altering the protein at the genetic level. It was noted many years ago that occasionally a site-directed mutant of an interesting protein was more easily crystallizable than wild-type protein; this effect was shown to be due to the presence of the relevant site in a lattice contact region. Currently, a number of methods have been shown to produce high-quality crystals, the most common of which is simply to crystallize a construct which represents a subset of the entire wild-type sequence, and which constitutes a domain containing the important functionalities of the intact protein. Approximately 10–20% of all new structures being reported in the literature are of this variety.

The clearest example of this strategy is the crystallization of the components of the pol I transcription complex, particularly the MAX transcriptional activator in complex with its cognate DNA (Ferre-D'Amare *et al.*, 1993). This protein was only crystallized after extensive analysis of solution aggregaton and dispersion of over a dozen constructs of various length using dynamic light-scattering assays. The most promising candidate was crystallized and the structure solved in a period of less than 6 months after its identification. A large number of other structures were elucidated in 1990–91 through a similar fashion, particularly functional domains of members of two protein classes which are traditionally difficult to crystallize: membrane-bound receptors proteins (bacterial aspartate chemotaxis receptor, Milburn *et al.*, 1991; human growth hormone receptor, De Vos, Ultsch and Kossiakoff, 1992; LDL receptor, Wilson *et al.*, 1991; CD4, Ryu *et al.*, 1990 and Wang *et al.*, 1990; glucocorticoid receptor, Luisi *et al.*, 1991) and multidomain transcriptional

Table 2. Crystallization methodologies and representative structures

	Technique	Some recent structures
I.	Purification	All species
II.	Genetic manipulation	
	A. Expression of independent domains	MAX b-zip domain/DNA complex
		DNA gyrase B
		Glucocorticoid receptor
		Heat shock protein ATP domain
		CD4
		Aspartate receptor ligand domain
		Variable surface glycoprotein
		Klenow fragment
		HGH receptor
	B. Site mutants to improve lattice contacts	H-ferritin
	C. Artificial disulphide cross-links	Aspartate receptor ligand domain
		Anti-chymotrypsin (serpin inhibitor)
	D. 'Internal' site-directed mutations	Aspartate amino transferase
		Azurin
		Fibroblast growth factor
III.	Determination of aggregation state and solution behaviour	MAX
		USF
		TATA-binding protein
		αl helical protein
		Gene 5 protein
		Ferritin
IV.	Temperature control and variation	10-THFS
		Myosin regulatory domain
		Xylose isomerase
V.	Chemical modification	Mysosin head protein

regulators (MAX, Ferre-D'Amare *et al.*, 1993; GCN4, Ellenberger *et al.*, 1992).

In addition to the wholesale reduction of intact wild-type protein molecules to functional domains, which is a 'divide and conquer' approach to crystalization, more subtle changes in primary sequence can often improve results dramatically. As shown in *Table 2*, changes in internal or inter-subunit cross-linking, single point mutations at the site of a lattice contact, and the alteration of a single site of a post-translational modification (particularly phosphorylation, methylation and glycosylation) can improve crystallization results (Milburn *et al.*, 1991; Lawson *et al.*, 1991; Ago *et al.*, 1991).

CRYSTALLIZABILITY VERSUS SOLUTION AGGREGATION BEHAVIOUR

A number of investigators are now discovering that along with the degree of purity and nature of the genetic construct used to produce the target protein, there is an extremely strong correlation between the degree of non-specific aggregation in solution (polydispersion) and crystallizability, with those molecules which display clean oligomerization and aggregation behaviour (forming a mondisperse solution) often being much more easily crystallized (Burley and Roeder, 1992; Ferre-D'Amare *et al.*, 1993). The most commonly reported technique for assessing aggregation is photon correlation spectroscopy, also known as dynamic or quasi-elastic light scattering. This method is currently being used by a handful of crystallography laboratories around the world, and is currently primarily responsible for the elucidation of approximately a dozen novel structures in the past couple of years. It is clear that routine application of this technology, in conjunction with genetic and purification methodologies mentioned above, will have an enormous impact on protein crystallization in the future.

TECHNIQUES FOR RESOLUTION IMPROVEMENT

Organizations and investigators have recently backed away from initial predictions that space-based macromolecular crystallization will allow the initial elucidation of structures that cannot otherwise be determined, primarily because microgravity-based crystallization experiments have not produced any specimens in a unique morphology, space group, or dramatically improved scattering power as compared to standard crystallization techniques. As a result, it is now clear that it is far more efficient and cost-effective to discover a new route to a desired novel crystalline species through the application of standard laboratory methodologies, than it is to attempt to use the microgravity environment to achieve the same goal.

Therefore, most claims on the side of microgravity-based protein crystallization are now found at the doorstep of 'improved resolution' or scattering power from many space-grown crystals (DeLucas and Bugg, 1990, 1991; McPherson, Greenwood and Day, 1991; DeLucas *et al.*, 1991; Stoddard *et al.*, 1992; Asano *et al.*, 1992). The average improvement in resolution for these species averages between 0.1 and 0.4 Å, and is presumably caused by improved internal lattice formation. As mentioned above, there are four commonly used ground-based methods for producing similar resolution improvements. The first is improved genetic constructs which eliminate disordered regions (particularly *N*- and *C*-termini in the target proteins). Sensitive area detector systems and synchrotron radiation, taken together, may produce substantial improvements in diffraction limits for a given crystal. Finally, ultra-low temperature data collection, which can often allow collection of much higher resolution data simply by extending the length of scan time across reflection profiles sufficiently to dramatically improve counting statistics for weak data (Hope, 1990) as described below.

The cryo-preservation of crystals at temperatures approaching –190°C can improve the observed diffraction data by as much as that seen with microgravity crystallization. Crystals of biological macromolecules are sensitive to damage caused by X-ray irradiation. The result is an often severe degradation in the quality of the data as they are collected. A technique pioneered by Hope (reviewed 1990) immortalizes crystals by rapidly reducing their temperature to that of liquid nitrogen. When successful, the method results in a sample that is preserved as a vitreous freeze; as long as the crystal is maintained at low temperature, it experiences no damage due to irradiation. The advantage is two-fold: data no longer suffer the inaccuracies introduced by radiation decay, and higher resolution data may be collected by exposing cryo-preserved crystals to much higher X-ray fluxes, allowing weak, high-resolution terms to be measured accurately. The technique has been instrumental in the recent elucidation of several important structures (Teng, 1990; Joshua-Tor *et al.*, 1988; Brown *et al.*, 1993). The resolution enhancement observed can be as high as 1.0 Å (Pamela J. Bjorkman, pers. comm.).

Conclusions

A theoretical treatment of the conditions which affect macromolecular crystallization indicates that the microgravity environment should be conducive to lattice formation and growth, as discussed in the first section of this review. However, 10 years of microgravity crystallization experiments, covering over 100 protein species and using several different microgravity platforms, indicates that there has yet to be a single crystallization which has yielded any major improvement, leading to significant structural information, which could not be attained through ground-based methodologies at lower cost. Indeed, those cases which have been reported as the most 'successful' microgravity crystallizations (isocitrate lyase, aspartate ammonia lyase, 10-THFS) have all been shown to either have been caused by experimental artifacts, or to have subsequently improved dramatically by simple modifications of standard crystallization trials on Earth. It is therefore worthwhile to perform a cost–benefit analysis of microgravity-based crystallization.

The cost of a single launch of a crystallization payload which runs on either *Mir* or the US Shuttle varies between US$100 000 and well over US$1 million by the time all factors have been accounted for. In some cases, these expenses are not actually shouldered by the individual academic laboratory, but rather by the taxpayer through the support of the relevant space agency, but the final price is the same nonetheless. In comparison, for US$100 000 an individual laboratory could purchase dynamic light-scattering and automated chromatography systems (roughly US$25 000 and US$35 000, respectively) and pay a full-time senior technician or two postdoctoral fellows a year's salary to conduct somewhere between 2000 and 8000 h of purification, cloning and crystallization trials. For the upper price tag, a pair of crystallography laboratories could be fully outfitted with the best data collection hardware available.

Microgravity crystallization is, without question, one method to improve the quality of protein crystals. However, it is by no means the most efficacious method developed in the past decade, and still remains a technique available only to those researchers who are able to work on the US Space Shuttle, or who are rich enough to afford to pay the commercial launch costs which INTOSPACE or Payload Systems charge. Currently, realistic, unsubsidized launch costs begin in the hundreds of thousands of dollars, and go up from there. The cost of a shuttle mission, in its entirety, has been estimated at hundreds of millions of dollars, the proposed space station is estimated to cost in the tens of *billions* of dollars. The methods that we have briefly outlined have been developed in average laboratories, and are available for use by any researcher so inclined. Barring some breakthrough in current results, we would consider microgravity crystallization a poor justification for building a space station. A vastly greater impact on our understanding of disease and the fundamental processes of life, the development of new drugs and therapeutic agents, could be made instead by investing just 1% of the cost of the proposed space station in basic biomedical research.

The take-home message of this analysis is clear: space-based research, while important in its own right as a driving force for technological innovation, cannot and should not be justified by relationships with well-established research areas, such as structural biology, which actually have few benefits derived from manned space programmes. Additionally, it is clear that in an age of vanishing monetary resources for basic research, it is possible for critically important research areas which are composed solely of many small academic and industrial research groups (such as structural biology) to be adversely affected by the funding of a massive programme in a separate technology simply because both are broadly classified as 'science' and connected to one another in competition for federal support. Finally, it is clear that the greater the potential cost of a research programme to a nation, the greater the need for impartial peer review, criticism and policy advisement from a broad spectrum of the most well-respected and productive principle investigators in that particular research field, in order to maintain accountability among all involved agencies and investigators.

References

AGO, H., KITAGAWA, Y., FUJISHIMA, A., MATSUURA, Y. AND KATSUBE, Y. (1991). Crystal structure of basic fibroblast growth factor at 1.6 Å resolution. *Journal of Biochemistry* **110**, 360–363.

ASANO, K., FUJITA, S., SENDA, T. AND MITSUI, Y. (1992). Crystal growth of ribonuclease S under microgravity. *Journal of Crystal Growth* **122**, 323–329.

BROWN, J.H., JARDETZKY, T.S., GORGA, J.C., STERN, L.J., URBAN, R.G., STROMINGER, J.L. AND WILEY, D.C. (1993). Three-dimensional structure of the human class II histocompatibility antigen HLA-DR1. *Nature* **364**, 33–39.

D'ARI, L., BOLDUC, J., RABINOWITZ, J. AND STODDARD, B.L. (submitted). Purification, crystallization, and preliminary characterization of 10-tetrahydrofolate synthetase. *Journal of Molecular Biology*.

DAY, J. AND MCPHERSON, A. (1992). Macromolecular crystal growth experiments on International Microgravity Laboratory-1. *Protein Science* **1**, 1254–1260.

DELUCAS, L.J. AND BUGG, C.E. (1990). Protein crystal growth procedure for the U.S. Space Shuttle. *Methods: A Companion to Methods in Enzymology* **1**, 105–109.

DELUCAS, L.J. AND BUGG, C.E. (1990). Protein crystal growth in space. *Advances in Space Biolody and Medicine* **1**, 249–278.

DELUCAS, L.J., SUDDATH, F.L., SNYDER, R., NAUMANN, R., BROOM, M.B., PUSEY, M., YOST, V., HERREN, B., CARTER, D., NELSON, B., MEEHAN, E.J., MCPHERSON, A. AND BUGG, C.E. (1986). Preliminary investigation of protein crystal growth using the space shuttle. *Journal of Crystal Growth* **76**, 681–693.

DELUCAS, L.J., SMITH, C.D., SMITH, W., VIJAY-KUMAR, S., SENADHI, S.E., EALICK, S.E., CARTER, D.C., SNYDER, R.S., WEBER, P.C., SALEMME, F.R., OHLENDORF, D.H., EINSPAHR, H.M., CLANCY, L.L., NAVIA, M.A., MCKEEVER, B.M., NAGABHUSHAN, T.L., NELSON, G., MCPHERSON, A., KOSZELAK, S., TAYLOR, G., STAMMERS, D., POWELL, K., DARBY, G. AND BUGG, C.E.

(1991). Protein crystal growth results for shuttle flights STS-26 and STS-29. *Journal of Crystal Growth* **110**, 302–311.

DEMATTEI, R.C., FEIGELSON, R.S. AND WEBER, P.C. (1992). Factors affecting the morphology of isocitrate lyase crystals. *Journal of Crystal Growth* **122**, 152–160.

DE VOS, A.M., ULTSCH, M. AND KOSSIAKOFF, A.A. (1992). Human growth hormone and extracellular domain of its receptor: Crystal structure of the complex. *Science* **255**, 306–312.

EALICK, S.E., COOK, W.J., VIJAY-KUMAR, S., CARSON, M., NAGABHUSHAN, T.L., TROTTA, P.P. AND BUGG, C.E. (1991). Three-dimensional structure of recombinant human interferon-γ. *Science* **252**, 698–702.

ELLENBERGER, T.E., BRANDL, C.J., STRUHL, K. AND HARRISON, S.C. (1992). The GCN4 basic region leucine zipper binds DNA as a dimer of uninterrupted α helices: Crystal structure of the protein-DNA complex. *Cell* **71**, 1223–1237.

FERRE-D'AMARE, A.R., PRENDERGAST, G.C., ZIFF, E.B. AND BURLEY, S.K. (1993). Recognition by Max of its cognate DNA through a dimeric b/HLH/Z domain. *Science* **363**, 38–44.

HENDRICKSON, W. AND WUTTRICH, K. (Eds) (1991). *Macromolecular Structures 1990: Atomic Structures of Biological Macromolecules Reported During 1990*. Current Biology Ltd, London.

HENDRICKSON, W. AND WUTTRICH, K. (Eds) (1992). *Macromolecular Structures 1991: Atomic Structures of Biological Macromolecules Reported During 1991*. Current Biology Ltd, London.

HOPE, H. (1990). Crystallography of biological macromolecules at ultra-low temperatures. *Annual Review of Biophysics and Biophysical Chemistry* **19**, 107–126.

JOSHUA-TOR, L., RABINOVICH, D., HOPE, H., FROLOW, F., APPELLA, E. ABD SUSSMAN, J.L. (1988). The three-dimensional structure of a DNA duplex containing looped-out bases. *Nature* **334**, 82–86.

KATZ, B.S., DARIN, S., WEI, A., ZHONG, Q., RUBIN, H., COOPERMAN, B.S. AND CHRISTIANSON, D.W. (1993). Crystallization and atomic X-ray diffraction analysis of antichymotrypsin variants. *BBRC* **196**, 752–757.

KENDREW, J.C., DICKERSON, R.E., STRANDBERT, B.E., HART, R.G., DAVIES, D.R., PHILLIPS, D.C. AND SHORE, V.C. (1960). Structure of myoglobin: a three-dimensional Fourier synthesis at 2 Å resolution. *Nature* **185**, 422–427.

KOSZELAK, S., AND MCPHERSON, A. (1988). Time lapse microphotography of protein crystal growth using a color VCR. *Journal of Crystal Growth* **90**, 340–343.

KOSZELAK, S., DODDS, J.A. AND MCPHERSON, A. (1989). Preliminary analysis of crystals of satellite tobacco mosaic virus (STMV). *Journal of Molecular Biology* **208**, 323–326.

LAWSON, D.M., ARTYMIUK, P.J., YEWDALL, S.J., SMITH, J.M.A., LIVINGSTONE, J.C., TREFFRY, A., LUZZAGO, A., LEVI, S., AROSIO, P., CESARENI, G., THOMAS, C.D., SHAW, W.V. AND HARRISION, P.M. (1991). Solving the structure of human H ferritin by genetically engineering intermolecular crystal contacts. *Nature* **349**, 541–544.

LEWINSKI, K., HUI, Y., JAKOB, C.G., LOVELL, C.R. AND LEBIODA, L. (1993). Crystallization and preliminary crystallographic data for formyltetrahydrofolate synthetase from *Clostridium thermoaceticum*. *Journal of Molecular Biology* **229**, 1153–1156.

LITTKE, W. AND JOHN, C. (1984). Protein single crystal growth under microgravity. *Science* **225**, 203–204.

LUISI, B.F., XU, W.X., OTWINOWSKI, Z., FREEDMAN, L.P., YAMAMOTO, K.R. AND SIGLER, P.B. (1991). Crystallographic analysis of the interaction of the glucocorticoid receptor with DNA. *Nature* **352**, 497–505.

MCPHERSON, A. (1982). *Preparation and Analysis of Protein Cyrstals*. Krieger, Malabar, FL.

MCPHERSON, A., GREENWOOD, A. AND DAY, J. (1991). The effect of microgravity on protein crystal growth. *Advances in Space Research* **11(7)**, 343–356.

MILBURN, M.V., PRIVE, G.G., MILLIGAN, D.L., SCOTT, W.G., YEH, J., JANCARIK, J., KOSHLAND, D.E. JR AND KIM, S.-h. (1991). Three-dimensional structures of the ligand-binding domain of the bacterial aspartate receptor with and without a ligand. *Science* **254**, 1342–1347.

MILLER, T.Y., HE, X. AND CARTER, D.C. (1992). A comparison between protein crystals grown with vapor diffusion methods in microgravity and protein crystals using a gel liquid–liquid diffusion ground-based method. *Journal of Crystal Growth* **122**, 306–309.

NEIDHARDT, D.J., DISTEFANO, M.D., TANIZAWA, K., SODA, K., WALSH, C.T. AND PETSKO, G.A. (1987). X-ray crystallographic studies of the alanine-specific racemase from *Bacillus stearothermophilus. Journal of Biological Chemistry* **262** (32), 15323–15326.

NICOLOV, D.B., HU, S.-H., LIU, J., GASCH, A., HOFFMAN, A., HORIKOSHI, M., CHUA, N.-H., ROEDER, R.G. AND BURLEY, S.K. (1992). Crystal structure of TFIID TATA-box binding protein. *Nature* **360**, 40–46.

PUSEY, M. AND NAUMANN, R. (1986). Growth kinetics of tetragonal lysozyme crystals. *Journal of Crystal Growth* **76**, 593–599.

RYU, S.-e., KWONG, P.D., TRUNEH, A., PORTER, T.G., ARTHOS, J., ROSENBERG, M., DAI, X., XUONG, N.-H., AXEL, R., SWEET, R.W. AND HENDRICKSON, W.A. (1990). Crystal structure of an HIV-binding recombinant fragment of human CD4. *Nature* **348**, 419–426.

SHI, W., KIDD, R., GIORGIANNI, F., SCHLINDLER, J.F., VIOLA, R.E. AND FARBER, G.K. (in press). Crystallization and preliminary X-ray studies of L-aspartase from *Escherichia coli. Journal of Molecular Biology*.

SHLICHTA, P.J. (1986). Feasibility of mapping solution properties during the growth of protein crystals. *Journal of Crystal Growth* **76**, 656.

STODDARD, B.L., STRONG, R.K., ARROTT, A. AND FARBER, G.K. (1992). *Mir* for the crystallographer's money. *Nature* **26**, 293–294.

STRONG, R.K., STODDARD, B.L., ARROTT, A. AND FARBER, G.K. (1992). Long duration growth of protein crystals in microgravity aboard the *Mir* space station. *Journal of Crystal Growth* **119**, 200–214.

TENG, T.-Y. (1990). Mounting of crystals for macromolecular crystallography in a free-standing thin film. *Journal of Applied Crystallography* **23**, 387–391.

TILLER, W.A. (1991). *The Science of Crystallization: Microscopic Interfacial Phenomena.* Cambridge University Press, Cambridge.

TRAKHANOV, S.D., GREBENKO, A.I., SHIROKOV, V.A., GUDKOV, A.V., EGOROV, A.V., BARMIN, I.N., VAINSTEIN, B.K. AND SPIRIN, A.S. (1991). Crystallization of protein and ribosomal particles in microgravity *Journal of Crystal Growth* **110**, 317–321.

WANG, J., YAN, Y., GARRETT, T.P.J., KIU, J., RODGERS, D.W., GARLICK, R.I., TARR, G.E., HUSAIN, Y., REINHERZ, E.L. AND HARRISON, S.C. (1990). Atomic structure of a fragment of human CD4 containing two immunoglobulin-like domains. *Nature* **348**, 411–418.

WILSON, C., WARDELL, M.R., WEISGRABER, K.H., MAHLEY, R.W. AND AGARD, D.A. (1991). Three-dimensional structure of the LDL receptor-binding domain of human apolipoprotein E. *Science* **252**, 1817–1822.

4
Biotechnology of Cereals

PAUL A. LAZZERI[1] AND PETER R. SHEWRY[2]

[1] *AFRC Institute of Arable Crops Research, Rothamsted Experimental Station, Harpenden* and [2] *Department of Agricultural Sciences, University of Bristol, Long Ashton Research Station, Long Ashton, Bristol, UK*

Introduction: Cereal production and utilization

Cereals are the most important crops in the world in terms of area under cultivation, yield and contribution to the diet of man and his livestock. In 1991, over 700 million hectares were cultivated, with total yields of about 1900 million tonnes. Only three species account for about 80% of the total: wheat (550 million tonnes), rice (520 million tonnes) and maize (480 million tonnes) (FAO, 1988). Other widely grown cereals are (in order of decreasing production) barley, sorghum, oats, millets (various species, the most important of which is pearl millet) and rye. The great success of cereals is due to a number of factors, including their high yields, ease of harvest, technological properties and the availability of a range of species and cultivars to suit almost all climatic conditions. Their tolerance of a wide range of water status is illustrated by Tef, which is adapted to the semi-arid conditions of Ethiopia, and deep water rice, which will grow in water several metres deep. Similarly, their tolerance of temperature ranges from rye, which is particularly suited to the continental climate of Northern Europe and Canada, to a range of minor millets and rice, which flourish in the tropics.

The major harvested organ of cereals is, of course, the grain, although the leaves and stems are harvested for silage and forage in some countries. The seed is, in botanical terms, a caryopsis, in which the seed wall (testa) becomes fused with the fruit wall (pericarp). It consists of two organs, the triploid endosperm and diploid embryo. The endosperm is the major storage organ,

---

Abbreviations: ACCase, acetyl CoA carboxylase; ADP, adenosine diphosphate; ATP, adenosine triphosphate; CaMV, cauliflower mosaic virus; CAT, chloramphenicol acetyltransferase; cDNA, complementary DNA; CMS, cytoplasmic male sterility; es, embryo-derived suspension; ftp, fertile transgenic plants; HMG, 3 hydroxy 3 methylglutaryl; IEF, isoelectric focusing; is, inflorescence-derived suspension; ISL, inflorescence sheath leaf; LUX, luciferase; mc, mesophyll cells; M_r, relative molecular mass; mRNA, messenger RNA; ms, microspore-derived suspension; PAT, phosphinothricin phosphotransferase; PEG, polyethylene glycol; PVC, polyvinyl chloride; RFLP, restriction fragment length polymorphism; sc, stably transformed callus; SDS-PAGE, sodium dodecyl sulphate polyacrylamide gel electrophoresis; T-DNA, transferred DNA; T_R-DNA, right border of the DNA; te, transient expression; tp, transgenic plants.

Biotechnology and Genetic Engineering Reviews — Vol. 11, December 1993
0264–8725/93/11/79–145 $20.00 + $0.00

and differentiates to form two tissues – a central starchy endosperm surrounded by an aleurone layer. The starchy endosperm is a dead tissue in the mature grain, while the aleurone layer (which consists of one layer of cells in wheat but up to three in barley) remains live and acts as a source of hydrolytic enzymes during germination. The embryo consists of the embryonic axis (including the plumule and radicle), which will grow to form the seedling, and a single storage cotyledon (scutellum). In the case of wheat, the starchy endosperm forms the white flour, and the aleurone and embryo the bran and germ, respectively. A longitudinal section of a typical cereal grain (wheat) is shown in *Figure 1*.

Plant seeds contain three major groups of storage compounds, the proportions of which vary between and within different botanical groups. In the cereals, the major storage compound is always starch, which usually accounts for 60–70% of the grain dry weight. The other two groups, protein and lipids, are relatively minor components, accounting for about 8–12% and 2–8% of the dry weight, respectively (Bewley and Black, 1978). The variation in the content of lipid, from about 2% in wheat and barley to 8% in oats and 5–9% in maize, is of considerable interest, as it indicates the potential for manipulating the balance of carbon partitioning between starch and lipid. Maize is already grown as an oil crop, and increasing the proportions of oil in other cereals could lead to a change in the balance of utilization.

Although proteins only account for a relatively small proportion of the cereal grain, they have been of overwhelming importance in determining the pattern of utilization. The total amount of protein present in the cereal grains harvested by humankind far exceeds that in the more protein-rich (up to 40% dry weight) legume seeds, and consequently they form the major source of dietary protein for humans and their livestock. In addition, they have a major impact on the properties of the grain for utilization in the food industry, and in particular in the properties of wheat for making bread and baked goods, pasta and noodles, and other food products.

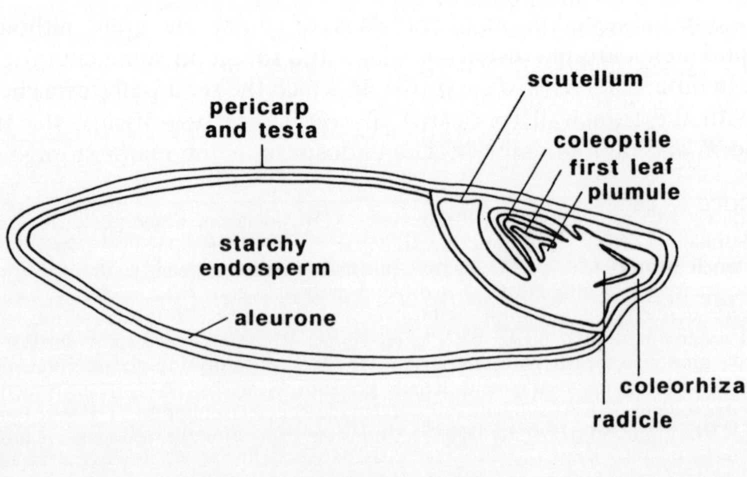

Figure 1. Longitudinal section of a typical cereal grain (wheat).

Targets for manipulation

Cereals have been grown by humans for about 9000 years (Harlan, 1978) and have undoubtedly been selected for improved characteristics (either consciously or unconsciously) over most of this period. In addition, the application of the scientific principles of plant breeding over the last century has led to tremendous improvements in yield, agronomic performance and aspects of quality. However, improvement using this classical approach is limited by the available range of genetic variation in the crop itself or in related species that can be readily crossed. Biotechnology offers the opportunity to introduce variation beyond that normally available, as well as providing tools to maximize the efficiency of classical plant breeding. However, before looking at aspects of cereal biotechnology in detail, we will first briefly discuss the main targets for manipulation.

GRAIN YIELD

Increases in yield have always been a major aim of plant breeders. Since the mature grain consists mainly of starch, attention has been focused on the pathway of starch synthesis and its control (see Preiss, 1988). However, we do not fully understand the role of source–sink relations in determining cereal grain yield (Wardlaw, 1990), and biotechnology offers an opportunity to explore this in detail (see, for example, the elegant studies of Willmitzer and co-workers on source–sink relations in tobacco: Stitt, von Schaewen and Willmitzer, 1990; Sonnewald *et al.*, 1991). A further factor which may also affect yield is endosperm cell number. Chojecki, Bayliss and Gale (1986) demonstrated that the yield of wheat was correlated with endosperm DNA content and cell number, confirming previously reported correlations between kernel size and endosperm cell number in wheat (Brochlehurst, 1977; Radley, 1978; Singh and Jenner, 1982) and barley (Schacherer and Beringer, 1984). It is therefore of considerable importance to understand the control of cell division in the developing cereal endosperm, and in particular the transition between the phases of cell division and cell expansion (the latter phase being associated with storage compound synthesis).

RESISTANCE TO BIOTIC STRESSES

Resistance to pests and pathogens is a major target for plant breeders, with attention focused on a range of fungal, bacterial and viral diseases, and on insect and nematode pests. This is important both for intensive arable agriculture in the developed world and for lower-input farming systems in less developed countries. In the former, the growth of cereal monocultures and intensive use of chemical control methods facilitates the development and spread of new strains and species of pathogen, leading to a constant battle between the crop and plant breeder on the one hand and the pest or pathogen on the other. The exploitation of natural resistance mechanisms is of equal importance in lower-input systems where economic and other factors preclude the extensive use of agrochemicals. In addition, one should not ignore the environmental benefits of decreased pesticide use resulting from the exploitation of natural and engineered defence mechanisms, especially in high-input systems.

We still understand very little about the natural mechanisms of plant resistance to pests and pathogens, and in particular about gene-for-gene resistance to fungal pathogens. Nevertheless, strategies have been used with varying degrees of success to confer resistance in transgenic plants. These include the use of viral coat protein genes (Beachy, 1988) or satelite DNA (Harrison, Mayo and Baulcombe, 1987; Gerlach, Llewellyn and Haselhoff, 1987) to confer resistance to viruses, and of bacterial genes encoding insecticidal proteins (*Bacillus thuringiensis* toxins) to confer insect resistance (Koziel *et al.*, 1993).

In addition, a number of plant proteins, which may be expressed constitutively or induced on infection or feeding, have been demonstrated to confer resistance to pests or pathogens either *in vitro* or in transgenic plants. These include inhibitors of hydrolytic enzymes (α-amylases and proteases), β-glucanases, endochitinases, protein synthesis inhibitors, lectins, thionins and membrane-permeabilizing proteins. The first such protein to be used in transgenic plants, and the most well known, is the cowpea trypsin inhibitor, which was shown to confer resistance to tobacco budworm in transgenic tobacco (Hilder *et al.*, 1987). Other proteins may confer resistance to fungal or bacterial pathogens, or even broad-spectrum resistance to a range of such organisms. Some of these proteins are derived from cereals and are therefore attractive targets for eventual incorporation into transgenic cereal plants.

An alternative approach to confering resistance in transgenic plants is to transfer genes for the production of defence-related chemicals. Because such chemicals are synthesized by multigene pathways, this is a more formidable challenge than the transfer of single genes for disease resistance proteins as discussed above. However, Hain *et al.* (1993) have recently reported enhanced disease resistance of tobacco to the fungal pathogen *Botrytis cinerea* by inserting two stilbene synthase genes which resulted in the synthesis of phytoalexins.

These and other strategies for enhancing crop resistance to microbial pathogens are discussed in a recent review article by Lamb *et al.* (1992). However, there is no reason to assume that such engineered sources of pest and pathogen control will prove to be any more durable than many chemical control methods used at present, as they will almost certainly lead to the evolution of resistant forms of pests and pathogens (May, 1993).

RESISTANCE TO ABIOTIC STRESSES

In global terms, the most important abiotic stress is mineral toxicity, although this is not always appreciated in developed countries. It has been estimated that nearly a quarter of the world's soils are affected by mineral stress (Clark, 1982), the major problems being salinity, aluminium, manganese and boron. Heavy metal toxicity also occurs under some circumstances, notably copper (see Manyowa and Miller, 1991). The most successful approach to this problem in cereals is to exploit natural variation in tolerance, either in the crop species itself or in wild relatives (see Manyowa and Miller, 1991). An interesting illustration of the potential of genetic engineering to improve stress tolerance is the recent report by Tarczynski, Jensen and Bohnert (1993) that the introduction of bacterial mannitol-1-phosphate dehydrogenase into tobacco plants results in the accumulation of mannitol and increased tolerance of high salinity.

Tolerance to high and low temperatures (the latter including frost damage) and to waterlogging are also of some importance, and have been studied in cereals. High temperature effects in cereals include the appearance of 'heat shock' proteins as in other systems. In addition, high temperatures during grain filling in wheat result in decreased quality for bread-making, which may result in part from specific effects on gluten protein gene expression (Blumenthal *et al.*, 1990).

NUTRIENT USE EFFICIENCY

The current concern about the environmental consequences of intensive agricultural practices includes recognition of the need to reduce the use of fertilizers and the levels of nitrate in groundwaters. This has stimulated interest in nitrogen use efficiency, and in the mechanisms of nitrate uptake, assimilation and utilization. Cloned genes for nitrate assimilation (nitrate reductase, nitrite reductase, glutamine synthetase, glutamate synthase) are available from several species and are currently being used to explore the control of nitrogen metabolism in transgenic plants (see Vaucheret *et al.*, 1992). Similar studies could lead to improved nitrogen use efficiency in cereals.

GRAIN QUALITY AND COMPOSITION

There has been increased emphasis on grain composition and quality over the past decade, partly because the increased yields of cereals in the developed world, and the spread of wheat cultivation within the European Community, have led to new opportunities for utilization. Recent studies have indicated that it should be possible to manipulate many aspects of grain composition and quality, some of which are discussed in detail later.

HERBICIDE RESISTANCE

Engineered resistance to herbicides is perhaps the most widely discussed and emotive issue in plant genetic engineering, because it is seen as establishing dependence on the intensive use of herbicides. In fact, if managed properly, it should ultimately reduce the herbicide dosage by allowing more efficient use. Resistance to several major classes of herbicide has been achieved by two basic strategies: by manipulating the amount or sensitivity of the target enzyme, or by introducing genes for detoxification. Furthermore, the latter approach has been successfully used with genes from bacterial as well as plant sources. This work has, of course, been facilitated by detailed studies of the target enzymes and of the detoxification pathways. It has been discussed in a number of recent articles (see, for example, Gasser and Fraley, 1989; Hartnett *et al.*, 1991; Botterman *et al.*, 1991; O'Keefe, Lenstra and Omer, 1992), and will not be elaborated here.

So far, the only herbicide-resistant cereal to be marketed is maize, and this has been produced by mutant selection rather than genetic engineering. However, the development of reliable methods of cereal transformation will undoubtedly lead to rapid exploitation of the strategies that have been successfully developed for other plants.

Development and application of methods

Three strategies have been adopted to improve cereals. The first is plant breeding, which involves crossing contrasting genotypes to generate new combinations of genes, followed by selection of the most suitable of these. In addition, exotic genotypes or closely related (i.e. crossable) wild species may be used as sources of specific genes (for example, for resistance to pests or abiotic stress). This approach does not involve biotechnology, and will not be discussed further.

The second strategy is still based on classical plant breeding, but elements of biotechnology (tissue culture, regeneration) are used to make the process more efficient and to allow genes to be incorporated from wild relatives that cannot readily be crossed. The third strategy is genetic engineering: the insertion of specific genes to improve aspects of resistance, agronomic performance, quality or yield.

In the following sections, we will discuss progress in the latter two strategies, highlighting achievements and identifying problems and limitations.

TISSUE AND CELL CULTURE

Plant tissue culture is based on two principles: first, that plant cells may be cultured axenically *in vitro* and, second, that they are inherently totipotent, so that given the correct conditions they are able to regenerate to form new plants. In herbaceous dicot species, regenerable tissue cultures can be established from a variety of plant parts, including relatively mature organs. However, in most monocots the process is more difficult, as only immature tissues which are to some degree still meristematic can be induced to proliferate *in vitro* (Vasil, 1987). This poor capacity for dedifferentiation in culture is correlated with a wound response in which damaged areas are 'sealed-off' rather than being 'repaired' by *de novo* cell division forming a wound callus (Kahl, 1982). Among the monocots, members of the grass family (*Poaceae*) are especially determinate, their cells becoming committed to a particular developmental pathway soon after emergence from the meristem (see Morrish, Vasil and Vasil, 1987). These characteristics have necessitated the development of specific methods for the culture and manipulation of cereals *in vitro* (Vasil, 1988; Lazzeri, Kollmorgen and Loerz, 1990).

Meristem culture

As with other plant species, it is possible to culture isolated shoot apices of cereals and to propagate them indefinitely (Dalton and Dale, 1985). In most cereals there is limited potential for vegetative multiplication, as they are efficiently propagated by seed, but there may be occasions where it is of interest to clone individual genotypes *in vitro* (e.g. the multiplication of transformants before transfer to soil, or of lines with reduced fertility). Additionally, *in vitro* tiller cultures may be maintained at low temperatures for germplasm storage (Dale and Webb, 1985).

Callus culture

Cereal callus cultures are initiated by plating explanted tissue on medium containing a synthetic auxin, usually 2,4-D. Cell proliferation is induced and a heterogeneous callus is typically produced, containing both embryogenic tissue competent for regeneration and non-embryogenic tissue. The former is characteristically slow-growing, compact and distinctly organized, whereas the latter tends to be faster-growing, more friable and disorganized. Regeneration may occur spontaneously from embryogenic tissues, but is usually stimulated by a reduction in the auxin content of the medium, and possibly by the addition of cytokinin.

The choice of explant is the major factor determining culture success. The most responsive tissues are immature embryos/scutella (before the onset of starch deposition), immature inflorescences, apical meristems and seedling leaf base meristems. In contrast with most dicots, cereal leaf tissues are generally determinate and non-responsive *in vitro*. One exception is inflorescence sheath leaves (ISLs), which appear to retain more developmental plasticity (Barcelo *et al.*, 1991). ISL cultures also demonstrate the rule that in cereal tissue cultures, as in other systems, increasing levels of cellular competence are required for proliferation, root formation and embryogenesis/regeneration, with these different competencies being successively lost as tissues mature (Barcelo *et al.*, 1992).

Apart from explant type and ontogeny, several other factors are important in determining response in culture. A major element is donor plant genotype. There is great variation in the culturability of different genotypes within all cereal species (Duncan *et al.*, 1985; Luehrs and Loerz, 1988; Vasil, 1987). Culture response is clearly under genetic control and there has been some success in breeding for improved performance *in vitro* (Armstrong, Romero-Severson and Hodges, 1992 and references therein). As well as genotype, donor plant conditions (and thus the physiological status of the explant), culture medium composition and culture environment also influence response.

Regeneration from cereal tissue cultures occurs primarily by embryogenesis, with somatic embryos, similar in morphology to zygotic embryos, being formed from the surfaces of embryogenic callus. In some cases, however, regeneration appears to occur via organogenesis, with shoot apical meristems being formed superficially on callus tissues. As organogenesis is usually also associated with embryogenesis, the shoots observed may actually develop from the apical meristems of 'disorganized' somatic embryos.

Cereal cultures generally show the greatest capacity for plant regeneration during the first passage in culture, with this capacity declining over subsequent subcultures. This effect may to some extent be overcome by visual selection for embryogenic tissues at each subculture (Redway, Vasil and Vasil, 1990), but remains a limitation to the use of long-term cultures. Loss of regeneration competence is frequently correlated with increasing disorganization of cultured tissues and with the accumulation of cytological aberrations, but the underlying causes of the process are not fully understood (see pp. 107–108). Regenerable callus cultures can now be produced in all the important cereal species and the majority of genotypes can be cultured and regenerated, albeit with varying efficiencies.

The major uses for cereal callus cultures are as starting material for suspension and protoplast cultures and as regenerable target tissues for transformation. Callus cultures may also be used for the *in vitro* selection of mutants, and for the generation of somaclonal variants.

Suspension culture

Cell suspensions are produced by transferring callus tissue to liquid medium and shaking the culture to ensure aeration. Although the process is simple, primary embryogenic cereal callus does not adapt readily to liquid conditions and it is generally necessary to select for specific callus phenotypes which allow more efficient suspension initiation (Shillito *et al.*, 1989; Redway, Vasil and Vasil, 1990). In order to avoid long-term selection for embryo-derived callus suitable for suspension establishment, some workers have used microspore-derived callus as starting material (Datta, Datta and Potrykus, 1990; Jaehne, Lazzeri and Loerz, 1991). However, this approach requires an efficient anther or microspore culture system (see below). Early-stage cereal suspensions are typically slow-growing and heterogeneous, with large cell aggregates. They become finer and more homogeneous during the establishment phase and stable suspensions are usually composed of small aggregates (0.1–2.00 mm in diameter) of isodiametric, cytoplasm-rich cells.

Cereal suspensions do not produce somatic embryos in the liquid medium, but must be plated onto solid medium to test their regeneration capacity, where they first form embryogenic callus, which then develops somatic embryos. With their highly disorganized nature and fast growth rate (doubling times typically 4–7 days), suspension cultures tend to lose regeneration capacity relatively quickly (see below), and it is unusual for a suspension to remain regenerable for more than 1 year. This problem can to some extent be overcome by cryopreserving embryogenic lines and re-initiating them when the 'parent' line is no longer regenerable (Shillito *et al.*, 1989), but in most cases it is still necessary to establish new cell lines regularly.

Embryogenic cell suspensions capable of plant regeneration have been produced in all the major cereals (*see Table 1*). The process is most efficient in japonica rice, and probably least efficient in wheat and barley. In addition, there is considerable genotypic variation within each species in the ease and reproducibility of suspension establishment.

Cell suspensions are attractive experimental tools, as they are fast-growing, easy to maintain, have an indefinite life-span and can provide large numbers of easily accessible, relatively homogeneous cells. Their major disadvantage, as discussed above, is that they have a relatively short 'window' of competence for plant regeneration. Today, the most important applications of cereal suspensions are as sources of protoplasts (see below) and as targets for transformation (see below). They may also be useful for *in vitro* selection purposes (see below) and in physiological, biochemical and molecular studies. However, because cell suspensions do not have the structural organization of the parent plant and are grown under highly artificial conditions, they may not always be good models for investigating plant processes. Nevertheless, where cell suspensions do retain important characteristics of the parent plant/tissue (e.g. normal or mutant storage protein profiles: Tewes *et al.*, 1991; Schaeffer, Sharpe and Dudley, 1992), they can prove valuable as it is possible to make highly controlled manipulations of nutritional and growth factors and to study their effects.

Protoplast culture

Protoplasts can be isolated from cereal cells using the same methods as those used for herbaceous dicot species, but it is much more difficult to induce these protoplasts to divide. Regeneration from cereal protoplasts was a major objective in plant cell culture for some 15 years and it was not until 1986 that the production of fertile cereal (rice) plants from protoplasts was reported

Table 1. Regenerable suspension cultures of cereals

Species	Common name	Reference[a]
Hordeum vulgare	Barley	Jaehne, Lazzeri and Loerz (1991) Funatsuki, Loerz and Lazzeri (1992)
Hordeum chilense × *Triticum durum*	Tritordeum	Barcelo *et al.* (in press)
Oryza sativa ssp. *indica*	Indica rice	Lee *et al.* (1989) Datta *et al.* (1990)
Oryza sativa ssp. *japonica*	Japonica rice	Fujimura, Sakurai and Akagi (1985) Kyozuka, Mayasami and Shimamoto (1987)
Pennisetum americanum	Pearl millet	Vasil and Vasil (1980)
Sorghum vulgare	Sorghum	Wei and Xu (1990)
× *Triticosecale*	Triticale	Stolarz and Loerz (1986)
Triticum aestivum	Bread wheat	Wang and Nguyen (1990) Redway, Vasil and Vasil (1990)
Zea mays	Maize	Prioli and Sondahl (1989) Morocz *et al.* (1990)

[a] Selected references, not necessarily first reports.

(Toriyama, Hinata and Sasaki, 1986). The fact that protoplasts isolated from differentiated cereal tissues very rarely divide reflects the extreme determinacy of cereal cells, as discussed above. Furthermore, it is difficult to obtain adequate yields of protoplasts from meristematic tissues and those protoplasts which are obtained are fusogenic and difficult to handle. The preferred source of dividing cereal protoplasts is cell suspensions or, less frequently, fast-growing friable callus. The most efficient protoplast isolations and the highest plating efficiencies are obtained from long-established suspensions which grow very fast and have friable cell aggregates. Such suspensions have, however, usually lost the ability to regenerate plants, and embryogenic suspensions are required for the isolation of competent protoplasts. The development of methods for the production of embryogenic suspensions of various cereal species has been followed by reports of the regeneration of plants from protoplasts derived from them (*see Table 2*). The efficiency of protoplast regeneration in a particular species is dependent on the ease of establishing suspensions, which means that at present the process is only widely reproducible in japonica rice, and relatively few laboratories world-wide have produced fertile plants from maize, wheat or barley protoplasts. As experience grows, however, it is clear that techniques are being improved and that reproducible systems will be developed for more species.

The use of techniques such as alginate embedding (Tricoli, Hein and Carnes, 1986) and the use of nurse cultures (Funatsuki, Loerz and Lazzeri,

Table 2. Regeneration from cereal protoplasts

Species	Common name	References[a]	Source[b]
Hordeum vulgare	Barley	Jaehne, Lazzeri and Loerz (1991)	ms
		Funatsuki, Loerz and Lazzeri (1992)	es
Hordeum chilense × *Triticum durum*	Tritordeum	Barcelo *et al.* (in press)	es
Oryza sativa ssp. *indica*	Indica rice	Lee (1989)	es
		Su, Rudert and Hodges (1992)	ms
Oryza sativa ssp. *japonica*	Japonica rice	Kyozuka, Mayasami and Shimamoto (1987)	es
		Wu and Zapata (1992)	pc
		Gupta and Pattanayak (1993)	mc
Pennisetum americanum	Pearl millet	Vasil and Vasil (1980)	is
Sorghum vulgare	Sorghum	Wei and Xu (1990)	is
Triticum aestivum	Bread wheat	Vasil, Redway and Vasil (1990)	es
		Ahmed and Sagi (1993)	es
Triticum durum	Durum wheat	Yang *et al.* (1989)	es
Zea mays	Maize	Shillito *et al.* (1989)	es
		Prioli and Sondahl (1989)	es

[a] Selected references, reporting fertile plant regeneration where achieved.
[b] ms, microspore-derived suspension; es, embryo-derived suspension; mc, mesophyll cells; is, inflorescence-derived suspension.

1992) can improve cereal protoplast culture efficiency significantly and some workers have re-tested the possibility of using these methods for culturing protoplasts isolated directly from cereal organs. It has been shown that it is possible to obtain macroscopic calli from mesophyll protoplasts of oat (Hahne, Fleck and Hahne, 1989) and of *Tritordeum* (Barcelo, pers. comm.), and recently plants have been regenerated from mesophyll protoplasts isolated from rice leaf base and sheath (Gupta and Pattanayak, 1993).

Cereal protoplasts are used quite extensively in physiological and biochemical studies, but their main importance in biotechnology is as tools for genetic manipulation. Before cell wall reformation occurs, they are easily fused, either by chemical or electrical stimuli, allowing a range of somatic cell genetic techniques, including hybridization, organelle transfer or partial genome transfer (see below). In addition, protoplasts may easily be induced to take up DNA from surrounding solutions, again in response to electrical or chemical stimuli. Protoplast transformation is discussed below.

HAPLOIDY

Among the different *in vitro* genetic manipulation techniques used in plants, haploid induction and embryo rescue have the most direct applications in breeding programmes. Haploid plants themselves are not of practical use, but doubled haploids, resulting from spontaneous or induced chromosome duplication, constitute a single step to homozygosity and have value in breeding and genetic research.

Doubled haploids allow breeders to produce completely homozygous genotypes from heterozygous parents in a single generation. The uses of double haploid production in breeding depend largely on whether a particular crop is in- or out-breeding. Varieties of inbreeding species, including small-grain cereals such as wheat, barley and rice, are usually single homozygous genotypes. In this case, doubled haploids could become new cultivars in a single step by 'fixing' the product of a cross without the need for backcrossing. Alternatively, the doubled haploid may be used in further crosses.

In out-breeding crops, such as maize, doubled haploid lines may be used as the parents for hybrid production instead of the true-breeding lines conventionally produced by several generations of inbreeding (for discussions of haploid breeding systems, see Snape, 1989; Petolino, 1992; Pickering and Devaux, 1992).

There are three major routes for the production of doubled haploid plants. The first is the use of genetic systems, such as the haploid initiator *hap* gene in barley (Hagberg and Hagberg, 1987), which produces haploids through disturbances in zygote development. These systems will not be discussed as they do not involve biotechnological techniques. The second route, which is used principally in barley but also functions in wheat, is the *bulbosum* system (Kasha and Kao, 1970), in which a cross between *Hordeum bulbosum* and a cultivar is made, followed by rescue of the embryo (see below) which becomes haploid due to the elimination of the *H. bulbosum* chromosomes after fertilization. The third route is via the culture of gametophytic tissues (i.e. anthers or microspores, ovaries or ovules), which under appropriate conditions will proliferate and regenerate haploid or doubled haploid plants. In the cereals, anther/microspore culture is generally practised rather than ovary/ovule culture, as the latter require much more labour to dissect them out of the donor plant and the efficiency of haploid production is low (Pickering and Devaux, 1992; Castillo and Cistué, 1993).

The production of plants from microspores involves a drastic developmental switch in which the normal course of microspore development towards a pollen grain is diverted into the formation of an embryo or callus tissue. The precise trigger(s) causing this diversion are not understood, but it is postulated that the isolation and culture procedures lead to the depletion of factors required to maintain normal pollen development, allowing the switch from the gametophytic to a sporophytic pathway (Dunwell, 1992).

The first cereal haploids were produced in the early 1970s from anther cultures, but efficiencies were extremely low and high frequencies of albino plants occurred (see Dunwell, 1985, for a review). In the intervening years, considerable effort has been devoted to improving the methodology, particularly in the major crops, and today haploid techniques are routinely applied in maize, rice, wheat and barley breeding (Petolino, 1992; Zhang, 1989; Devaux, 1992).

Although, as previusly discussed, the underlying mechanisms of microspore embryogenesis have yet to be understood, many of the major factors influencing microspore response *in vitro* have been identified. Donor plant genotype is of primary importance. Response is clearly under genetic control and it is possible to transfer associated genes via crossing (e.g. Petolino and Thompson, 1987). Recently, genes affecting androgenetic response have been mapped by restriction fragment length polymorphism (RFLP) analysis in order to improve haploid breeding efficiency (Cowen *et al.*, 1992). After the genotype, the physiological status of the donor plants and the pretreatments (temperature, preculture) of the spikes containing microspores are the most important factors which influence microspore development and subsequent response in culture. Similarly, the stage at which spikes are taken for culture is significant, microspores at different developmental stages responding differently and yielding different frequencies of albino and haploid versus doubled haploid plants. In addition to these biological parameters, medium composition (particularly carbon source and growth regulators) and the physical culture environment have profound effects. Apart from the basic response, that is, the diversion to an embryogenic or callus formation pathway, other components of the process, such as chromosome doubling, plantlet formation and the ratio of green : albino plants, are all influenced by these parameters, giving in total a highly complex and interactive system, albeit one which is used successfully. Pickering and Devaux (1992) have made a comprehensive review of the factors affecting barley anther culture response, and a similar spectrum of factors operates in other cereal species.

Although anther culture remains the most widely used dihaploid production technique, it is also possible to produce plants from the culture of isolated microspores (Kasha, Cho and Ziauddin, 1992). The method is less labour-intensive than plating individual anthers and offers greater opportunities to understand the pathway of microspore embryogenesis without the influence of anther (tapetal) tissue. In addition, the technique allows the manipulation of large populations of uniform haploid cells, making it attractive for transformation, *in vitro* mutation and selection. Although high levels of response are currently only being obtained in model genotypes, efficiencies are constantly being improved.

Cultivars derived from doubled-haploid techniques have now been released in all of the major cereals (for examples, see Zhang, 1989; Pickering and Devaux, 1992). As the methodology improves, the efficiency will increase and the costs should decrease in comparison with more conventional techniques, leading to wider applications.

EMBRYO RESCUE/WIDE CROSSING

Plant breeders have produced improved cereals by combining desirable genes from the natural gene pool. However, the variability within several of the major cereals has been reduced as landraces have been replaced by pure line varieties (Frankel, 1970). In order to offset this effect and to provide supplementary genetic variability, breeders now look to the incorporation of genes from wild progenitor species or from other genera. Where wild species are closely related to the cultivated species (e.g. *Hordeum vulgare* and *H. spontaneum*), standard sexual crosses may be made to introgress genetic diversity for agronomic characteristics (Nevo, 1992). Where wild species are more distantly related or where attractive genetic variation exists in other genera, then standard crosses will not be possible and enabling technology will be needed for gene transfer. The best established technique for alien gene transfer (wide crossing) is that of hybrid embryo rescue, but two more recent techniques which may have applications are *in vitro* fertilization and somatic hybridization (protoplast fusion).

Embryo rescue

Barriers to the crossability of two species may exist at a number of levels, either pre- or post-fertilization (see Kalloo and Chowdhury, 1992). In cereals, it is often desirable to make wide crosses between distantly related species within a genus, or between species in related genera (Islam and Shepherd, 1991). In these cases, it is common for fertilization to take place, but for endosperm development to fail so that the hybrid embryo starves. In such situations, the hybrid plant may be recovered by embryo rescue, in which the developing embryo is excised from the caryopsis and cultured *in vitro*. In this procedure, the female parent is emasculated before pollen maturity and pollen from the male parent is applied when the stigmas are receptive. The efficiency of seed set may be improved by the addition of GA_3 at intervals after pollination, or by the application of 2,4-D as a pre-treatment before pollination. Immature embryos are isolated from enlarged caryopses as shortly before endosperm degeneration as possible, and are then transferred either to culture medium or to a nurse endosperm (Khush and Brar, 1992).

Athough the major factors controlling seed set in wide crosses are genetic, other factors such as the growth environment and physiological status of the parental plants are also of importance. Furthermore, frequencies of seed set may be good while plant recovery frequencies are low (Bothmer *et al.*, 1989), due to poor embryo germination. This may result from the operation of a number of post-fertilization barriers, and several procedures such as alteration of the genomic ratios of the parents, or the induction of chromosomal exchanges, have been developed to overcome such blocks (Khush and Brar, 1992).

Embryo rescue techniques have allowed the introgression of many important agronomic characteristics from wild species into cultivated cereals, and have great potential. Within the Triticeae, new genomic combinations have been made to produce fertile synthetic amphiploids (homozygous products of chromosome doubling of interspecific or intergeneric hybrids). In most cases, amphiploids derived from cultivated species show inferior performance to the cultivated parent, but two exceptions are *triticale* and *tritordeum*. Hexaploid *triticale* (AABBRR) is an amphiploid between tetraploid wheat and rye. It has a number of superior agronomic characteristics and is an accepted cereal crop, particularly in Eastern Europe. *Tritordeum* (HHAABB) is a more recently developed amphiploid between *Hordeum chilense* and tetraploid wheat (Martin and Sanchez-Monge, 1982). It has several interesting characteristics, including high protein content, pathogen resistances and drought and salinity tolerance (see Barcelo *et al.*, 1993).

Rather than transferring whole genomes, the purpose of most wide crosses is the transfer of small chromosomal segments or single genes from a wild relative into a crop species. This can be achieved by 'chromosome engineering' techniques, which start with hybridization and then involve a series of cytogenetic manipulations to give an elite line containing an alien segment.

In wheat, wide crossing has been used to transfer numerous genes from other cereals or wild relatives, including genes conferring resistances to pests and pathogens, tolerance to abiotic stresses such as salinity, and improved nutrient use efficiency (see Gale and Miller, 1987; Islam and Shepherd, 1991). Similar transfers have been performed in barley, although to a lesser extent (see Fedak, 1992).

A limitation in many of these gene transfer experiments has been the lack of markers on the cereal or alien chromosomes. The development of RFLP maps for the major cereals is changing this situation and will increase the efficiency of introgression of useful alien genes into cereal crops.

In vitro *fertilization*

Where pre-fertilization barriers to wide hybridization exist, a number of techniques may be applied, such as removal of the stigma followed by direct application of pollen to the style, or *in vitro* pollination of isolated ovaries (Zenkteler, 1990; Khush and Brar, 1992). The logical extension to these techniques is to carry out the fertilization *in vitro*, using isolated sperm and egg cells. This has long been a goal of plant cell biologists and has recently been achieved in maize (Kranz, Bautor and Loerz, 1991). The process involves the enzymic isolation of egg cells, which are then fused with mechanically isolated sperm cells using an electric pulse. The fusion products are cultivated in a feeder-cell system and will develop into multicellular structures. Further development was limited in the early experiments, but fertile hybrid maize plants have now been recovered (Kranz and Loerz, 1993). In this system, the frequency of egg + sperm cell fusions is high (~80%), as is the frequency of zygotic cell division (80%+) (Kranz, Bautor and Loerz, 1991). Furthermore, high fusion frequencies have also been achieved between maize egg cells and *Coix* pollen (E. Kranz, pers. comm.), indicating that the technique may have further applications in making wide crosses. At present, it offers a powerful research tool to study the fertilization process in plants.

Somatic hybridization

The final technique to overcome crossability barriers, particularly sexual incompatibility, is somatic cell fusion, using isolated protoplasts. The fusion of protoplasts is achieved either by the use of polyethylene glycol (PEG) under conditions of high pH and high calcium (Ozias-Akins, Ferl and Vasil, 1986), or by electrofusion using a short DC pulse after the alignment of protoplast pairs in an AC field (Terada *et al.*, 1987). Protoplast fusion offers the opportunity to make genetic manipulations which are difficult to achieve by other means. First, it is possible to make 'full' somatic hybrid plants from nuclear and cytoplasmic fusion of the two partners. Second, it is possible to create 'cybrid' plants (i.e. cytoplasm transfer) by inactivating the cytoplasm of one partner (via treatment with iodoacetamide) and the nucleus of the other partner (via X-irradiation), followed by fusion. Third, it is possible to achieve partial genome transfer (asymmetric hybridization) by both inactivating the cytoplasm and pulverizing the nucleus of the donor cells. The products are then selected for the transfer of fragments the donor genome into the recipient nucleus. To recover somatic hybrid plants, it must be possible to regenerate from callus produced from protoplast fusion. As discussed above, protoplast regeneration is efficient in only a few cereals and this has limited the application of somatic hybridization.

Somatic hybrid cell lines have been produced from a number of intergeneric and interspecific combinations (e.g. *Pennisetum* × *Panicum*, Ozias-Akins, Ferl and Vasil, 1986; *Triticum* × *Pennisetum*, Vasil, Ferl and Vasil, 1988; *Zea* × *Triticum*, Wang *et al.*, 1993; *Hordeum vulgare* × *H. bulbosum*, H. Funatsuki, pers. comm.; *H. vulgare* × *H. marinum*, X.-H. Wang, pers. comm.), but hybrid plants have only been recovered in rice.

Thus somatic hybrid plants have been obtained between rice and barnyard grass (*Echinochloa* sp.): Terada *et al.*, 1987), between rice and several wild *Oryza* species (Hayashi, Kyozuka and Shimamoto, 1988) and between rice cultivars (Toriyama *et al.*, 1988). Rice hybrid plants have also been produced, with the purpose of transferring cytoplasmic male sterility (CMS) cytoplasms into cultivars for hybrid seed production (Akagi *et al.*, 1989; Yang *et al.*, 1989; Kyozuka, Kaneda and Shimamoto, 1989). These experiments demonstrate that protoplast fusion can potentially be used to manipulate agronomic characters, but cereal protoplast culture techniques will need to be improved for these methods to have wider application.

TRANSFORMATION

For genetic engineering to be possible, we need to be able to introduce and stably integrate foreign genes which will modify the characteristics of the recipient plant. In most dicotyledonous species, this can be achieved by *Agrobacterium* transformation in which the T-DNA of the bacterium acts as the vector for the gene transfer (see Ooms, 1992, for review). Cereals are not natural hosts for *Agrobacterium* and consequently there is the need to apply alternative transformation techniques. A spectrum of different methods have been tried and three direct gene transfer methods – cell electroporation, protoplast transformation and particle bombardment – have now yielded fertile transgenic plants.

In this section, we will first survey the various techniques used for cereal transformation (*see Table 3*), then discuss the DNA constructs used, selection for transformed cells, transgene integration and stability in cereals, and finally consider the current situation regarding application of the technology.

Whole plant transformation methods

The attraction of whole plant methods is that an *in vitro* regeneration step is avoided and the technology is relatively simple, requiring no specialized equipment, but only a supply of plasmid DNA. The two approaches which have received most attention are macroinjection and pollen transformation.

De la Pena, Lorz and Schell (1987) injected plasmid DNA containing the *nptII* gene into immature floral tillers of rye and recovered three kanamycin-resistant plants which, in Southern blots, appeared to contain the appropriate

Table 3. Transformation methods tested in cereals

Method	Results
Whole plant level; macroinjection, pollen tube pathway, topical DNA applications, etc.	No good evidence for stable, integrative transformation
DNA uptake into imbibing seeds, embryos	Transient gene expression
Agrobacterium transformation, agroinfection	Transient gene expression, systemic virus multiplication
Microinjection	Transient gene expression, stably transformed somatic cells
Silicon carbide fibre injection	Stably transformed tissues
Cell electroporation	Fertile transformed plants
Protoplast transformation	Fertile transformed plants
Particle bombardment	Fertile transformed plants

npt fragments. However, transmission of the transgene to progeny was not reported. This first report prompted many groups to test the macroinjection method, but although there were several reports of the detection of sequences hybridizing to the input DNA, both in primary transformants and progeny, transgene expression was often undetectable. Furthermore, there was no clear proof for the stable integration of these sequences into genomic DNA, so that mechanisms for extrachromosomal maintenance of input DNA or endophyte transformation have been invoked (Rogers and Rogers, 1992; Konstantinov, Mladenovic and Denic, 1991).

Pollen transformation methods aim to transform pollen cells either before fertilization or shortly after fertilization when the gametic cells or zygote have no cell walls and may therefore be accessible to exogenous DNA. There has been a long-standing interest in gametophytic transformation of cereals (De Wet *et al.*, 1986) and in the mid- to late 1980s several workers claimed success using such methods in maize (Ohta, 1986), wheat (Picard *et al.*, 1988) and rice (Luo and Wu, 1988). These reports generated much interest and many laboratories attempted to repeat the experiments. The initial results proved dificult to reproduce, however, and no study has produced clear evidence for stable transformation using these methods. A frequent observation has been that the very high nuclease activity associated with germinating pollen degrades the applied DNA within a few minutes of application (Roeckel *et al.*, 1988; Booy, Krens and Huizing, 1989).

A common feature of the 'successful' gametophytic transformation experiments has been that Southern analyses suggest the presence of sequences corresponding to the input DNA, but that expression of the marker genes cannot be detected. This observation is consistent with degradation of the plasmid DNA before uptake and integration and suggests that pollen transformation could become a viable technique if a suitable DNA protection method can be devised.

DNA uptake into seeds/embryos

The cell membranes of dry tissues are disorganized and show increased permeability. During the process of imbibition, a very large water potential gradient exists between the hydrating tissue and the external solution and the rapid cell expansion leads to cell wall disruption. These conditions could be expected to facilitate the uptake of macromolecules into cells, and consequently a number of workers have examined the imbibition of dry tissues in DNA solutions as a potential transformation method. In early experiments, foreign DNA uptake could only be demonstrated by detecting its physical presence, leading to ambiguous results. The advent of heterologous marker genes has made detection far more sensitive and in 1989 Toepfer and co-workers demonstrated that dry cereal embryos could take up plasmid DNA from an imbibition buffer and that transient gene expression could be measured. Embryos are still viable after treatment and will germinate to produce normal plants. Although the method would therefore appear to have potential for stable transformation, this has not yet been achieved despite large-scale experiments. It may be that exogenous DNA is only taken up by damaged cells that ultimately do not survive, but it is more likely that uptake is confined to somatic cells of the embryos, which will not proliferate further, and that meristematic cells are not accessible to transformation. It has been shown that desiccated somatic embryos can also be transiently transformed by the imbibition method (Senaratna *et al.*, 1991), but there are no reports of stable transformation.

Agrobacterium *transformation/agroinfection*

For most plants, the most efficient transformation method is the use of the *Agrobacterium* vector system (Ooms, 1992). Cereals are classed as non-hosts to *Agrobacterium*, as they are not infected *in vivo* and show no response to inoculation *in vitro*. However, a series of investigations has shown that *Agrobacterium* can attach to cereal cells (Mooney and Goodwin, 1991), that cereal cells produce factors which induce *Agrobacterium* virulence genes (Schlaeppi and Hohn, 1992), and that the bacterium can transfer its T-DNA to cereal cells, leading to expression of genes contained within the T-DNA sequence.

Despite these indications of cereal × *Agrobacterium* compatibility, a reproducible method for stable transformation has yet to be developed. Attempts have been made to transform cereals at the whole plant level by pipetting Agrobacteria into flowers at the time of fertilization, but with mixed results. Hess, Dressler and Nimmrichter (1990) reported inheritance of the transferred genes in progeny of wheat treated using this method, while Langridge *et al.* (1992) made similar experiments in wheat, barley and maize but found no evidence for transmission of transgenes. They suggested that this result and the anomalous integration patterns observed could be explained by the transformation of endophytic bacteria present in the tissues of treated plants.

Relatively more effort has been applied to the transformation of cereal tissues with *Agrobacterium* under *in vitro* conditions. A variety of different culture systems have been used, including sterile seedlings (Graves and Goldman, 1986), shoot apices (Gould *et al.*, 1991) and embryos (Raineri *et al.*, 1990; Mooney *et al.*, 1991). All of these reports presented biochemical evidence for transformation, and most also presented molecular evidence for transgene integration. However, in no case was transmission of transgenes to progeny clearly demonstrated, and the experiments have not been repeated by independent workers. There is, therefore, still some doubt about the validity of this approach.

While the stable transformation of cereals by *Agrobacterium* is still questionable, the evidence that gene transfer occurs is now very good, largely as a result of the use of the agroinfection technique (Grimsley *et al.*, 1987), in which viral sequences capable to replication in cereal cells are cloned into the *Agrobacterium* T-DNA. This creates an extremely sensitive assay for T-DNA transfer, as a single viral molecule has the potential to replicate and establish a systemic infection which may easily be detected. Agroinfection has enabled factors affecting T-DNA transfer to cereal cells to be investigated. It has been shown that the *Agrobacterium* strain (Marks *et al.*, 1989), bacterial virulence genes (Grimsley *et al.*, 1989), recipient cell type (Grimsley *et al.*, 1988) and developmental stage (Schlaeppi and Hohn, 1992) are all important factors.

As the viral molecules transferred via agroinfection do not integrate into the host cell chromosomes, the method does not lead to stable transformation. However, it may be possible to use the system to transfer a transposable element which might subsequently transfer the gene of interest into chromosomal DNA. The first step in the process, the excision of a transposable element from a T-DNA-introduced viral vector, has recently been described (Shen and Hohn, 1992).

A recent positive development in the *Agrobacterium*-mediated transformation of cereals is the finding that bacterial strains containing multiple copies of *virG* genes transiently transform rice seedling tissues at enhanced frequencies (Li *et al.*, 1992). Although the authors do not report stable transformation, the results suggest that it may be possible to engineer *Agrobacterium* to improve its compatibility with cereal cells.

Microinjection

Microinjection of DNA into embryos or germline cells is an efficient transformation method in a number of animal systems and there have been repeated attempts to apply this approach to plants, in particular to cereals. To transform plant cells via microinjection, it must be possible to manipulate and culture single cells at high efficiency, the cell wall must be penetrated by the microcapillary needle, DNA delivery must avoid the vacuole, there must be a high frequency of DNA integration, and the injected cells must survive and subsequently proliferate and regenerate. Such conditions are satisfied by few plant cell culture systems, and in cereals it is particularly dificult to culture isolated cells at low density. These technical difficulties have meant that there has been little success in stably transforming cereals by microinjection despite major efforts and early optimism (Potrykus, 1989, 1990). The technique may, however, have applications in studying the developmental biology of cereals, as it is relatively easy to achieve transient gene expression after microinjection (e.g. Toyoda *et al.*, 1990). For example, it is possible to study cell lineages from maize apical meristems by following the transient expression of marker genes which can be retained over several cell divisions (G. Neuhaus, pers. comm.). Similarly, the newly developed *in vitro* fertilization technique for maize (Kranz and Loerz, 1993), the developing zygote is amenable to microinjection, which might allow the 'marking' of cells to study early events in embryo development.

DNA delivery via silicon carbide fibres

A variation on injection techniques for cell transformation is the use of microscopic fibres which can be coated with DNA and then penetrate a cell, thus effecting transformation. This method has been demonstrated in various animal systems and in yeast cells using asbestos or silicon carbide fibres and has recently been shown to function in plant cells (Kaeppler *et al.*, 1990). Fibres of diameter ~0.6 μm and length 10–80 μm are used, and transformation is dependent on vortexing the fibre/DNA/cell mixture to achieve cell penetration. In the grasses, the method has resulted in transient gene expression in maize and *Agrostis* suspension cells (Kaeppler *et al.*, 1990; Asano, Otsuki and Ugaki, 1991) and in stable transformation of maize suspension cells (Kaeppler *et al.*, 1990). The attraction of the technique is its simplicity, and it could become an important tool in cereal biotechnology if it proves applicable to regenerable tissues as suggested by Kaeppler *et al.* (1992).

Cell electroporation

Transformation via electroporation of intact cells (electroinjection) was initially attempted by many workers, and was generally concluded not to work. However, there is now clear evidence for cereal transformation by this method. Dekeyser *et al.* (1990) reported transient gene expression in leaf base tissues of rice, wheat and barley using a procedure involving prolonged incubation of the explants in buffer containing plasmid DNA before electroporation. Subsequently, transient expression has been reported from electroporated microspores, suspension cells (Fennel and Hauptmann, 1992) and immature embryos (Songstad *et al.*, 1993) of maize. The utility of the technique for stable transformation has been demonstrated by D'Halluin *et al.* (1992), who electroporated immature maize embryos or embryogenic callus in the presence of a plasmid containing the *neo* gene and obtained fertile kanamycin-resistant plants. Transmission and Mendelian segregation of the transgene in progeny was confirmed. Electroinjection has also been used to obtain stably transformed rice plants by the treatment of germinating seeds (Li *et al.*, 1991).

The mechanism of transformation by electroporation is not yet understood. It is possible that the electric pulse makes both the cell wall and plasma membrane porous to DNA, or that DNA uptake through the wall (via pre-existing pores or areas of damage) is passive and that the electric pulse simply perforates the membrane. The latter mechanism is supported by the observation that pre-incubation of tissues with the plasmid DNA is important, and by the report that in rice small cell groups may also be transformed by PEG treatment (Lee *et al.*, 1991).

The advantages of cell electroporation are that regenerable tissues such as immature scutella and embryogenic callus are accessible to transformation and that the procedure does relatively little damage to treated cultures. It is now of great interest to determine whether the technique can be applied to other cereals such as wheat and barley.

Protoplast transformation

Protoplast transformation was the first direct gene transfer method to be used in plants (Paszkowski *et al.*, 1984), and the first method used to transform a member of the grass family (Potrykus *et al.*, 1985). The principle is simple: isolated protoplasts are suspended in a buffer solution containing plasmid DNA and the protoplast plasma membranes transiently disrupted by a chemical (PEG) or electrical (electroporation) stimulus, allowing DNA uptake. The protoplasts are then either cultured for a few days before harvesting to assess transient gene expression, or are allowed to proliferate before selection with a toxin to identify transformed lines expressing a marker gene conferring resistance.

The advantages of protoplast transformation are that very large numbers ($>10^7$) of cells can be handled with ease, the transformation process (whether mediated by PEG or electroporation) is simple, with similar protocols being suitable for a wide range of species, and the selection of small protoplast-derived microcalli (typically 10–100 cells) is efficient.

The main disadvantage of protoplast transformation is that regeneration from protoplasts in many species, and particularly cereals, is difficult (see above). A second problem for use of the technique in cereals is that cell suspensions must be used as source material to obtain protoplasts with plating efficiencies high enough for transformation to be feasible. These cultures (and thus their protoplasts) lose regeneration potential relatively quickly and are also prone to cytological aberrations (i.e. somaclonal variation), which may result in variant plants being regenerated (see below). However, despite the difficulties associated with cereal protoplast culture, stably transformed cell lines have been produced via protoplast transformation in all the major

Table 4. Protoplast-mediated transformation in cereals

Species	Common name	References[a]	Level of transformation[b]
Hordeum vulgare	Barley	Lazzeri *et al.* (1991)	sc
Hordeum chilense × *Triticum durum*	Tritordeum	Barcelo *et al.* (in press)	te
Oryza sativa ssp. *indica*	Indica rice	Datta *et al.* (1989)	ftp
		Rathore, Chowdhury and Hodges (1993)	ftp
Oryza sativa ssp. *japonica*	Japonica rice	Shimamoto *et al.* (1989)	ftp
		Battraw and Hall (1992)	ftp
Pennisetum americanum	Pearl millet	Hauptmann *et al.* (1987)	te
Sorghum vulgare	Sorghum	Battraw and Hall (1991	sc
Triticum aestivum	Bread wheat	Mueller *et al.* (1992)	sc
Zea mays	Maize	Rhodes *et al.* (1988)	tp
		Omirulleh *et al.* (1993)	ftp

[a] Selected references, not necessarily first reports.
[b] te, transient expression; sc, stably transformed callus; tp, transgenic plants; ftp, fertile transgenic plants.

cereals, and the method has yielded fertile transgenic plants of maize and of japonica and indica types of rice (*see Table 4*).

Both of the techniques for inducing DNA uptake by protoplasts, PEG treatment and electroporation, have been used successfully in cereals and neither is consistently superior. PEG transformation is cheap, as it requires no specialized equipment, but rigorous washing of sensitive protoplasts may be necessary to achieve good division frequencies. Electroporation usually employs buffer solutions having lower toxicity to protoplasts (although toxic ions may be liberated during the pulse discharge: Tada, Sakamoto and Fujimura, 1990), but some protoplasts are damaged by the voltages needed to achieve membrane poration. In addition, the technique requires special equipment for pulse generation.

The efficiency of stable transformation of cereal protoplasts may be high (e.g. $1.0–4.5 \times 10^{-4}$: Hayashimoto, Li and Murai, 1990), but the production of transgenic plants is invariably limited by the frequency of regeneration from transgenic calli. At present, protoplast transformation can be considered a usable and routine procedure only in rice (japonica varieties, and in fewer indica varieties) and in a single highly-selected maize line (Omirulleh *et al.*, 1993). Significant improvements must be made in the culture of protoplasts of other cereals for the method to have wider applicability.

Particle bombardment

The development of particle bombardment (biolistic) transformation technology (Klein *et al.*, 1987) has been of major significance for plant genetic engineering, as it has enabled the transformation of species which are not infected by *Agrobacterium* and for which protoplast regeneration procedures are not efficient.

The principle of the method is that dense microscopic particles, usually tungsten or gold, are coated with DNA and then accelerated into plant cells, through cell walls, allowing transient gene expression, or possibly stable transformation, to take place. The first generation of 'particle guns' used gunpowder charges as the propulsive force (Klein *et al.*, 1987), but although these devices could be effective there were problems with reproducibility due to variation in the charges used. The second generation of particle guns includes electrostatic, pneumatic and compressed gas propulsion systems. These newer guns give more controlled and reproducible particle delivery and have improved transformation efficiency (for reviews on transformation by particle bombardment, see Klein *et al.*, 1992; Christou, 1992).

In particle bombardment, the DNA transfer process is essentially genotype- and tissue-independent, giving the method great flexibility. It can be used to obtain transient gene expression in any plant tissue which can be introduced into the bombardment chamber. To obtain transgenic plants, however, particles must be targeted to regenerable cells, which must survive the delivery and proliferate allowing the transformed cells to be selected from non-transformed tissues. To achieve this at acceptable efficiency requires co-optimization of the physical parameters of particle delivery and the regeneration of the target tissue. Several physical and chemical factors, such as particle type, size and density, DNA precipitation method, particle acceleration speed, etc, influence transformation efficiency (see Morrish *et al.*, 1992), but the critical factor is that the DNA is delivered to cells which are competent for regeneration. In cereals, such cells are found principally in embryogenic cell cultures. In the first reports of cereal transformation via bombardment, cell suspensions were targeted (Klein *et al.*, 1988). Subsequently, several groups produced transgenic maize plants by bombarding embryogenic suspensions (Fromm *et al.*, 1990; Gordon-Kamm *et al.*, 1990). This approach is, however, limited by the need to establish embryogenic suspensions and the fact that suspension-derived regenerants are prone to be variant. More recently, attention has focused on the bombardment of callus cultures (wheat: Vasil *et al.*, 1992), or of primary explants such as embryos (rice: Christou, Ford and Kofron, 1992; wheat: Weeks, Anderson and Blechl, 1993) or inflorescences (*tritordeum*: Barcelo *et al.*, in press) (*Figure 2*). Primary cultures have the advantages that they are highly embryogenic and that the short culture periods involved should limit the incidence of somaclonal variation. There are also other target tissues for bombardment in cereals, such as exposed meristems or microspore cultures, but these are technically more difficult to prepare than primary explants such as embryos and there is as yet no report of transgenic plants from these sources.

Today, particle bombardment is the most efficient of the various methods

Table 5. Particle gun-mediated transformation in cereals

Species	Common name	References[a]	Level of transformation[b]
Avena sativa	Oats	Somers *et al.* (1992)	ftp
Hordeum vulgare	Barley	Mendel *et al.* (1989)	te
		Kartha *et al.* (1989)	te
Hordeum chilense × *Triticum durum*	Tritordeum	Barcelo *et al.* (in press)	ftp
Oryza sativa ssp. *indica*	Indica rice	Christou, Ford and Kofron (1991)	ftp
		Christou, Ford and Kofron (1992)	ftp
Oryza sativa ssp. *japonica*	Japonica rice	Cao *et al.* (1992)	ftp
		Christou, Ford and Kofron (1992)	ftp
Pennisetum glaucum	Pearl millet	Taylor and Vasil (1991)	te
Sorghum vulgare	Sorghum	Hagio, Blowers and Earle (1991)	sc
Triticum aestivum	Bread wheat	Vasil *et al.* (1992)	sc
		Vasil *et al.* (1992)	ftp
		Weeks, Anderson and Blechl (1993)	ftp
Zea mays	Maize	Klein *et al.* (1988)	sc
		Gordon-Kamm *et al.* (1990)	ftp

[a] Selected references, not necessarily first reports.
[b] te, transient expression; sc, stably transformed callus; ftp, fertile transgenic plants.

(a)

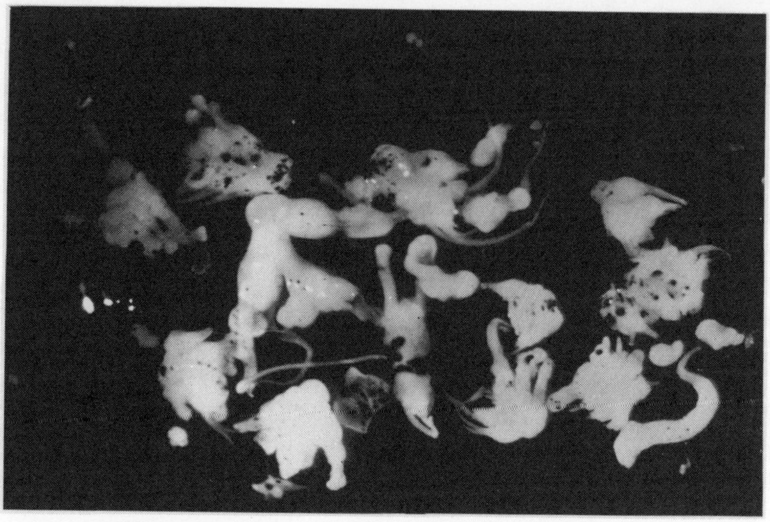

(b)

Figure 2. (a) Transient expression of the *gus* reporter gene (seen as dark spots) in bombarded immature inflorescences of *tritordeum*. (b) Mature and fully fertile transformed plants of *tritordeum*.

for cereal transformation. It has already been used to produce fertile transgenic plants in several of the major cereals (*Table 5*), and it is certain that more species will shortly be added to this list.

DNA constructs

The plasmid vectors used for direct gene transfer to plants typically contain an expression cassette comprising a promoter region, a transcription iniation site and a multilinker sequence with unique restriction sites, allowing the insertion of the desired coding sequences (selectable and/or scorable markers, genes of interest), followed by a polyadenylation signal. The components of these vectors which have received most attention in attempts to improve transformation efficiency are promoter sequences, and the selectable and scorable markers.

To achieve high-level expression of inserted transgenes, it is necessary to use promoters which are active in the recipient cells. A number of bacterial and viral promoters have been used for transformation of dicot species, for example, the *Agrobacterium* opine and T_R-DNA gene 5′ sequences or the cauliflower mosaic virus (CaMV) 35S sequence. Of the numerous promoters tested, the CaMV 35S promoter has given high-level constitutive expression in many species and this promoter is also the most commonly used in cereals: maize (Rhodes *et al.*, 1988), rice (Shimamoto *et al.*, 1989), barley (Lazzeri *et al.*, 1991) and wheat (Vasil *et al.*, 1991). It is observed, however, that the levels of expression from the CaMV 35S promoter are generally lower in cereals than in dicot species, especially in transient expression assays (e.g. Hauptmann *et al.*, 1987). Consequently, efforts have been made to identify or construct more active promoters for cereals.

In transient expression assays, the introduction of monocot intron sequences into the transcriptional units of constructs has been shown to increase activity. Several different introns incuding those from the maize *adh1* (alcohol dehydrogenase) gene (Callis, Fromm and Walbot, 1987), the maize *shrunken1* gene (Vasil *et al.*, 1989) and the rice *act1* (actin) gene (McElroy *et al.*, 1990) have been shown to be effective.

The choice of a 'strong' promoter for stable transformation is less clear cut and the effect of the insertion site within the recipient genome (position effects) becomes a major factor deciding transgene expression level. Furthermore, there are clearly differences in the levels of expression of the promoters currently available between different cereal species. For example, in rice the maize *adh1* promoter gives good levels of expression (Zhang and Wu, 1988), whereas it is only weakly active in wheat (Vasil *et al.*, 1992). Typically, expression levels are lowest in barley and wheat, are higher in maize and are highest in rice (authors' personal observations). One promoter which has given high levels of expression in a number of different cereals is the rice actin 1 promoter/actin 1 intron sequence (McElroy *et al.*, 1990).

There has been some success in constructing composite transcriptional units to give improved expression in cereals. Last *et al.* (1991) combined an anaerobic response element from the maize ADH gene with multiple enhancer elements from the *Agrobacterium* octopine synthase gene, the *adh1* intron and a truncated *adh1* promoter, and obtained levels of expression in wheat up to 50 times those obtained from the CaMV 35S promoter. More recently, Omirulleh *et al.* (1993) have combined two CaMV 35S enhancer elements with a truncated wheat α-amylase promoter to achieve high levels of expression in transgenic maize plants.

Various different scorable marker genes have been used in plant systems, including chloramphenicol acetyltransferase (CAT) (Fromm, Taylor and Walbot, 1985, radiochemical assay), luciferase (LUX) (Ow, Jacobs and Howell, 1987, luminescence assay) and β-glucuronidase (*uidA*, GUS) (Jefferson, 1987, histochemical or fluorometric assays). The latter marker is the most popular because it is easy to assay and allows the visualization of transgene expression *in situ*. Selectable marker genes are considered below.

Selection of transformed tissues

An ideal selection system should give good discrimination between transformed and non-transformed cells, allowing few escapes, and should have no influence on regeneration from transformed tissues. A number of compounds, and their associated resistance genes, have been tested in plants, but only a few, generally antibiotic or herbicide markers, are in common use (*see Table 6*). Cereals, particularly in culture, show higher levels of resistance to selection agents than most dicot species, meaning that high concentrations of the compounds must be used and that the timing and maintenance of selection pressure are critical. In recent experiments, the most popular selectable markers for cereals have been the *nptII* (neomycin phosphotransferase) gene (kanamycin or G418 selection) or the *bar* (phosphinothricin phosphotransferase, PAT) gene (phosphinothricin, Basta or Bialaphos selection). The latter system has the advantage that the herbicides Basta or Bialaphos may be used for selection at the whole plant level, facilitating the selection of plants containing the transgene in segregating progeny generations.

Transgene expression and stability

As the production of transgenic cereal plants is a relatively new development, there are limited published data on the stability and heritability of transgenes in plants. Several studies made at the callus level and on primary (T_0) plants have, however, given information on integration and expression patterns. As observed in other species, direct gene transfer to cereals appears to involve random integration into the recipient genome. It is common for multiple copies of the transforming plasmid to be integrated, which may vary from low numbers (e.g. 1–2: maize, D'Halluin *et al.*, 1992; 1–6: rice, Battraw and Hall, 1992; 1–10: rice, Hayashimoto, Li and Murai, 1990) to high numbers (e.g. 1–20+: oats, Somers *et al.*, 1992; 50–100: rice, Datta *et al.*, 1990). Plasmids are commonly integrated as head-to-tail concatamers of two or more molecules and there are frequently sequence rearrangements which give restric-

Table 6. Selection systems used in cereal transformation

Selection agent	Resistance gene	Reference[a]	Species
Kanamycin	*nptII* Neomycin phosphotransferase	D'Halluin *et al.* (1992)	Maize
		Battraw and Hall (1992)	Rice
G418	*nptII*	Battraw and Hall (1992)	Rice
		Barcelo *et al.* (in press)	Tritordeum
Hygromycin B	*hpt.aphIV* Hygromycin phosphotransferase	Hayashimoto, Li and Murai (1990)	Rice
		Walters *et al.* (1992)	Maize
Phosphinothricin (Basta, Bialaphos)	*bar* Phosphinothricin Acetyltransferase	Gordon-Kamm *et al.* (1990)	Maize
		Vasil *et al* (1992)	Wheat
		Rathore, Chowdhury and Hodges (1993)	Rice
Chlorsufuron	*ILV2, ilvG*	Fromm *et al.* (1990)	Maize
Glyphosate	*EPSP synthase* 5-enolpyruvylshikimate phosphate synthase	Vasil *et al.* (1991)	Wheat

[a] Selected references.

tion fragments larger or smaller than expected (Gordon-Kamm *et al.*, 1990; Battraw and Hall, 1992; D'Halluin *et al.*, 1992). Integration sites are generally at single or closely linked loci (e.g. Spencer *et al.*, 1992; D'Halluin *et al.*, 1992), but integrations at multiple and unlinked loci are also found (Hayashimoto *et al.*, 1990: Rathore, Chowdhury and Hodges, 1993). The factors determining the complexity of integration pattern are not understood. Variables such as culture system, DNA conformation and delivery method may have an influence, but the observation that a range of integration patterns (varying from single copies to multiple insertions and multiple copies with rearrangements) may be found within a single experiment (Rathore, Chowdhury and Hodges, 1993), suggests that cell-to-cell variation in factors such as cell cycle stage, activity of DNA repair mechanisms or endogenous nuclease activity may also be involved.

When levels of transgene activity are assessed in independent transformants, there is typically wide variation (e.g. Omirulleh *et al.*, 1993), but this variation is rarely correlated with the number of copies integrated (e.g. Gordon-Kamm *et al.*, 1990). Variation in expression levels may be explained by the influence of the surrounding genomic environment, and the lack of correlation between copy number and expression levels suggests a significant frequency of silent transgenes. The occurrence of inactive insertions becomes apparent in co-transformation experiments where two independent transgenes are supplied on different plasmids. It is common for co-integration frequencies to be high (e.g. 77%, Gordon-Kamm *et al.*, 1990), but co-expression frequencies are variable (e.g. 18%, Gordon-Kamm *et al.*, 1990; 30%, Rathore, Chowdhury and Hodges, 1993; 70%, Barcelo *et al.*, in press; 75%, Somers *et al.*, 1992). In some cases, transgene inactivation may result from rearrangements occurring during transformation or from mechanisms such as promoter methylation (Matzke *et al.*, 1989) and co-suppression by multiple transgene copies (DeCarvalho *et al.*, 1992).

Several studies have now examined the heritability of transgenes in maize and rice, the cereals in which transformation procedures are most developed. In maize, transgenes usually segregate as single dominant alleles, even when multiple copies are present, indicating either linkage of the copies or that only single copies are active (Spencer *et al.*, 1992; Walters *et al.*, 1992; D'Halluin *et al.*, 1992). However, instances of non-Mendelian inheritance were reported in all of these studies, with some integrated copies of the transgenes appearing to be unstable. Similarly, in rice, Rathore, Chowdhury and Hodges (1993) found both Mendelian and non-Mendelian segregation ratios in T_1 and T_2 populations.

Transgene integration via direct gene transfer is at present an uncontrolled process and this has implications for transgene expression and stability. Further work is needed to understand its mechanisms in order to be able to manipulate the process to produe plants with predictable and stable transgene expression.

An observation common to virtually all the studies on transgenic cereals cited is the occurrence of phenotypically abnormal and infertile plants. These aberrations are not associated with the transformation process, but result from culture-induced variation. This phenomenon is discussed below.

Applications

Cereal transformation is too recent a procedure for there to be many examples of practical application. The majority of experiments performed to date have been directed at establishing gene transfer techniques and at identifying suitable selectable and scorable markers and promoter sequences, but there is increasing use of cereal transformation as a research tool. Transient expression systems, either via protoplast transformation or particle bombardment, are particularly suitable for gene expression analysis, with bombardment giving the possibility of examining tissue-specific expression in intact organs. Examples of these applications include analyses of intron sequence function (Callis, Fromm and Walbot, 1987), organelle targeting with heterologous sequences (Teeri *et al.*, 1989), anthocyanin biosynthesis genes (Klein, Roth and Fromm, 1989), α-amylase promoter specificity (Lee *et al.*, 1991) and regulation (Salmenkallio *et al.*, 1990; Jacobsen and Close, 1991), and analysis of phytochrome gene photoregulation (Bruce *et al.*, 1989).

There is limited scope for analyses of gene expression in stably transformed callus, as the undifferentiated callus state is an artificial condition which does not constitute a good model for the organized plant. The ability to produce transgenic plants, however, offers a great opportunity for analysing gene function via the introduction of heterologous genes or the over-expression or suppression of homologous sequences. Several studies have investigated spatial expression patterns of heterologous and homologous promoters in cereals (Zhang and Wu, 1988; Battraw and Hall, 1990; Terada, Nakayama and Iwabuchi, 1993; Omirulleh *et al.*, 1993; Zhang *et al.*, 1991; Tada *et al.*, 1991) using fusions with the GUS gene.

The first examples of applied genetic engineering of cereals are now starting to appear. Much of this work is carried out by the major agrochemical and plant breeding companies, so the actual state of progress is significantly in advance of that which has been published. There are several reports of herbicide-resistant cereals: rice (Datta *et al.*, 1992; Cao *et al.*, 1992; Rathore, Chowdhury and Hodges, 1993), maize (Gordon-Kamm *et al.*, 1990; Fromm *et al.*, 1990), wheat (Vasil *et al.*, 1992) and oats (Somers *et al.*, 1992). In most cases, the herbicide resistance genes have been introduced to act as selection markers, although there are a number of situations where herbicide-resistant cereals may solve agronomic problems, such as where weeds have developed resistance to conventional selective herbicides, or where engineered resistant plants would allow the use of a 'benign' molecule such as Glyphosate instead of toxic molecules such as Atrazine (Gressel, 1992).

A rather less controversial application of transformation technology is the engineering of pest resistance. The field testing of transgenic maize expressing a *Bacillus thuringiensis* crystal protein gene has recently been reported (Koziel *et al.*, 1993; see also Chapter 6, this volume). In this work, an elite maize inbred was transformed, via particle bombardment, with an engineered crystal protein gene under the control of either the CaMV 35S promoter or of two tissue-specific promoters from maize (minimizing transgene expression in seeds). The transgenic plants were protected in the field against very high infestations of the European corn borer, a major pest of maize in North America and Europe.

This experiment demonstrates that cereal genetic engineering is now a reality, and in the next few years many of the targets identified in other sections of this review will be approached using this strategy.

SOMACLONAL VARIATION

A feature of plant cell and tissue culture is that regenerated plants may be variant, that is, they may differ in some characteristics from the parent plant. This variation may be stable and heritable or may be unstable and show non-Mendelian inheritance ('epigenetic' variation). This phenomenon is termed 'somaclonal variation' (Larkin and Scowcroft, 1981) and has been observed in all groups of plants which have been regenerated *in vitro* (for reviews, see Karp, 1991; Peschke and Phillips, 1992). The causes of somaclonal variation are not fully understood, but a number of the factors involved have been identified.

Variation is associated with *de novo* regeneration from dedifferentiated and disorganized tissue; plants regenerated from cultured meristems are rarely variant, indicating that the normal controls which maintain genetic fidelity in the whole plant operate *in vitro* in isolated meristems. Some variation is undoubtedly present in the explants used to establish cultures, as plant somatic tissues commonly contain polyploid or aneuploid cells. However, the frequency of variation increases with culture age. This may result simply from the accumulation of mutations/epigenetic changes over time, or from faster proliferation of variant cells so that they form an increased proportion of the culture. The variation increases as cultures become progressively less organized and grow rapidly, so that established fast-growing cell suspensions typically have high frequencies of aberrant cells.

A number of different mechanisms of somaclonal variation have been identified, including chromosomal aberrations such as polyploidy and aneuploidy, chromosomal breakage, activation of transposable elements, DNA methylation changes, single base pair changes and gene amplifications and deletions (for bibliography, see Karp, 1991; Peschke and Phillips, 1992).

Somaclonal variation has both positive and negative aspects for cereal biotechnology. In transformation, all the currently proven methods involve an *in vitro* culture step (see above) and somaclonal variation is a problem in these systems. The ideal transformation system is one in which elite cultivars can be modified by the introduction of specified genes, but remain otherwise unchanged. The most obvious marker for the occurrence of variation *in vitro* is loss of regeneration competence, which correlates closely with the accumulation of cytological aberration. Such cultures regenerate fewer plants and those that are produced become progressively more variant with lower fertility (see Morrish, Vasil and Vasil, 1987). The impact of this process is seen in several of the first reports of cereal transformation, where regeneration from transgenic cell lines was inefficient and the transgenic plants were frequently infertile (e.g. Fromm *et al.*, 1990; Gordon-Kamm *et al.*, 1990; Vasil *et al.*, 1992). To avoid these problems, primary cultures have been transformed in more recent experiments (Koziel *et al.*, 1993; Barcelo *et al.*, in press).

There has been some interest in the harnessing of somaclonal variation as a source of novel variation for cereal improvement (Larkin and Scowcroft, 1981), but the general view today is more cautious. However, there are several reports of useful variation arising from tissue culture, such as improved protein content (Schaeffer, Sharpe and Cregan, 1984; Ryan and Scowcroft, 1987), dwarf stature (Galiba *et al.*, 1985) and modified flowering time (Sun *et al.*, 1983; Larkin *et al.*, 1984). In addition, there has also been some success with *in vitro* selection experiments which have, for example, yielded plants resistant to pathotoxins (Rines and Luke, 1985; Pauly, Shane and Gengenbach, 1987), resistant to an insect toxin (Zemetra *et al.*, 1993), with increased freezing tolerance (Lazar *et al.*, 1988) and with increased levels of specific amino acids (Miao, Zhuang and Hu, 1988) (see review in Lazzeri *et al.*, 1991). Overall, however, attempts to exploit somaclonal variation in cereals have not been very successful. Most of the variation observed is deleterious and that which is useful is frequently neither novel nor genetically stable, and therefore difficult to use in seed-propagated crops.

Although somaclonal variation is today generally seen in a negative light, it is still a subject worthy of study. With a better understanding of its underlying mechanisms, it should be possible to reduce its impact where it is unwanted and even to use the process to direct genetic change.

Application of biotechnology to grain composition and quality

The quality of a cereal grain is, of course, related to the end use. The major traditional end uses are as food or feed, either consumed directly or after processing. The quality for these uses is largely determined by the grain proteins, so we will initially discuss the properties of the individual proteins and strategies for their manipulation. In this we will focus on the use of barley for livestock feed and of wheat for bread-making, although we will also discuss other uses of these two cereals. We will then discuss wider opportunities for manipulating the proportions and compositions of the starch and lipid fractions of the grain, and for producing high-value chemicals for pharmaceutical and other purposes.

MANIPULATING THE PROTEIN QUALITY OF BARLEY AND WHEAT

Cereal grain proteins

Cereal grain proteins can be broadly divided into two types: storage and non-storage. The latter include structural and metabolic proteins, such as enzymes and components of cell walls and membranes. Storage proteins can be defined as those that are synthesized specifically to provide a store of nitrogen, sulphur and carbon for seed germination and seedling growth. They account in wheat and barley for 30–60% of the total grain nitrogen (depending on nutrient availability), and are stored in dense membrane-bound deposits called protein bodies. In most cereals (with the exception of oats and rice), the major storage protein fraction is insoluble in water or dilute salt solutions but soluble in alcohol–water mixtures.

These alcohol-soluble proteins (called prolamins) are characterized by high levels of proline and glutamine, about 20 mol % and 30 mol % respectively in total prolamin fractions from barley and wheat, and about 20 mol % and 10 mol % respectively in maize and sorghum. In all cereals, the prolamins are complex polymorphic mixtures of proteins, which range widely in their M_rs (from about 30 000 to 90 000 in barley and wheat and 14 000 to 27 000 in maize) and amino acid compositions (the combined proportions of proline and glutamine ranging from about 25 to 70 mol %). However, in all cases, they are low in one or more nutritionally essential amino acids. This results in low nutritional quality of the whole grain when used as food for humans or as feed for monogastric livestock (such as pigs and poultry). The first limiting amino acid in barley, wheat and maize is lysine, followed by threonine in wheat and barley and tryptophan in maize.

The prolamins are synthesized only in the starchy endosperm of the developing grain. Although they form the major protein fractions in most species, other storage proteins may also be present, either in the starchy endosperm or the aleurone and embryo, or both. These include albumins and globulins related to those stored in seeds of many dicotyledonous species.

The nutritional quality of barley

The nutritional quality of barley is limited by the levels of lysine and threonine: about 3.1 and 3.3 g per 100 g of protein, respectively, compared with WHO recommended levels of 5.5 and 5.0 g per 100 g, respectively (Ewart, 1967; FAO, 1973). The significance of this deficiency for human nutrition is uncertain, but it is probably unlikely to pose problems except perhaps for rapidly growing children living solely on cereals. The situation with livestock is more clear-cut, and it is necessary to supplement cereals with sources of lysine and threonine in order to obtain maximum growth of rapidly growing pigs and poultry. These sources include fish meal, pure amino acids or soybean meal. There is, of course, no problem with ruminants, as they can digest the bacterial flora of the rumen to provide essential amino acids.

The prolamin storage protein fraction of barley is called hordein. It is a complex mixture of over 20 proteins, which vary in precise number, proportions and properties (i.e. M_r and pI) between different genotypes (*Figures 3, 4*). These components are classified into two major groups (B and C hordeins) and two minor groups (D and γ-hordeins) (*Figure 3*). All of these groups are severely deficient in lysine, although D hordein is unusually rich in threonine (~8 mol %) (*Figure 3*). B and C hordeins account for about 80% and 10–20% of the total hordein fraction, respectively (Kirkman, Shewry and Miflin, 1982). Both consist of a number of proteins, which are encoded by families of about 20–30 closely linked genes at the *Hor* 1 (C hordein) and *Hor* 2 (B hordein) loci located on chromosome 5 (1H) (see Shewry *et al.*, 1990). Extensive restriction fragment polymorphism is observed, reflecting the protein polymorphism (*Figure 4*).

D and γ-hordeins each account for a small proportion of the total hordein fraction. D hordein is a single band on SDS-PAGE separations, and appears

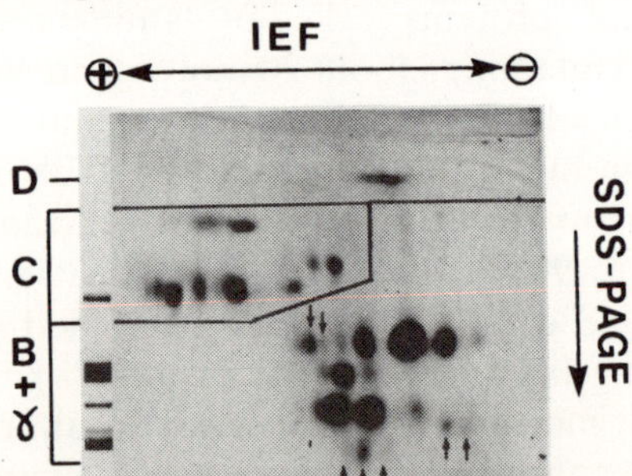

Group	Composition (mol%)				
	Glx	Pro	Cys	Lys	Thr
B	35	21	2·5	0·5	2·1
C	41	31	0	0·2	1·0
D	30	12	1·5	1·1	8·1
γ	32	16	2·7	1·6	3·1

Figure 3. Two-dimensional IEF/SDS-PAGE of barley hordeins, and the proportions of glutamate + glutamine (Glx), proline (Pro), cysteine (Cys), lysine (Lys) and threonine (Thr) in the individual groups. The electrophoretic separation of hordeins from cv Carlsberg II is taken from Kreis *et al.* (1983). The arrows indicate the γ-type hordeins which have similar M_rs to the B hordeins.

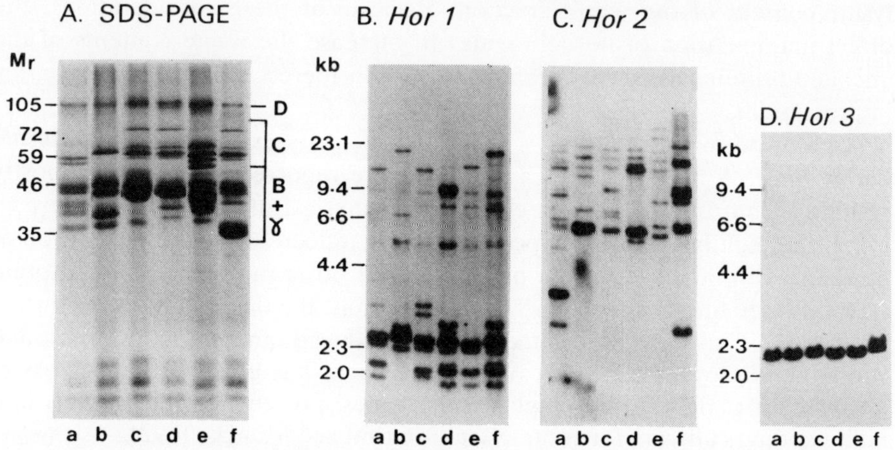

Figure 4. Polymorphism in hordein polypeptides and genes. (A) SDS-PAGE analyses of total reduced and pyridylethylated hordein fractions from the cultivars Athos (a), Keg (b), Jupiter (c), Hoppel (d), Igri (e) and Sundance (f). (B, C and D) Total genomic DNAs from the same six cultivars, digested with *Hin*dIII and probed with cDNA clones related to C hordein (B), B hordein (C) and D hordein (D). Rearranged from Bunce *et al.* (1986) and taken from Shewry (1992).

to be encoded by a single copy gene (*Figure 4*) at the *Hor* 3 locus on the long arm of chromosome 5 (1H) (Shewry *et al.*, 1983; Bunce *et al.*, 1986). γ-Hordeins consist of several minor bands with similar M_rs to B hordeins (*Figures 3,4*), and are encoded by a small multigene family, probably located at the *Hor* 5 locus which is closely linked to *Hor* 1 (Shewry and Parmar, 1987; Shewry *et al.*, 1990).

A number of partial and complete cDNAs and/or genes for hordeins have been isolated. These demonstrate a high degree of structural homology between members of a group, but greater differences between the groups. However, it is clear from sequence comparisons that all hordeins have some degree of structural relatedness (as discussed by Shewry, 1993), indicating a common evolutionary origin.

We have discussed the complexity of the hordein fraction in some detail, as this has important implications for selecting strategies for improvement. First, the low contents of lysine in the individual hordein polypeptides may be significant in relation to the packaging of the proteins as hydrated solid within the protein bodies. It is not known whether the lysine content could be increased sufficiently to have an impact on the nutritional quality of the grain without affecting this or other important biological properties. However, studies of maize zeins indicate that at least some additional lysine residues may be tolerated (Lending, Wallace and Larkins, 1992). Second, the existence of complex multigene families means that any inserted hordein genes will be expressed in the presence of many other actively expressed hordein genes. Consequently, the inserted genes would need to have exceptionally strong promoters or be expressed in lines in which other hordein genes are inactivated or deleted in order to have a significant impact on the lysine content of the whole fraction. Because of these considerations, the direct manipulation of hordein genes to increase the lysine contents of the encoded proteins has never been seriously considered as a strategy for barley improvement.

A more promising strategy is suggested by analyses of mutant high lysine barley lines. A number of such mutants were reported in the 1970s, mostly resulting from treatment with chemical or physical mutagens. In almost all cases, the high lysine phenotype results from a decrease in the proportion of hordein, with increases in the proportions of other more lysine-rich protein fractions (see Shewry *et al.*, 1987b). In addition, the decreased proportion of prolamins is usually associated with a reduced amount of starch, and consequently lower yield. Because of this there has been limited success in incorporating the mutant high lysine genes present in these lines into high-yielding cultivars for commercial use (but see Munck, 1992).

Although most high lysine mutants were produced by mutagenesis, the first to be described was a spontaneous mutant identified by screening the world barley collection held by the USDA (Munck *et al.*, 1970). This is a primitive line (land race) from Ethiopia, and was named Hiproly. Although it has not been possible to incorporate the high lysine gene from Hiproly (designated *lys*) into high-yielding cultivars, molecular and biochemical analyses of the mutant phenotype have proved to be very valuable.

Hiproly has about 30% more lysine than normal cultivars, accompanied by a modest decrease in the proportion of hordein (~10–20%: see Shewry *et al.*, 1987b). The latter is clearly not sufficient to account for the increased lysine content, and the answer lies in the presence of increased amounts of specific salt-soluble proteins. Four such proteins account for about half of the increased lysine content, and for 17% of the total grain lysine in Hiproly compared with about 7% in normal cultivars (Hejgaard and Boisen, 1980). These are β-amylase (5.0 g % lys), protein Z (7.1 g % lys) and chymotrypsin inhibitors CI-1 (9.5 g % lys) and CI-2 (11.5 g % lys). CI-1 and CI-2 are clearly of most interest because of their very high contents of lysine, and Hejgaard and Boisen (1980) showed that the amounts of these two proteins were increased about 6- to 8-fold in high lysine lines derived from Hiproly, as compared with the normal parents.

CI-1 and CI-2 were initially purified at the Carlsberg Laboratory and complete or partial amino acid sequences determined (Svendsen, Boisen and Hejgaard, 1982; Svendsen, Martin and Jonassen, 1980). They showed that both proteins exist in isoforms, and this was confirmed by cDNA cloning which demonstrated that both were encoded by two sub-families of mRNAs and genes (Williamson *et al.*, 1987; Williamson, Forde and Kreis, 1988). The major isoform of CI-2 consists of 84 residues with an M_r of about 9400, while two equally abundant isoforms of CI-1 consist of 83 residues with M_rs of about 8800 and 9000 (*Figure 5*). Neither protein contains cysteine residues, or appears to be synthesized with a signal peptide. Nevertheless, there is clear evidence from immunogold labelling that CI-2 is located in protein bodies and in vesicles derived from the Golgi apparatus (Rasmussen, Munck and Ullrich, 1990). Both proteins are encoded by small multigene families (3–4 copies)

Amino Acid Sequences of
CI-1 and CI-2 of Barley

```
CI - 1A   M S S M E G S V L K Y P E P T E G S I G
     1B   M R S M E G S V P K Y P E P T E G S I G
CI - 2A   M S S V E - - - - K K P E G V N T G A G
     2B                                 * D C L C

CI - 1A   A S S A - K T S W P E V V G M S A E K A
     1B   A S G A - K R S W P E V V G M S A E K A
CI - 2A   D R H N L K T E W P E L V G K S V E E A
     2B   D C Q N Q K T E W P E L V E K S V E E A

CI - 1A   K E I I L R D K P N A Q V E V I P V D A
     1B   K E I I L R D K P D A Q I E V I P V D A
CI - 2A   K K V I L Q D K P E A Q I I V L P V G T
     2B   K K V I L Q D K P E A Q I I V L P V G T

CI - 1A   M V H L N F D P N R V F V L V - - - - A
     1B   M V P L D F N P N R I F I L V - - - - A
CI - 2A   I V T M E Y R I D R V R L F V D K L D N
     2B   I V T M E Y R I D R V R L F V D R L D N

CI - 1A   V A R T P T V G
     1B   V A R T P T V G
CI - 2A   I A Q V P R V G
     2B   I A Q V P R V G

*    sequence incomplete
```

Figure 5. The amino acid sequences deduced from cDNA clones for barley chymotrypsin inhibitors CI-1 and CI-2. In each case, two sub-families of clones were identified, and were designated A and B. Note the high contents of lysine residues (K) and the absence of cysteine. Based on data in Williamson *et al.* (1987) and Williamson, Forde and Kreis (1988). Standard single letter abbreviations for amino acids are used.

located on the short arm of chromosome 5 (1H) (Cannell *et al.*, 1992).

Because of its small size and lack of disulphide bonds, CI-2 has been an attractive target for structural and protein engineering studies. Its three-dimensional structure has been determined (alone and in complex with subtilisin) by X-ray crystallography and nuclear magnetic resonance (NMR) spectroscopy, and its inhibitory constant and specificity have been altered by engineering of the reactive loop region that interacts with serine proteases (see Campbell, 1992).

The biological roles of CI-1 and CI-2 are not known, but they may play a role in natural defence against pests and pathogens. This suggestion is supported by the fact that related inhibitors present in tomato are induced by damage or feeding (Graham *et al.*, 1985).

An analysis of Hiproly demonstrates that substantial increases in the amounts of CI-1 and CI-2, which are sufficient to have an impact on the lysine content of the whole grain, can be tolerated without adverse effects on grain development or digestibility. Also, it is possible that such increases may also provide enhanced protection against pests and pathogens. In Hiproly, these increases are achieved by the action of a *trans*-acting 'regulatory' gene, and the incorporation of this gene into cultivars has been limited by pleiotropic effects on grain yield. An alternative strategy would be to increase the amount of CI-2 by inserting additional gene copies, either under control of their own promoter or, preferably, a stronger promoter which also confers seed specificity.

A gene for CI-2 has been isolated, and its control of expression studied using chimaeric genes with the GUS reporter (Peterson *et al.*, 1991; Shewry and Kreis, 1992; Kreis and Shewry, 1992). This has shown that 1100 bp of 5′ upstream sequence is sufficient to confer tissue-specific expression in transgenic tobacco plants, but that 263 bp of 5′ sequence is insufficient. It is unlikely, however, that the CI-2 promoter would be used to drive expression if additional CI-2 genes were inserted into barley. This is because it would be advantageous to use a stronger promoter, for example from the D hordein gene of barley or from one of the homologous HMW subunit genes of wheat. D hordein accounts for about 2–4% of the total hordein fraction, and is encoded by a single gene (see above). Similarly, each HMW subunit is encoded by a single gene and accounts (on average) for about 2% of the total extractable grain protein (Halford *et al.*, 1992b). D hordein gene expression has not been studied in detail, but several studies of HMW subunit gene expression have been made using transgenic tobacco and transient expression in protoplast systems. These have shown that only 280 bp of DNA immediately upstream of the transcription start site is sufficient to confer temporal and endosperm-specific gene expression (Halford *et al.*, 1989). It should, therefore, be possible to fuse the 5′ upstream sequence from an HMW subunit or D hordein gene with the coding sequence of CI-2 in order to obtain high levels of tissue-specific expression in transgenic barley plants. It is difficult to calculate how many such copies of CI-2 genes would need to be inserted to have a significant impact on grain quality, as it is not possible to predict the precise levels of expression that would be achieved.

We consider that increasing the level of expression of lysine-rich proteins such as CI-2 is the most promising approach to increasing the lysine content of cereals such as barley, wheat and maize, and that this will be more readily achieved by the insertion of additional gene copies under the control of strong promoters than by attempting to increase the levels of expression of endogenous genes. The latter will depend on identifying the molecular basis for the control of gene expression, and in particular the role of *trans*-acting factors that are presumably produced by regulatory genes such as the *lys* gene of Hiproly.

The malting quality of barley

The major food use of barley is for malting, and the subsequent use of the malt for brewing or distilling. Malting quality is therefore a prime consideration for plant breeders, but only gradual improvements have been made using classical approaches. This is probably because malting quality is a complex character, depending on the interactions of a number of individual characters expressed both during grain development and during the malting process. The former result in differences in endosperm structure and composition (e.g. amounts of β-glucans, hordein storage proteins and β-amylase), while the latter relate mainly to the production of hydrolytic enzymes (e.g. α-amylases, β-glucanases and proteases) and their effects on the endosperm cells and their storage components. These characters and their role in malting quality are still incompletely understood, and we will not attempt a detailed discussion here. Instead, we will focus on one character whose role has been the subject of considerable discussion and speculation. This is the amount, composition and spatial distribution of the hordein storage proteins.

It has long been known that high levels of grain protein are disadvantageous for malting, due to a negative correlation with hot water extract (Bishop, 1930a). Bishop (1930a) also showed that this was not solely due to mere dilution of starch, but that increases in respiration and rootlet growth also contributed to malting losses. Since the major effect of high grain nitrogen is an increase in the proportion of hordein storage proteins (Bishop, 1930b; Kirkman, Shewry and Miflin, 1982), this implies that a high level of hordein is a disadvantage. However, Bishop (1930a) and subsequent workers also appreciated that the magnitude of this effect varied with the variety. It was not until the late 1970s that improved methodology and increased knowledge of the structure and genetics of hordeins allowed this genetic effect to be examined in more detail.

The first studies of this type were reported by Baxter and Wainwright (1979a), who compared 16 varieties which varied in malting quality. They suggested that varieties with better quality had lower proportions of fast-moving (on electrophoresis at low pH) B hordein polypeptides. However, Shewry *et al.* (1980) were unable to confirm these observations using SDS-PAGE analyses of a wider range of malting and non-malting varieties of barley (28 in total). Similarly, Riggs *et al.* (1983) found no clear correlation between hordein pattern and malting quality in a collection of 84 cultivars, although they used an acid pH electrophoresis system similar to that used by Baxter and Wainwright (1979a).

Baxter and Wainwright (1979b) also drew attention to a further characteristic of hordein that could be relevant to malting quality – the proportion of the proteins present in high M_r polymers stabilized by disulphide bonds. We now know that these polymers consist predominately of D and B hordeins, but that some γ-hordeins may also be present. Baxter and Wainwright (1979b) extracted hordeins in two fractions: with 70% ethanol followed by the same solvent with a reducing agent (2-mercaptoethanol). They showed that the proportions of hordein present in these two fractions varied with genotype and environment, and that the second fraction (which presumably consisted mainly of polymers) was degraded less during malting than the first (mainly monomeric) fraction. Furthermore, the spent grains contained a higher proportion of polymeric hordeins than malt, suggesting that some hordeins became associated by disulphide bonds during mashing. The authors suggested that these polymeric hordeins could entrap starch granules resulting in poor extraction. Shewry *et al.* (1981) also showed that the proportion of aggregated hordeins was slightly greater in cultivars of low malting quality than in good-quality cultivars, while Smith and Lister (1983) showed a clear inverse correlation between malting quality and the amount of gel protein (which consists predominately of polymers of D and B hordeins). They also suggested that the amount of gel protein was determined by the proportion of D hordein, which was synthesized at a disproportionately higher level in poor-quality cultivars as the grain nitrogen increased. This implies that an improvement in quality could be achieved by down-regulating the synthesis of D hordein, which appears a realistic target since only one D hordein gene per haploid genome appears to be present.

A further factor which may influence malting quality is the spatial distribution of hordein within the endosperm. Millet, Montembault and Autran (1991) compared the distribution of monomeric and polymeric hordeins in pearling fractions prepared from one good-quality and one poor-quality cultivar of barley, and showed that a higher proportion of the hordein was present in the peripheral layers of the poor-quality cultivar. In addition, a higher proportion of the hordein was present in polymers stabilized by disulphide bonds. It is not yet known whether similar differences are present between other good- and poor-quality cultivars, but it raises the intriguing possibility that hordeins may limit the rate of endosperm modification during malting of poor-quality cultivars by forming an insoluble gel around the periphery of the endosperm, thus hindering the diffusion of α-amylase and other hydrolytic enzymes.

So far, hordein has been discussed in relation to adverse, or at best neutral, effects on the quality for malting, brewing and distilling. However, there is one respect in which hordein may play a positive role, in conferring foam retention and cling (adhesion to the glass) to beer. These are important quality parameters for the consumer, and appear to be related to the presence of hydrophobic peptides (Bamforth, 1985a,b) that have similar amino acid compositions to hordeins (Dale, Young and Brewer, 1989; Dale and Young, 1989). However, their precise origins and structures remain to be determined.

It is clear from this brief discussion that the relationship between hordein and the quality of barley for malting, brewing and distilling is complex, and may involve the interactions of genotypic and environmental factors operating during grain development and germination with chemical reactions occurring during the subsequent technological processes (e.g. mashing). In terms of hordein composition, these factors may affect the proportions and properties of the individual polypeptides, their assembly into disulphide-stabilized polymers and their spatial distribution within the grain. Furthermore, it is probable that the same proteins have negative effects on some aspects of quality (e.g. modification and hot water extract) and positive effects on others (foam stability and cling). Since hordein is only one of a number of factors which affect quality for malting, brewing and distilling, it is not surprising that only slow progress has been made in effecting improvements by plant breeding. Genetic engineering will undoubtedly provide opportunities to manipulate some of the individual characters that contribute to quality (e.g. the proportion of D hordein), but we still have insufficient understanding of these characters at the biochemical and molecular levels to assess which approaches would be most likely to succeed.

The bread-making quality of wheat

Bread-making is one of humankind's oldest technologies, dating back to at least ancient Egyptian times. In addition, wheat gluten (corresponding to the prolamin fraction) was one of the earliest plant proteins to be studied, being first isolated in 1745 (Beccari, 1745), and has been the subject of a vast volume of research over the past century. Despite this effort, we have only recently started to gain an understandng of the detailed structures of the gluten proteins and their role in grain technology.

The wheat prolamins are deposited in the developing grain in protein bodies, which become disrupted during the later stages of endosperm maturation to form a proteinaceous matrix surrounding the starch granules. When a dough is made, these proteins form a cohesive network which can be isolated in a form that is substantially pure by washing to remove starch granules and water-soluble components. Gluten plays a key role in bread-making, by enabling the entrapment and retention of the carbon dioxide produced by fermentation to give a light porous crumb structure which is fixed by baking. The ability to do this depends on a combination of two physical properties: elasticity and extensibility (viscous flow). A precise balance of these is essential, and the poor quality of many wheats (including some UK wheats) is related to insufficient elasticity.

The gluten proteins of wheat are usually classified into two groups: the gliadins are single monomeric proteins, while the glutenins consist of individual subunits that are assembled into high M_r polymers stabilized by inter-chain disulphide bonds. The individual gliadin monomers and glutenin polymers also interact by strong non-covalent forces, notably hydrogen bonds and hydrophobic interactions. These two fractions also have functional significance, as the glutenins are associated with elasticity and the gliadins with extensibility.

Electrophoretic analyses of total prolamin fractions from bread wheat show a very high degree of polymorphism, with at least 50 individual proteins revealed by two-dimensional IEF/SDS-PAGE under reducing conditions (*Figure 6*). The degree of polymorphism is considerably greater than that observed in barley (cf. *Figures 3 and 6*), which is at least partly due to the allohexaploid nature of bread wheat. Bread wheat has three genomes (called A, B and D), each consisting of seven pairs of chromosomes. Because these genomes are derived from related species of wild grass, there is much repetition of genetic information, with related genes often present at similar positions on the homoeologous (homologous but non-pairing) chromosomes of the three genomes. Thus in bread wheat the gliadins and glutenin subunits are encoded by loci on all three pairs of the group 1 (1A, 1B, 1D) and group 6 (6A, 6B, 6D) chromosomes (see Payne, 1987). It is of interest that the group 1 chromosomes are homeologues of chromosome 5 (1H) of barley, which is the location of all the hordein genes (Shewry *et al.*, 1990).

As in barley, it is possible to classify the individual gluten proteins into several groups. The gliadins are classified into three groups, α-, γ- and ω-, in order of decreasing mobility when separated by electrophoresis at low pH.

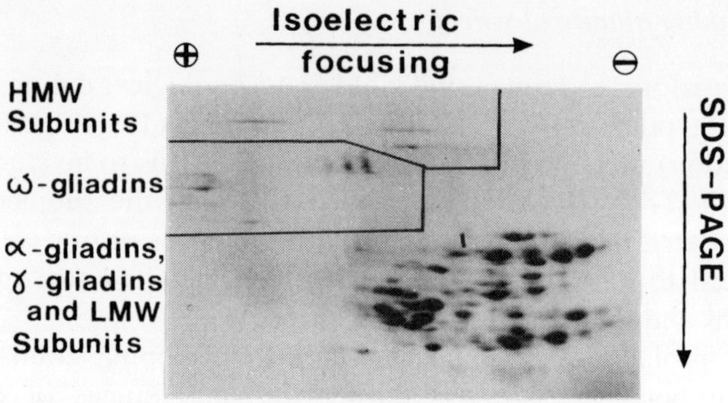

Figure 6. Two-dimensional IEF/SDS-PAGE of total prolamins from wheat cv Chinese Spring. The electrophoretic separation is from Shewry *et al.* (1987a).

The ω- and γ-gliadins are homologues of C and γ-hordeins, respectively, and are encoded by genes on the group 1 chromosomes. In contrast, the α-gliadins have no precise homologues in barley, and are encoded by genes on the group 6 chromosomes. The glutenin subunits are similarly divided into groups, but on the basis of mobility by SDS-PAGE. The HMW subunits are homologues of barley D hordein, while the major group of LMW subunits is most closely related to B hordeins. These two groups of proteins are also encoded by genes on the group 1 chromosomes.

There is no doubt that all these proteins contribute to the structure and functionality of wheat gluten, but two lines of evidence indicate that one group is particularly important in determining dough quality. This is the HMW subunits of glutenin. Cultivars of bread wheat contain three, four or five individual HMW subunits, which are divided into high M_r x-types and low M_r y-types (Payne, Holt and Law, 1981). They are encoded by single loci on the long arms of chromosomes 1A, 1B and 1D, each locus consisting of two genes: one x-type and one y-type (Harberd, Bartels and Thompson, 1986). Specific silencing of one, two or three of these genes results in the observed variation in HMW subunit number (*Figure 7*). Thus cultivars always contain 1Dx, 1Dy and 1Bx subunits, sometimes 1By and 1Ax subunits, but never 1Ay subunits. In addition to variation in subunit number, there is also allelic variation in the structures of the individual subunits, which is observed as differences in their mobilities in SDS-PAGE. A number of correlative studies have shown that this allelic variation in subunit number and mobility is correlated with differences in bread-making quality, providing the first line of evidence that the HMW subunits play a major role in determining gluten functionality (see Payne, 1987). On the basis of detailed comparisons, it has been possible to assign 'bread-making scores' to all the individual HMW subunits (Payne *et al.*, 1987). However, the most striking correlations are with the presence or absence of a 1Ax subunit, and with allelic variation in the subunit pair encoded by chromosome 1D (1Dx + 1Dy).

The second line of evidence for the importance of the HMW subunits comes from biochemical studies. It has long been known that good quality is associated with a high proportion of high M_r (above about 1 million) glutenin

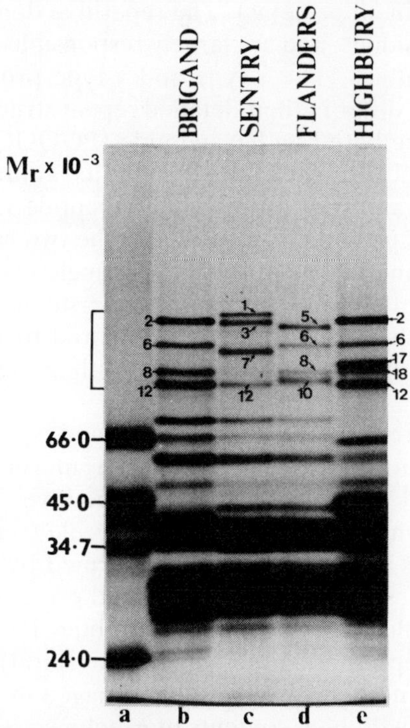

Figure 7. SDS-PAGE of total prolamins of four cultivars of wheat, illustrating variation in the number and mobilities of the HMW subunits (numbered according to Payne and Lawrence, 1983). Taken from Shewry and Tatham (1989).

polymers, and biochemical studies demonstrate that such polymers are enriched in HMW subunits (Field *et al.*, 1983). In contrast, the LMW subunits are present in polymers of high and low M_r.

In order to understand the role of the HMW subunits in gluten structure and functionality, it is necessary to determine the structures of the individual proteins, their interactions with other glutenin subunits to form polymers, and the properties and interactions of the polymers that are formed. Although our knowledge is still incomplete, studies carried out over the last decade are now providing valuable information on these subjects (see Shewry, Halford and Tatham, 1989, 1992).

A total of nine genes encoding different HMW subunits have been isolated and characterized (see Shewry *et al.*, 1989; Halford *et al.*, 1992). These include the complete family of six genes from cv Cheyenne: expressed genes encoding subunits 1Ax2*, 1Bx7, By9, 1Dx5 and 1Dy10, and a silent gene of the 1Ay type (*Figure 8*). In addition, genes for the allelic subunits 1Ax1, 1Dx2 and 1Dy12 have been isolated from other cultivars.

Analysis of the sequences of the encoded proteins show that they have similar structures, consisting of between about 480 and 700 residues, with M_rs of about 67 500–88 000 (see Shewry, Halford and Tatham, 1989). All consist of three structural domains – a long repetitive central domain which is flanked by shorter non-repetitive domains at the *N*-terminus (81–104 residues) and

C-terminus (42 residues in all proteins). The repetitive domains range in size from about 480 to 700 residues, and are largely responsible for the differences in M_rs of the whole proteins. The x-type and y-type proteins have similar structures (*Figure 8*), but differ in their detailed repeat structure. Whereas the y-type subunit repeats are based on hexapeptide (PGQGQQ) and nonapeptide (GYYPTSLQQ) motifs, the x-type subunits contain an additional tripeptide motif (GQQ) and the consensus nonapeptide motif differs in the substitution of P for L at position 7. In addition, the two subunit types differ in their number of cysteine residues. Whereas a single cysteine is present in the *C*-terminal domains of all subunits, only three cysteines are present in the *N*-terminal domains of the x-type subunits compared to five in the y-type. Additional cysteine residues are also present within the repeats of some subunits only (*Figure 9*).

A combination of predictive hydrodynamic and spectroscopic studies indicate that the repetitive sequences form an unusual supersecondary structure, a loose spiral structure based on repetitive β-turns, giving an extended rod-like conformation to the whole molecule (see Shewry, Halford and Tatham, 1989). This has recently been confirmed by the application of scanning tunnelling microscopy to image a purified protein fraction (Miles *et al.*, 1991). This showed aligned rods of diameter about 19.5 Å, with diagonal striations (presumably representing the turns of the spiral) with a periodicity of about 15 Å. The *N*- and *C*-termini were not imaged in this study, but are predicted to be globular with high contents of α-helix.

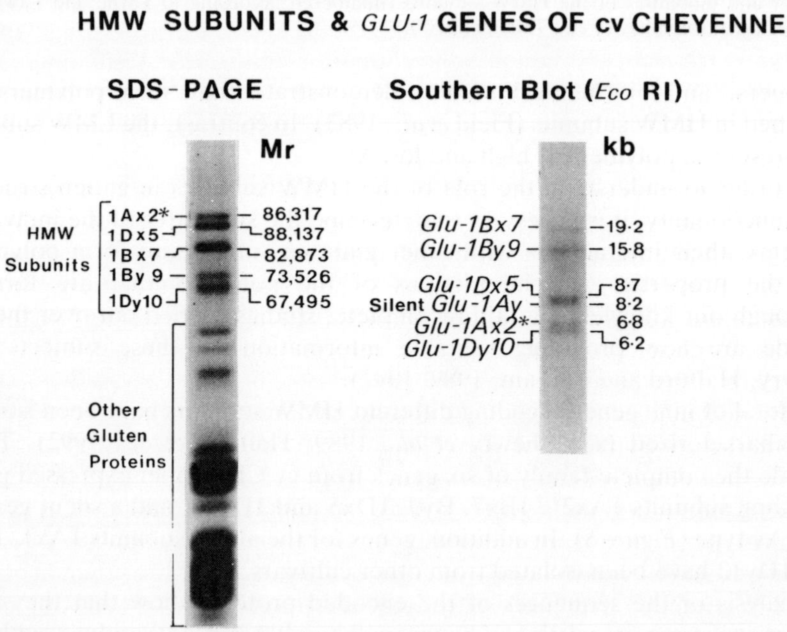

Figure 8. The HMW subunit proteins, separated by SDS-PAGE, and the *Glu-1* genes, determined by Southern blotting, of wheat cv Cheyenne. Taken from Shewry, Halford and Tatham (1989).

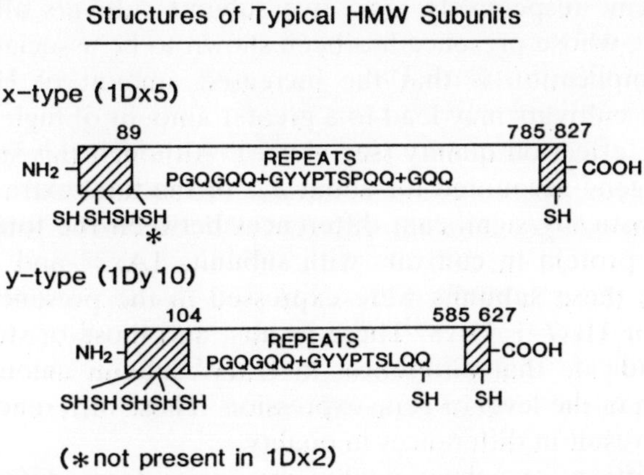

Figure 9. The structures of typical x-type and y-type HMW subunits. Based on sequences reported in Anderson *et al.* (1989).

What are the implications of HMW subunit structure for the formation and properties of glutenin polymers? An important factor is, of course, the number and distribution of cysteine residues. The presence of most of the cysteine residues in the *N*- and *C*-terminal domains would be expected to lead to predominantly head-to-tail polymers, with cross-links and branches arising from the additional cysteines present in the *N*-terminal domains and in the repetitive domains of some subunits. Despite impressive attempts to unravel the disulphide structure of wheat glutenin (Köhler, Belitz and Wieser, 1991, 1993; Tao, Adalsteins and Kasarda, 1992), we still know very little about the precise patterns of disulphide bond formation of the HMW subunits, and in particular about which cysteine residues form intra-chain or inter-chain disulphide bonds. There is, of course, the added complication that it is theoretically possible for the same cysteine residue to form an intra-chain bond (as reported by Köhler, Beltiz and Wieser, 1993), or an inter-chain bond with an identical subunit (as reported by Köhler, Belitz and Wieser, 1991), with a different type of subunit (x or y: as reported by Tao, Adalsteins and Kasarda, 1992), with a different subunit of the same type, or with an LMW subunit. It is probable that the number and distribution of such bonds will affect the elasticity of the glutenin polymers, in the same way that the degree of cross-linking determines the elastic modulus of rubbers (Flory, 1953). Although it has been proposed that the β-spiral conformation also contributes to glutenin elasticity (Tatham, Shewry and Miflin, 1984), recent studies cast doubt on this (Belton *et al.*, in press).

Assuming that glutenin cross-linking is the major determinant of glutenin elasticity (and hence bread-making quality), what can we say about the quality differences associated with allelic variation in HMW subunit composition? We believe that these arise from quantitative and qualitative effects. The former will be related to differences in the number of expressed genes, and in their relative levels of expression. Halford *et al.* (1992b) compared the total amounts of HMW subunit protein in cultivars with four and five expressed genes, showing that this represented about 8 and 10% of the total extracted protein, respectively. The five subunit cultivars all had subunit 1Ax1 or 1Ax2*, whose presence has been shown to be associated with good quality. The implication is that the increased amount of HMW subunit protein in these cultivars may lead to a greater amount of high M_r polymers, with a resulting effect on quality (see above). Although this study indicated that each 1Ax gene accounted for about 2% of the total extracted proteins, there were statistically significant differences between the total amounts of HMW subunit protein in cultivars with subunits 1Ax2* and 1Ax1, and in cultivars where these subunits were expressed in the presence of subunits 1Bx7 + 1By9 or 1Bx7 + 1By8. These results, and those of studies in other laboratories, indicate that differences in HMW subunit amount also occur due to variation in the level of gene expression. These differences would also be expected to result in differences in quality.

Qualitative effects on wheat quality may arise from differences in the structures of allelic subunits. The most notable variation of this type is between the allelic pairs of subunits 1Dx5 + 1Dy10 (good quality) and 1Dx2 + 1Dy12 (poor quality). Because recombination between the genes encoding 1Dx and 1Dy subunits is very rare, it is not possible to conclude whether the quality correlations are due to differences between the allelic 1Dx subunits (1Dx2 and 1Dx5), the allelic 1Dy subunits (1Dy10 and 1Dy12) or the allelic pairs. However, comparisons of the amino acid sequences of these four subunits show only one striking difference between the two pairs of alleles. This is the presence of a cysteine residue towards the *N*-terminal end of the repetitive domain of subunit 1Dx5 (*see Figure 9*), which is not present in subunit 1Dx2. This could lead to a higher degree of cross-linking of the glutenin polymers, and hence greater gluten elasticity.

It is clear from the brief summary presented here that we now know a considerable amount about the structures of the individual HMW subunits, and their potential roles in gluten structure and functionality. It should, therefore, be relatively straightforward to design strategies to manipulate gluten structure to improve quality. The two obvious strategies to increase the elasticity of gluten would be to manipulate the amounts or structures (i.e. cross-linking) of the high M_r glutenin polymers, by increasing the total amount of HMW subunit protein or by introducing additional cross-linking sites (i.e. cysteine residues). However, more subtle alterations to the structure of glutenin could also be made to optimize the quality for other end uses (e.g. noodles, biscuits, wafers and cakes), by manipulating the amounts and/or properties of other gluten proteins such as the gliadins or LMW subunits.

Other uses of wheat and gluten

Wheat is, of course, used for many other food products in addition to bread. These include baked goods (biscuits, cakes and pastries), noodles and pasta, and various batters, breadings and coatings. In addition, isolated gluten is used as a binder and filler, for example in processed meat products and pet foods. All of these uses depend on the unusual physical properties of the gluten proteins, but the precise balance of properties that is required differs. For example, less elastic gluten is required for biscuits, pastries and cakes than for bread-making, and it is usual to make these products with weak varieties.

A considerable volume of research has been carried out on the quality of wheats for pasta-making, including the identification of protein markers which can be followed in plant-breeding programmes (similar to the HMW subunits of glutenin and bread-making quality) (see MacRitchie, DuCros and Wrigley, 1990). The quality requirements for other food applications are less well understood, and vary considerably between different products (for example, different types of biscuit).

Understanding the molecular basis for the physical properties of gluten, and in particular for the balance between elasticity and extensibility, will undoubtedly facilitate the use of wheats for all these products, and will almost certainly also reduce the reliance on chemical improvers to produce the required properties. However, as with cereal starches (see below), the most important impact may well be on the development of new uses, particularly for isolated gluten.

Wheat gluten is readily separated from the other major grain component, starch, in industrial separation plants. However, whereas the starch finds a number of uses (see below), most of the gluten is used in baking, either to fortify poor-quality flours or for speciality products that require a high content of gluten. The latter include wholemeal and high-fibre breads, and hamburger buns. The other major use of isolated gluten is as a binder, most notably in pet foods.

At present, the supply of gluten and the demand from the bread-making and other industries are pretty well in balance, but several factors could result in a surplus of gluten for other uses. For example, further improvement in the intrinsic bread-making quality of European wheats would require less gluten for fortification, while more gluten would also be produced as a by-product of the substitution of wheat starch for imported maize starch. This could lead to the use of gluten to produce textured protein products, or to produce protein emulsifiers or foaming agents. Once again, a detailed understanding of gluten structure and functionality will be critical for the development of such uses, whether based on genetic engineering to produce specific cultivars or on downstream processing of isolated gluten (for example, by partial enzymic digestion or partial deamidation by acid hydrolysis).

Novel protein products

There has been considerable discussion about using plant storage organs to produce high-value chemicals, including biologically active peptides and proteins. The feasibility of this has been demonstrated by the production of functional antibodies in the leaves of transgenic tobacco plants (obtained by crossing plants independently transformed with genes encoding gamma or kappa immunoglobulin chains) (Hiatt, Cafferkey and Bowdish, 1989), and by the elegant studies of Krebbers, Vandekerckhove and co-workers at Ghent (see Krebbers, Bosch and Vandekerckhove, 1992; Krebbers *et al.*, 1993). The Ghent group have used a 2S albumin storage protein from *Arabidopsis* as host for two biologically active peptides: the six residue peptide Leu-enkephalin and magainin, a 28 residue antibacterial peptide from *Xenopus*. This was done by engineering the cDNA to replace a variable loop region with sequence encoding the biological peptide, flanked by cleavage sites to facilitate purification. The cDNAs were then expressed in seeds of transgenic *Arabidopsis* or oilseed rape, the 2S albumins purified and the peptide excised. Yields of up to 200 nmol per g seed were obtained which would correspond to about 1 kg of a 25 residue peptide per hectare of oilseed rape (Krebbers *et al.*, 1993). However, the authors consider that yields up to five times greater could be obtained using stronger promoters.

Although Krebbers and co-workers have used oilseed rape for their studies, it is clear that a similar approach could be used with cereals. In fact, cereals would probably be the preferred hosts because of their high yields and the ease of harvest and processing (compared to oilseed rape, where up to about 30% of the seeds can be lost before or during harvest). However, the extent to which this approach is finally adopted will probably depend on regulatory and economic factors rather than biological constraints.

MANIPULATING STARCH CONTENT AND COMPOSITION

Starch composition and synthesis

Because starch is the major storage compound in cereal grains, it is an obvious target for manipulation, in order to increase its total amount (and hence yield) or its composition and properties.

Starch is not a single compound but a mixture of two complex polymers called amylose and amlyopectin. Amylose is a linear chain of between 1000 and 10 000 D-glucopyranosyl units, with α1-4 linkages. Amylopectin is also composed of D-glucopyranosyl residues, but has a highly branched structure due to the presence of α1-6 linkages as well as α1-4 linkages. The average chain length between branches is about 20–25 and the total degree of polymerization about 1000. The properties of starches are determined to a large extent by the proportions of these polymers (which vary between ~70–85% amylopectin and ~15–30% amylose in cereals), their precise structures (i.e. chain lengths and degrees of branching) and their organization within the granule. These properties include the density, gelatinization temperature and crystallinity index, and affect the utilization in the food and other industries. The basic enzymology of starch synthesis has been reviewed by several authors (Preiss, 1988; Smith and Martin, in press) and is summarized in *Figure 10*.

It appears that ADP glucose is the major building block for starch synthesis, being produced by the reaction of glucose-1-phosphate with ATP. This reaction, catalysed by ADP glucose pyrophosphorylase (an enzyme consisting of two different subunits), is the first committed reaction in starch synthesis and is considered to be an important regulatory step. The glucosyl

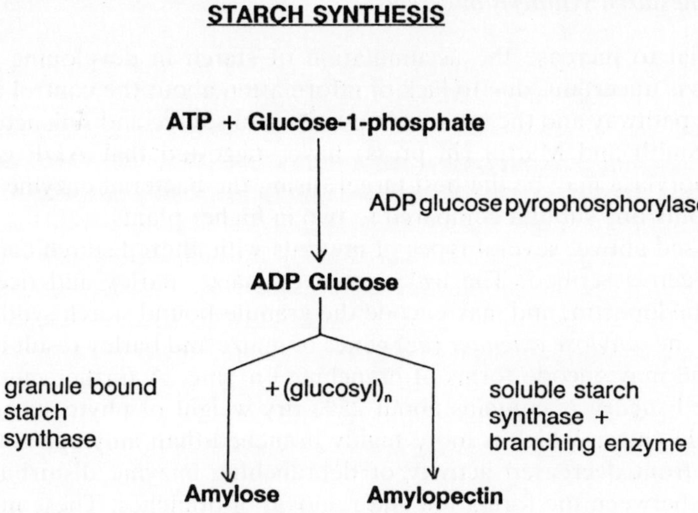

Figure 10. Summary of the enzymology of starch synthesis.

unit is then transferred to an α-glucan primer, forming an α1-4 linkage. This reaction is catalysed by starch synthases which occur in two classes (soluble and granule-bound) and, in most species, in two forms within each class. Analyses of mutant lines (see below) indicate that the bound forms are mainly responsible for the synthesis of amylose and the soluble forms for the synthesis of amylopectin. Finally, the branch points in amylopectins are introduced by branching enzyme (which again exists in several forms), which hydrolyses α1-4 bonds and transfers short oligosaccharides to form α1-6 linkages. Recent studies have also shown that debranching enzyme is active in developing endosperms of maize, and it has been suggested that the precise structure of amylopectin results from an equilibrium between the activities of branching and debranching enzymes (Preiss, 1988).

Genes and/or cDNAs have been cloned for several of these enzymes, either from cereals or from other species. The cereal enzymes which have been cloned include the two subunits of ADP glucose pyrophosphorylase (rice, wheat) and forms of granule-bound starch synthase (maize, barley and wheat) (see Smith and Martin, in press). Mutant genes encoding some starch biosynthetic enzymes have also been identified, and these result in changes in the composition of starch and a reduction in the total amount. These mutations also result in compensatory increases in soluble sugars, and could provide an opportunity for the diversion of carbon into alternative pathways, for example lipid synthesis. The feasibility of blocking starch synthesis by genetic engineering has been elegantly demonstrated by Willmitzer and colleagues, who used antisense technology to eliminate the activity of one of the two subunits of ADP glucose pyrophosphorylase from potato tubers (see Smith and Martin, in press). The result was an absence of starch and the accumulation of sucrose.

Manipulating starch synthesis in cereals

The potential to increase the accumulation of starch in developing cereal endosperms is uncertain, due to lack of information about the control of flux through the pathway and the relative importance of source and sink activities. However, Smith and Martin (in press) have suggested that ADP glucose pyrophosphorylase may be the best target, using the bacterial enzyme which consists of only one subunit compared to two in higher plants.

As discussed above, several types of mutants with altered starch composition have been described. The *waxy* genes of maize, barley and rice have increased amylopectin, and may encode the granule-bound starch synthase I. In contrast, the *amylose extender* (*ae*) genes of maize and barley result in high amylose, and may encode forms of branching enzyme. A further mutant of maize, called *sugary 1*, contains about 25% dry weight of phytoglycogen, a water-soluble glucan which is more highly branched than amylopectin. This may result from decreased activity of debranching enzyme disturbing the equilibrium between the formation and removal of branches. These mutants are important in relation to genetic engineering, as they demonstrate that changes in the amounts of single enzyme proteins can have major effects on starch composition. This is despite the fact that most of the enzymes occur in more than one form.

It is anticipated, therefore, that the composition of cereal starches could be altered by the insertion of genes to either increase the activities of intrinsic enzymes or introduce novel activities, or by the use of antisense technology to reduce the activities of other enzymes. For example, the proportion of amylopectin could be increased by reducing granule-bound starch synthase and/or introducing glycogen branching enzyme, while the proportion of amylose could be increased by reducing branching enzyme and/or increasing debranching enzyme. The consequences of such changes for seed development and germination are not known, but the occurrence of *waxy, amylose extender* and *sugary* mutants demonstrates that wide variation in starch composition can be tolerated.

Current and novel end uses of cereal starches

Assuming, therefore, that the composition of starch could be readily altered by genetic engineering, it is of interest to consider what types of alterations would be required. At present, about two-thirds of the starch produced in Western Europe is used in the food and drink industries, and most of this for enzymic hydrolysis to give syrups for confectionery and soft drinks. In this case, engineering could possibly be used to make the polymer more susceptible to enzymic digestion (Kennedy, Cabalda and White, 1988). Starch with a normal amylose : amylopectin ratio also finds a range of uses in the food industry, both in the unmodified state and after chemical modification (e.g. oxidation, esterification and phosphorylation), while starches from *waxy* and *amylose extender* types are also produced for specific uses (Dunn, 1985). The functional properties of these starches are determined by their physical properties such as gelatinization temperature, density, crystallinity index, freeze–thaw stability and resistance to water separation. These properties could possibly be optimized by genetic engineering, reducing the need for chemical modifications with associated legal constraints concerned with food safety (Galliard, 1985).

Although the food industry is currently the major consumer of starch, there is no doubt that the greatest opportunity for increased consumption lies in non-food products. At present, these account for only about a third of starch production, with the paper and board and the chemical and pharmaceutical industries being the main users. Munck, Rexen and Haastrup (1988) identified three areas in which increased use of starch can be expected.

The first is in the cellulose industry. The growing demand for paper and packaging will lead to increased use of non-traditional fibres (e.g. from recycled products and straw) which are shorter than the traditional hardwood fibres. Laboratory tests show that up to 30% starch can be incorporated to increase the strength and maintain a low cost. The second area is in the production of chemicals and pharmaceuticals. In this case, starch will be used as a substrate for fermentation, and similar considerations apply as for production of syrups for the food and drink industry. Some bulk chemicals are already produced by starch fermentation (e.g. ethanol, glycerol, acetone), but the range can be expected to increase. The third, and perhaps the most exciting, of the potential new outlets is in the production of synthetic polymers. Starch can be used as a filler (e.g. in PVC) or as a graft co-polymer with synthetic polymers. It can also be used for the production of phenol resin as a substitute for formaldehyde, and to produce polyols which find uses in various products including polyurethane foams.

The extent to which such applications take off will, of course, depend on a number of factors – environmental as well as economic. They have the advantages that they will allow the utilization of agricultural surpluses and conserve fossil fuels in favour of a renewable resource, while the products and by-products will be more biodegradable and biocompatible. In addition, the cosmetic, pharmaceutical and chemical products can be expected to be non-irritant and non-allergenic, with the consumer appeal of being 'natural'. Developing a sound understanding of the control of starch synthesis and composition in cereals will undoubtedly make an important contribution to such developments.

REDESIGNING GRAIN COMPOSITION

Despite the wide range of morphological variation in cereals and their adaptation to a range of climatic zones and environmental extremes, there is very little variation in grain composition. Thus protein accounts on average for about 11–12% of the grain dry weight, and starch for about 75% of the dry weight in barley, rye, wheat and maize, but slightly less in sweetcorn and oats. These minor differences in starch content are associated with differences in total lipids, from about 2–3% dry weight in barley, wheat and rye to 5% in maize, 8% in oats and 9% in sweetcorn. Although these proportions vary to a certain extent within species (for example, in relation to nitrogen nutrition), this is limited when compared with the range of variation between seeds of other species. For example, protein can account for up to almost 40% of dry weight in soybean and cotton, and lipid for up to 50% in peanut, oilseed rape and oilpalm, and over 60% in castor bean and brazil nut. Similarly, the proportion of starch varies from less than 10% of dry weight in brazil nut, to 15–30% in temperate oilseeds (rape, cotton, soybean) and over 75% in cereals (see Bewley and Black, 1978). At present, maize is the only cereal that is used to produce oil, although oats could also be used if problems of stability could be eliminated.

The protein content of cereals is regulated to some extent by nutrition, as higher levels of nitrogen fertilization result in increased levels of grain protein. For example, the total nitrogen content of field-grown barley in the UK has been reported to vary between 1.27 and 2.0% dry weight (Kirkman, Shewry and Miflin, 1982). However, even excessive fertilization would not result in the production of protein to a level of 40% of dry weight, as in soybean. It appears, therefore, that there are two levels of control on cereal grain protein content: an environmental control operating within genetically determined limits. Recent studies indicate that the expression of cereal prolamin genes is controlled by the levels of free amino acids in the endosperm, so we may need to increase the transport of amino acids into the grain if we wish to increase the protein content beyond the current levels (Balconi *et al.*, 1991; Müller and Knudsen, 1993). This would clearly require an understanding of the nitrogen economy of the whole plant, including the mechanisms of uptake, assimilation and long-range transport. It may therefore still be some time before we have sufficient understanding to manipulate cereal grain protein content. However, of much greater potential value, and of more immediate interest, is the manipulation of cereals to produce larger amounts of lipids.

At present, a range of plant lipids are required for specific food and non-food uses, coming principally from two types of plant: the tropical palms (oilpalm and coconut) and cocoa, and a range of temperate species (e.g. sunflower, cotton, soybean, oilseed rape, safflower, linseed and various minor species). A wide range of species are used because of their characteristic oils and their adaptation to different climates. The palms are used in particular to provide short chain fatty acids for soaps and detergents, and cocoa to provide specific types of triacylglycerol for chocolate and confectionery, but both have a limited geographical range. Similarly, the major temperate oilseeds also have limited ranges, and do not produce the full spectrum of oils required for food and industrial applications. Finally, a number of minor species are currently of interest in that they produce oils with specific pharmaceutical or industrial applications, but these have poor agronomic characteristics and low yields (Röbbelen, 1991). The production of such oils in cereals would open new opportunities for the industrial use of plant products. Thus it would be of value to produce cereals with a range of lipids for specific end uses as well as high levels of total lipid.

Triacylglycerols (triglycerides) consist of three fatty acid molecules esterified to the three OH groups of glycerol. Their properties depend on the structures of the individual fatty acids (chain length, number and positions of double bonds, presence of other groups such as hydroxyl and epoxy) and on their spatial distribution on the glycerol molecule. Most triacylglycerols from plant sources have high proportions of C16 and C18 fatty acids, with zero or one double bond in the former and zero to three in the latter. However, fatty acids of longer or shorter chain length and with more double bonds or other modifications occur in some seed oils and have specific industrial or food applications.

The first reaction in fatty acid synthesis is the carboxylation of acetyl CoA to give malonyl CoA, catalysed by acetyl CoA carboxylase (ACCase). This is a key regulatory step in animal systems, the enzyme activity being controlled by phosphorylation and dephosphorylation (Davies *et al.*, 1992). The phosphorylation reaction is catalysed by a specific protein kinase which also regulates the activity of HMG CoA reductase, which catalyses the first reaction in terpenoid synthesis (Davies *et al.*, 1992). In plants, ACCase and the initial elongation (by the addition of 2 carbon units) and desaturation reactions occur in the plastid, to give fatty acids with up to 18 carbon atoms and one double bond. Further elongation and desaturation reactions, other modifications (e.g. hydroxylation) and triacylglycerol assembly then take place in the endoplasmic reticulum (Slabas *et al.*, 1993; Stymne and Stobart, 1993).

ACCase is an obvious target for attempts to increase the flux of carbon into fatty acids, and several laboratories are making good progress in characterizing the enzyme and isolating corresponding DNA clones from oilseeds. However, an alternative approach is to assume that the flux is controlled not by the total amount of enzyme protein, but by regulation of its activity by reversible phosphorylation. MacKintosh *et al.* (1992) have recently purified the plant HMG CoA kinase and shown that this also phosphorylates human ACCase but not plant ACCase. The regulation of both enzymic activities by a single kinase would in any case be unlikely, since ACCase and HMG Co reductase have different subcellular locations in plants (in the plastid and endoplasmic reticulum, respectively). However, the presence of a separate and specific ACCase kinase in plants cannot be ruled out.

At Long Ashton, we are studying a family of plant protein kinases which may also play a regulatory role in carbon metabolism in the developing endosperm. These are related to the *Sucrose Non-Fermenting 1 (SNF1)* gene of yeast, which controls a carbon catabolite repression system (Celenza and Carlson, 1986). Thus in the presence of glucose, the *SNF1* gene acts in concert with other regulatory genes to repress the synthesis of a number of enzymes including invertase, and *snf1* mutants cannot grow on substrates such as sucrose, maltose, galactose, melibiose, glycerol and ethanol. We have shown that a cDNA prepared using rye endosperm mRNA can functionally complement the *snf1* mutant of yeast, enabling it to grow with glycerol as a substrate (Alderson *et al.*, 1991). In addition, barley contains at least two families of *SNF1*-related genes, one of which appears to be expressed in all tissues, and the other only in aleurones and starchy endosperms (Halford *et al.*, 1992a; Hannappel *et al.*, in prep.). The roles of these genes are not known, but the endosperm-specific form may well play a role in regulating carbon metabolism in the developing seed.

Any attempt to increase lipid synthesis, for example by increasing the synthesis and/or activation of ACCase, may well need to be accompanied by a decrease in the activity of starch synthesis by using an available mutant or down-regulating ADP glucose pyrophosphorylation) in order to provide an adequate supply of carbon skeletons. In addition, it is not known whether the activities of the elongases, desaturases, etc, that catalyse subsequent steps in fatty acid synthesis, would also need to be increased. Finally, there may also be an opportunity to produce lipids rich in specific fatty acids by incorporating enzyme systems from other species, for example the 14-hydroxylase from castor bean to synthesize ricinoleic acid for industrial use, or the $\triangle 6$ desaturase from evening primrose or borage to synthesize γ-linolenic acid for pharmaceutical use. These and other opportunities are discussed in a number of recent reviews of oilseed manipulation (see, e.g. Stobart, Stymne and Shewry, 1992; Slabas *et al.*, 1993; Safford, 1993; Lapinskas, 1993).

Acknowledgement

We are grateful to Dr Pilar Barcelo (Universität Hamburg) for providing the photographs used in *Figure 2*.

References

AHMED, K.Z. AND SAGI, F. (1993). Culture and fertile plant regeneration from regenerable embryogenic suspension cell-derived protoplasts of wheat (*Triticum aestivum* L.). *Plant Cell Reports* **12**, 175–179.

AKAGI, H., SAKAMOTO, M., NEGISHI, T. AND FUJIMURA, T. (1989). Construction of rice cybrid plants. *Molecular and General Genetics* **215**, 501–506.

ALDERSON, A., SABELLI, P.A., DICKINSON, J.R., COLE, D., RICHARDSON, M., KREIS, M., SHEWRY, P.R. AND HALFORD, N.H. (1991). Complementation of *snf1*, a mutation affecting global regulation of carbon metabolism in yeast, by a plant protein kinase cDNA. *Proceedings of the National Academy of Sciences, USA* **88**, 8602–8605.

ANDERSON, O.D., GREENE, F.C., YIP, R.E., HALFORD, N.G., SHEWRY, P.R. AND MALPICA-ROMERO, J.-M. (1989). Nucleotide sequences of the two high-molecular-weight glutenin genes from the D-genome of a hexaploid bread wheat, *Triticum aestivum* L. cv. Cheyenne. *Nucleic Acids Research* **17**, 461–462.

ARMSTRONG, C.L., ROMERO-SEVERSON, J. AND HODGES, T.K. (1992). Improved tissue culture response of an elite maize inbred through backcross breeding, and identification of chromosomal regions important for regeneration by RFLP analysis. *Theoretical and Applied Genetics* **84**, 755–762.

ASANO, Y., OTSUKI, Y. AND UGAKI, M. (1991). Electroporation-mediated and silicon carbide fibre-mediated DNA delivery in *Agrostis alba* L. (Redtop). *Plant Science* **79**, 247–252.

BALCONI, C., RIZZI, E., MANZOCCHI, L., SOAVE, C. AND MOTTO, M. (1991). Analysis of *in vivo* and *in vitro* grown endosperms of high and low protein strains of maize. *Plant Science* **73**, 1–9.

BAMFORTH, C.W. (1985a). Cambridge Prize Lecture: Biochemical approaches to beer quality. *Journal of the Institute of Brewing* **91**, 154–160.

BAMFORTH, C.W. (1985b). The foaming properties of beer. *Journal of the Institute of Brewing* **91**, 370–383.

BARCELO, P., LAZZERI, P.A., MARTIN, A. AND LOERZ, H. (1991). Competence of cereal leaf cells. I. Patterns of proliferation and regeneration capability *in vitro* of the influorescence sheath leaves of barley, wheat and tritordeum. *Plant Science* **77**, 243–251.

BARCELO, P., LAZZERI, P.A., MARTIN, A. AND LOERZ, H. (1992). Competence of cereal leaf cells. II. Influence of auxin, ammonium and explant age on regeneration. *Journal of Plant Physiology* **139**, 448–454.

BARCELO, P., LAZZERI, P.A., HERNANDEZ, P., MARTIN, A. AND LOERZ, H. (1993). Morphogenic cell and protoplast cultures of *tritordeum*. *Plant Science* **88**, 209–218.

BARCELO, P., HAZEL, C., BECKER, D., MARTIN, A. AND LOERZ, H. (in press). Fertile transgenic cereal (tritordeum) plants obtained by particle bombardment of inflorescence tissue. *The Plant Journal*.

BATTRAW, M.J. AND HALL, T.C. (1990). Histochemical analysis of CaMV35S promoter-β-glucuronidase gene expression in transgenic rice plants. *Plant Molecular Biology* **15**, 527–538.

BATTRAW, M.J. AND HALL, T.C. (1991). Stable transformation of *Sorghum bicolor* protoplasts with chimeric neomycin phosphotransferase II and β-glucuronidase genes. *Theoretical and Applied Genetics* **82**, 161–168.

BATTRAW, M.J. AND HALL, T.C. (1992). Expression of a chimeric neomycin phosphotransferase II gene in first and second generation transgenic rice plants. *Plant Science* **86**, 191–202.

BAXTER, E.D. AND WAINWRIGHT, T. (1979a). Hordein and malting quality. *Journal of the American Society of Brewing Chemists* **37**, 8–12.

BAXTER, E.D. AND WAINWRIGHT, T. (1979b). The importance in malting and mashing of hordein proteins with a relatively high sulphur content. *Proceedings of the EBC Congress, Berlin*, 131–143.

BEACHY, R.N. (1988). Virus cross-protection in transgenic plants. In *Plant Gene Research: Temporal and Spatial Regulation of Plant Genes* (D.P.S. Verma and R.B. Goldberg, Eds), pp. 313–27. Springer-Verlag, New York.

BECCARI, J.B. (1745). De frumento. *De Bononiensi Scientiarum et Artium atque Acade mia Commentarii, Tomi Secundi*, Bononia.

BELTON, P.S., COLQUHOUN, I.J., FIELD, J.M., GRANT, A., SHEWRY, P.R. AND TATHAM, A.S. (in press). ^{1}H and ^{2}H NMR relaxation studies indicate that the elasticity of the HMW subunits of glutenin is not elastin-like. *Journal of Cereal Science*.

BEWLEY, J.D. AND BLACK, M. (1978). *Physiology and Biochemistry of Seeds in Relation to Germination. 1. Development, Germination and Growth*. Springer-Verlag, Berlin.

BISHOP, L.R. (1930a). The institute of brewing research scheme. 1. The prediction of extract. *Journal of the Institute of Brewing* **36**, 421–434.

BISHOP, L.R. (1930b). The nitrogen content and quality of barley. *Journal of the Institute of Brewing* **36**, 352–369.

BLUMENTHAL, C.S., BATEY, I.L., BEKES, F., WRIGLEY, C.W. AND BARLOW, E.W.R. (1990). Gliadin genes contain heat-shock elements: Possible relevance to heat-induced changes in grain quality. *Journal of Cereal Science* **11**, 185–187.

BOOY, G., KRENS, F.A. AND HUIZING, H.J. (1989). Attempted pollen-mediated transformation of maize. *Journal of Plant Science* **135**, 319–324.

BOTHMER, R. VON, CLAESSON, L., FLINK, J. AND LINDE-LAURSEN, I. (1989). Triple hybridisation with cultivated barley (*Hordeum vulgare* L.). *Theoretical and Applied Genetics* **78**, 818–824.

BOTTERMAN, J., D'HALLUIN, K., DEBLOCK, M., DEGREEF, W. AND LEEMANS, J. (1991). Engineering of glufosinate resistance and evaluation under field conditions. In *Herbicide Resistance in Weeds and Crops* (JC Caseley, G.W. Cussans and R.K. Atkin, Eds), pp. 355–363. Butterworth-Heinemann, Oxford.

BROCHLEHURST, P.A. (1977). Factors controlling grain weight in wheat. *Nature* **266**, 348–349.

BRUCE, W.B., CHRISTENSEN, A.H., KLEIN, T., FROMM, M.E. AND QUAIL, P.H. (1989). Photoregulation of a phytochrome gene promoter from oat transferred into rice by particle bombardment. *Proceedings of the National Academy of Sciences, USA* **86**, 9692–9696.

BUNCE, N.A.C., FORDE, B.G., KREIS, M. AND SHEWRY, P.R. (1986). DNA restriction fragment length polymorphism at hordein loci: Application to identifying and fingerprinting barley cultivars. *Seed Science and Technology* **14**, 419–429.

CALLIS, J., FROMM, M. AND WALBOT, V. (1987). Introns increase gene expression in cultured maize cells. *Genes and Development* **1**, 1183–1200.

CAMPBELL, A.F. (1992). Protein engineering of the barley chymotrypsin inhibitor 2. In *Plant Protein Engineering* (P.R. Shewry and S. Gutteridge, Eds), pp. 257–268. Cambridge University Press, Cambridge.

CANNELL, M., KARP, A., ISAACS, P. AND SHEWRY, P.R. (1992). Chromosomal assignment of genes in barley using telosomic wheat–barley addition lines. *Genome* **35**, 17–23.

CAO, J., DUAN, X., MCELROY, D. AND WU, R. (1992). Regeneration of herbicide resistant transgenic plants following microprojectile-mediated transformation of suspension culture cells. *Plant Cell Reports* **11**, 586–591.

CASTILLO, A.M. AND CISTUÉ, L. (1993). Production of gynogenic haploids of *Hordeum vulgare* L. *Plant Cell Reports* **12**, 139–143.

CELENZA, J.L. AND CARLSON, M. (1986). A yeast gene that is essential for release from glucose repression encodes a protein kinase. *Science* **233**, 1175–1180.

CHOJECKI, A.J.S., BAYLISS, M.W. AND GALE, M.D. (1986). Cell production and DNA accumulation in the wheat endosperm, and their association with grain weight. *Annals of Botany* **58**, 809–817.

CHRISTOU, P. (1992). Genetic transformation of crop plants using microprojectile bombardment. *The Plant Journal* **2**, 275–281.

CHRISTOU, P., FORD, T.L. AND KOFRON, M. (1991). Production of transgenic rice (*Oryza sativa* L.) plants from agronomically important Indica and Japonica varieties via electric discharge particle acceleration of exogenous DNA into immature zygotic embryos. *Bio/Technology* **9**, 957–962.

CHRISTOU, P., FORD, T.L. AND KOFRON, M. (1992). The development of a variety-independent gene-transfer method for rice. *Tib Tech* **10**, 239–246.

CLARK, R.B. (1982). Plant response to mineral element toxicity and deficiency. In *Breeding Plants for Less Favourable Environments* (MN Christiansen and C.F. Lewis, Eds), pp. 71–142. John Wiley, New York.

COWEN, N.M., JOHNSON, C.D., ARMSTRONG, K., MILLER, M., WOOSLEY, A., PESCITELLI, S., SKOKUT, M., BELMAR, S. AND PETOLINO, J.F. (1992). Mapping genes conditioning *in vitro* androgenesis in maize using RFLP analysis. *Theoretical and Applied Genetics* **84**, 720–724.

DALE, C.J. AND YOUNG, T.W. (1989). Fractionation of small peptides from beer. *Journal of the Institute of Brewing* **95**, 35–41.

DALE, C.J., YOUNG, T.W. AND BREWER, S. (1989). Amino acid analysis of beer polypeptides. *Journal of the Institute of Brewing* **95**, 89–97.

DALE, P.J. AND WEBB, K.J. (1985). Germplasm storage and micropropagation. In *Cereal Tissue and Cell Culture* (S.W.J. Bright and M.G.K. Jones, Eds), pp. 79–96. Martinus Nijhoff/Junk, Dordrecht.

DALTON, S.J. AND DALE, P.J. (1985). The application of *in vitro* tiller induction in *Lolium multiflorum*. *Euphytica* **34**, 897–904.

DATTA, S.K., DATTA, K. AND POTRYKUS, I. (1990). Fertile indica rice plants regenerated from protoplasts isolated from microspore derived cell suspensions. *Plant Cell Reports* **9**, 253–256.

DATTA, S.K., PETERHANS, A., DATTA, K. AND POTRYKUS, I. (1990). Genetically engineered fertile Indica rice recovered from protoplasts. *Bio/Technology* **8**, 736–740.

DATTA, S.K., DATTA, K., SOLTANIFAR, N., DONN, G. AND POTRYKUS, I. (1992). Herbicide-resistant Indica rice plants from IRRI breeding line IR72 after PEG-mediated transformation of protoplasts. *Plant Molecular Biology* **20**, 619–629.

DAVIES, S.P., CARLING, D., MUNDAY, M.R. AND HARDY, D.G. (1992). Diurnal rhythm of phosphorylation of rat liver acetyl-CoA carboxylase by the AMP-activated protein kinase, demonstrated using freeze-clamping effects of high fat diets. *European Journal of Biochemistry* **203**, 615–623.

DECARVALHO, F., GHEYSEN, G., KUSHNIR, S., VAN MONTAGU, M., INZE, D. AND CASTRESANA, C. (1992). Suppression of β-1,3-glucanase transgene expression in homozygous plants. *EMBO Journal* **11**, 2595–2602.

DEKEYSER, R.A., CLAES, B., DERYCKE, R.M.U., HABETS, M.E., VAN MONTAGU, M.C. AND CAPLAN, A.B. (1990). Transient gene expression in intact and organized rice tissues. *The Plant Cell* **2**, 591–602.

DE LA PENA, A., LORZ, H. AND SCHELL, J. (1987). Transgenic rye plants obtained by injecting DNA into young floral tillers. *Nature* **325**, 274–276.

DEVAUX, P. (1992). Haploidy in barley and wheat improvement. In *Reproductive Biology and Plant Breeding* (Y. Dattée, C. Dumas and A. Gallais, Eds), pp. 139–151. Springer-Verlag, Berlin.

DE WET, J.M.J., DE WET, A.E., BRINK, D.E., HEPBURN, A.G. AND WOODS, J.H. (1986). Gametophyte transformation in maize (*Zea mays*, Gramineae). In *Biotechnology and Ecology of Pollen* (D.C. Mulcahy, Ed.), pp. 59–64. Springer-Verlag, Heidelberg.

D'HALLUIN, K., BONNE, E., BOSSUT, M., DE BEUCKELEER, M. AND LEEMANS, J. (1992). Transgenic maize plants by tissue electroporation. *The Plant Cell* **4**, 1495–1505.

DUNCAN, D.R., WILLIAMS, M.E., ZEHR, B.E. AND WIDHOLM, J.M. (1985). The production of callus capable of plant regeneration from immature embryos of numerous *Zea mays* genotypes. *Planta* **165**, 32–332.

DUNN, C.F. (1985). High amylose starches. *Food Technology in Australia* **37**, 110–111.

DUNWELL, J.M. (1985). Anther and ovary culture. In *Cereal Tissue and Cell Culture* (S.W. Bright and M.G.K. Jones, Eds), pp. 1–44. Martinus Nijhoff, Dordrecht.

DUNWELL, J.M. (1992). Mechanisms of microspore embryogenesis. In *Reproductive Biology and Plant Breeding* (Y. Dattée, C. Dumas and A. Gallais, Eds), pp. 121–130. Springer-Verlag, Berlin.

EWART, J.A.D. (1967). Amino acid analyses of cereal flour proteins. *Journal of the Science of Food and Agriculture* **18**, 548–552.

FEDAK, G. (1992). Intergeneric hybrids with *Hordeum*. In *Barley: Genetics, Biochemistry, Molecular Biology and Biotechnology* (P.R. Shewry, Ed.), pp. 45–70. CAB International, Wallingford.

FENNEL, A. AND HAUPTMANN, R. (1992). Electroporation and PEG delivery of DNA into maize microspores. *Plant Cell Reports* **11**, 567–570.

FIELD, J.M., SHEWRY, P.R. AND MIFLIN, B.J. (1983). Solubilization and characterization of wheat gluten proteins: correlations between the amount of aggregated proteins and baking quality. *Journal of Agricultural and Food Chemistry* **39**, 390–377.

FLORY, P.J. (1953). Principles of polymer chemistry. In *Rubber Elasticity*, pp. 432–494. Cornell University Press, Ithaca, NY.

FOOD AND AGRICULTURE ORGANIZATION (1973). *Energy and Protein Requirements*. FAO Nutritional Meeting Report Series No. 52, WHO Technical Report Series No. 522. FAO, Rome.

FOOD AND AGRICULTURE ORGANIZATION (1988). *Production*, Vol. 42. FAO, Rome.

FRANKEL, O.H. (1970). Genetic conservation in perspective. In *Genetic Resources in*

Plants: Their Exploration and Conservation (O.H. Frankel and E. Bennett, Eds), pp. 469–489. Blackwell Scientific, Oxford.

FROMM, M., TAYLOR, L.P. AND WALBOT, V. (1985). Expression of genes transferred into monocot and dicot plant cells by electroporation. *Proceedings of the National Academy of Sciences, USA* **82**, 5824–5828.

FROMM, M.E., MORRISH, F., ARMSTRONG, C., WILLIAMS, R., THOMAS, J. AND KLEIN, T.M. (1990). Inheritance and expression of chimeric genes in the progeny of transgenic maize plants. *Bio/Technology* **8**, 833–839.

FUJIMURA, T., SAKURAI, M. AND AKAGI, H. (1985). Regeneration of rice plants from protoplasts. *Plant Tissue Culture Letters* **2**, 74–75.

FUNATSUKI, H., LOERZ, H. AND LAZZERI, P.A. (1992). Use of feeder cells to improve barley protoplast culture and regeneration. *Plant Science* **85**, 179–187.

GALE, M.D. AND MILLER, T.E. (1987). The introduction of alien genetic variation in wheat. In *Wheat Breeding: Its Scientific Basis* (F.G.H. Lupton, Ed.), pp. 173–210. Chapman and Hall, London.

GALIBA, G., KERTESZ, Z., UTKA, J. AND SAGI, L. (1985). Differences in somaclonal variation in three winter wheat (*Triticum aestivum*) varieties. *Cereal Research Communications* **13**, 343–350.

GALLIARD, T. (1985). Bulk chemicals from plants: Starch and starch-derived products. In *Plant Products and the New Technology* (K.W. Fuller and J.R. Gallon, Eds), pp. 113–115. Oxford University Press, Oxford.

GASSER, C.S. AND FRALEY, R.T. (1989). Genetically engineered plants for crop improvement. *Science* **244**, 1293–1299.

GERLACH, W.L., LLEWELLYN, D. AND HASELHOFF, J. (1987). Construction of a plant disease resistance gene from the satellite RNA of tobacco ringspot virus. *Nature* **328**, 802–5.

GORDON-KAMM, W.J., SPENCER, T.M., MANGANO, M.L., ADAMS, T.R., DAINES, R.J., START, W.G., O'BRIEN, J.V., CHAMBERS, S.A., ADAMS, W.R.JR, WILLETTS, N.G., RICE, T.B., MACKEY, C.J., KRUEGER, R.W., KAUSCH, A.P. AND LEMAUX, P.G. (1990). Transformation of maize cells and regeneration of fertile transgenic plants. *The Plant Cell* **2**, 603–618.

GOULD, J., DEVEY, M., HASEGAWA, O., ULIAN, E.C., PETERSON, G. AND SMITH, H. (1991). Transformation of *Zea mays* L. using *Agrobacterium tumefaciens* and the shoot apex. *Plant Physiology* **95**, 426–434.

GRAHAM, J., PEARCE, G., MERRYWEATHER, J., TITAN, K., ERICSSON, L. AND RYAN, C. (1985). Wound-induced proteinase inhibitors from tomato leaves. The cDNA-deduced primary structure of preinhibitor I and its post-translational processing. *Journal of Biological Chemistry* **260**, 6555–6560.

GRAVES, A.C.F. AND GOLDMAN, S.L. (1986). The transformation of *Zea mays* seedlings with *Agrobacterium tumefaciens*. *Plant Molecular Biology* **7**, 43–50.

GRESSEL, J. (1992). Genetically engineered herbicide-resistant crops: A moral imperative for world food production. *Agro Food Industry, Hi-tech* **3**, 3–7.

GRIMSLEY, N., HOHN, T., DAVIES, J.W. AND HOHN, B. (1987). *Agrobacterium*-mediated delivery of infectious maize streak virus into maize plants. *Nature* **325**, 177–179.

GRIMSLEY, N.H., RAMOS, C., HEIN, T. AND HOHN, B. (1988). Meristematic tissues of maize plants are most susceptible to agroinfection with maize streak virus. *Bio/Technology* **6**, 185–189.

GRIMSLEY, N., HOHN, B., RAMOS, C., KADO, C. AND ROGOWSKY, P. (1989). DNA transfer from *Agrobacterium* to *Zea mays* or *Brassica* by agroinfection is dependent on bacterial virulence functions. *Molecular and General Genetics* **217**, 309–316.

GUPTA, H.S. AND PATTANAYAK, A. (1993). Plant regeneration from mesophyll protoplasts of rice (*Oryza sativa* L.). *Bio/Technology* **11**, 90–94.

HAGBERG, A. AND HAGBERG, G. (1987). Production of spontaneously doubled haploids in barley using a breeding system with marker genes and the '*hap*'-gene. *Biologisches Zentralblatt* **106**, 53–58.

HAGIO, T., BLOWERS, A.D. AND EARLE, E.D. (1991). Stable transformation of sorghum cell cultures after bombardment with DNA-coated microprojectiles. *Plant Cell Reports* **10**, 260–264.

HAHNE, B., FLECK, J. AND HAHNE, G. (1989). Colony formation from mesophyll protoplasts of a cereal, oat. *Proceedings of the National Academy of Sciences, USA* **86**, 6154–6160.

HAIN, R., REIF, H.-J., KRAUSE, E., LANGEBARTELS, R., KINDL, H., VORNAM, B., WIESE, W., SCHMELZER, E., SCHREIER, P.H., STÖCKER, R.H. AND STENZEL, K. (1993). Disease resistance results from foreign phytoalexin expression in a novel plant. *Nature* **361**, 153–156.

HALFORD, N.G., FORDE, J., SHEWRY, P.R. AND KREIS, M. (1989). Functional analysis of the upstream regions of a silent and an expressed member of a family of wheat seed protein genes in transgenic tobacco. *Plant Science* **62**, 207–216.

HALFORD, N.G., VICENTE-CARBAJOSA, J., SABELLI, P.A., SHEWRY, P.R., HANNAPPEL, U. AND KREIS, M. (1992a). Molecular analysis of a barley multigene family related to *SNF1*, a yeast gene encoding a protein kinase responsible for global regulation of carbon metabolism. *The Plant Journal* **2**, 791–797.

HALFORD, N.G., FIELD, J.M., BLAIR, H., URWIN, P., MOORE, K., ROBERT, L., THOMPSON, R., FLAVELL, R.B., TATHAM, A.S. AND SHEWRY, P.R. (1992b). Analysis of HMW glutenin subunits encoded by chromosome 1A of bread wheat (*Triticum aestivum* L.) indicates quantitative effects on grain quality. *Theoretical and Applied Genetics* **83**, 373–378.

HANNAPPEL, U., VICENTE-CARBAJOSA, J., BARKER, J.H.A., SHEWRY, P.R. AND HALFORD, N.G. (in prep.). Differential tissue-specific expression of members of a family of *SNF1*-related barley genes encoding protein kinases.

HARBERD, N.P., BARTELS, D. AND THOMPSON, R.D. (1986). DNA restriction fragments variation in the gene family encoding high molecular weight (HMW) glutenin subunits of wheat. *Biochemical Genetics* **24**, 579–596.

HARLAN, J.R. (1978). The origins of cereal agriculture in the old world. In *Origins of Agriculture* (C.A. Reeds, Ed.), pp. 357–383. Mouton, The Hague.

HARRISON, B.D., MAYO, M.A. AND BAULCOMBE, D.G. (1987). Virus resistance in transgenic plants that express cucumber mosaic virus satellite RNA. *Nature* **328**, 799–802.

HARTNETT, M.E., CHUI, C.-F., FALCO, S.C., KNOWLTON, S., MAUVAIS, C.J. AND MAZUR, B.J. (1991). Molecular analysis of sulfonylurea herbicide resistant ALS genes. In *Herbicide Resistance in Weeds and Crops* (J.C. Caseley, G.W. Cussans and R.K. Atkin, Eds), pp. 343–535. Butterworth-Heinemann, Oxford.

HAUPTMANN, R.M., OZIAS-AKINS, P., VASIL, V., TABAEIZADEH, Z., ROGERS, S.G., HORSCH, R.B., VASIL, I.K. AND FRALEY, R.T. (1987). Transient expression of electroporated DNA in monocotyledonous and dicotyledonous species. *Plant Cell Reports* **6**, 265–270.

HAYASHI, Y., KYOZUKA, J. AND SHIMAMOTO, K. (1988). Hybrids of rice (*Oryza sativa* L.) and wild *Oryza* species obtained by cell fusion. *Molecular and General Genetics* **214**, 6–10.

HAYASHIMOTO, A., LI, Z. AND MURAI, N. (1990). A polyethylene glycol-mediated protoplast transformation system for production of fertile transgeneric rice plants. *Plant Physiology* **93**, 857–863.

HEJGAARD, J. AND BOISEN, S. (1980). High lysine proteins in Hiproly barley breeding: Identification, nutritional significance and new screening methods. *Hereditas* **93**, 311–20.

HESS, D., DRESSLER, K. AND NIMMRICHTER, R. (1990). Transformation experiments by pipetting *Agrobacterium* into the spikelets of wheat (*Triticum aestivum* L.). *Plant Science* **72**, 233–244.

HIATT, A., CAFFERKEY, R. AND BOWDISH, K. (1989). Production of antibodies in transgenic plants. *Nature* **342**, 76–8.

HILDER, V.A., GATEHOUSE, A.M.R., SHEERMAN, S.E., BARKER, R.F. AND BOULTER, D. (1987). A novel mechanism of insect resistance engineered into tobacco. *Nature* **300**, 160–3.

ISLAM, A.K.M.R. AND SHEPHERD, K.W. (1991). Alien genetic variation in wheat improvement. In *Chromosome Engineering in Plants: Genetics, Breeding, Evolution, Part A* (P.K. Gupta and T. Tsuchiya, Eds), pp. 291–312. Elsevier, Amsterdam.

JACOBSEN, J.V. AAND CLOSE, T.J. (1991). Control of transient expression of chimaeric genes by giberellic acid and abscisic acid in protoplasts prepared from mature barley aleurone layers. *Plant Molecular Biology* **16**, 713–724.

JAEHNE, A., LAZZERI, P.A. AND LOERZ, H. (1991). Regeneration of fertile plants from protoplasts derived from embryogenic cell suspensions of barley (*Hordeum vulgare* L.). *Plant Cell Reports* **10**, 1–6.

JEFFERSON, R.A. (1987). Assaying chimeric genes in plants: The GUS gene fusion system. *Plant Molecular Biology* **5**, 387–405.

KAEPPLER, H.F., GU, W., SOMERS, D.A., RINES, H.W. AND COCKBURN, A.F. (1990). Silicon carbide fibre-mediated DNA delivery into plant cells. *Plant Cell Reports* **9**, 415–418.

KAEPPLER, H.F., SOMERS, D.A., RINES, H.W. AND COCKBURN, A.F. (1992). Silicon carbide fiber-mediated stable transformation of plant cells. *Theoretical and Applied Genetics* **84**, 560–566.

KAHL, G. (1982). Molecular biology of wound healing: The conditioning phenomenon. In *Molecular Biology of Plant Tumors* (G. Kahl and J. Schell, Eds), pp. 211–267. Academic Press, New York.

KALLOO, G. AND CHOWDHURY, J.B. (EDS) (1992). *Distant Hybridisation of Crop Plants*. Springer-Verlag, Berlin.

KARP, A. (1991). On the current understanding of somaclonal variation. In *Oxford Surveys of Plant Molecular and Cell Biology* (B.J. Miflin, Ed.), Vol. 7, pp. 1–58. Oxford University Press, Oxford.

KARTHA, K.K., CHIBBAR, R.N., GEORGES, F., LEUNG, N., CASWELL, K., KENDAL, L.E. AND QURESHI, J. (1989). Transient expression of chloramphenicol acetyltransferase (CAT) gene in barley cell cultures and immature embryos through microprojectile bombardment. *Plant Cell Reports* **8**, 429–432.

KASHA, K.J. AND KAO, K.N. (1970). High frequency haploid production in barley (*Hordeum vulgare*). *Nature* **225**, 874–876.

KASHA, K.J., CHO, U.-H. AND ZIAUDDIN, A. (1992). Application of microspore cultures. In *Barley Genetics VI* (L. Munck, Ed.), Vol. II, pp. 793–806. Munksgaard International Publishers Ltd, Copenhagen.

KENNEDY, J.F., CABALDA, V.M. AND WHITE, C.A. (1988). Enzymic starch utilization and genetic engineering. *Tib Tech* **6**, 184–189.

KHUSH, G.S. AND BRAR, D.S. (1992). Overcoming the barriers in hybridisation. In *Distant Hybridisation of Crop Plants* (G. Kalloo and J.B. Chowdhury, Eds), pp. 47–61. Springer-Verlag, Berlin.

KIRKMAN, M.A., SHEWRY, P.R. AND MIFLIN, B.J. (1982). The effect of nitrogen nutrition on the lysine content and protein composition of barley seeds. *Journal of the Science of Food and Agriculture* **33**, 115–127.

KLEIN, T.M., ROTH, B.A. AND FROMM, M.E. (1989). Regulation of anthocyanin biosynthetic genes introduced into intact maize tissues by microprojectiles. *Proceedings of the National Academy of Sciences, USA* **85**, 6681–6685.

KLEIN, T.M., WOLF, E.D., WU, R. AND SANFORD, J.C. (1987). High-velocity microprojectiles for delivering nucleic acids into living cells. *Nature* **327**, 70–73.

KLEIN, T.M., FROMM, M., WEISSINGER, A., TOMES, D., SCAAF, S., SLETTEN, M. AND SANFORD, J.C. (1988). Transfer of foreign genes into intact maize cells with high-velocity microprojectiles. *Proceedings of the National Academy of Sciences, USA* **85**, 4305–4309.

KLEIN, T.M., KORNSTEIN, L., SANFORD, J.C. AND FROMM, M. (1989). Genetic transformation of maize cells by particle bombardment. *Plant Physiology* **91**, 440–444.

KLEIN, T.M., ARENTZEN, R., LEWIS, P.A. AND FITZPATRICK-MCELLIGOTT, S. (1992). Transformation of microbes, plants and animals by particle bombardment. *Bio/Technology* **10**, 286–291.

KÖHLER, P., BELITZ, H.-D. AND WIESER, H. (1991). Disulphide bonds in wheat gluten: Isolation of a cysteine peptide from glutenin. *Zeitschrift für Lebensmitteluntersuchung und -Forschung* **192**, 234–239.

KÖHLER, P., BELITZ, H.-D. AND WIESER, H. (1993). Disulphide bonds in wheat gluten: Further cysteine peptides from high molecular weight (HMW) and low molecular weight (LMW) subunits of glutenin and from γ-gliadins. *Zeitschrift für Lebensmittel-untersuchung und -Forschung* **196**, 239–247.

KONSTANTINOV, K., MLADENOVIC, S. AND DENIC, M. (1991). Recombinant DNA technology in maize breeding. V. Plant and indigenous bacterial strains transformation by the gene controlling kanamycin resistance. *Genetika (Beograd)* **23**, 121–135.

KOZIEL, M.G., BELAND, G.L., BOWMAN, C., CAROZZI, N.B., CRENSHAW, R., CROSSLAND, L., DAWSON, J., DESAI, N., HILL, M., KADWELL, S., LAUNIS, K., LEWIS, K., MADDOX, D., MCPHERSON, K., MEGHJI, M.R., MERLIN, E., RHODES, R., WARREN, G.W., WRIGHT, M. AND EVOLA, S.V. (1993). Field performance of elite transgenic maize plants expressing an insecticidal protein derived from *Bacillus thuringiensis*. *Bio/Technology* **11**, 194–200.

KRANZ, E. AND LOERZ, H. (1993). *In vitro* fertilization with isolated, single gametes results in zygotic embryogenesis and fertile maize plants. *The Plant Cell* **5**, 739–746.

KRANZ, E., BAUTOR, J. AND LOERZ, H. (1991). *In vitro* fertilization of single, isolated gametes of maize mediated by electrofusion. *Sexual Plant Reproduction* **4**, 12–16.

KREBBERS, E., BOSCH, D. AND VANDERKERCKHOVE, J. (1992). Prospects and progress in the production of foreign proteins and peptides in plants. In *Plant Protein Engineering* (P.R. Shewry and S. Gutteridge, Eds), pp. 315–326. Cambridge University Press, Cambridge.

KREBBERS, E., DA SILVA, CONCEICAO, A., DENIS, M., D'HONDT, K. AND VANDERKERCKHOVE, J. (1993). Modification of plant seed storage proteins. In *Plant Storage Compounds: Biosynthesis, Interactions and Manipulation* (P.R. Shewry and A.K. Stobart, Eds) pp. 317–324. Oxford University Press, Oxford.

KREIS, M. AND SHEWRY, P.R. (1992). The control of protein synthesis in developing barley seeds. In *Barley: Genetics, Biochemistry, Molecular Biology and Biotechnology* (P.R. Shewry, Ed.), pp. 319–333. CAB International, Wallingford.

KREIS, M., SHEWRY, P.R., FORDE, B.G., RAHMAN, S. AND MIFLIN, B.J. (1983). Molecular analysis of a mutation conferring the high lysine phenotype on the grain of barley (*Hordeum vulgare*). *Cell* **34**, 161–167.

KYOZUKA, J., KANEDA, T. AND SHIMAMOTO, K. (1989). Production of cytoplasmic male sterile rice (*Oryza sativa* L.) by cell fusion. *Bio/Technology* **7**, 1171–1174.

KYOZUKA, J., MAYASAMI, Y. AND SHIMAMOTO, K. (1987). High frequency of plant regeneration from rice protoplasts by novel nurse culture methods. *Molecular and General Genetics* **206**, 408–413.

LAMB, C.J., RYALS, J.A., WARD, E.R. AND DIXON, R.A. (1992). Emerging strategies for enhancing crop resistance to microbial pathogens. *Bio/Technology* **10**, 1436–1445.

LANGRIDGE, P., BRETTSCHNEIDER, R., LAZZERI, P. AND LOERZ, H. (1992). Transformation of cereals via *Agrobacterium* and the pollen pathway: A critical assessment. *The Plant Journal* **2**, 631–638.

LAPINSKAS, P. (1993). Oil crops for the pharmaceutical industry. In *Plant Storage*

Compounds: Biosynthesis, Interactions and Manipulation (P.R. Shewry and A.K. Stobart, Eds)pp. 332–342. Oxford University Press, Oxford.

LARKIN, P.J. AND SCOWCROFT, W.R. (1981). Somaclonal variation: A novel source of variability from cell cultures for plant improvement. *Theoretical and Applied Genetics* **60**, 197–214.

LARKIN, P.J., RYAN, S.A., BRETTELL, R.I.S. AND SCOWCROFT, W.R. (1984). Heritable somaconal variation in wheat. *Theoretical and Applied Genetics* **60**, 197–214.

LAST, D.I., BRETTELL, R.I.S., CHAMBERLAIN, D.A., CHAUDHURY, A.M., LARKIN, P.J., MARSH, E.L., PEACOCK, W.J. AND DENNIS, E.S. (1991). pEmu: An improved promoter for gene expression in cereal cells. *Theoretical and Applied Genetics* **81**, 581–588.

LAZAR, M.D., CHEN, T.H.H., GUSTA, L.V. AND KARTHA, K.K. (1988). Somaclonal variation for freezing tolerance in a popularion derived from norstar winter wheat. *Theoretical and Applied Genetics* **75**, 480–484.

LAZZERI, P.A., KOLLMORGEN, J. AND LOERZ, H. (1990). *In vitro* technology. In *Reproductive Versatility in the Grasses* (G.P. Chapman, Ed.), pp. 182–219. Cambridge University Press, Cambridge.

LAZZERI, P.A., BRETTSCHNEIDER, R., LUEHRS, R. AND LOERZ, H. (1991). Stable transformation of barley via PEG-induced DNA uptake into protoplasts. *Theoretical and Applied Genetics* **18**, 437–444.

LEE, B., MURDOCH, K., TOPPING, J., KREIS, M. AND JONES, M.G.K. (1989). Transient gene expression in aleurone protoplasts isolated from developing caryopses of barley and wheat. *Plant Molecular Biology* **13**, 21–39.

LEE, N., WANG, Y., YANG, J., GE, K., HUANG, S., TAN, J. AND TESTA, D. (1991). Efficient transformation and regeneration of rice small cell groups. *Proceedings of the National Academy of Sciences, USA* **88**, 6389–6393.

LENDING, C.R., WALLACE, J.C. AND LARKINS, B.A. (1992). Synthesis of zeins and their potential for amino acid modification. In *Plant Protein Engineering* (P.R. Shewry and S. Gutteridge, Eds), pp. 209–218. Cambridge University Press, Cambridge.

LI, B.-J., XU, X.-P., SHI, H.-P. AND KE, S.-Y. (1991). Introduction of foreign genes into the seed embryo cells of rice by electroinjection and the regeneration of transgenic rice plants. *Science in China* **34**, 923–931.

LI, X.-Q., LIU, C.-N., RITCHIE, S.W., PENG, J.-Y., GELVIN, S.B. AND HODGES, T.K. (1992). Factors influencing *Agrobacterium*-mediated transient expression of gusA in rice. *Plant Molecular Biology* **20**, 1037–1048.

LUEHRS, R. AND LOERZ, H. (1988). Initiation of morphogenic cell suspension and protoplast cultures of barley (*Hordeum vulgare* L.). *Planta* **175**, 71–81.

LUO, Z. AND WU, R. (1988). A simple method for the transformation of rice via the pollen-tube pathway. *Plant Molecular Biology Reporter* **6**, 165–174.

MACKINTOSH, R.W., DAVIES, S.P., CLARKE, P.R., WEEKES, J., GILLESPIE, J.G., GIBB, B.J. AND HARDIE, D.G. (1992). Evidence for a protein kinase cascade in higher plants: 3-Hydroxy-3-methylglutaryl-CoA reductase kinase. *European Journal of Biochemistry* **209**, 923–931.

MACRITCHIE, F., DUCROS, D.L. AND WRIGLEY, C.W. (1990). Flour polypeptides related to wheat quality. In *Advances in Cereal Science and Technology* (Y. Pomeranz, Ed.), Vol. X, pp. 79–145. American Association of Cereal Chemists, Inc, St Paul, MN.

MANYOWA, N.M. AND MILLER, T.E. (1991). The genetics of tolerance to high mineral concentrations in the tribe Triticeae: A review and update. *Euphytica* **57**, 175–185.

MARKS, M.S., KEMP, J.M., WOOLSTON, C.J. AND DALE, P.J. (1989). Agroinfection of wheat: A comparison of *Agrobacterium* strains. *Plant Science* **63**, 247–256.

MARTIN, A. AND SANCHEZ-MONGE, E. (1982). Cytology and morphology of the amphiploid *Hordeum chilense* × *Triticum turgidum* conv. *durum Euphytica* **31**, 261–267.

MATZKE, M.A., PRIMIG, M., TRNORSKY, J. AND MATZKE, A.J.M. (1989). Reversible methylation and inactivation of marker genes in sequentially transformed tobacco plants. *EMBO Journal* **8**, 643–649.

MAY, R.M. (1993). Resisting resistance. *Nature* **361**, 593–4.

MCELROY, D., ZHANG, W., CAO, J. AND WU, R. (1990). Isolation of an efficient actin promoter for use in rice transformation. *The Plant Cell* **2**, 163–171.

MENDEL, R.R., MUELLER, B., SCHULZE, J., KOLESNIKOV, V. AND ZELENIN, A. (1989). Delivery of foreign genes into intact barley cells by high-velocity microprojectiles. *Theoretical and Applied Genetics* **78**, 31–34.

MIAO, Z., ZHUANG, J. AND HU, H. (1988). Expression of various gametic types in pollen plants regenerated from hybrids between *Triticum–Agropyron* and wheat. *Theoretical and Applied Genetics* **75**, 485–491.

MILES, M.J., CARR, H.J., MCMASTER, T., BELTON, P.S., MORRIS, V.J., FIELD, J.M., SHEWRY, P.R. AND TATHAM, A.S. (1991). Scanning tunnelling microscopy of a wheat gluten protein reveals details of a spiral supersecondary structure. *Proceedings of the National Academy of Sciences USA* **88**, 68–71.

MILLET, M.-O., MONTEMBAULT, A. AND AUTRAN, J.-C. (1991). Hordein compositional differences in various anatomical regions of the kernel between two different barley types. *Sciences des Alimentations* **11**, 155–161.

MOONEY, P.A. AND GOODWIN, P.B. (1991). Adherence of *Agrobacterium tumefaciens* to the cells of immature wheat embryos. *Plant Cell, Tissue and Organ Culture* **25**, 199–208.

MOONEY, P.A., GOODWIN, P.B., DENNIS, E.S. AND LLEWELLYN, D.J. (1991). *Agrobacterium tumefaciens*-gene transfer into wheat tissues. *Plant Cell, Tissue and Organ Culture* **25**, 209–218.

MOROCZ, S., DONN, G., NEMETH, J. ABD DUDITS, D. (1990). An improved system to obtain fertile regenerants via maize protoplasts isolated from a highly embryogenic suspension culture. *Theoretical and Applied Genetics* **80**, 721–726.

MORRISH, F., VASIL, V. AND VASIL, I.K. (1987). Developmental morphogenesis and genetic manipulation in tissue and cell cultures of the Gramineae. *Advances in Genetics* **24**, 431–499.

MORRISH, F., SONGSTAD, D.D., ARMSTRONG, C.L. AND FROMM, M. (1992). Microprojectile bombardment: A method for the production of transgenic cereal crop plants and the functional analysis of genes. In *Transgenic Plants: Fundamentals and Applications* (A. Hiatt, Ed.), pp. 133–171. Marcel Dekker, New York.

MUELLER, E., SCHMITT, M., LAZZERI, P.A. AND LOERZ, H. (1992). Stable transformation fo wheat cultures (*T. aestivum*). In *Meeting of the German Section of the IAPTC*, 18–19 September 1992, University of Hamburg, Abstract No 34.

MÜLLER, M. AND KNUDSEN, S. (1993). The nitrogen response of a barley C-hordein promoter is controlled by positive and engative regulation of the GCN4 and endosperm box. *The Plant Journal* **4**, 343–355.

MUNCK, L. (1992). The case of high-lysine barley breeding. In *Barley: Genetics, Biochemistry, Molecular Biology and Biotechnology* (P.R. Shewry, Ed.), pp. 573–601. CAB International, Wallingford.

MUNCK, L., KARLSSON, K.-E., HAGBURG, A. AND EGGUM, B.O. (1970). Gene for improved nutritive value in barley seed protein. *Science* **168**, 985–987.

MUNCK, L., REXEN, F. AND HAASTRUP, L. (1988). Cereal starches within the European community: Agricultural production, dry and wet milling and potential use in Industry. *Starch* **40**, 81–87.

NEVO, E. (1992). Origin, evolution, population genetics and resources for breeding of wild barley, *Hordeum spontaneum*, in the fertile crescent. In *Barley: Genetics, Biochemistry, Molecular Biology and Biotechnology* (P.R. Shewry, Ed.), pp. 19–43. CAB International, Wallingford.

OHTA, Y. (1986). High-efficiency genetic transformation of maize by a mixture of pollen and exogenous DNA. *Proceedings of the National Academy of Sciences, USA* **83**, 715–719.

O'KEEFE, D.P., LENSTRA, R. AND OMER, C.A. (1992). An enzymatic basis for herbicide resistance: Cytochrome P450 mono-oxygenases. In *Plant Protein Engineering* (P.R. Shewry and S. Gutteridge, Eds), pp. 281–292. Cambridge University Press, Cambridge.

OMIRULLEH, S., ABRAHAM, M., GOLOVKIN, M., STEFANOV, I., KARABAEV, M.K., MUSTARDY, L., MOROCZ, S. AND DUDITS, D. (1993). Activity of a chimeric promoter with the doubled CaMV 35S enhancer element in protoplast-derived cells and transgenic plants in maize. *Plant Molecular Biology* **21**, 415–428.

OOMS, G. (1992). Genetic engineering of plants and cultures. In *Plant Biotechnology* (M.W. Fowler and G.S. Warren, Eds), pp. 223–257. Pergamon Press, Oxford.

OW, D.W., JACOBS, J.D. AND HOWELL, S.H. (1987). Functional regions of the cauliflower mosaic virus 35S RNA promoter determined by use of the firefly Luciferase gene as a reporter of promoter activity. *Proceedings of the National Academy of Sciences, USA* **84**, 4870–4874.

OZIAS-AKINS, P., FERL, R.J. AND VASIL, I.K. (1986). Somatic hybridisation in the Gramineae: *Pennistum americanum* L.K. Schum. (Pearl millet) + *Panicum maximum* Jacq. (Guinea grass). *Molecular and General Genetics* **203**, 365–370.

PASZKOWSKI, J., SHILLITO, R.D., SAUL, M., MANDAK, V., HOHN, T., HOHN, B. AND POTRYKUS, I. (1984). Direct gene transfer to plants. *EMBO Journal* **3**, 2717–2722.

PAULY, M.H., SHANE, W.W. AND GENGENBACH, B.G. (1987). Selection for bacterial blight phytotoxin resistance in wheat tissue culture. *Crop Science* **27**, 340–344.

PAYNE, P.I. (1987). Genetics of wheat storage proteins and the effect of allelic variation on breadmaking quality. *Annual Review of Plant Physiology* **38**, 141–153.

PAYNE, P.I. AND LAWRENCE, G.D. (1983). Catalogue of alleles for the complex gene loci, *Glu-A1, Glu-B1* and *Glu-D1* which code for high molecular weight subunits of glutenin in hexaploid wheat. *Cereal Research Communications* **11**, 29–35.

PAYNE, P.I., HOLT, L.M. AND LAW, C.N. (1981). Structural and genetical studies on the high molecular weight subunits of wheat glutenin. *Theoretical and Applied Genetics* **60**, 229–236.

PAYNE, P.I., NIGHTINGALE, MA., KRATTIGER, A.F. AND HOLT, L.M. (1987). The relationship between HMW glutenin subunit composition and the breadmaking quality of British grown wheat varieties. *Journal of the Science of Food and Agriculture* **40**, 51–65.

PESCHKE, V.M. AND PHILLIPS, R.L. (1992). Genetic implications of somaclonal variation in plants. *Advances in Genetics* **30**, 41–75.

PETERSON, D.M., FORDE, J., WILLIAMSON, M.S., RHODE, W. AND KREIS, M. (1991). The nucleotide sequence of a chymotrypsin inhibitor-2 gene of barley (*Hordeum vulgare*). *Plant Physiology* **96**, 1389–90.

PETOLINO, J.F. (1992). The use of androgenesis in maize breeding. In *Reproductive Biology and Plant Breeding* (Y. Dattée, C. Dumas and A. Gallais, Eds), pp. 131–138. Springer-Verlag, Berlin.

PETOLINO, J.F. AND THOMPSON, S.A. (1987). Genetic analysis of anther culture response in maize. *Theoretical and Applied Genetics* **74**, 284–286.

PICARD, E., JACQUEMIN, J.M., GRAINER, F., BOBIN, M. AND FORGEOIS, P. (1988). Genetic transformation of wheat (*Triticum aestivum*) by plasmid DNA uptake during pollen tube germination. In *Proceedings of the 7th International Wheat Genetics Symposium*, pp. 779–787. IPSR, Cambridge.

PICKERING, R.A. AND DEVAUX, P. (1992). Haploid production: Approaches and use in plant breeding. In *Barley: Genetics, Biochemistry, Molecular Biology and Biotechnology* (P.R. Shewry, Ed.), pp. 519–548. CAB International, Wallingford.

POTRYKUS, I. (1989). Gene transfer to cereals: An assessment. *Tib Tech* **7**, 269–273.

POTRYKUS, I. (1990). Gene transfer to cereals: An assessment. *Bio/Technology* **8**, 535–542.

POTRYKUS, I., SAUL, M., PETRUSKA, J., PASZKOWSKI, J. AND SHILLITO, R.D. (1985). Direct gene transfer into protoplasts of a graminaceous monocot. *Molecular and General Genetics* **199**, 183–188.

PREISS, J. (1988). Biosynthesis of starch and its regulation. In *The Biochemistry of Plants, Vol. 14: Carbohydrates* (J. Preiss, Ed.), pp. 181–254. Academic Press, San Diego, CA.

PRIOLI, L.M. AND SONDAHL, M.R. (1989). Plant regeneration and recovery of fertile plants from protoplasts of maize (*Zea mays* L.). *Bio/Technology* **7**, 589–594.

RADLEY, M. (1978). Factors affecting grain enlargement in wheat. *Journal of Experimental Botany* **29**, 919–934.

RAINERI, D.M., BOTTINO, P., GORDON, M.P. AND NESTER, E.W. (1990). *Agrobacterium*-mediated transformation of rice (*Oryza sativa* L.). *Bio/Technology* **8**, 33–38.

RASMUSSEN, U., MUNCK, L. AND ULLRICH, S.E. (1990). Immunogold localisation of chymotrypsin inhibitor-2, a lysine-rich protein, in developing barley endosperm. *Planta* **180**, 272–277.

RATHORE, K.S., CHOWDHURY, V.J. AND HODGES, T.K. (1993). Use of *bar* as a selectable marker gene and for production of herbicide-resistant rice plants from protoplasts. *Plant Molecular Biology* **21**,'871–884.

REDWAY, F.A., VASIL, V. AND VASIL, I.K. (1990). Characterisation and regeneration of wheat (*Triticum aestivum* L.) embryogenic cell suspension cultures. *Plant Cell Reports* **8**, 714–717.

RHODES, C.A., PIERCE, D.A., METTLER, I.J., MASCARENHAS, D. AND DETMER, J.J. (1988). Genetically transformed maize plants from protoplasts. *Science* **240**, 204–207.

RIGGS, T.J., SANADA, M., MORGAN, A.F. AND SMITH, D.F. (1983). Use of acid gel electrophoresis in the characterisation of 'B' hordein protein in relation to malting quality and mildew resistance in barley. *Journal of the Science of Food and Agriculture* **34**, 576–586.

RINES, H.W. AND LUKE, H.H. (1985). Selection and regeneration of toxin-insensitive plants from tissue cultures of oats (*Avena sativa*) susceptible to *Helminthosporium victoriae. Theoretical and Applied Genetics* **74**, 31–37.

RÖBBELEN, G. (1991). The genetic improvement of seedoil. *Chemistry and Industry* 713–716.

ROECKEL, P., HEIZMANN, P., DUBOIS, M. AND DUMAS, C. (1988). Attempts to transform *Zea mays* via pollen grains: Effect of pollen and stigma nuclease activities. *Sexual Plant Reproduction* **1**, 156–163.

ROGERS, S.W. AND ROGERS, J.C. (1992). The importance of DNA methylation for stability of foreign DAN in barley. *Plant Molecular Biology* **18**, 945–961.

RYAN, S.A. AND SCOWCROFT, W.R. (1987). A somaclonal variant of wheat with additional α-amylase isosymes. *Theoretical and Applied Genetics 273*, 459–464.

SAFFORD, R. (1993). Opportunities for genetic engineering of oil production. In *Plant Storage Compounds: Biosynthesis, Interactions and Manipulation* (P.R. Shewry and A.K. Stobart, Eds) pp. 343–358. Oxford University Press, Oxford.

SALMENKALLIO, M., HANNUS, R., TEERI, T.H. AND KAUPPINEN, V. (1990). Regulation of α-amylase promoter by giberellic acid and abscisic acid in barley protoplasts transformed by electroporation. *Plant Cell Reports* **9**, 352–355.

SCHACHERER, A. AND BERINGER, H. (1984). Zahl und Gröβe von Endospermzellen in wachsenden Getreidekorn als Indikator der Speicherkapazität. *Berichte Deutsche Botanische Gesellschaft* **97**, 183–195.

SCHAEFFER, G.W., SHARPE, F.T., JR AND CREGAN, P.B. (1984). Variation for improved protein and yield from rice anther culture. *Theoretical and Applied Genetics* **67**, 383–389.

SCHAEFFER, G.W., SHARPE, F.T. JR AND DUDLEY, J.T. (1992). Rice protein mutant expressed in liquid suspension cultures: Chitinases, β-glucanases and other proteins. *Theoretical and Applied Genetics* **84**, 26–32.

SCHLAEPPI, M. AND HOHN, B. (1992). Competence of immature maize embryos for *Agrobacterium*-mediated gene transfer. *The Plant Cell* **4**, 7–16.

SENARATNA, T., MCKERSIE, B.D., KASHA, K.J. AND PROCUNIER, J.D. (1991). Direct DNA uptake during the imbibition of dry cells. *Plant Science* **79**, 223–228.

SHEN, W.-H. AND HOHN, B. (1992). Excision of a transposable element from a viral vector introduced into maize plants by agroinfection. *The Plant Journal* **2**, 35–42.

SHEWRY, P.R. (1992). Barley seed storage proteins – structure, synthesis and deposition. In *Nitrogen Metabolism of Plants* (D.J. Pilbeam and K. Mengel, Eds) pp. 201–227. Oxford University Press, Oxford.

SHEWRY, P.R. (1993). Biological and evolutionary aspects of cereal seed storage proteins. In *Plant Storage Compounds: Biosynthesis, Interactions and Manipulation* (P.R. Shewry and A.K. Stobart, Eds) pp. 53–78. Oxford University Press, Oxford.

SHEWRY, P.R. AND KREIS, M. (1992). Tissue-specific gene expression in the developing barley seed. In *Barley Genetics VI* (L. Munck, Ed.), Vol. 2, pp. 725–735. Proceedings of the 6th International Barley Genetics Symposium. Munksgaard International Publishers, Copenhagen.

SHEWRY, P.R. AND PARMAR, S. (1987). The *Hrd F* locus encodes γ-type hordeins. *Barley Genetics Newsletter* **17**, 32–34.

SHEWRY, P.R. AND TATHAM, A.S. (1989). New light on an old technology: The structure of wheat gluten and its role in breadmaking. *Outlook on Agriculture* **18**, 65–71.

SHEWRY, P.R., HALFORD, N.G. AND TATHAM, A.S. (1989). The high molecular weight subunits of wheat, barley and rye: Genetics, molecular biology, chemistry and role in wheat gluten structure and functionality. In *Oxford Surveys of Plant Molecular and Cell Biology* (B.J. Miflin, Ed.), Vol. 6, pp. 163–219. Oxford University Press, Oxford.

SHEWRY, P.R., HALFORD, N.G. AND TATHAM, A.S. (1992). The high molecular weight subunits of wheat glutenin. *Journal of Cereal Science* **15**, 105–120.

SHEWRY, P.R., PARMAR, S. AND PAPPIN, D.J.C. (1987). Characterization and genetic control of the prolamins of *Haynaldia villosa*: Relationship to cultivated species of the Triticeae (rye, wheat and barley). *Biochemical Genetics* **25**, 309–325.

SHEWRY, P.R., WILLIAMSON, M.S. AND KREIS, M. (1987). Effects of mutant genes on the synthesis of storage components in developing barley endosperms. In *Mutant Genes that Affect Plant Development* (H. Thomas and D. Grierson, Eds), pp. 95–118. Cambridge University Press, Cambridge.

SHEWRY, P.R., FAULKS, A.J., PARMAR, S. AND MIFLIN, B.J. (1980). Hordein polypeptide pattern in relation to malting quality and the varietal identification of malted grain. *Journal of the Institute of Brewing* **86**, 138–141.

SHEWRY, P.R., WOLFE, M.S., SLATER, S.E., PARMAR, S., FAULKS, A.J. AND MIFLIN, B.J. (1981). Barley storage proteins in relation to varietal identification, malting quality and mildew resistance. In *Barley Genetics IV*, pp. 596–603. Edinburgh University Press, Edinburgh.

SHEWRY, P.R., FINCH, R., PARMAR, S., FRANKLIN, J. AND MIFLIN, B.J. (1983). Chromosomal location of *Hor 3*, a new locus governing storage proteins in barley. *Heredity* **50**, 179–189.

SHEWRY, P.R., PARMAR, S., FRANKLIN, J. AND BURGESS, S.R. (1990). Analysis of a rare recombination event within the multigenic *Hor 2* locus of barley (*Hordeum vulgare* L.). *Genetical Research* **55**, 171–176.

SHILLITO, R.D., CARSWELL, G.K., JOHNSON, C.M., DIMAIO, J.S. AND HARMS, C.T. (1989). Regeneration of fertile plants from plant protoplasts of elite inbred maize. *Bio/Technology* **7**, 581–587.

SHIMAMOTO, K., TERADA, R., IZAWA, T. AND FUJIMOTO, H. (1989). Fertile rice plants regenerated from transformed protoplasts. *Nature* **338**, 274–276.

SINGH, B.K. AND JENNER, C.F. (1982). Association between concentrations of organic nutrients in the grain, endosperm cell number and grain dry weight within the ear of wheat. *Australian Journal of Plant Physiology* **9**, 83–95.

SLABAS, A.R., ELBOROUGH, K., CHASE, D., SINDEN, B., MARTINEZ-RIVAS, J.M., BROWN, A.P., SWINHOE, R., SIMON, B. AND FAWCETT, T. (1993). Fatty acid synthesis in oilseeds. In *Plant Storage Compounds: Biosynthesis, Interactions and Manipulation* (P.R. Shewry and A.K. Stobart, Eds) pp. 81–95. Oxford University Press, Oxford.

SMITH, A.M. AND MARTIN, C. (in press). Starch biosynthesis and the potential for its manipulation. In *Biosynthesis and Manipulation of Plant Products* (D. Grierson, Ed.). Plant Biotechnology Vol. 3. Blackie and Son, Glasgow.

SMITH, D.B. AND LISTER, P.R. (1983). Gel forming proteins in barley grain and their relationship with malting quality. *Journal of Cereal Science* **1**, 229–239.

SNAPE, J.W. (1989). Doubled haploid breeding: Theoretical basis and practical applications. In *Review of Advances in Plant Biotechnology*, 1985–88. (A. Mujeeb-Kasi and L.A. Sitch, Eds), pp. 19–29. Second International Symposium on Genetic Manipulation in Crops. CIMMYT and IRRI, Mexico and Manila.

SOMERS, D.A., RINES, H.W., GU, W., KAEPPLER, H.F. AND BUSHNELL, W.R. (1992). Fertile transgenic oat plants. *Bio/Technology* **10**, 1589–1594.

SONGSTAD, D.D., HALAKA, F.G., DEBOER, D.L., ARMSTRONG, C.L., HINCHEE, M.A.W., FORD-SANTINO, C.G., BROWN, S.M., FROMM, M.E. AND HORSCH, R.B. (1993). Transient expression of GUS and anthocyanin constructs in intact maize immature embryos following electroporation. *Plant, Cell Tissue and Organ Culture* **33**, 195–201.

SONNEWALD, U., BRAUER, M., VON SCHAEWEN, A., STITT, M. AND WILLMITZER, L. (1991). Transgenic tobacco plants expressing yeast-derived invertase in either the cytosol, vacuole or apoplast: A powerful tool for studying sucrose metabolism and sink/source interactions. *The Plant Journal* **1**, 95–106.

SPENCER, T.M., O'BRIEN, J.V., START, W.G., ADAMS, T.R., GORDON-KAMM, W.J. AND LEMAUX, P.G. (1992). Segregation of transgenes in maize. *Plant Molecular Biology* **18**, 201–210.

STITT, M., VON SCHAEWEN, A. AND WILLMITZER, L. (1990). 'Sink' regulation of photosynthetic metabolism in transgenic tobacco plants expressing yeast invertase in their cell wall involves a decrease of the Calvin-cycle enzymes and an increase of glycolytic enzymes. *Planta* **183**, 40–50.

STOBART, A.K., STYMNE, S. AND SHEWRY, P.R. (1992). Manipulation of cereal protein and oilseed quality. In *Control of Plant Gene Expression* (D.P. Verma, Ed.), pp. 499–533. CRC Press, Boca Raton, FL.

STOLARZ, A. AND LOERZ, H. (1986). Somatic embryogenesis, *in vitro* multiplication and plant regeneration from immature embryo explants of hexaploid triticale (× *Triticosecale* Wittmark). *Zeitschrift für Planzenzüchtung* **96**, 353–362.

STYMNE, S. AND STOBART, A.K. (1993). Triacylglycerol biosynthesis. In *Plant Storage Compounds: Biosynthesis, Interactions and Manipulation* (P.R. Shewry and A.K. Stobart, Eds) pp. 96–114. Oxford University Press, Oxford.

SU, R.-C., RUDERT, M.L. AND HODGES, T.K. (1992). Fertile indica and japonica rice plants regenerated from protoplasts isolated from embryogenic haploid suspension cultures. *Plant Cell Reports* **12**, 45–49.

SU, Z.-X., CHEN-ZHANG, Z., KANG-LE, Z., XIU-FANG, Q. AND YA-PING, F. (1983). Somaclonal genetics of rice, *Oryza sativa* L. *Theoretical and Applied Genetics* **67**, 67–73.

SVENDSEN, I., BOISEN, S. AND HEJGAARD, J. (1982). Amino acid sequence of serine protease inhibitor CI-1 from barley: Homology with barley inhibitor CI-2 potato inhibitor I, and leech eglin. *Carlsberg Research Communications* **47**, 45–53.

SVENDSEN, I., MARTIN, H.B. AND JONASSEN, I. (1980). Characteristics of Hiproly barley III. Amino acid sequences of two lysine-rich proteins. *Carlsberg Research Communications* **45**, 79–85.

TADA, Y., SAKAMOTO, M. AND FUJIMURA, T. (1990). Efficient gene introduction into rice by electroporation and analysis of transgenic plants: Use of electroporation buffer lacking chloride ions. *Theoretical and Applied Genetics* **80**, 475–480.

TADA, Y., SKAMOTO, M., MATSUOKA, M. AND FUJIMURA, T. (1991). Expression of a monocot LHCP promoter in transgenic rice. *EMBO Journal* **10**, 1803–1808.

TAO, H.P., ADALSTEINS, A.E. AND KASARDA, D.D. (1992). Intermolecular disulfide bonds link specific high-molecular-weight glutenin subunits in wheat endosperm. *Biochimica et Biophysica Acta* **1159**, 13–21.

TARCZYNSKI, M.C., JENSEN, R.G. AND BOHNERT, H.J. (1993). Stress protection of transgenic tobacco by production of the osmolyte mannitol. *Science* **259**, 508–510.

TATHAM, A.S., SHEWRY, P.R. AND MIFLIN, B.J. (1984). Wheat gluten elasticity: A similar molecular basis to elastin? *FEBS Letters* **177**, 205–208.

TAYLOR, M.G. AND VASIL, I.K. (1991). Histology of, and physical factors affecting, transient GUS expression in pearl millet (*Pennisetum glaucum* (L.) R.Br.) embryos following microprojectile bombardment. *Plant Cell Reports* **10**, 120–125.

TEERI, T.H., PATEL, G.K., ASPEGREN, K. AND KAUPPINEN, V. (1989). Chloroplast targeting of neomycin phosphotransferase II with a pea transit peptide in electroporated barley mesophyll protoplasts. *Plant Cell Reports* **8**, 187–190.

TERADA, R., NAKAYAMA, T. ND IWABUCHI, M. (1993). A wheat histone H3 promoter confers cell division-dependent and -independent expression of the gus A gene in transgenic rice plants. *The Plant Journal* **3**, 241–252.

TERADA, R., KYOZUKA, J., NISHIBAYASHI, S. AND SHIMAMOTO, K. (1987). Plantlet regeneration from somatic hybrids of rice (*Oryza sativa* L.) and barnyard grass (*Ecinochloa oryzicola* Vasing). *Molecular and General Genetics* **210**, 39–43.

TEWES, A., MANTEUFFEL, R., ADLER, K., WEBER, E. AND WOBUS, U. (1991). Long-term cultures of barley synthesize and correctly deposit seed storage proteins. *Plant Cell Reports* **10**, 467–470.

TOEPFER, R., GRONENBORN, B., SCHELL, J. AND STEINBISS, H.H. (1989). Uptake and transient expression of chimaeric genes in seed-derived embryos. *Plant Cell* **1**, 133–139.

TORIYAMA, K. AND HINATA, K. (1988). Diploid somatic-hybrid plants regenerated from rice cultivars. *Theoretical and Applied Genetics* **76**, 665–668.

TORIYAMA, K., HINATA, K. AND SASAKI, T. (1986). Haploid and diploid plant regeneration from protoplasts of anther callus in rice. *Theoretical and Applied Genetics* **70**, 505–509.

TORIYAMA, K., ARIMOTO, Y., UCHIMAYA, H. AND HINATA, K. (1988). Transgenic rice plants after direct gene transfer into protoplasts. *Bio/Technology* **6**, 1072–1074.

TOYODA, H., YAMAGA, T., MATSUDA, Y. AND OUCHI, S. (1990). Transient expression of the β-glucuronidase gene introduced into barley coleoptile cells by microinjection. *Plant Cell Reports* **9**, 299–302.

TRICOLI, D.M., HEIN, M.B. AND CARNES, M.G. (1986). Culture of soybean mesophyll protoplasts in alginate beads. *Plant Cell Reports* **5**, 334–337.

VASIL, I.K. (1987). Developing cell and tissue culture systems for the improvement of cereal and grass crops. *Journal of Plant Physiology* **128**, 193–218.

VASIL, I.K. (1988). Progress in the regeneration and genetic manipulation of cereal crops. *Bio/Technology* **6**, 397–402.

VASIL, I.K. AND VASIL, V. (1980). Isolation and culture of cereal protoplasts: Embryogenesis and plantlet formation from protoplasts of *Pennisetum americanum*. *Theoretical and Applied Genetics* **56**, 97–99.

VASIL, I.K. AND VASIL, V. (1992). Advances in cereal protoplast research. *Physiologia Plantarum* **85**, 279–283.

VASIL, V., FERL, R.J. AND VASIL, I.K. (1988). Somatic hybridisation in the Gramineae: *Triticum monococcum* L. (Einkorn) + *Pennisetum americanum* (L.) K. Schum. (Pearl millet). *Journal of Plant Physiology* **132**, 160–163.

VASIL, V., REDWAY, F. AND VASIL, I.K. (1990). Regeneration of plants from embryogenic suspension culture protoplasts of wheat (*Triticum aestivum* L.). *Bio/Technology* **8**, 429–434.

VASIL, V., CLANCY, M., FERL, R.J., VASIL, I.K. AND HANNAH, L.C. (1989). Increased gene expression by the first intron of maize *shrunken-1* locus in grass species. *Plant Physiology* **91**, 1575–1579.

VASIL, V., BROWN, S.M., RE, D., FROMM, M.E. AND VASIL, I.K. (1991). Stably transformed callus lines from microprojectile bombardment of cell suspension cultures of wheat. *Bio/Technology* **9**, 743–747.

VASIL, V., CASTILLO, A.M., FROMM, M.E. AND VASIL, I.K. (1992). Herbicide resistant fertile transgenic wheat plants obtained by microprojectile bombardment of regenerable embryogenic callus. *Bio/Technology* **10**, 667–675.

VAUCHERET, H., KRONENBERGER, J., LEPINGLE, A., VILAINE, F., BOUTIN, J.-P. AND CABOCHE, M. (1992). Inhibition of tobacco nitrite reductase activity by expression of antisence RNA. *The Plant Journal* **2**, 559–569.

WALTERS, D.A., VETSCH, C.S., POTTS, D.E. AND LUNDQUIST, R.C. (1992). Transformation and inheritance of a hygromycin phosphotransferase gene in maize plants. *Plant Molecular Biology* **18**, 189–200.

WANG, T.B., NIIZEKI, M., HARADA, T., ISHIKAWA, R., QIAN, Y.Q. AND SAITO, K. (1993). Establishment of somatic hybrid cell lines between *Zea mays* L. (maize) and *Triticum* sect. *trititrigia* MacKey (Trititrigia). *Theoretical and Applied Genetics* **86**, 371–376.

WANG, W.C. AND NGUYEN, H.T. (1990). A novel approach for efficient plant regeneration from long-term suspension culture of wheat. *Plant Cell Reports* **8**, 639–642.

WARDLAW, I.F. (1990). The control of carbon partitioning in plants. *New Phytologist* **116**, 341–381.

WEEKS, J.T., ANDERSON, O.D. AND BLECHL, A.E. (1993). Rapid production of multiple independent lines of fertile transgenic wheat. *Plant Physiology* **102**, 1077–1084.

WEI, Z. AND XU, Z. (1990). Regeneration of fertile plants from embryogenic suspension culture protoplasts of *Sorghum vulgare*. *Plant Cell Reports* **9**, 51–53.

WILLIAMSON, M.S., FORDE, J. AND KREIS, M. (1988). Molecular cloning of two isoinhibitor forms of chymotrypsin inhibitor 1 (CI-1) from barley endosperm and their expression in normal and mutant barleys. *Plant Molecular Biology* **10**, 521–35.

WILLIAMSON, M.S., FORDE, J., BUXTON, B. AND KREIS, M. (1987). Nucleotide sequence of barley chymotrypsin inhibitor-2 (CI-2) and its expression in normal and high-lysine barley. *European Journal of Biochemistry* **165**, 99–106.

WU, C. AND ZAPATA, F.J. (1992). Plant regeneration from protoplasts isolated from primary callus of four japonica rice (*Oryza sativa* L.) varieties. *Plant Science* **86**, 83–87.

YANG, Z.-Q., SHIKANAI, T., MORI, K. AND YAMADA, Y. (1989). Plant regeneration from cytoplasmic hybrids of rice (*Oryza sativa* L.). *Theoretical and Applied Genetics* **77**, 305–310.

ZEMETRA, R.S., SCHOTZKO, D.J., SMITH, C.M. AND LAUVER, M. (1993). *In vitro* selection for Russian wheat aphid (*Diuraphis noxiz*) resistance in wheat (*Triticum aestivum*). *Plant Cell Reports* **12**, 312–315.

ZENKTELER, M. (1990). *In vitro* fertilisation and wide hybridisation in higher plants. *Critical Reviews in Plant Science* **9**, 267–279.

ZHANG, J.-L., XIAO-MING, L., RUI-ZHU, C., RUI-XIN, H. AND MENG-MIN, H. (1991). Transposition of maize transposable element Activator in rice. *Plant Science* **73**, 191–198.

ZHANG, W. AND WU, R. (1988). Efficient regeneration of transgenic plants from rice protoplasts and correctly regulated expression of the foreign gene in the plants. *Theoretical and Applied Genetics* **76**, 835–840.

ZHANG, Z.H. (1989). The practicability of anther culture breeding in rice. In *Review of Advances in Plant Biotechnology*, 1985–88 (A. Mujeeb-Kasi and L.A. Sitch, Eds), pp. 31–41. Second International Symposium on Genetic Manipulation in Crops.

5

Progress in the Development of New Barley, Hop and Yeast Variants for Malting and Brewing

JOHN R.M. HAMMOND AND CHARLES W. BAMFORTH

BRF International, Nutfield, Surrey, UK

Introduction

This review reports recent genetical developments related to humankind's oldest biotechnology, the brewing of beer. It will, hopefully, become evident to the reader that, despite the longevity of the processes, malting and brewing are anything but stagnant. The review focuses on progress in the development of new and improved barleys, hops and yeasts. In particular, it emphasizes the potential for modern gene technology in these improvements. Readers with no fundamental knowledge of the malting and brewing processes are referred to Hough (1985) for a useful overview. The simplest outlines of the processes involved are given by Bamforth (1993).

Barley and malt

Barley remains the principal cereal crop, employed worldwide for the production of beer (Bamforth and Barclay, 1993). The vast majority of the barley is prior-malted, in order to mobilize the starch reserves of the endosperm for ready hydrolysis in mashing.

Barley varieties differ in their propensity to 'modify' during the germination phase of malting, viz. to have their cell walls and proteins enzymically hydrolysed, thereby releasing starch. The more readily modified cultivars (so-called 'malting' varieties) are progressively turned over commercially as newer, improved lines emerge through conventional breeding programmes (Sage, 1992). Since the advent of Plant Variety Rights in 1964, there has been a notable surge in the numbers of malting varieties coming to the market-place. Whereas once varieties might have had a lifetime of 100 years, they may now become outclassed in 4–6 years (Smeaton, 1990).

Abbreviations: ABA, abscisic acid; ABRE, ABA response element; ALDC, acetolactate decarboxylase; GA, gibberellic acid; GARE, GA-responsive element; mRNA, messenger RNA; rDNA, recombinant DNA; RFLP, restriction fragment length polymorphism.

Biotechnology and Genetic Engineering Reviews — Vol. 11, December 1993
0264–8725/93/11/147–169 $20.00 + $0.00 © Intercept Ltd, P.O. Box 716, Andover, Hampshire SP10 1YG, UK

Breeding of barleys by traditional techniques is, however, a protracted and imprecise affair, taking perhaps 10–15 years, from the initial cross until a malting variety is fully accepted for commercial use in brewing. Considerable interest exists in identifying rapid methods, applicable to strictly limited quantities of grain, which can be used to select promising malting quality lines very early in a breeding operation (Home and Elamo, 1993). It is clearly advisable to have assessment protocols at all filial stages, which ensure that the barley is not only going to satisfy the farmer, in terms of its agronomic performance, but also the maltster and, ultimately, brewer in terms of the quality (and quantity) of malt which it will yield.

The techniques involved in conventional plant breeding are comprehensively explained elsewhere (Smith, Austin and Haslemore, 1986). Increasing attention, however, is being paid to newer opportunities for more selectively developing barleys with defined and specific characteristics, through modern techniques of genetic modification (Holm *et al.*, 1992). The rate-limiting step in such operations for monocotyledonous plants is the ability to regenerate plants from protoplasts. In the case of barley, this is compounded by an imperfect understanding of the molecular mechanisms which control gene expression. It is useful to address the latter issue first.

THE CONTROL OF GENE EXPRESSION IN BARLEY

The expression of genes in barley seems largely to involve the hormones gibberellin and abscisic acid (ABA). There is a diversity of gibberellins, but as the majority of studies have involved gibberellic acid (GA), it is convenient to concentrate on this molecule.

The levels of GA peak during the development of seeds (Slominski, Rgowski and Nowak, 1979) whereas ABA peaks later, around the onset of seed desiccation (King, 1976). ABA or osmotic stress induces expression of the so-called *rab* (*r*esponsive to *AB*A) genes (Mundy and Chua, 1988). These genes are valuable models for the investigation of ABA-regulated gene expression (Mundy, Yamaguchi-Shinozaki and Chua, 1990). Such studies, in common with others reported later in this section, involved experiments with so-called transient gene expression in which exogenous DNA remains in a non-integrated form in the host nucleoplasm but can, nonetheless, be expressed for short periods in a manner equivalent to that of integrated genes.

ABA may have a role in regulating dormancy (Fong, Smith and Koehler, 1983). Non-dormant mutants, which lack ABA or are unable to respond to it, do not desiccate properly, and neither do they accumulate *rab* gene products (Koornneef, 1986). The RAB proteins, translated from *rab* genes, peak later in seed formation, but disappear early in germination (Close, Kortt and Chandler, 1989; Mundy and Chua, 1988). The precise function of the RAB proteins is unknown.

Early in germination, the scutellum synthesizes hydrolytic enzymes, although the aleurone soon adopts a priority role in this regard (Mundy and Munck, 1985). Detailed investigations in several laboratories have revealed something of the influence of ABA and GA on the synthesis of enzymes in germinating barley (*Table 1*).

GA is known to trigger the accumulation of messenger RNA molecules coding for hydrolases critical to endosperm conversion, for example α-amylase (Higgins, Jacobsen and Zwar, 1982), β-glucanase (Mundy and Fincher, 1986) and endo-peptidases (Rogers, Dean and Heck, 1985). GA, or a factor dependent on GA, activates transcription of the principle α-amylase isozyme of barley (Jacobsen and Beach, 1985). Skriver *et al.* (1991) delineated both a *GA-r*esponsive *e*lement (GARE) in the α-amylase promoter and an *ABA r*esponsive *e*lement (ABRE) in the promoter of a *rab* gene. They showed that ABA represses expression from the GARE. These elements inserted into other promoters directed strong, hormone-responsive expression of genes, showing that they could be used as regulatory cassettes for 'plugging into' any gene. The α-amylase gene promoter is particularly relevant in the context of genes, the expression of which is only desired during malting and not in raw barley.

It appears that sequence motifs in the 3′ untranslated tails of GA-responsive genes have a key role in the stability and processing of mRNA in barley (Khursheed and Rogers, 1989). Secretion of hydrolases in germinating barley is dictated by calcium (Deikman and Jones, 1986).

As can be seen in *Table 1*, ABA antagonizes GA in the expression of genes (Mozer, 1980). ABA, for instance, induces factors capable of either inhibiting the transcription of α-amylase genes or destabilizing the mRNA encoding α-amylase (Nolan and Ho, 1988).

GENETIC TRANSFORMATION OF BARLEY

The production of transgenic plants, originally tobacco in the early 1980s, was facilitated by the use of the soil pathogen *Agrobacterium tumefaciens*, which is capable of transferring genes into the plant genome. While dicotyledonous plants are readily transformed by vector mediation, such transformation of monocotyledonous plants has proved much more problematic. It seems that differentiated cells of cereals have a poor capacity to dedifferentiate and become embryogenic, viz to eventually reform a plant (Potrykus, 1989). Furthermore, exogenous DNA introduced via *Agrobacterium* may not integrate into the host genome (Grimsley *et al.*, 1987). Accordingly, a range of

Table 1. Hormonal regulation of proteins in barley seeds

Gene	Effect of GA	Effect of ABA	Reference
β(1→3)(1→4)-glucanase	+++	–	Mundy and Fincher (1986)
Thiol protease	+++	–	Rogers, Dean and Heck (1985)
Carboxypeptidase	+	–	Hammerton and Ho (1986)
α-amylase 1	+++	–	Higgins, Jacobsen and Zwar (1982)
α-amylase 2	+++	–	Higgins, Jacobsen and Zwar (1982)
α-amylase inhibitor	–	+	Mundy *et al.* (1986)

+, Increased expression; –, decreased expression.

direct gene transfer techniques has been explored as an alternative (*Table 2*). However, there is no evidence for stable transformation from any of these techniques.

The most favoured technique seems to be that of biolistic attack on embryogenic tissues. However, it does appear that there is an associated problem with mutations emerging through the regeneration phase, causing albinism, growth retardation and sterility.

The successful regeneration of barley protoplasts has been reported (Lazzeri and Lorz, 1988) and it has recently been claimed that green barley plants containing an integrated gene for β-glucanase for *Trichoderma reesei* have been produced (Mannonen *et al.*, 1993).

TARGETS FOR GENETIC TRANSFORMATION OF BARLEY

Successful application of modern gene technology to barley will facilitate a broad range of benefits (*Table 3*). Improvements may lie in manipulation of genes native to barley or by introducing exogenous genes from diverse sources. *Table 3* suggests that progress is limited. This certainly seems to be the case from a survey of the published literature, but there is little doubt that significant progress is being made in various locations, although this activity is shrouded in commercial secrecy. It should be realized that few brewers can have the same direct influence on barley genetic manipulation as they can, for example, on modification of yeast. All brewers have their own yeast strains which, subject to acquiring the necessary legislative permission, they are free to modify as they require. Relatively few brewers produce their own malt, and even fewer are vertically integrated to the extent of growing and, ultimately, breeding their own barleys. Thus, a myriad of commercial decision points divide breeders, with their ability to consider genetical modification of barley, and end-users, that is maltsters and brewers. It is important, then, to have mechanisms to bring the various parties into one place (for example, the Institute of Brewing Barley Committee; see Smeaton, 1990). In the future, one might envisage much more direct control by end-users over raw material supply, for example, contract growing.

Table 2. Alternative strategies for introducing exogenous DNA into plants

Method	Reference
Injection of DNA into young floral tillers	de la Pena, Lorz and Schell (1987)
Transfer of DNA through the pollen tube into the egg	Luo and Wu (1988)
Imbibition of DNA by desiccated embryos	Topfer *et al.* (1989)
Microinjection of DNA	Neuhaus *et al.* (1987)
Electroporation of DNA	Lindsey and Jones (1987)
Bombardment with tungsten/gold particles coated with DNA ('biolistics')	McCabe *et al.* (1988)

Table 3. Goals for genetic modification of malting barley

Goal	Rationale	Status	Reference
Disease resistance	Avoidance of use of pesticides ('clean labelling')	Current strategy is growth of mixtures of barley varieties, differing in disease resistance.	Wolfe, Bennett and Jenkins (1981)
		Report of transfer of gene for powdery mildew resistance from *H. bulbosum* to *H. vulgare*	Xu and Kasha (1992)
Improved yield	Decreased use of fertilizers – reduced nitrogen enabling increased starch	No published reference	
Reduced fibre	Fewer processing problems with non-starchy polysaccharides	Low β-glucan barley	Aastrup, Erdal and Munck (1985)
			Molina-Cano *et al.* (1989)
High fibre-conversion potential	Increased enzyme levels to deal with non-starchy polysaccharides	Claim for integration of heat-stable β-glucanase	Mannonen *et al.* (1993)
High starch-conversion potential	Increased yield of fermentable sugar per unit weight of barley: enhance limit dextrinase and/or optimize amylose/amylopectin ratios	No published reference	
High foam potential	Quality improvements in beer	No published reference	
Low haze potential	Quality improvements in beer	Low anthocyanogen barley – improvements needed to agronomic properties	von Wettstein *et al.* (1985)
Dormancy and vigour control	Ability to use barley in maltings, as and when required	No published reference – fundamental physiology not fully evaluated	
Modifiability	Even and rapid modification of endosperm during germination	No published reference – fundamental physiology not fully evaluated	
Low protein accumulation	Low protein, therefore high starch	No published reference	
High flavour stability product	Quality improvements in beer	No published reference – fundamental physiology not fully evaluated	
Flavour control	Elimination of precursors of possibly undesirable flavours (e.g. dimethyl sulphide)	No published reference	

Hops

Although a relatively small contributor to the cost of beer production, hops have a disproportionate influence on quality, notably in terms of bitterness and aroma. As a consequence, the majority of brewers are conservative with regard to their 'hop bill' and generally accept new varieties with reluctance.

Notwithstanding this, there are several distinct improvements which brewers would like to see in their preferred varieties. Such characteristics cannot be introduced by traditional breeding techniques without severe risk to perceived quality attributes. The only hope has been selection of variants which display the new properties, while retaining all desirable features (so-called 'clonal variation'). There is clearly a substantial chance element to this approach, and for this reason, modern gene technology holds exciting possibilities for hops.

A principal aim is the introduction into hops of resistance to infection and infestation. The cultivated hop (*Humulus lupulus*) is susceptible to powdery mildew (*Pseudoperononspora humuli*) and to the damson hop aphid (*Phorodon humuli*). Its Far Eastern counterpart *H. japonicus* is not. Consequently, there is a considerable interest in transferring the relevant resistance genes. However, while *H. japonicus* can pollinate *H. lupulus*, the resultant embryos abort (Winge, 1914). Nevertheless, there would be some promise if an appropriate culture medium could be identified on which to activate immature 'cross'-embryos, such as was first demonstrated for flax (*Linum*; Laibach, 1925). Having said this, certain varieties of *H. lupulus*, such as Wye Target, contain dominant genes, notably R2, which confer resistance to powdery mildew, so that this may be a useful alternative.

Protoplast formation, fusion and genetic modification by electroporation have all been demonstrated for *H. lupulus*. As yet it has not proved possible to regenerate plants beyond the microcallus stage. However, it is known that hop plants can be regenerated from callus cultures (Heale *et al.*, 1989).

The problem is rather one of which callus lines to select. Evaluation of new varieties is a prolonged affair, which must take into account the considerable impact which growth site and conditions have on expression. Not only this, there is immense undesirability in the methods which need to be used to select for disease resistance, viz growing in strictly quarantined, highly infectious conditions.

As yet there are no suitable selection markers available which can be used as indicators of desired characters. Perhaps restriction fragment length polymorphisms (RFLPs) provide the likeliest opportunity (Gunn, 1989). Certainly, RFLP analysis did detect clear polymorphisms in isolates of *Verticillium albo-atrum*, the organism which causes hop wilt (Heale *et al.*, 1989).

The bitterness components of beer originate from the α-acids, which are normally found only in the cones of female plants. Crossing experiments have confirmed the genetic potential for maximizing levels of α-acids in hops (Gunn, 1988). The genes for α-acid production are found in the male plant as well; the future goal must surely be to identify markers for those genes in order to facilitate crossing of male and female with high α-acid potential.

Brewers' yeasts

The basic purpose of fermentation has not changed over the centuries that beer has been produced: it is the conversion of wort, chiefly into an alcoholic, pleasant tasting beer. This is brought about by the activity of brewers' yeast, which converts sugars into ethanol and carbon dioxide, together with small quantities of highly flavoured compounds which give beers their characteristic flavours. The basic biochemistry of these processes is well understood, although the detailed mechanisms underlying the production of many flavour compounds is still a fruitful source of much basic research. Similarly, the factors determining flocculation of yeast from fermenting beer, to facilitate clarification, are only now being elucidated. The current state of knowledge of yeast physiology has been described in some detail in a previous volume in this series (Boulton, 1991), and so the present authors will attempt not to tread the same ground.

Brewers' yeasts all belong to the species *Saccharomyces cerevisiae*. However, they are quite distinct from the well-characterized laboratory strains of the same species. Unlike the latter, which have been bred for their ease of handling in laboratories, brewers' yeasts have been selected for their performance in brewery fermenters. Little effort has been devoted to rendering them amenable to genetic analysis and so laboratory-based breeding programmes have not, until recently, featured in the work schedules of brewing laboratories. This has changed with the advent of 'genetic engineering' and many laboratories have embarked upon work to develop new strains of brewing yeast with defined properties.

THE SCOPE FOR IMPROVING BREWERS' YEASTS

The properties which are of most importance for a good brewers' yeast strain have been described by Tubb and Hammond (1987). They inclue the ability to ferment wort sugars to ethanol rapidly and efficiently, without excessive yeast growth. They need to be able to withstand the high osmotic pressure and significant levels of carbon dioxide and ethanol encountered in brewery fermenters. They must also produce the required levels of flavour compounds, while being easy to separate from the beer at the end of fermentation. Once separated, they must remain healthy during storage before they are re-used. There is considerable scope for improving the behaviour of traditional brewing yeasts in a number of these areas, either by modifying their physiology or by extending their biochemical capabilities.

TRADITIONAL BREEDING METHODS

The breeding of laboratory strains of yeast involves sporulation and hybridization of cells. For many years, this was not possible with brewers' yeasts, since they sporulate poorly and many of the spores that are produced are non-viable (Anderson and Martin, 1975). Recently, techniques have been developed to improve the sporulation characteristics of brewers' yeasts (Gjermansen and Sigsgaard, 1981; Bilinski, Russell and Stewart, 1986, 1987a), but they still sporulate poorly when compared with laboratory strains. Using these techniques, meiotic segregants of brewers' yeasts have been mated and a number of hybrids obtained (Von Wettstein, 1983; Bilinski, Russell and Stewart, 1987b). Although these approaches have proved successful, much work is involved in selecting hybrids with suitable brewing properties and, accordingly, few laboratories are pursuing this approach.

Direct selection of mutants with or without prior mutagenesis has been used to obtain yeasts with altered flocculation and flavour production characteristics (Molzahn, 1977). The selection of mutants resistant to 2-deoxyglucose has yielded derepressed mutants which are able to ferment maltose and glucose simultaneously, increasing the speed and efficiency of fermentation (Jones, Russell and Stewart, 1986).

By mixing large numbers of yeast cells together and by applying a strong positive selection pressure, it is possible to produce hybrids between brewers' yeasts and other yeast strains. The technique is known as rare-mating (Gunge and Nakatomi, 1972) and has been used to produce brewers' yeasts with the ability to ferment dextrins (Tubb *et al.*, 1981) and strains with anti-contaminant properties (Young, 1981, 1983; Hammond and Eckersley, 1984). The technique has been largely superseded by more specific recombinant DNA methods.

Spheroplast fusion provides an alternative means for overcoming the mating difficulties experienced with brewers' yeasts. The cell walls of yeasts are removed enzymatically and the resulting spheroplasts induced to fuse either chemically (van Solingen and van der Plaat, 1977) or electrically (Halfmann *et al.*, 1982). Spheroplast fusion has been used to produce dextrin-fermenting brewers' yeasts (Barney, Jansen and Helbert, 1980; Hansen, Rocken and Emeis, 1990), yeasts with both anti-yeast and anti-bacterial activities (Sasaki *et al.*, 1984), as well as yeasts with changed flocculation character (Urano, Nishikawa and Kamimura, 1990). One problem often experienced with fusion hybrids is that they have poor brewing properties because of the contribution to their genome from the non-brewing strain. Considerable work is required to overcome this and, again, this is a technique largely superseded by more specific methods.

GENETIC ENGINEERING

The use of DNA molecules to bring about a change in a recipient organism bypasses the mating system and so can be readily applied to brewers' yeasts. With the development of recombinant DNA techniques in the 1970s, such transformation procedures became feasible and have been widely applied to the breeding of new strains of brewers' yeasts. It is normal when carrying out transformation experiments with haploid or diploid yeasts to employ strains carrying auxotrophic markers in order to simplify the identification of transformed cells among the background of parental organisms. However, since brewers' yeasts are polyploid, it is extremely difficult to produce auxotrophic mutants and so the plasmids used for transformation of brewers' yeasts normally carry dominant selection markers.

For brewing applications, the most widely used selection marker is *CUP1*, which confers copper resistance upon cells harbouring multiple copies of the gene (Henderson, Cox and Tubb, 1985). Almost all brewers' yeasts are sensitive to copper and become resistant when transformed with plasmids containing the *CUP1* gene. Their fermentation behaviour is unaltered and the beer produced is identical to that normally produced (Meaden and Tubb, 1985). Another useful dominant resistance marker is *SMR1*, which encodes resistance to the herbicide sulphometuron-methyl (Casey, Xiao and Rank, 1988a). This is a mutant allele of the yeast *ILV2* gene, which encodes the enzyme acetolactate synthase. Since the *ILV2* gene is a normal part of the genome of a brewers' yeast, the mutant version can not only act as a selectable marker but can also be used as a way of directing integration of DNA into the yeast chromosome (Casey, Xiao and Rank, 1988b). Both the *CUP1* and *SMR1* genes are derived from *Saccharomyces cerevisiae* and so are not 'foreign' to brewers' yeasts, an important consideration when modified yeasts are being considered for commercial production and approval for their use is being sought.

Two other dominant selectable markers widely used are those coding for resistance to geneticin (G418) (Jimenez and Davies, 1980) and to chloramphenicol (Hadfield, Cashmore and Meacock, 1986). Both are derived from bacteria and so, once they have been used for selection purposes, must be jettisoned from a transformed yeast before it can be considered for commercial use. The unaltered bacterial G418 resistance gene is expressed in yeasts but, for efficient expression, it is necessary to place the gene under the control of a yeast promoter. Resistance to chloramphenicol is a mitochondrial function and so can only be used when transformed yeast cells can be grown on a non-fermentable substrate. Despite these complications, both resistance genes have been successfully employed in a number of breeding programmes with brewers' yeasts.

Most early transformation experiments with brewers' yeasts employed multicopy plasmids which led to many copies of the gene of interest being inserted into the host organism. This usually resulted in particularly good expression of the inserted gene. Such over-expression did however, on occasion, have deleterious effects on the fermentation behaviour of the transformed yeast (Perry and Meaden, 1988). The genetic stability of yeasts transformed with multi-copy plasmids is not good, the new property gradually being lost with successive fermentations (Meaden and Tubb, 1985; Cantwell *et al.*, 1985).

More recently, transformation of brewers' yeasts has been carried out by integration of the gene of interest, either into a chromosome (Enari *et al.*, 1987) or into the endogenous 2 μm plasmids (Hinchliffe, Fleming and Vakeria, 1987). Chromosomal integrants are stable but, because of the lower copy number, the level of expression can be less than with multi-copy transformants (Enari *et al.*, 1987). This can be partially overcome by integrating the gene of interest at several sites within the yeast chromosomes. Alternatively, a highly efficient yeast promoter, typically one involved in controlling a gene from the glycolytic pathway, can be employed to increase the level of expression. This latter approach has now been almost universally adopted.

DIACETYL-LESS BREWERS' YEASTS

Diacetyl is a major off-flavour in finished beers. It is formed by chemical oxidation of α-acetolactate, which diffuses from yeast cells during fermentation. The need to remove diacetyl is the major reason for the long flavour maturation period at the end of fermentation. In order to eliminate this costly process, various genetic approaches have been proposed to prevent or control diacetyl formation. The rate-limiting step is conversion of α-acetolactate to diacetyl, and so most solutions to the problem have involved either removing α-acetolactate before it is converted to diacetyl or preventing its formation in the first place.

The removal of α-acetolactate can be brought about by converting it to acetoin using the bacterial enzyme acetolactate decarboxylase (ALDC). The gene encoding this enzyme has been cloned from several bacterial species. In initial trials, the gene from *Enterobacter aerogenes* was expressed off the yeast *ADH1* promoter. When transformed brewers' yeasts were used for fermentation, reduced levels of diacetyl were produced compared with control fermentations (Shimizu, Sone and Inoue, 1989). There appeared to be no effect on other aspects of yeast metabolism. In subsequent experiments, the ALDC gene was integrated into the rDNA genes of brewers' yeasts. Because of the many copies of rDNA present in the yeast genome, the ALDC gene was integrated more than 20 times into each cell and consequently all transformants showed a high enzyme activity (Fujii *et al.*, 1990). Again, beers made in laboratory-scale fermentation trials had very low diacetyl contents. The work was extended to investigate the effects of yeast promoters on gene expression and genes from other bacteria were included. Cloned ALDC genes from two strains of *E. aerogenes* and from one *Klebsiella terrigena* strain were placed under the control of both *ADH1* and *PGK* promoters and terminators and were used to transform brewers' yeasts by integration into their *rDNA*, *ADH1* and *PGK* genes. All the transformants fermented well and two, in which the *ALDC* gene had been integrated at the *PGK* site under the control of the *PGK* promoter and terminator, produced beer in which the diacetyl level after primary fermentation was below the taste threshold. The finished beer was of good quality and the transformed yeasts were completely stable (Suihko *et al.*, 1989, 1990; Blomqvist *et al.*, 1991).

Modification of the amino-acid biosynthetic pathway leading to the formation of α-acetolactate (*Figure 1*) is another approach which has been used in the control of diacetyl production. The *ILV* genes coding for the various enzymes have been cloned (Petersen *et al.*, 1983) and attempts have been made to regulate beer diacetyl levels by manipulating these genes in brewing organisms. Mutant yeasts, lacking the enzymes threonine deaminase and acetolactate synthase, produce no diacetyl but ferment very poorly, presumably because of their inability to synthesize several amino acids (Ryder and Masschelein, 1983). This difficulty can be overcome by only partially reducing the activity of acetolactate synthase. Dominant mutations of the *ILV2* gene are known which prevent sulphometuron methyl inhibiting acetolactate synthase activity. They do this by reducing the binding affinity of the enzyme for the herbicide (Falco and Dumas, 1985). Some of the resistant forms of the enzyme are less active and so should produce less α-acetolactate and hence less diacetyl. Spontaneous dominant mutants of this type have been isolated but reductions in diacetyl levels were disappointingly small (Gjermansen *et al.*, 1988; Galvan *et al.*, 1987). More recently, a much more complex procedure has been developed for the production of recessive mutations in the *ILV2* locus of brewing yeasts (Kielland-Brandt *et al.*, 1989). Strains have been produced having very low acetolactate synthase activities by screening mutagenized, spontaneous herbicide-resistant, meiotic segregants from brewing yeasts for slow growth on media lacking isoleucine and valine. Hybrids, produced by mating these strains, ferment well, produce low levels of diacetyl and yield acceptable beer (P. Sigsgaard, pers. comm.). Another approach to reducing the activity of acetolacetate synthase has involved transforming brewers' yeast with a fragment of the *ILV2* gene in such a way that anti-sense mRNA was made (Vakeria, Box and Hinchliffe, 1991). The level of normal acetolactate synthase mRNA was thereby reduced, and less enzyme was synthesized. Unfortunately, although less diacetyl was produced, the fermentation performance of the transformed yeast was poor.

Another possible way of reducing diacetyl levels is by increasing the flux through the pathway leading to valine production, which would result in any excess α-acetolactate being consumed. This has been attempted by transforming yeasts with multi-copy plasmids containing copies of either *ILV3* or *ILV5* genes. Yeasts transformed with *ILV5* had increased activity of acetolactate reductoisomerase and produced considerably less diacetyl than did parent yeasts (Dillemans *et al.*, 1987; Villanueba, Goossens and Masschelein, 1990; Goossens *et al.*, 1991, 1993). Those transformed with *ILV3* produced just as much diacetyl as did their parents, despite a considerable increase in the corresponding enzyme activity (Goossens *et al.*, 1987).

It is clear that a number of different approaches to producing diacetyl-less brewers' yeasts have been used with different degrees of success. Technically, the ALDC-producing strains are probably the most successful, but their application in breweries must await clarification of the legislative position.

AMYLOLYTIC BREWERS' YEASTS

Normal brewing yeasts cannot utilize the dextrins, which represent about 25% of wort sugars. Yeasts able to ferment these additional sugars are more efficient than brewing yeasts and can be used to produce both low carbohydrate diabetic beers and lower calorie 'light' beers. At present, these types of beers are often made by the addition of fungal glucoamylases derived from *Aspergillus niger*. Amylolytic brewing yeasts offer a number of advantages,

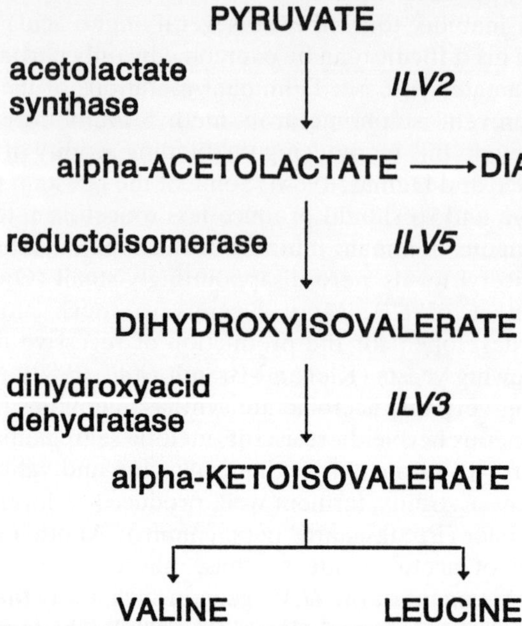

Figure 1. Diacetyl formation and amino acid biosynthesis. Diacetyl is formed as a by-product of the biosynthetic pathway for valine and leucine. The *ILV* genes and the enzymes coded by them are indicated on the pathway.

since mistakes in enzyme addition are eliminated and other deleterious activities which are often present in commercial enzyme preparations are avoided.

Various approaches to the production of amylolytic brewers' yeasts have been adopted. A glucoamylase gene (*DEX1* or *STA2*) has been cloned from *Saccharomyces cerevisiae* var. *diastaticus* (Meaden *et al.*, 1985) and inserted into a multi-copy plasmid. Both ale and lager strains have been transformed with this plasmid using *CUP1* as the selectable marker (Meaden and Tubb, 1985). The transformants produced extracellular glucoamylase but, because yeast glucoamylase lacks α-1,6 debranching activity, complete superattenuation was not achieved. The plasmid-based transformants were not stable and both growth and fermentation rates were slower than in the parent organisms (Perry and Meaden, 1988). Plasmid instability was much improved by the development of an all-yeast multi-copy plasmid (Vakeria and Hinchliffe, 1989), but the fermentation rate of the transformant was still slow. The *STA1* glucoamylase gene has also been cloned from *S. cerevisiae* var. *diastaticus* and used to transform brewing yeasts (Sakai *et al.*, 1989; Park *et al.*, 1990). Poor plasmid stabilities and slower fermentations were again observed. Only when the gene was integrated into the yeast chromosome was a stable transformant obtained (Park *et al.*, 1990).

The lack of α-1,6 debranching activity in glucoamylases derived from *S. cerevisiae* var. *diastaticus* is a major problem that can only be overcome by using genes for glucoamylases from other microbial sources. The *GA* gene from *Aspergillus niger* has been cloned and inserted into an integrating vector (Yocum, 1986). This gene encodes a glucoamylase with both α-1,4 and α-1,6 activities. The vector used for transformation of brewers' yeast was designed to achieve integration of the *GA* gene into the yeast genome while excluding bacterial DNA sequences (*Figure 2*). To avoid instability problems caused by recombination, three copies of the *GA* gene were integrated into three different copies of the yeast *HO* gene. Gene expression and protein secretion were optimized by the use of a yeast glycolytic enzyme promoter and a secretion signal sequence, respectively. Fermentation trials in 100 litre fermenters have been carried out successfully: high levels of extracellular glucoamylase have been obtained and good-quality 'light beer' has been produced (Gopal and Hammond, 1992).

The glucoamylase from *A. niger* is stable at pasteurization temperatures, retaining its activity in packaged beer. Therefore, if any undigested dextrin remains in the beer, this can cause difficulties, the beer gradually becoming sweeter as glucose is produced. The glucoamylase from *Schwanniomyces occidentalis*, unlike the *Saccharomyces* enzyme, possesses debranching activity and is more heat-labile than is the fungal enzyme. It is not, however, commercially available, and so is not currently used for the production of low-carbohydrate beers. The gene for this enzyme has been cloned and placed under the control of the *Saccharomyces ADH1* promoter (Lancashire *et al.*, 1989). Both integrated and multi-copy transformants have been produced, which successfully fermented brewery worts to a lower gravity than normal and produced acceptable beers.

GLUCANOLYTIC BREWERS' YEASTS

Barley β-glucanase is unstable at the temperatures found during malt kilning and so β-glucans can sometimes be found in beer due to the low level of enzyme activity during wort production. Large molecular weight glucans in beer can cause filtration difficulties and give rise to sediments, gels and hazes. Microbial β-glucanases are often added to mashes or worts to prevent these problems and so there has been much interest in transferring into brewing yeasts the gene for endo-1,3-1,4-β-D-glucanase. This gene has been cloned from *Bacillus subtilis* (Cantwell and McConnell, 1983; Hinchliffe and Box, 1984; Lancashire and Wilde, 1987), *Trichoderma reesei* (Knowles *et al.*, 1985) and barley (Jackson, Ballance and Thomsen, 1986). At first, only low levels of intracellular enzyme activity were obtained from yeasts transformed with the *B. subtilis* gene (Hinchliffe and Box, 1985). When the β-glucanase gene was placed under the control of the yeast *ADH1* promoter (Cantwell *et al.*, 1985), high levels of β-glucanase were produced intracellularly by transformed brewing yeasts but only very low levels were detected in the beer. Significant levels of extracellular β-glucanase were finally obtained when the bacterial gene was placed under the control of the yeast α-mating factor signal sequence (Lancashire and Wilde, 1987). The experimental beers produced using these last transformants had much lower β-glucan contents and viscosities than did control beers and their filterabilities were much improved.

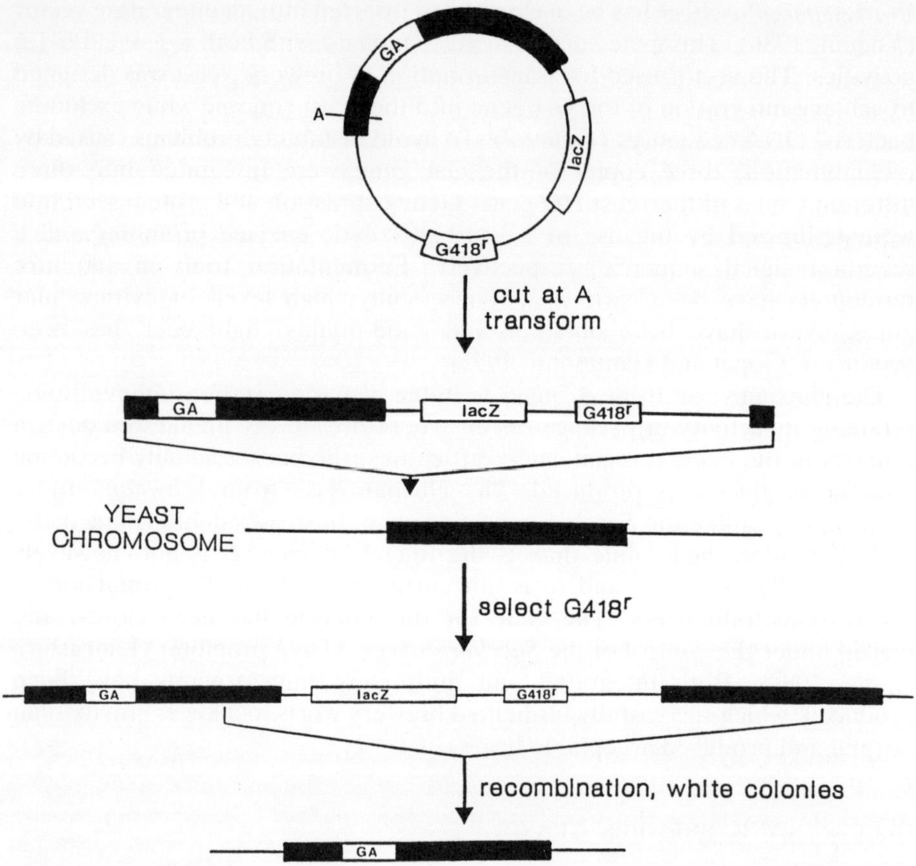

Figure 2. Transformation protocol used to achieve stable integration of the *Aspergillus niger* glucoamylase gene into brewers' yeast. The *GA* gene is inserted into a copy of the yeast *HO* gene contained on a plasmid, which also contains copies of the *LacZ* and G418 resistance genes from bacteria. The plasmid is cut at A and used to transform brewers' yeast. Transformants are selected by their ability to grow on plates containing G418. Spontaneous recombination between the two copies of the *HO* gene can lead to loss of the bacterial DNA, while the *GA* gene is retained. This recombination can be detected by virtue of the simultaneous loss of the *LacZ* gene. The desired recombinants can then be detected by screening for glucoamylase activity.

When the β-glucanase gene from *T. reesei* was cloned and used for transformation of brewing yeasts, it was placed under the control of the yeast *PGK* promoter. This was then inserted into both multi-copy and integrating plasmids (Enari *et al.*, 1987). A yeast secretion signal sequence was unnecessary since, unlike their bacterial equivalents, fungal extracellular enzymes are efficiently excreted by yeasts. With both multi-copy and integrated transformants, wort β-glucans were efficiently degraded and beer filterability was much improved, while fermentation and beer characteristics were largely unaltered (Enari *et al.*, 1987; Penttila *et al.*, 1987).

The β-glucanase from *B. subtilis* shows highest enzyme activity at a pH value of 6.7, whereas the barley enzyme is most active at pH 4.7, a value which more closely resembles those encountered during the brewing process.

The barley enzyme would therefore appear attractive for use in transformed yeasts. The barley gene has been cloned by joining two separate DNA sequences isolated from barley aleurone, which together encompassed the barley β-glucanase gene. The gene was then fused to a mouse α-amylase secretion signal sequence and inserted into an expression vector between yeast *ADH1* promoter and terminator sequences (Jackson, Ballance and Thomsen, 1986). Yeast transformants all produced extracellular β-glucanase (Thomsen, Jackson and Brenner, 1988). A similarly transformed brewers' yeast has been used to produce beer in a pilot plant. The β-glucan content of the beer was much lower than usual, flavour and foam stability were unaffected, and filtration rates were markedly increased (Berghof and Stahl, 1991).

As discussed earlier, however, a much more attractive option is the insertion of a heat-stable β-glucan into barley. The earlier the enzyme is introduced, the more benefits it has. β-Glucanase introduced into yeast can have no benefits in wort separation, for instance. Furthermore, there are clear attractions to having foreign enzymes destroyed in the kettle boil.

FLOCCULENT BREWERS' YEASTS

The factors controlling flocculation of brewing yeasts during fermentation are still far from clear (Boulton, 1991) and the genes involved have not been identified with any certainty. A number of genes (*FLO1*, *FLO5*, *FLO8* and *tup1*) are known to affect flocculation, but their presence in brewers' yeasts is still the subject of much research. Brewers' yeasts appear to flocculate by two different mechanisms (Stratford and Assinder, 1991). The flocculation behaviour of bottom-fermenting lager strains resembles that of laboratory strains containing *FLO* genes and is different from that of top-fermenting ale strains. This suggests that flocculation of ale strains may be caused by something other than the *FLO* genes. The involvement of *FLO1* in the flocculation of bottom-fermenting yeasts has been supported by hybridization studies (Watari *et al.*, 1991a). Recently, the *FLO1* gene product has been tentatively identified as a hydrophobic cell wall protein (van der Aar, Straver and Teunissen, 1993).

The *FLO1* gene has been cloned (Watari *et al.*, 1989) and used to transform both top- and bottom-fermenting non-flocculent yeasts. The transformed yeasts were more flocculent than their parents (Watari *et al.*, 1991b) and fermented satisfactorily (Watari *et al.*, 1991a). As far as lager yeasts are concerned, it would seem that transformation provides a mechanism for manipulating flocculation characteristics.

THE GENETIC MAKE-UP OF A BREWERS' YEAST

The development of techniques for the modification of brewers' yeasts has taught us much of the structure of the genome of these organisms. The Carlsberg lager yeast strain M244 has been examined in considerable detail and much fascinating information obtained. This has been made possible by the technique of single chromosome transfer (Nilsson-Tillgren *et al.*, 1980), whereby single chromosomes have been transferred from meiotic segregants of the lager yeast into laboratory strains, thereby allowing detailed genetic analysis of the brewers' yeast chromosomes to be carried out. Chromosomes III, V, X, XII and XIII have been studied in this way.

Within the lager yeast there are two forms of chromosome III: one is functionally and structurally identical to the chromosome III found in laboratory strains, whereas the other is functionally homologous to the laboratory-strain chromosome but is structurally different. Recombination between the two chromosome types only occurs in certain regions (Nilsson-Tillgren *et al.*, 1981; Holmberg, 1982). Similarly, two forms of chromosome V have been identified: one is both structurally and functionally homologous to the chromosome of *Saccharomyces cerevisiae*, while the other is functionally homologous but fails to recombine at all (a homeologous chromosome) (Nilsson-Tillgren *et al.*, 1986).

Three forms of chromosome X have been identified, two of which are not homologous to chromosome X from *S. cerevisiae*, recombination occurring along only part of the chromosome (Casey, 1986a,b). The third chromosome, although the same length as that from laboratory strains, has not been examined in any detail. Chromosome XII occurs in two forms, one homologous and the other homeologous to the equivalent chromosome from laboratory strains (Petersen *et al.*, 1987). In contrast, three different chromosomes XIII exist, one showing homology to that of *S. cerevisiae*, a non-recombining homeologous chromosome and a mosaic-type chromosome exhibiting partial recombination (Petersen *et al.*, 1987).

The chromosome structure of lager yeast M244 is clearly extremely complicated and it appears to have two recombination-incompatible genomes. Direct examination of the number of copies of the *ILV2* gene in the Carlsberg lager yeast suggests that it contains two copies of each of the two verions of the *ILV2* region (Gjermansen *et al.*, 1988). If this is true of all genes, then the brewing yeast would appear to be allotetraploid, although the alloploidy is obviously irregular because of the occurrence of mosaic chromosomes. It also appears that some chromosomes lack certain genes (Nilsson-Tillgren *et al.*, 1986; Casey, 1986b). If all brewing yeasts are allotetraploids with irregular chromosome structures and missing or defective genes, this may explain their poor sporulation behaviour.

The future

The techniques of modern yeast genetics have been applied to both unravelling the detailed structure of the brewers' yeast genome and to the construction of new yeast strains for the more efficient production of beer. Many of these new strains have been technically very successful and have been used to produce beer in pilot plants. Similar advances have been made in developing new barley and hop varieties. The major barrier to the commercial exploitation of these new plants and yeasts is the classification of the beers produced by them as 'novel foods', together with a requirement for government approval for their use (Hammond, 1991). There is a pressing need for brewing technologists to devote time and effort to ensuring that this barrier is overcome, and that genetic modification comes to be seen as a normal part of the commercial development of the brewing process. If this does not happen, the application of genetic modification techniques in the brewing industry will be severely curtailed and the benefits available will be lost.

Acknowledgement

The Director-General of BRF International is thanked for permission to publish.

References

AASTRUP, S., ERDAL, K. AND MUNCK, L. (1985). Low β-glucan barley mutants and their malting behaviour. In *Proceedings of the European Brewery Convention Congress*, Helsinki, pp. 387–393.

ANDERSON, E. AND MARTIN, P.A. (1975). The sporulation and mating of brewing yeasts. *Journal of the Institute of Brewing* **81**, 242–247.

BAMFORTH, C.W. (1993). Brewing distilled for the biochemist. *The Biochemist* **15**, 2–6.

BAMFORTH, C.W. AND BARCLAY, A.H.P. (1993). Malting technology and the uses of malt. In *Barley: Chemistry and Technology* (A.W. MacGregor and R.S. Bhatty, Eds) pp. 297–354. American Association of Cereal Chemists, St Paul, Minnesota.

BARNEY, M.C., JANSEN, G.P. AND HELBERT, J.R. (1980). Use of spheroplast fusion and genetic transformation to introduce dextrin utilization into *Saccharomyces uvarum. Journal of the American Society of Brewing Chemists* **38**, 1–5.

BERGHOF, K. AND STAHL, U. (1991). Improving the filterability of beer by the use of β-glucanase active brewers' yeast. *BioEngineering* **7**, 27–32.

BILINSKI, C.A., RUSSELL, I. AND STEWART, G.G. (1986). Analysis of sporulation in brewer's yeast: Induction of tetrad formation. *Journal of the Institute of Brewing* **92**, 594–598.

BILINSKI, C.A., RUSSELL, I. AND STEWART, G.G. (1987a). Physiological requirements for induction of sporulation in lager yeast. *Journal of the Institute of Brewing* **93**, 216–219.

BILINSKI, C.A., RUSSELL, I. AND STEWART, G.G. (1987b). Cross-breeding of *Saccharomyces cerevisiae* and *Saccharomyces uvarum* (*carlsbergensis*) by mating of meiotic segregants: Isolation and characterization of species hybrids. *Proceedings of the European Brewery Convention Congress*, Madrid, pp. 497–504.

BLOMQVIST, K., SUIKHO, M.-L., KNOWLES, J. AND PENTTILA, M. (1991). Chromosomal integration and expression of two bacterial α-acetolactate decarboxylase genes in brewer's yeast. *Applied and Environmental Microbiology* **57**, 2796–2803.

BOULTON, C.A. (1991). Developments in brewery fermentations. *Biotechnology and Genetic Engineering Reviews* **9**, 355–410.

CANTWELL, B.A. AND MCCONNELL, D.J. (1983). Molecular cloning and expression of a *Bacillus subtilis* β-glucanase gene in *Escherichia coli. Gene* **23**, 211–219.

CANTWELL, B., BRAZIL, G., HURLEY, J. AND MCCONNELL, D. (1985). Expression of the gene for the endo-β-1,3-1,4-glucanase from *Bacillus subtilis* in *Saccharomyces cerevisiae*. In *Proceedings of the European Brewery Convention Congress*, Helsinki, pp. 259–266.

CASEY, G.P. (1986a). Cloning and analysis of two alleles of the *ILV3* gene from *Saccharomyces carlsbergenis. Carlsberg Research Communications* **51**, 327–341.

CASEY, G.P. (1986b). Molecular and genetic analysis of chromosomes X in *Saccharomyces carlsbergensis. Carlsberg Research Communications* **51**, 343–362.

CASEY, G.P., XIAO, W. AND RANK, G.H. (1988a). A convenient dominant selection marker for gene transfer in industrial strains of *Saccharomyces yeast: SMR1* encoded resistance to the herbicide sulfometuron methyl. *Journal of the Institute of Brewing* **94**, 93–97.

CASEY, G.P., XIAO, W. AND RANK, G.H. (1988b). Construction of α-galactosidase-positive strains of industrial baker's (*Saccharomyces cerevisiae*) yeasts. *Journal of the American Society of Brewing Chemists* **46**, 67–71.

CLOSE, T.J., KORTT, A.A. AND CHANDLER, P.M. (1989). A cDNA-based comparison of dehydration-induced proteins (dehydrins) in barley and corn. *Plant Molecular Biology* **13**, 95–108.

DEIKMAN, J. AND JONES, R.L. (1986). Regulation of the accumulation of mRNA for α-amylase isoenzymes in barley aleurone. *Plant Physiology* **80**, 672–675.

DE LA PENA, A., LORZ, H. AND SCHELL, J. (1987). Transgenic rye plants obtained by injecting DNA into young floral tillers. *Nature* **325**, 274–276.

DILLEMANS, M., GOOSSENS, E., GOFFIN, O. AND MASSCHELEIN, C.A. (1987). The amlification effect of the *ILV5* gene on the production of vicinal diketones in *Saccharomyces cerevisiae. Journal of the American Society of Brewing Chemists* **45**, 81–84.

ENARI, T.M., KNOWLES, J., LEHTINEN, U., NIKKOLA, M., PENTTILA, M., SUIHKO, M.L., HOME, S. AND VILPOLA, A. (1987). Glucanolytic brewer's yeast. In *Proceedings of the European Brewery Convention Congress*, Madrid, pp. 529–536.

FALCO, S.C. AND DUMAS, K.S. (1985). Genetic analysis of mutants of *Saccharomyces cerevisiae* resistant to the herbicie sulfometuron methyl. *Genetics* **109**, 21–35.

FONG, F., SMITH, J.D. AND KOEHLER, D.E. (1983). Early events in maize seed development: 1-Methyl-3-phenyl-5-(3-[trifluoromethyl]phenyl)-4-(1H)-pyridinone induction of vivipacy. *Plant Physiology* **73**, 899–901.

FUJII, T., KONDO, K., SHIMIZU, F., SONE, H., TANAKA, J.-I. AND INOUE, T. (1990). Application of a ribosomal DNA integration vector in the construction of a brewer's yeast having α-acetolactate decarboxylase activity. *Applied and Environmental Microbiology* **56**, 997–1003.

GALVAN, L., PEREZ, A., DELGADO, M. AND CONDE, J. (1987). Diacetyl production by sulfometuron resistant mutants of brewing yeast. In *Proceedings of the European Brewery Convention Congress*, Madrid, pp. 385–392.

GJERMANSEN, C. AND SIGSGAARD, P. (1981). Construction of a hybrid brewing strain of *Saccharomyces carlsbergensis* by mating of meiotic segregants. *Carlsberg Research Communications* **46**, 1–11.

GJERMANSEN, C., NILSSON-TILLGREN, T., PETERSEN, J.G.L., KIELLAND-BRANDT, M.C., SIGSGAARD, P. AND HOLMBERG, S. (1988). Towards diacetyl-less brewers' yeast: Influence of *ilv2* and *ilv5* mutations. *Journal of Basic Microbiology* **28**, 175–183.

GOOSSENS, E., DILLEMANS, M., DEBOURG, A. AND MASSCHELEIN, C.A. (1987). Control of diacetyl formation by the intensification of the anabolic flux of acetohydroxyacid intermediates. In *Proceedings of the European Brewery Convention Congress*, Madrid, pp. 553–560.

GOOSSENS, E., DEBOURG, A., VILLANUEBA, K.D. AND MASSCHELEIN, C.A. (1991). Decreased diacetyl production by site directed integration of the *ILV5* gene into chromosome XIII of *Saccharomyces cerevisiae*. In *Proceedings of the European Brewery Convention Congress*, Lisbon, pp. 289–296.

GOOSSENS, E., DEBOURG, A., VILLANUEBA, K.D. AND MASSCHELEIN, C.A. (1993). Decreased diacetyl production in lager brewing yeast by integration of the *ILV5* gene. In *Proceedings of the European Brewery Convention Congress*, Oslo, pp. 251–258.

GOPAL, C.V. AND HAMMOND, J.R.M. (1992). Use of genetically-modified yeasts for beer production. In *Proceedings of the 5th International Brewing Technology Conference*, Harrogate, pp. 297–306. Brewing Technology Services Ltd, London.

GUNGE, N. AND NAKATOMI, Y. (1972). Genetic mechanisms of rare matings of the yeast *Saccharomyces cerevisiae* heterozygous for mating type. *Genetics* **70**, 41–58.

GUNN, R.E. (1988). Hop breeding in England. *Monograph XIII, European Brewery Convention*, pp. 46–53.

GUNN, R.E. (1989). The application of biotechnology to hops. *Monograph XV, European Brewery Convention*, pp. 59–69.

HADFIELD, C., CASHMORE, A.M. AND MEACOCK, P.A. (1986). An efficient chloramphenicol-resistance marker for *Saccharomyces cerevisiae* and *Escherichia coli. Gene* **45**, 149–158.

HALFMANN, H.J., ROCKEN, W., EMEIS, C.C. AND ZIMMERMANN, U. (1982). Transfer of mitochondrial function into a cytoplasmic respiratory-deficient mutant of *Saccharomyces* yeast by electro-fusion. *Current Genetics* **6**, 25–28.

HAMMERTON, R.W. AND HO, T.-H.D. (1986). Hormonal regulation of the development of protease and carboxypeptidase activity in barley. *Plant Physiology* **80**, 692–697.

HAMMOND, J.R.M. (1991). The development of brewing processes: The impact of European biotechnology regulations. In *Proceedings of the European Brewery Convention Congress*, Lisbon, pp. 393–400.

HAMMOND, J.R.M. AND ECKERSLEY, K.W. (1984). Fermentation properties of brewing yeast with killer character. *Journal of the Institute of Brewing* **90**, 167–177.

HANSEN, M., ROCKEN, W. AND EMEIS, C.C. (1990). Construction of yeast strains for the production of low-carbohydrate beer. *Journal of the Institute of Brewing* **96**, 125–129.

HEALE, J.B., LEGG, T., BRAR, J., FABB, A. AND BAINBRIDGE, B. (1989). Application of plant tissue culture and molecular biology techniques to 'progressive' wilt of hops caused by *Verticillium albo-atrium. Monograph XV, European Brewery Convention*, pp. 70–83.

HENDERSON, R.C.A., COX, B.S. AND TUBB, R.S. (1985). The transformation of brewing yeasts with a plasmid containing the gene for copper resistance. *Current Genetics* **9**, 133–138.

HIGGINS, T.J.V., JACOBSEN, J.V. AND ZWAR, J.A. (1982). Gibberellic acid and abscisic acid modulate protein synthesis and mRNA levels in barley aleurone layers. *Plant Molecular Biology* **1**, 191–203.

HINCHLIFFE, E. AND BOX, W.G. (1984). Expression of the cloned endo-1,3-1,4-β-glucanase gene of *Bacillus subtilis* in *Saccharomyces cerevisiae. Current Genetics* **8**, 471–475.

HINCHLIFFE, E. AND BOX, W.G. (1985). Beer enzymes and genes: The application of a concerted approach to β-glucan degradation. In *Proceedings of the European Brewery Convention Congress*, Helsinki, pp. 267–274.

HINCHLIFFE, E., FLEMING, C.J. AND VAKERIA, D. (1987). A novel 'stable' gene maintenance system for brewing yeast. In *Proceedings of the European Brewery Convention Congress*, Madrid, pp. 505–512.

HOLM, P., KNUDSEN, S., OLSEN, F.L., JENSEN, J. AND MULLER, M. (1992). Genetics transformation of barley for quality traits. *Ferment* **5**, 417–424.

HOLMBERG, S. (1982). Genetic differences between *Saccharomyces carlsbergensis* and *S. cerevisiae*. II: Restriction endonuclease analysis of genes in chromosome III. *Carlsberg Research Communications* **47**, 233–244.

HOME, S. AND ELAMO, E. (1993). Evaluation of malting potential in barley breeding programmes. *Monatsschift für Brauwissenschaft* **6**, 216–220.

HOUGH, J.S. (1985). The Biotechnology of Malting and Brewing. Cambridge University Press, Cambridge.

ICEC, H., HORSCH, R. AND ROGERS, S. (1987). *Agrobacterium*-mediated plant transformation and its further applications to plant biology. *Annual Review of Plant Physiology* **38**, 467–486.

JACKSON, E.A., BALLANCE, G.M. AND THOMSEN, K.K. (1986). Construction of a yeast vector directing the synthesis and release of barley 1-3,1-4-β-glucanase. *Carlsberg Research Communications* **51**, 445–458.

JACOBSEN, J.V. AND BEACH, L.R. (1985). Control of transcription of α-amylase and rDNA genes in barley aleurone protoplasts by gibberellin and abscisic acid. *Nature* **316**, 275–277.

JIMENEZ, A. AND DAVIES, J. (1980). Expression of a transposable antibiotic resistance element in *Saccharomyces*. *Nature* **287**, 869–871.

JONES, R.M., RUSSELL, I. AND STEWART, G.G. (1986). The use of catabolite derepression as a means of improving the fermentation rate of brewing yeast strains. *Journal of the American Society of Brewing Chemists* **44**, 161–166.

KHURSHEED, B. AND ROGERS, J.C. (1989). Barley α-amylase genes and the thiol protease gene Aleurain: Use of a single Poly(A) addition signal associated with a conserved pentanucleotide at the cleavage site. *Proceedings of the National Academy of Sciences, USA* **86**, 3987–3991.

KIELLAND-BRANDT, M.C., GJERMANSEN, C., NILSSON-TILLGREN, T. AND HOLMBERG, S. (1989). Yeast breeding. In *Proceedings of the European Brewery Convention Congress*, Zurich, pp. 37–47.

KING, R.W. (1976). Abscisic acid in developing wheat grains and its relationship to grain growth and maturation. *Planta* **132**, 43–51.

KNOWLES, J.K.C., PENTTILA, M., TEERI, T.T., ANDRE, L., SALOVUORI, I. AND LEHTOVAARA, P. (1985). Transfer of genes coding for fungal glucanases into yeast. In *Proceedings of the European Brewery Convention Congress*, Helsinki, pp. 251–258.

KOORNNEEF, M. (1986). Genetic aspects of abscisic acid. In *Plant Gene Research: A Genetic Approach to Plant Biochemistry* (R. Blonstein and B. King, Eds), pp. 35–54. Springer-Verlag, Berlin.

LAIBACH, F. (1925). Das Taubwerden von Bastardsamen und die Künstliche Aufzucht früh absterbender Bastardembryonen. *Zeitschrift Botanische* **17**, 417–459.

LANCASHIRE, W.E. AND WILDE, R.J. (1987). Secretion of foreign proteins by brewing yeasts. In *Proceedings of the European Brewery Convention Congress*, Madrid, pp. 513–520.

LANCASHIRE, W.E., CARTER, A.T., HOWARD, J.J. AND WILDE, R.J. (1989). Superattenuating brewing yeast. In *Proceedings of the European Brewery Convention Congress*, Zurich, pp. 491–498.

LAZZERI, P.A. AND LORZ, H. (1988). Production of regenerable suspensions and protoplast cultures of barley (*Hordeum vulgae*). In *Seventh International Congress on Plant Tissue and Cell Culture*, Amsterdam.

LINDSEY, K. AND JONES, M.G.K. (1987). The permeability of electroporated cells and protoplasts of sugar beet. *Planta* **172**, 346–355.

LUO, Z. AND WU, R. (1988). A simple method for transformation of rice via the pollen-tube pathway. *Plant Molecular Biology Reports* **6**, 165–174.

MCCABE, D.E., SWAIN, W.F., MARTINELL, B.J. AND CHRISTOU, P. (1988). Stable transformation of soybean (*Glycine max*) by particle acceleration. *Biotechnology* **6**, 924–926.

MANNONEN, L., KURTEN, V., RITALA, A., SALMENKALLIO-MARTTILA, M., HANNUS, R., KAUPPINEN, V., ASPERGREN, K. AND TEERI, T. (1993). Biotechnology in improvement of malting barley. *Proceedings of the European Brewery Convention Congress*, Oslo, pp. 85–93.

MEADEN, P.G. AND TUBB, R.S. (1985). A plasmid vector system for the genetic manipulation of brewing strains. In *Proceedings of the 20th Congress of the European Brewery Convention*, Helsinki, pp. 219–226.

MEADEN, P., OGDEN, K., BUSSEY, H. AND TUBB, R.S. (1985). A *Dex* gene conferring production of extracellular amyloglucosidase on yeast. *Gene* **34**, 325–334.

MOLINA-CANO, J.L., ROCA DE TOGORES, F., ROYO, C. AND PEREZ, A. (1989). Fast-germinating low beta-glucan mutants induced in barley with improved malting quality and yield. *Theoretical and Applied Genetics* **78**, 748–754.

MOLZAHN, S.W. (1977). A new approach to the application of genetics to brewing yeast. *Journal of the American Society of Brewing Chemists* **35**, 54–59.

MOZER, T.J. (1980). Control of protein synthesis in barley aleurone layers by the plant hormones gibberellic acid and abscisic acid. *Cell* **20**, 479–485.

MUNDY, J. AND CHUA, N.-H. (1988). Abscisic acid and water-stress induce the expression of a novel rice gene. *EMBO Journal* **7**, 2279–2286.

MUNDY, J. AND FINCHER, G.B. (1986). Effects of gibberellic acid and abscisic acid on levels of translatable mRNA (1-3,1-4)β-D-glucanase in barley aleurone. *FEBS Letters* **198**, 349–352.

MUNDY, J. AND MUNCK, L. (1985). Synthesis and regulation of hydrolytic enzymes in germinating barley. In *New Approaches to Research in Cereal Carbohydrates* (R.D. Hill and L. Munck, Eds), pp. 139–148. Elsevier, Amsterdam.

MUNDY, J., YAMAGUCHI-SHINOZAKI, K. AND CHUA, N.-H. (1990). Nuclear proteins bind conserved elements in the abscisic acid-responsive promoter of a rice *rab* gene. *Proceedings of the National Academy of Sciences, USA* **87**, 1406–1410.

MUNDY, J., HEJGAARD, J., HANSEN, A., HALLGREN, L., JORGENSEN, K.G. AND MUNCK, L. (1986). Differential synthesis *in vitro* of barley aleurone and starchy endosperm proteins. *Plant Physiology* **81**, 630–636.

NEUHAUS, G., SPANGENBERG, G., MITTELSTEIN, SCHEID, O. AND SCHWEIGER, H.G. (1987). Transgenic rape seed plants obtained by the micro-injection of DNA into microspore-derived embryoids. *Theoretical and Applied Genetics* **75**, 30–36.

NILSSON-TILLGREN, T., PETERSEN, J.G.L., HOLMBERG, S. AND KIELLAND-BRANDT, M.C. (1980). Transfer of chromosome III during *kar* mediated cytoduction in yeast. *Carlsberg Research Communications* **45**, 113–117.

NILSSON-TILLGREN, T., GJERMANSEN, C., KIELLAND-BRANDT, M.C., PETERSEN, J.G.L. AND HOLMBERG, S. (1981). Genetic differences between *Saccharomyces carlsbergensis* and *S. cerevisiae*: Analysis of chromosome III by single chromosome transfer. *Carlsberg Research Communications* **46**, 65–76.

NILSSON-TILLGREN, T., GJERMANSEN, C., HOLMBERG, S., PETERSEN, J.G.L. AND KIELLAND-BRANDT, M.C. (1986). Analysis of chromosome V and the *ILV1* gene from *Saccharomyces carlsbergensis. Carlsberg Research Communications* **51**, 309–326.

NOLAN, R.-C. AND HO, T.-H.D. (1988). Hormonal regulation of α-amylase expression in barley aleurone layers: The effects of gibberellic acid removal and abscisic acid and phaseic acid treatments. *Plant Physiology* **88**, 558–593.

PARK, C.S., PARK, Y.J., LEE, Y.H., PARK, K.J., KANG, H.S. AND PEK, U.H. (1990). The novel genetic manipulation to improve the plasmid stability and enzyme activity in the recombinant brewing yeast. *Technical Quarterly of the Master Brewers Association of the Americas* **27**, 112–116.

PENTTILA, M.E., SUIHKO, M.L., LEHTINEN, U., NIKKOLA, M. AND KNOWLES, J.K.C. (1987). Construction of brewer's yeasts secreting fungal endo-β-glucanase. *Current Genetics* **12**, 413–420.

PERRY, C. AND MEADEN, P. (1988). Properties of a genetically-engineered dextrin-fermenting strain of brewers' yeast. *Journal of the Institute of Brewing* **94**, 64–67.

PETERSEN, J.G.L., HOLMBERG, S., NILSSON-TILLGREN, T., KIELLAND-BRANDT, M.C. AND GJERMANSEN, C. (1983). Approaches to genetic improvement of brewers' yeast: Cloning of the isoleucine-valine genes. *European Brewery Convention Monograph* **IX**, 30–39.

PETERSEN, J.G.L., NILSSON-TILLGREN, T., KIELLAND-BRANDT, M.C., GJERMANSEN, C. AND HOLMBERG, S. (1987). Structural heterozygosis at genes *ILV2* and *ILV5* in *Saccharomyces carlsbergensis. Current Genetics* **12**, 167–174.

POTRYKUS, I. (1989). Gene transfer to cereals: An assessment. *Trends in Biotechnology* **7**, 268–273.

ROGERS, J.C., DEAN, D. AND HECK, G.R. (1985). Aleurain: A barley thiol protease closely related to mammalian cathepsin H. *Proceedings of the National Academy of Sciences, USA* **82**, 6512–6516.

RYDER, D.S. AND MASSCHELEIN, C.A. (1983). Aspects of metabolic regulatory systems and physiological limitations with a view to the improvement of brewing yeast performance. *European Brewery Convention Monograph* **IX**, 2–29.

SAGE, G.C.M. (1992). Current and future breeding of barley for malting and brewing. *Brewers Guardian* **121**, 24–29.

SAKAI, K., FUKUI, S., YABUUCHI, S., AOYAGI, S. AND TSUMURA, Y. (1989). Expression of the *Saccharomyces diastaticus STA1* gene in brewing yeasts. *Journal of the American Society of Brewing Chemists* **47**, 87–91.

SASAKI, T., WATARI, J., KOHGO, M., NISHIKAWA, N. AND MATSUI, Y. (1984). Breeding of a brewer's yeast possessing anti-contaminant properties. *Journal of the American Society of Brewing Chemists* **42**, 164–166.

SHIMIZU, F., SONE, H. AND INOUE, T. (1989). Brewing performance of a genetically transformed yeast with acetolactate decarboxylase activity. *Technical Quarterly of the Master Brewers Assocation of the Americas* **26**, 47–50.

SKRIVER, K., OLSEN, F.L., LEAH, R. AND MUNDY, J. (1991). Gene promoter analysis of malting enzymes. In *Proceedings of the European Brewery Convention Congress*, Lisbon, pp. 117–124.

SLOMINSKI, B., RGOWSKI, A. AND NOWAK, J. (1979). Abscisic acid and gibberellic acid contents in ripening barley seeds. *Physiological Plant and Molecular Pathology* **45**, 167–169.

SMEATON, R.H. (1990). New barley materials. *Ferment* **3**, 122–126.

SMITH, D.B., AUSTIN, R.B. AND HASLEMORE, R.M. (1986). Genetic manipulation of barley: Practice and prospects. In *Proceedings of the Second Aviemore Conference on Malting, Brewing and Distilling* (I. Campbell and F.G. Priest, Eds), pp. 62–79. Institute of Brewing, London.

STRATFORD, M. AND ASSINDER, S. (1991). Yeast flocculation: Flo1 and NewFlo phenotypes and receptor structure. *Yeast* **7**, 559–574.

SUIHKO, M.-L., PENTTILA, M., SONE, H., HOME, S., BLOMQVIST, K., TANAKA, J., INOUE, T. AND KNOWLES, J. (1989). Pilot-brewing with α-acetolactate decarboxylase active yeasts. In *Proceedings of the European Brewery Convention Congress*, Zurich, pp. 483–490.

SUIHKO, M.-L., BLOMQVIST, K., PENTTILA, M., GISLER, R. AND KNOWLES, J. (1990). Recombinant brewer's yeast strains suitable for accelerated brewing. *Journal of Biotechnology* **14**, 285–300.

THOMSEN, K.K., JACKSON, E.A. AND BRENNER, K. (1988). Genetic engineering of yeast: construction of strains that degrade β-glucans with the aid of a barley gene. *Journal of the American Society of Brewing Chemists* **46**, 31–36.

TOPFER, R., GRONENBORN, B., SCHELL, J. AND STEINBISS, H.-H. (1989). Uptake and transient expression of chimeric genes in seed-derived embryos. *Plant Cell* **1**, 133–139.

TUBB, R.S. AND HAMMOND, J.R.M. (1987). Yeast genetics. In *Brewing Microbiology* (F.G. Priest and I. Campbell, Eds), pp. 47–82. Elsevier, London.

TUBB, R.S., SEARLE, B.A., GOODEY, A.R. AND BROWN, A.J.P. (1981). Rare mating and transformation for construction of novel brewing yeasts. In *Proceedings of the European Brewery Convention Congress*, Copenhagen, pp. 487–496.

URANO, N., NISHIKAWA, N. AND KAMIMURA, M. (1990). Electrofusion of a brewer's yeast: Construction of fusants between a brewer's yeast and *Saccharomyces cerevisiae (flo5)*. In *Proceedings of the Institute of Brewing (Australia and New Zealand Section) Convention*, Auckland, pp. 154–157.

VAKERIA, D. AND HINCHLIFFE, E. (1989). Amylolytic brewing yeast: Their commercial and legislative acceptability. In *Proceedings of the European Brewery Convention Congress*, Zurich, pp. 475–482.

VAKERIA, D., BOX, W.G. AND HINCHLIFFE, E. (1991). The control of gene expression in brewing yeast with anti-sense RNA. In *Proceedings of the European Brewery Convention Congress*, Lisbon, pp. 305–312.

VAN DER AAR, P.C., STRAVER, M.H. AND TEUNISSEN, W.R.H. (1993). Flocculation of brewers' lager yeast. In *Proceedings of the European Brewery Convention Congress*, Oslo, pp. 259–266.

VAN SOLINGEN, P. AND VAN DER PLAAT, J.B. (1977). Fusion of yeast spheroplasts. *Journal of Bacteriology* **130**, 946–947.

VILLANUEBA, K.D., GOOSSENS, E. AND MASSCHELEIN, C.A. (1990). Sub-threshold vicinal diketone levels in lager brewing yeast fermentations by means of *ILV5* gene amplification. *Journal of the American Society of Brewing Chemists* **48**, 111–114.

VON WETTSTEIN, D. (1983). Emil Christian Hansen Centennial Lecture: From pure yeast culture to genetic engineering of brewers' yeast. In *Proceedings of the European Brewery Convention Congress*, London, pp. 97–119.

VON WETTSTEIN, D., NILAN, R.A., AHRENT-LARSEN, B., ERDAL, K., INGVERSEN, J., JENDE STRID, B., KRISTIANSEN, K.N., CARSEN, J., OUTTRUP, H. ABD ULLRICH, S.E. (1985). Proanthocyanidin-free barley for brewing: Progress in breeding for high yield and research tool in polyphenol chemistry. *Technical Quarterly of the Master Brewers Association of the Americas* **22**, 41–52.

WATARI, J., TAKATA, Y., OGAWA, M., NISHIKAWA, N. AND KAMIMURA, M. (1989). Molecular cloning of a flocculation gene in *Saccharomyces cerevisiae. Agricultural and Biological Chemistry* **53**, 901–903.

WATARI, J., TAKATA, Y., MURAKAMI, J. AND KOSHINO, S. (1991a). Breeding of flocculent brewer's yeast by genetic engineering. In *Proceedings of the European Brewery Convention Congress*, Lisbon, pp. 297–304.

WATARI, J., TAKATA, Y., OGAWA, M., MURAKAMI, J. AND KOSHINO, S. (1991b). Breeding of flocculent industrial *Saccharomyces cerevisiae* strains by introducing the flocculation gene *FLO1*. *Agricultural and Biological Chemistry* **55**, 1547–1552.

WINGE, Ö. (1914). Investigations on hops (*Humulus lupulus* L.) III. The pollination and fertilization processes in *Humulus lupulus* L. and *H. japonicus* Sieb. et Zucc. *Comptes-rendus des travaux du Laboratoire de Carlsberg* **11**, 1–44.

WOLFE, M.S., BENNETT, J.A. AND JENKINS, J.E.E. (1981). The use of cultivar mixtures for disease control. In *Strategies for the Control of Cereal Diseases* (J.F. Jenkyn and R.T. Plumb, Eds), pp. 73–80. Blackwell, Oxford.

XU, J. AND KASHA, K.J. (1992). Transfer of a dominant gene for powdery mildew resistance and DNA from *Hordeum bulbosum* into cultivated barley (*H. vulgare*). *Theoretical and Applied Genetics* **84**, 771–777.

YOCUM, R.R. (1986). Genetic engineering of industrial yeasts. *Proceedings of BioExpo 86*, pp. 171–180.

YOUNG, T.W. (1981). The genetic manipulation of killer character into brewing yeast. *Journal of the Institute of Brewing* **87**, 292–295.

YOUNG, T.W. (1983). The properties and brewing performance of brewing yeasts possessing killer character. *Journal of the American Society of Brewing Chemists* **41**, 1–4.

6
The Insecticidal Crystal Proteins of *Bacillus thuringiensis*: Past, Present and Future Uses

MICHAEL G. KOZIEL, NADINE B. CAROZZI, THOMAS C. CURRIER, GREGORY W. WARREN AND STEPHEN V. EVOLA

Ciba Agricultural Biotechnology Research Unit, Research Triangle Park, North Carolina, USA

Introduction

Bacillus thuringiensis (Bt) is a Gram-positive, spore-forming bacterium, which produces a parasporal crystal during sporulation. These crystals are predominantly comprised of one or more proteins, called δ-endotoxins or insecticidal crystal proteins, known to possess insecticidal activity when ingested by certain insects. Numerous strains of Bt are currently known. Each strain produces differing numbers of δ-endotoxins with various insecticidal activities. A given δ-endotoxin is typically insecticidal towards a narrow spectrum of insect targets. Bt has been used since 1938 to produce an insecticidal spray for the control of certain insect pests. These sprays are known for their environmental safety and safety towards non-target insects as well as other animals. As agricultural practices demand ever-increasing safety for pest control practices, interest in Bt has increased. Originally, Bt was considered to be useful only against a few insect species in the order Lepidoptera, but expanded screening efforts have yielded several strains with a number of different activities, including activities against non-insect pests (Feitelson, Payne and Kim, 1992). As the spectrum of activity for Bt as a class of pest control agents increases, so too does its use. New technologies to aid strain discovery have hastened the identification of new strains harbouring novel activities. New technologies in the area of molecular biology are providing the means to construct strains of Bt with desired insecticidal

Abbreviations: AcMNPV, *Autographa californica* multiple nuclear polyhedrosis virus; AcNPV, *Autographa californica* nuclear polyhedrosis virus; Aq, *Agmenellum quadruplicatum*; BBMV, brush border membrane vesicle; Bt, *Bacillus thuringiensis*; CaMV, cauliflower mosaic virus; CPB, Colorado potato beetle; *Cxc, Clavibacter xyli* subsp. *cynodontis*; ECB, European corn borer; mRNA, messenger RNA; Pc, *Pseudomonas cepacia*; PCR, polymerase chain reaction; PEPC, phosphoenolpyruvate carboxylase; SDS-PAGE, sodium dodecyl sulphate polyacrylamide get electrophoresis; UV, ultraviolet.

Biotechnology and Genetic Engineering Reviews — Vol. 11, December 1993
0264–8725/93/11/171–228 $20.00 + $0.00 © Intercept Ltd, P.O. Box 716, Andover, Hampshire SP10 1YG, UK

activities or to overcome or prevent resistance to these proteins by the target insect. Further, these new tools allow the introduction and expression of the δ-endotoxins in new host organisms, thereby increasing the possible uses for these insecticidal proteins. Their attributes of being direct, single gene products make them attractive for such engineering endeavours. This review focuses on the traditional uses of the δ-endotoxins as the active components of microbial sprays and the future uses of such sprays in light of new technologies and an increased understanding of the biology of these proteins. It also looks at the uses of these proteins in various new host organisms, and examines the issue of resistance management. The δ-endotoxins can be viewed as pesticidal proteins whose applications are broad, and rapidly expanding, and no longer limited to a particular use within the confines of one particular organism.

Traditional and current uses of Bt insecticides

Bacillus thuringiensis has been used to produce biological insecticides for foliar application for over 50 years. These products were, and are still, produced by fermentation of single Bt strains in crude, inexpensive media. As the Bt cells begin to exhaust the medium, they enter the sporulation phase of growth where one or more insecticidal proteins, or δ-endotoxins, are synthesized. The δ-endotoxins accumulate in morphologically distinct crystal(s) within the cells and are released into the medium with the spores when the cells ultimately lyse. Typically, the first step in the manufacturing process of a Bt bio-insecticide is to concentrate the fermentation solids (spores, δ-endotoxin crystals and unused particulate media components) by centrifugation. Sometimes tangential flow dialysis is also used to further concentrate the solids before spray drying. The dried material is then milled to a uniform size to produce a technical powder which can then be formulated in a variety of ways. The most common formulations are wettable powders or dispersible granules, but oil- and aqueous-flowable preparations have also been produced.

At least three major factors contribute to the effectiveness of Bt insecticides produced in this manner. First, a Bt strain must be obtained that produces δ-endotoxins which are effective on the target insects. Second, the strain must be able to produce large amounts of active δ-endotoxin during large-scale fermentations at a minimum cost. Third, a formulation that maximizes the longevity and effectiveness of the δ-endotoxins should be used.

Strain selection

The bacterial strains used for all of the early and some of the current Bt foliar insecticides are wild-type strains, that is, they were found in nature in the form in which they are used to produce the microbial spray. These strains have good activity on a variety of commercially important insects and they can be easily fermented and formulated. Current knowledge concerning the mechanism of action of Bt δ-endotoxins clearly indicates that the insecticidal activity of a strain is determined by several factors. The number, type and amount of δ-endotoxin(s) produced are the primary determinants, and for many insects this alone can fully account for a strain's activity. However, for a few insects the spores and the δ-endotoxins appear to act in synergy, possibly due to septicaemia and the production of degradative enzymes and/or other toxins in the insect's gut (Heimpel and Angus, 1959). Most wild-type Bt isolates produce multiple δ-endotoxin proteins, each having a characteristic insecticidal activity spectrum. It is generally believed that the overall activity spectrum of a strain is due to additive and/or synergistic interactions of the individual δ-endotoxins present in their proportional amounts. Both the total amount of δ-endotoxin produced and the relative amount of each is largely determined by the copy number of each gene and the extent to which it is expressed. Increasing copy number of a single gene or the presence of multiple δ-endotoxin gene types can increase the amount of δ-endotoxin produced up to about 30% of the dry weight of the cell (Lilley, Ruffell and Somerville, 1980). Thus, the maximum amount of a single δ-endotoxin protein would be obtained from a strain expressing a single δ-endotoxin gene at its maximum biosynthetic capacity. However, the insecticidal spectrum of such a strain would be relatively narrow. In contrast, a strain producing several δ-endotoxins would have a broader insecticidal spectrum, but the activity contribution of each δ-endotoxin would be decreased relative to the proportional amount produced.

When the first Bt products were developed, little was known about the determinants of toxicity or the δ-endotoxin composition of the strains. Analysis of these strains shows that they have good biosynthetic capabilities in crude fermentation media, they have relatively broad insecticidal activity, and they express at least four different δ-endotoxins (*Table 1*). For example, the broad insecticidal activity spectra of HD1 (Dipel®, Abbott) is reflected in the insecticidal activities of the individuals δ-endotoxins (*Table 2*). In addition, high toxicity to a limited number of insects is predicted by the presence of three individual δ-endotoxins with good activity on the same insect.

For many target crops, such as cole crops, only a few of the pests listed in *Table 2* are of importance. In such cases, it should be possible to obtain a better performing bio-insecticide based upon a Bt strain that produces only those toxins with good activity on the target insects. The realization that closely related δ-endotoxins could have quite different insecticidal activities and the discovery that most δ-endotoxin genes are located on plasmids made it possible to begin to rationally construct strains with increased activity on specific pests. Knowing the insecticidal spectrum of an δ-endotoxin and its plasmid location, it is possible to use the classical genetic techniques of plasmid curing and conjugal plasmid transfer to eliminate unnecessary δ-endotoxins and/or introduce desirable ones. Bt strains present in several currently available bio-insecticides were constructed using these techniques (*Table 1*). Although these genetic manipulation techniques have the capability to increase potency and control specificity, they are limited by three major factors: (1) not all plasmids containing δ-endotoxin genes can be easily cured;

Table 1. Gene composition of several currently available Bt insecticides

		cry Gene type								
Product[a]	Company	IA(a)	IA(b)	IA(c)	IB	IC	ID	IIA	IIB	IIIA
Agree®	Ciba-Geigy	+		+		+	+			
Xentari®	Abbott	+	+			+	+			
Dipel®	Abbott	+	+	+				+	+	
Biobit®	Novo	+	+	+				+	+	
Javelin®	Sandoz	+	+	+				+	+	
Condor®	Ecogen	+	+	+				+	+	
Cutlass®	Ecogen	+	+	+				+	+	
Foil®	Ecogen			+						+
Novodor®	Novo Nordisk									+

Note: These data were generated at Ciba using gene specific primers in the polymerase chain reaction (Carozzi *et al.*, 1991). A positive PCR reaction only indicates the likely presence of the gene. It does not imply gene expression. Thus the gene may be present in the strain but the δ-endotoxin may not be present in the product. In some cases, the same gene type may be present on two different δ-endotoxin plasmids.
[a] Agree® (Burgess and Jarrett, 1985), Condor® (Gonzalez and Macaluso, 1992), Cutlass® (Gonzalez and Macaluso, 1992) and Foil® (Donovan *et al.*, 1991) were constructed using plasmid curing and conjugation techniques.

(2) not all δ-endotoxin plasmids can be transferred by conjugation; and (3) some plasmids contain more than one δ-endotoxin gene. Because of these limitations, it may not be possible to construct a strain containing only the desired δ-endotoxin genes using classical genetic techniques.

The tools of genetic engineering can be used to overcome all of the limitations of classical genetics. However, it is desirable to use both classical and molecular approaches together if possible. The classical techniques of plasmid transfer and curing are easier and quicker than recombinant procedures and can be complemented by molecular techniques when necessary. In addition to being a useful complement to the classical techniques of strain construction, molecular techniques have the potential to further optimize δ-endotoxin production in several ways. For example, they could be used to control gene dosage by altering plasmid copy number or by integrating the δ-endotoxin gene into an existing plasmid or the bacterial chromosome. Copy number mutations of a Bt vector have been reported (Arantes and Lereclus, 1991) which can to some extent control δ-endotoxin production. In addition, gene expression can also be controlled by altering the elements regulating transcription and translation (promoters and/or ribosome binding sites) or by increasing the mRNA stability of the message by modifying its 3′ end. Also, the δ-endotoxin itself could be engineered to have a broader insecticidal activity spectrum. Additionally, molecular techniques have the potential to further optimize and enhance δ-endotoxin production. As additional information on the determinants of specificity and toxicity are obtained, it may be possible to more rationally make directed changes in these parameters.

Regardless of the techniques used for rational strain construction, the first and perhaps the most difficult step is to determine the desired insecticidal activities and to identify δ-endotoxins with these properties. Multiple δ-endotoxin genes could be introduced into a strain for one or more of the following reasons: to increase toxicity on a particular pest, to broaden the

Table 2. Relative activities of δ-endotoxin gene types on selected insects[a]

Endotoxin	Px sen	Px res	Se	Sl	Tn	Hv	Hz	On	Ms	Ldis	Pb	Ld
cryIA(a)			–	–	+	++		++	+++		+++	
cryIA(b)	+++	–	–	–	+++	++	++	++	+++	+++	+++	
cryIA(c)	+++		–	–	+++	+++	+	++	+++	+++	+++	
cryIB	+++	+++		–		++		[++]	–		+++	
cryIC	++	++	[++]	++		–			++		+++	
cryID	–	–		–		≤+			+++		≤++	
cryIE	–	–	(++)	++		–			++			
cryIF	++		–			++		++				
cryIIA					++	++	++	++	+++	+++		
cryIIB					++	++	++	+		+++		
cryIIC												
cryIIIA												++
cryIIIB												+
cryIIIB2												++
cryIIIC(a)												++
cryIIIC(b)												+
cryIIID												(+++)
cryV								++				++

Px sen, *Plutella xylostella* (diamondback moth) sensitive to Dipel; Px res, *Plutella xylostella* (diamondback moth) resistant to Dipel; Se, *Spodoptera exigua* (beet armyworm); Sl, *Spodoptera litoralis* (cotton leafworm); Tn, *Trichoplusia ni* (cabbage looper); Hv, *Heliothis virescens* (tobacco budworm); Hz, *Helicoverpa zea* (cotton bollworm, corn earworm); On, *Ostrinia nubilalis* (corn borer); Ms, *Manduca sexta* (tobacco hornworm); Ldis, *Lymantria dispar* (gypsy moth); Pb, *Pieris brassicae* (European cabbage worm); Ld, *Leptinotarsa decemlineata* (Colorado potato beetle).

[a] Relative activities are estimates made from published values or from our unpublished data [data in brackets]. The activities are defined as follows: +++ = $LC_{50} \leq 10$ ng cm^{-2} or for Pb < 1 μg ml^{-1}; ++ = 10 ng $cm^{-2} \leq LC_{50} \leq 100$ ng cm^{-2} or for Pb $1 \leq LC_{50} \leq 10$ μg ml^{-1}; + = 100 ng $cm^{-2} \leq LC_{50} \leq 1000$ ng cm^{-2} or for Pb $10 \leq LC_{50} \leq 100$ μg ml^{-1}; – = $LC_{50} \geq 1000$ ng cm^{-2} or for Pb > 100 μg ml^{-1}.
Relative activities given in parentheses indicate the reported activity was not given in ng cm^{-2}. In these cases, the relative activity was estimated by comparison of the δ-endotoxin's activity on a second insect whose activity was reported elsewhere in ng cm^{-2}.

host range in cases where one δ-endotoxin cannot be identified with acceptable toxicity on multiple insect targets, or to manage resistance by including δ-endotoxins that bind to different sites in the insect midgut. Whatever the desired goal, the effectiveness of the strain will depend on the biosynthetic capacity of the cell, because introducing more toxins to broaden specificity will usually reduce the amount of each δ-endotoxin present.

The task of identifying δ-endotoxins with new or more potent activities typically involves screening wild-type strains of Bt. The presence of a new activity is obvious, but to detect the presence of a more potent δ-endotoxin, bioassays must be normalized to the amount of δ-endotoxin protein in the assay. To find genes encoding proteins that are more potent on insects known to be susceptible to Bt δ-endotoxins, it is necessary to isolate genetically the δ-endotoxin gene by classical or molecular methods so that activity per unit of δ-endotoxin protein can be compared with the specific activities of known δ-endotoxins.

Molecular probes such as gene-specific polymerase chain reactions (PCRs; Carozzi *et al.*, 1991) or hybridization with gene-specific DNA sequences (Gaetner, Sick and Schwab, 1993) are useful for the identification of known or related genes in strains with interesting insecticidal properties. A general idea of the biological activity of a δ-endotoxin can be made if a known gene class is detected, but accurate prediction is not possible because specific activities of δ-endotoxins of a gene class can differ by as much as 10-fold on the same insect (Von Tersch *et al.*, 1991). Novel δ-endotoxins which cannot be identified using gene probes might be identified by biochemical or immunological means. Analysis of the solubilized proteins from crystal inclusion by SDS-PAGE can determine if proteins of an unexpected size are present. Differences of less than 1000 daltons can be resolved on such gels. Another method for finding unique proteins using antibodies directed against known δ-endotoxins was described by Geiser, Hartmann and Oddou (1993), which involves the binding of known δ-endotoxins to antibody affinity columns. A protein with insecticidal activity that does not bind may constitute a new δ-endotoxin class. After the genes of interest are identified, they must then be isolated by classical or molecular techniques and the activity spectrum and specific activities on individual insects determined. Finally, strains are constructed by combining the selected genes in a Bt strain suitable for fermentation.

Fermentation

The first Bt product was marketed in France in 1938. In 1960, there were five US companies marketing Bt bio-insecticides, but by 1979 only two remained. For a review of the historical development of commercial production of Bt bio-insecticides, the reader is referred to Rowe and Margaritis (1987). Both of these products were produced by growth of the bacterium in liquid medium in large mechanically agitated vessels sparged with air. This type of deep liquid fermentation is the preferred method of production for current Bt insecticides. The details of the fermentation, downstream processing and formulation of Bt products are generally kept as trade secrets. The development of a fermentation process is largely empirical and must be determined separately for each Bt strain selected. Unlike laboratory research where cost of media components, cell harvesting and sample processing are not the highest priority, the goal in commercial production is to make the maximum amount of active ingredient and final formulated product for the minimum cost. From the fermentation aspect, this necessitates the use of crude, inexpensive fermentation ingredients. These materials often contain a high proportion of insoluble protein and other materials. The insoluble protein is an inexpensive source of carbon and nitrogen and can be utilized because the Bt cells produce large amounts of extracellular enzymes in order to solubilize and degrade this material.

The suitability of these components depends on many factors. Clearly, the cost is important and the material must be readily available in large quantities whenever it is needed. In addition, the batch-to-batch uniformity of the fermentation substrates may be critical for development of an easily reproducible process. Since many crude protein sources are waste or by-products of other manufacturing processes, different production lots are not always of consistent nutritional value for fermentation. This inconsistency may result from the fact that the fermentation substrates are not always handled, stored or even processed in the same manner, or because the starting materials from which the fermentation substrates are obtained are not always the same. Obtaining a consistent supply of any required inorganic nutrients, such as phosphate and divalent ions, may also be a problem because there are many crude sources of the materials which could be contaminated with unwanted inorganic poisons or inhibitors.

The development of an economical fermentation process involves a great deal of experimentation and must be specifically designed for a particular strain of Bt. Thus the production of a new bio-insecticide using a new Bt strain requires the development of a different fermentation process. This may only require minor changes in an existing process or the physiological properties of the new strain may be different enough to require major changes in the entire manufacturing process.

If strains are constructed by genetic manipulation, it should be possible to avoid the development of a different production process for each new Bt strain. Genetically manipulated strains could be constructed by introducing the desired δ-endotoxin genes into a 'standard' recipient strain. Such a recipient would be a variant of a strain for which the fermentation and formulation technology have already been developed. This variant could be developed using plasmid-curing techniques to remove those δ-endotoxin genes that are not desired in the new product. The recipient strain might be completely cured of all δ-endotoxin genes, or one or more δ-endotoxin genes from the original parent strain might be retained. The degree of plasmid curing would depend on whether any δ-endotoxin genes from the parent were desired in the final product and whether the desired genes were on plasmids that were free of undesirable genes. A genetically modified strain would then be constructed by introducing the desired δ-endotoxin genes into the plasmid-cured recipient strain using plasmid transformation, sexual conjugation or a variety of recombinant DNA techniques.

It should also be possible to increase δ-endotoxin production by genetically modifying the fermentation strain or the regulatory elements controlling δ-endotoxin gene expression. In fact, strain selection for increased δ-endotoxin production has been accomplished with Novodor® (Novo Nordisk). The strain in this product is a mutant *B. thuringiensis* var. *tenebrionis* obtained by gamma irradiation. It produces about twice as much δ-endotoxin as its parent (Gurtler and Petersen, 1991), apparently as a result of a regulatory change which results in the production of δ-endotoxin prior to sporulation. In addition, the results of Mettus and Macaluso (1990) suggest an additive increase in δ-endotoxin production in a strain where a δ-endotoxin gene is expressed by both its native sporulation dependent promoter and by a strongly expressed vegetative promoter. Donovan (1991) was also able to express a silent *cryIIB* gene by fusing it to the *cryIIIA* promoter. Furthermore, Gamel and Piot (1992) fused the *cryIC* promoter to the *cryIIIA* gene and introduced this construct into a strain already containing CryIA(a), CryIA(b), CryIA(c) and CryIIA proteins. They state that the introduction of the newly introduced *cryIIIA* gene did not interfere with expression of the native genes. Together, these data show that δ-endotoxin expression can be regulated by placing δ-endotoxin genes under the control of promoters from either δ-endotoxin or non-endotoxin genes. This kind of regulatory change, together with changes in gene copy number and alterations at the ribosome binding site, should make it possible to further increase δ-endotoxin production.

Some other possibilities for increasing δ-endotoxin production include directing more of the cell's biosynthetic capacity to δ-endotoxin production or to prolong the time period allowed for biosynthesis. The major biosynthetic activity occurring during δ-endotoxin synthesis is the production of spores. This complex process consumes a substantial amount of biosynthetic capacity and for the most part spores are not required for insecticidal activity on many commercially important insects. The biosynthetic requirements for spore production might be diverted to δ-endotoxin production in strains where sporulation was aborted at an early stage. Spore-negative, δ-endotoxin-producing mutants of Bt have been reported (Johnson, 1981; Wakisaka *et al.*, 1982) and about as much or a little more δ-endotoxin was produced in the mutants as compared to their respective parental strains. Other spore-negative strains have been described where δ-endotoxin production was increased relative to the parental strains (Fitzjames, 1984; Herrnstadt and Gaertner, 1987). The greatest increase reported was about 1.5-fold better than the parental strain. Further improvements in this area might be realized as more information is obtained about the biosynthetic and regulatory relationships between δ-endotoxin synthesis and spore formation.

It might also be possible to isolate an overproducing mutant by extending the normal period for δ-endotoxin biosynthesis. Inhibiting or completely preventing cell lysis by elimination or inactivation of enzymes (Kingan and Ensign, 1968) responsible for cell wall degradation during sporulation could produce the desired phenotype.

Formulation and application

Bt bio-insecticides have been formulated in many ways since their introduction in 1938. Wettable powders, dispersible granules, dusts, microgranules, aqueous flowable liquids and oil-based flowable liquids that are dispersible in water have been developed. Early formulations were often difficult to apply and their performance was often poor or unreliable. Advances in application equipment and improvements in formulation technology have largely overcome application difficulties. However, different strains may well require different formulations because of differences in their respective fermentation media and their by-products. Also, batch-to-batch formulation problems can be encountered due to variable quality of the fermentation components.

Substantial improvements in the performance of Bt insecticides have been made through advances in formulation technology. Many new spreaders and stickers have been developed to help evenly distribute and bind the material over the leaf surface, which is often very waxy. Newer, more effective additives are continually being developed and further improvements in this area can be expected in the future.

The second major task of a good formulation is to aid in the persistence of the active ingredient. It was recognized very early that Bt insecticides do not persist for very long in the environment (Gelernter, 1990). Possible causes for loss of activity include inactivation by ultraviolet (UV) light, heat, leaf exudates and pH (Pozsgay *et al.*, 1987; Dulmage and Aizawa, 1982). The most important factors have not been identified. Some of the factors are clearly environmental and others, like leaf surface chemistry, are plant-dependent. At high pH, particularly in the presence of reductant, δ-endotoxin crystals are solubilized and thus become more accessible to proteolytic inactivation. High leaf pH might be important on cotton where the leaf surface can be in the range of pH 8–10 (Andrews and Sikorowski, 1973).

Formulations of Bt insecticides have addressed these problems by either including additives to help prevent loss of the active ingredient or encapsulating the material in such a way that the active ingredient is protected from inactivation. The addition of UV protectants might effectively reduce irradiation damage. Perhaps inactivation due to heat could be controlled by including an additive, but no materials of this type have been described. Another approach to control the inactivating effects of UV light and/or leaf exudates would be to encapsulate the material. Much of the information on specific additives and formulations can be found in the patent literature and will not be reviewed here. Encapsulation can be accomplished chemically (McGuire and Shasha, 1992) or biologically (Barnes and Cummings, 1987). Chemical encapsulation involves a process in which the active ingredient is coated with some type of polymer, while biological encapsulation involves the expression of the δ-endotoxin genes in a microbe. For biological encapsulation, the intact microbe containing the δ-endotoxin protein is used as the active ingredient in the formulation. Unfortunately, there is a limited amount of information in the public domain on the effectiveness of any of the above formulation approaches. All are described as beneficial, but there are few data available that accurately quantify their protective effects. Formulation is an active but rather proprietary area of research that will undoubtedly increase the effectiveness of Bt insecticides in the future.

Most commercially available Bt insecticides can be tank-mixed with a variety of other agricultural chemicals (*1993 Crop Protection Chemicals Reference*, Chemical and Pharmaceutical Press, New York). These include other pesticides, herbicides and additives intended to promote insect feeding and thereby increase the effectiveness of the Bt insecticide. There is a good deal of information available on various positive interactions of Bt with other insecticides and feeding stimulants. Regardless of the effectiveness of any formulation or mix with beneficial agrochemicals, the benefits must be economical. To be useful, they must provide a benefit justified by the added price.

Spectrum of activity

The various insecticidal crystal proteins have been classified based upon their spectrum of activity and sequence homology. The classification put forth by Hofte and Whiteley (1989) placed the then known insecticidal crystal proteins into four major classes. Generally, the major classes are defined by the spectrum of activity with the CryI proteins active against Lepidoptera, CryII proteins active against both Lepidoptera and Diptera, CryIII proteins active against Coleoptera, and CryIV proteins active against Diptera. Subsequently, a crystal protein was identified with activity against both Lepidoptera and Coleoptera (Tailor *et al.*, 1992) and the authors suggested proteins with such activity be placed in a new class, the CryV proteins. This protein has activity against European corn borer (*Ostrinia nubilalis*; Lepidoptera) and Colorado potato beetle (*Leptinotarsa decemlineata*; Coleoptera). Another group (Schnepf *et al.*, 1992) discovered Bt crystal proteins with activity against nematodes. These proteins were placed into two groups based upon their structures and these two groups were proposed to be called CryV and CryVI, thus conflicting with the other authors. The nomenclature system is still evolving, but it is providing a needed clarity and consistency in the naming of the various δ-endotoxins. A δ-endotoxin with activity against a Hymenoptera has also been found (Payne *et al.*, 1992) and is yet unclassified. The relationship, both in terms of protein sequence homology and structural homology, of these new δ-endotoxins with the better known δ-endotoxins classified by Hofte and Whiteley (1989) remains unknown. As yet, it is unclear whether these new toxins have the same mode of action or whether they represent new modes of action as well as new pesticidal activities.

Within each major class, the δ-endotoxins are grouped according to sequence homology. The CryI proteins are typically produced as 130–140 kDa proteins which are proteolytically activated to produce proteins about 60–70 kDa. The active portion of the δ-endotoxin resides in the NH_2 terminal portion of the full-length molecule. Hofte and Whiteley classified the then known CryI proteins into six groups: IA(a), IA(b), IA(c), IB, IC and ID. Since then, proteins classified as CryIE, CryIF, CryIG and CryIX have been characterized. Other lepidopteran active genes have also been identified, but have not been classified as yet. *Table 3* lists various crystal proteins produced by *B. thuringiensis*. New coleopteran-active proteins have also been identified since this classification scheme was put forth. Originally, there was but one characterized protein active against Coleoptera and this was classified as CryIIIA. CryIIIA has activity against mealworm (*Tenebrio molitar*) and Colorado potato beetle (*Leptinotarsa decemlineata*). Subsequently, CryIIIB, CryIIIB2, CryIIIC, CryIIIC(b) and CryIIID δ-endotoxins have been identified with activity against Coleoptera.

The spectrum of insecticidal activity of an individual δ-endotoxin from Bt tends to be quite narrow, with a given δ-endotoxin being active against only a few (known) insects. Specificity is the result of the efficiency of the various steps involved in producing an active toxin protein and its subsequent interaction with the epithelial cells in the insect digestive tract. It is also likely

Table 3. Current list of Bt δ-endotoxins and their activities

Endotoxin/strain	Size (kDa)	Specificity	Reference
CryIA(a)	133	Lep	Schnepf, Wong and Whiteley (1985)
CryIA(b)	131	Lep	Wabiko, Raymond and Bulla (1986)
CryIA(c)	133	Lep	Adang *et al.* (1985)
CryIA(c) variants	133	Lep	Dardenne, Seurinck and Peferoen (1990); Von Tersch *et al.* (1991)
CryIB	138	Lep,Col	Brizzard and Whiteley (1988)
CryIC	135	Lep	Honee, van der Salm and Visser (1988)
CryIC(b)	134	Lep	Bosse, Masson and Brousseau (1990)
CryID	133	Lep	Hofte *et al.* (1990)
CryIE	133	Lep	Visser *et al.* (1990); Masson *et al.* (1992)
CryIF	134	Lep	Chambers *et al.* (1991)
CryIG	130		Smulevitch *et al.* (1991); Gleave, Hedges and Broadwell (1992)
CryIX	81	Lep	Gawron-Burke, Chambers and Gonzalez (1991)
PS81RR1	133	Lep	Payne and Sick (1990)
OS81GG	133	Lep	Payne, Sick and Thompson (1992)
81IA2	133	Lep	Payne and Sick (1991a)
81IB	132	Lep	Payne and Sick (1991a)
81IB2	135	Lep	Payne and Sick (1991a)
81IA	134	Lep	Payne and Sick (1991a)
PS81F[b][cryIE(a)]	133	Lep	Payne and Sick (1991b)
34 kDa crystal protein	34	Lep	Brown and Whiteley (1992)
PS158C2	47,37,34 and 32	Lep	Payne, Cummings and Cannon (1993)
CryIIA	71	Lep,Dip	Donovan *et al.* (1988)
CryIIB	71	Lep,Dip	Widner and Whiteley (1989)
CryIIC	110	Lep,Dip	Wu *et al.* (1991)
CryIIIA	73	Col	Herrnstadt *et al.* (1987)
CryIIIB	74	Col	Sick, Gaertner and Wong (1990)
CryIIIB2	74	Col	Donovan *et al.* (1992a)
CryIIIC*	74.4	Col	Donovan *et al.* (1993)
CryIIIC*	129	Col	Lambert *et al.* (1992a)
CryIIIC(b)	70	Col	Donovan, Rupar and Slaney (1992)
CryIIID	73	Col	Lambert *et al.* (1992b)
43F (resembles IIC)	74	Col	Sick and Gilroy (1991)
PS50C	130	Col	Foncerrada, Sick and Payne (1992)
PS40D1		Col	Hickle, Bradfisch and Payne (1992)
NCIMB40152	63–70,32 and 14	Col	Cidaria *et al.* (1991)
PS86A1	58 and 45	Col	Bradfisch, Michaels and Payne (1993)
PS86Q3	155,135, 98,62 and 58	Col	Bradfisch, Michaels and Payne (1993)

Continued over

Table 3. cont.

Endotoxin/strain	Size (kDa)	Specificity	Reference
Strain 14-4		Col	Carozzi *et al.* (1993)
Strain 14-5		Col	Carozzi *et al.* (1993)
Strain 14-8		Col	Carozzi *et al.* (1993)
Strain 14-10		Col	Carozzi *et al.* (1993)
CryIVA	134	Dip	Ward and Ellar (1987)
CryIVB	128	Dip	Chunjatpornchai *et al.* (1988)
CryIVC	78	Dip	Thorne *et al.* (1986)
CryIVD	72	Dip	Donovan, Dankocsik and Gilbert (1988)
PS71M3-1	135	Dip	Sick (1991)
PS71M3-2	78	Dip	Sick (1991)
CryV[a]	81	Col,Lep	Tailor *et al.* (1992)
PS140E2[b]	78,70 and 35	Hym	Payne *et al.* (1992a)
PS86Q3[b]	58–155	Hym	Payne *et al.* (1992a)
PS211B2[b]	27–175	Hym	Payne *et al.* (1992a)
PS17a(CryV[a])	156	Nem	Schnepf *et al.* (1992)
PS17b(CryV[a])	146	Nem	Schnepf *et al.* (1992)
PS33F2(CryV[a])	142	Nem	Schnepf *et al.* (1992)
PS63B(CryV[a])	91	Nem	Schnepf *et al.* (1992)
PS52A1(CryVI)	54	Nem	Schnepf *et al.* (1992)
PS69D1(CryVI)	45	Nem	Schnepf *et al.* (1992)
PS81F	133	Protozoa	Thompson and Gaertner (1991)

* Two groups have used the CryIIIC name.
[a] Two groups have proposed using CryV to describe a new class of δ-endotoxins.
[b] The name CryIE(a) has been proposed for this gene.
Lep, Lepidoptera active; Col, Coleoptera active; Dip, Diptera active; Hym, Hymenoptera active; Nem, Nematode active. The Hymenoptera active strains produce a variety of inclusion proteins ranging from about 27 to 155 kDa. The activity of the individual proteins and precursor–product relationships are unknown at present.

to be a result of the limited testing these proteins undergo to identify their activity. Typically, only a few pests of commercial significance are used to determine the activity of new strains or proteins. It is possible that a given δ-endotoxin has a broader spectrum of activity than is known, but other target organisms simply have not been tested.

To be insecticidal, a δ-endotoxin must first be ingested by the insect and proteolytically activated to form an active toxin. Activation of the insecticidal crystal proteins is a multistep process. After ingestion, the crystals must first be solubilized in the insect gut. The crystals are typically insoluble at neutral pH and soluble at alkaline pH. Once solubilized, the δ-endotoxins are activated by specific proteolytic cleavages. The proteases in the insect gut can play a role in specificity by determining where the protoxin is cleaved. The activated toxins are typically quite stable in the presence of trypsin or other proteases. Once activated, the δ-endotoxins bind to specific proteins present on the surface of the gut epithelial cells. These binding proteins have been called receptors by some researchers, but it is not known whether they perform the function of a true receptor, so others prefer these proteins simply to be called binding proteins. The interaction of activated δ-endotoxins and the cellular binding proteins can be monitored using iodinated δ-endotoxin and purified brush border membrane vesicles (Hofmann *et al.*, 1988a,b),

Western blots of the brush border membrane vesicle proteins probed with activated δ-endotoxin (Oddou, Hartmann and Geiser, 1991), or can be visualized with immunosectioning techniques (Bravo *et al.*, 1992; Bravo, Jansens and Peferoen, 1992). In all cases examined, binding occurs whenever there is toxicity. In most, but not all cases, toxicity results when there is binding. After binding, the δ-endotoxin apparently inserts itself, or part of itself, into the cell membrane, creating a pore that ultimately results in the osmotic rupture of the cell. It appears that solubilization, activation and binding are steps which play roles in determining specificity of a δ-endotoxin. It is not known whether membrane insertion, or perhaps subsequent steps, also provide levels of specificity. Presently, it is simplest to conjecture that the largest part of specificity is determined by the binding properties of an activated δ-endotoxin, since this is one of the better characterized aspects of the chain of events that results in toxicity. There have been recent reviews dealing with the mode of action of these protein toxins (Gill, Cowles and Pietrantonio, 1992; English and Slatin, 1992), and the reader is directed elsewhere for a more detailed description of the current knowledge concerning the mode of action of the δ-endotoxins.

In certain cases, a δ-endotoxin that is capable of activity against a particular insect is not active because one or more of the above steps are not properly carried out in the gut. Solubilizing and activating δ-endotoxins prior to insect feeding can sometimes result in a broader spectrum of activity. For instance, CryIB is active against the European corn borer (ECB: *Ostrinia nubilalis*; Lepidoptera) but the full-length protein is not active against the Colorado potato beetle (CPB: *Leptinotarsa decemlineata*; Coleoptera). If it is solubilized and activated with proteases, it is still active against ECB, but it is now also active against CPB (Bradley *et al.*, 1992). The gut of CPB apparently does not solubilize and/or activate CryIB. The gut environment of CPB has a more neutral pH than the gut environment of lepidopterans and therefore the solubilization of CryIB may not occur in CPB.

The proteolytic activation of a solubilized δ-endotoxin can play a role in determining its specificity. The δ-endotoxin from *B. thuringiensis* var. *aizawa*, originally called the IC1 gene, has been classified as a *cryIA(b)* gene based upon its homology with other known *cryIA(b)* genes. However, the IC1 protein has activity against both lepidopteran and dipteran insects depending upon how it is activated (Haider, Knowles and Ellar, 1986). This δ-endotoxin is processed differently in the yellowfever mosquito (*Aedes aegypti*; Diptera) than it is in European cabbage worm (*Pieris brassicae*; Lepidoptera). In *Aedes* gut juice, a 53 kDa active δ-endotoxin is obtained, whereas in *Pieris* gut juice, a 55 kDa activated toxin is obtained. IC1 differs from the HD-1 CryIA(b) by only four amino acids, so gross changes in the binding region do not seem to account for the difference in activity. The different cleavages possibly allow the activated molecules to fold differently, thereby exposing different regions capable of binding to the insect gut. The choice of specificity of the cleavage, in this case, appears to reside with the gut proteases of the different insects. This raises the possibility that other δ-endotoxins may carry cryptic activities and that their spectrum of activity can be broadened by proper activation. An example of this is a *cryIIIC* gene described by Lambert *et al.* (1992a). This gene was isolated from Bt strain BTS137J and encodes a 129 kDa protein which is not active against any known insect unless it is proteolytically cleaved to produce an activated molecule prior to feeding. The activated 72 kDa protein has activity against Colorado potato beetle. If one supplies the proper proteolytic activity, the δ-endotoxin becomes toxic to at least one insect.

It is unclear what determines the spectrum of binding of the δ-endotoxin to the brush border of epithelial cells once it is properly activated. Some δ-endotoxins, such as CryIA(b), are active against a number of insects, while others appear to be active against only one (known) insect. The precise reasons for this are unclear at present. One possibility is that a single toxin, once activated, is able to bind gut proteins in a number of insects because of multiple binding sites within the binding domain. Domain II, postulated to be the binding domain for CryIIIA and likely for other δ-endotoxins also (Li, Carroll and Ellar, 1991), appears to be quite large and could possess more than one binding site. The argument for multiple binding sites on a single molecule would seem to be supported by the observation that some δ-endotoxins are active against multiple insects, sometimes even insects in different orders. The differential activity against the two orders, in some cases, results from differential cleavage of the same toxin in the gut juices from the different insects. The size of the region of a δ-endotoxin required for binding to the insect gut cell binding protein is unclear. Ge, Shivarova and Dean (1989) identified a region of CryIA(a) important for binding to the gut of the silkworm (*Bombyx mori*; Lepidoptera). This region lies between amino acids 332 and 450 of the δ-endotoxin, which is within Domain II as defined by Li, Carroll and Ellar (1991). CryIIA, which is toxic to both the lepidopteran tobacco hornworm (*Manduca sexta*) and the yellowfever mosquito (*Aedes aegypti*; Diptera), is 87% homologous with CryIIB, which has activity against only *Manduca* (Widner and Whiteley, 1989). By generating and characterizing hybrids between these two proteins, Widner and Whiteley (1990) showed that specificity towards *Aedes* larvae could be localized in 76 amino acids in the region between residues 307 and 382 of CryIIA. These two proteins differ by only 18 amino acids in this region, making it appear that relatively small changes in the proper region can drastically alter the spectrum of activity of a particular δ-endotoxin. In this case, the authors ruled out alternative processing in this region of the protein as an explanation for differential activity and concluded that differential binding was the cause of the different activity of the two proteins.

There may be stringent requirements for a specific interaction between the activated δ-endotoxin and the gut cell receptor proteins prior to cell lysis. Such requirements could influence the spectrum of activity. The orientation of the bound toxin relative to the cell membrane or the distance of the bound toxin from the cell surface could influence the efficiency with which the δ-endotoxin can insert itself into the membrane. The insect gut cell receptor, or binding protein, also plays a role in the relative activity of a given δ-endotoxin against a particular insect. Van Rie *et al.* (1989) have shown that the differential activities of CryIA(a), CryIA(b) and CryIA(c) against tobacco budworm (*Heliothis virescens*; Lepidoptera) are not due to differences in the binding affinity of the activated δ-endotoxins to the target cell, but instead are a reflection of the relative concentrations of the different receptors. That is, all three activated δ-endotoxins bind their respective receptors with about the same affinity, but the more active δ-endotoxin appeared to have a greater number of receptor sites and hence the target cells could bind more toxin molecules. In contrast, tobacco hornworm (*Manduca sexta*; Lepidoptera) brush border membrane vesicles bind all three toxins with about the same affinity and appear to have the same number of receptors for each of the three δ-endotoxins. As one might expect, all three δ-endotoxins are about equally active against *Manduca*. Thus, the number of receptor molecules on the target cells can at least influence the activity of a δ-endotoxin and perhaps, below a critical concentration, determine whether or not a given δ-endotoxin can be active at all. Presently, it is not known how many δ-endotoxin molecules are needed to bring about cell lysis.

Another possibility to account for activity of a given δ-endotoxin against insects in different orders is that the toxin-binding proteins in a number of different insects are structurally related and hence are capable of binding the same δ-endotoxin. If a δ-endotoxin bound to receptors in different insects using the same binding region, the receptors would have to be very similar, at least in the region bound by the δ-endotoxin. In the case of CryIIA, the receptor in both Lepidoptera and Diptera would have to be similar enough to bind the same protein, although it seems that changes in a few amino acids, and perhaps their influence on the folding of the protein, also play an important role in the toxic action of this δ-endotoxin. The search for the δ-endotoxin-binding protein(s) in the insect gut is currently an active area with a number of laboratories close to cloning an insect gut Bt δ-endotoxin receptor. Once some of these receptors are characterized in terms of their sequence, one will be able to determine if a δ-endotoxin that is active against a number of insects is active because of its ability to bind multiple sites or because its particular binding site is present on a number of different receptor molecules whose sequence is conserved among a number of different insects, perhaps even in different orders of insects. If the latter is actually the case, however, one might anticipate that the δ-endotoxins would have a broader spectrum of activity than that observed, since their conserved targets would be likely to be present in a number of insects, especially closely related insects. It is somewhat surprising that some δ-endotoxins are not, in fact, active against closely related species, supporting the argument that the δ-endotoxin-receptor interaction is quite specific.

The role of the δ-endotoxin receptor in the insect gut is unknown, but it is conjectured to play an important role in the insect. This is based on the observation that in one of the cases where an insect developed resistance to a given δ-endotoxin, the insect had an increased susceptibility to another δ-endotoxin. Van Rie *et al.* (1990a,b) demonstrated that resistance to CryIA(b) in the Indian meal moth (*Plodia interpunctella*; Lepidoptera) was accompanied by an increased sensitivity to CryIC. The decreased susceptibility to CryIA(b) was accompanied by a decreased binding of CryIA(b) to brush border membrane vesicles from the resistant insects. Increased susceptibility to CryIC was accompanied by an increase in the number of binding sites for this δ-endotoxin. CryIA(b) and CryIC did not compete for the same binding sites in these insects. Such compensation might indicate a critical role for the receptor protein in the insect gut. However, in the case of resistance to CryIA(b) in the diamondback moth (*Plutella xylostella*; Lepidoptera), there was a decreased binding of CryIA(b), but there was not a difference in the binding of, or susceptibility to, CryIB or CryIC (Ferre *et al.*, 1991).

Finding and engineering new activities

Use of *B. thuringiensis* δ-endotoxins for insect control is increasing. With increased use, there has been an increased effort in finding new strains with improved or novel activities. The last few years have seen a dramatic increase in the number of known strains and genes from Bt having insecticidal activities. Further, Bt strains and proteins with activities against other classes of organisms, such as nematodes, have also been discovered. The first coleopteran active strain was reported in 1983 by Krieg *et al.* There are now a number of known strains with coleopteran activity. This is the result of expanded screening efforts as well as expanded strain isolation efforts. There are various strategies for identifying strains with a novel activity, but each ultimately rests upon the relevant biological assay.

New strains can be isolated, all of which can be screened against the target pest(s). This is, however, a labour-intensive approach with many of the same strains, or known strains, being tested repeatedly. The source material for strain isolation can be random samples, such as soil or grain dust from storage elevators. There have been a number of strategies developed to try to enrich a strain population for the desired activity. One can try to use a directed screen based on ecological factors, such as habitat of the insect target or isolation of strains from diseased target insects. Such ecological methods of directing a strain search seem to be reasonable methods for reducing the number of strains one must investigate to find a desired activity, although such methods do not guarantee finding the desired activity. However, such searches have proved useful, especially in light of the fact that the first Bt strain was isolated from diseased silkworms by Ishiwata in 1901. Various other approaches to reduce screening efforts include the use of genes with the desired activity as hybridization probes to identify strains with the same or related genes (Visser, 1989) or using oligonucleotide probes from known genes (Prefontaine *et al.*, 1987); the use of monoclonal antibodies specific for various δ-endotoxins and also somewhat predictive of insecticidal activity (Hofte *et al.*, 1988); and the use of the PCR to identify strains carrying genes the same as or related to a given gene (Carozzi *et al.*, 1991). Such screens can identify strains carrying genes related to known genes and they can also identify strains carrying genes not related to currently characterized genes. By identifying strains not related to known strains, one may be able to enrich for strains carrying a novel activity.

While the traditional screening approach, with its multiple variations and refinements, continues to bear fruit, other approaches for producing δ-endotoxins with novel activities are also possible. As more knowledge is gained as to how the δ-endotoxins function, it becomes increasingly likely that one can engineer a δ-endotoxin to have a new activity. The silk moth (*Bombyx mori*; Lepidoptera) specificity domain from CryIA(a) has been moved to CryIA(c), thus imparting a new insecticidal activity to the resulting chimeric protein (Ge, Shivarova and Dean, 1989). The chimeric protein with *Bombyx* activity is also capable of binding *Bombyx* brush border membrane vesicles (Lee *et al.*, 1992). Similar chimeric proteins were constructed to better define the regions of CryIA(c) associated with determining specificity towards various insects (Ge, 1991). The region between amino acids 332–450 appears responsible for activity against cabbage looper (*Trichoplusia ni*; Lepidoptera), while the region between amino acids 335–615 was identified as the tobacco budworm (*Heliothis virescens*; Lepidoptera) specificity region. Transfer of amino acids 450–612 from CryIA(c) to CryIA(a) results, surprisingly, in a protein with about 30-fold more activity against *Heliothis* than either parental δ-endotoxin. How broadly such engineering of improved δ-endotoxins can be applied remains to be seen, since the rules governing such improvement remain to be elucidated. The region of CryIIA that allows it to be toxic towards mosquitoes has been identified by moving a 76 amino acid region to CryIIB (Widner and Whiteley, 1990). Li, Carroll and Ellar (1991) have resolved the structure of the CryIIIA molecule and have determined that it is composed of three distinct domains. As regions of the molecule responsible for binding to the insect gut receptors are better defined, it may be possible to alter them in a directed manner. Epitope mutagenesis (Scott and Smith, 1990) could be used to change the binding characteristics of a given δ-endotoxin and perhaps also alter its spectrum of activity. Such an approach would be dependent upon the development of good strategies to select the desired binding, since the molecules are quite large and the number of potential mutants too large to screen in a strictly random fashion. However, the ability to isolate brush border membrane vesicles from the desired target insect could facilitate the search for altered molecules able to bind the desired insect gut proteins.

Sivasubramanian and Federici (1991) have demonstrated that fusions between a δ-endotoxin and a viral-derived protein from *Autographa californica* multiple nuclear polyhedrosis virus (AcMNPV) can produce proteins with novel activity. In this case, the authors claim that altering the binding of the δ-endotoxin indeed alters its spectrum of activity. A fusion between the AcMNPV gp-64 protein and CryIIIA showed signs of activity against cabbage looper (*Trichoplusia ni*; Lepidoptera), as judged by a slight increase in mortality relative to a plasmid control where insects were fed *Escherichia coli* strains mised into their diet. This same fusion appeared to cause a decrease in weight gain when fed to tobacco budworm (*Heliothis virescens*; Lepidoptera) neonate larvae. The role of the receptor, or cellular 'toxin-binding protein', in this toxicity is unknown, so how broadly this type of engineering can be applied is uncertain. Further, such chimeric proteins appear to be unstable, so significant efforts in engineering stable fusion proteins may have to be made. More of these fusions will have to be attempted to ascertain the usefulness of this approach as compared with the random screening, or directed screening, of naturally occurring δ-endotoxins for new activities.

The steps of toxicity after binding are largely unknown. It is not known whether binding of the δ-endotoxin to the receptor results in a configurational change in the δ-endotoxin. If so, this could present another level of specificity. One can envision a δ-endotoxin binding to a cell surface protein in such a manner that a required configurational change does not occur properly and thus subsequent steps for toxicity are not possible. It is also possible that a specific interation between the cell receptor and the δ-endotoxin produces such a configurational change in the δ-endotoxin and this interaction could influence specificity. If such configurational changes do result from binding, the ability to engineer δ-endotoxins to alter specificity will be a more difficult task, since any changes in the δ-endotoxin which alter its binding ability could also deleteriously alter its ability to change configuration in response to binding.

It is unclear at this stage whether the receptor plays a passive role in events of toxicity or whether it plays an active role. If it plays a passive role, any epithelial cell surface protein may be able to serve as a receptor for an engineered δ-endotoxin, providing configurational changes mentioned above do not play a significant role in toxicity. The receptor may simply serve as an anchor near the cell surface, which allows a close interaction between the toxic region of the δ-endotoxin and the cell membrane. If the receptors play an active role in toxicity, such as triggering a cellular response, then specific receptors would have to be targeted. This would increase the difficulty of engineering a new host specificity onto an existing δ-endotoxin. When the insect cell's receptors are cloned and characterized, this issue will become less ambiguous. Until then, selection of new binding targets will have to rely on a more random approach.

Other host organisms

MICROORGANISMS

The first Bt δ-endotoxin gene was cloned and expressed in *Escherichia coli* in 1981 by Schnepf and Whiteley. Since then, many 'alternate hosts' have been transformed with δ-endotoxin genes. Examples of the micro-organisms other than *E. coli* that have been used as δ-endotoxin gene recipients are listed in *Table 4*. This list of alternate δ-endotoxin hosts can be grouped into four main categories: those used primarily to clone, or characterize specific δ-endotoxin genes; those used as fermentation hosts to increase production; those used to deliver the δ-endotoxin more effectively to the target insect pest; and insect pathogens which use δ-endotoxins to improve their effectiveness. This section will emphasize alternate hosts which have the potential of delivering the δ-endotoxin more effectively to the target insect and insect pathogens which use the δ-endotoxins to improve their effectiveness.

Although Bt is an effective microbial pesticide, a number of biological constraints limit its use. The Bt δ-endotoxin is short-lived on crops, thereby reducing its residual activity, necessitating the need for many applications during a growing season. Bt, as with other spray-on pesticides, is difficult to deliver to insect species which burrow into their host plant, hide under leaves or live primarily under the soil surface. In the case of aquatic insect control, maintaining the δ-endotoxin in the water at the level of insect feeding is difficult. Many of the problems and limitations of conventional Bt-based products are being addressed by molecular biologists using recombinant DNA technology. The isolation of δ-endotoxins has enabled scientists to clone and characterize the genes encoding these molecules. Each δ-endotoxin is encoded by a single gene; therefore, it can 'easily' be transferred to other Bts or to organisms such as plants, bacteria, fungi and algae. In the future, genetic engineering can be expected to result in the development of strains with improved potency, longer residual activity and broader host range, as well as novel delivery systems for δ-endotoxins, allowing better targeting and environmental persistence.

ALTERNATE DELIVERY IN EPIPHYTES AND ENDOPHYTES

Genetic engineering technology allows the use of micro-organisms that multiply on (epiphyte) or in (endophyte) plants to produce insecticidal proteins. This approach can overcome problems of insufficient coverage and persistence while providing season-long control.

One of the first attempts to improve the delivery and potency of δ-endotoxins against soil insect species was the work of Watrud *et al.* (1985) and Obukowicz *et al.* (1986a,b). The authors introduced a δ-endotoxin gene from *B. thuringiensis* var. *kurstaki* HD-1 into two corn root-colonizing strains of *Pseudomonas fluorescens*, using transposase deletion derivatives of the transposon Tn5 containing the δ-endotoxin gene. Their objective was to clone the δ-endotoxin gene into non-pathogenic soil microbes which showed a high degree of association with corn roots. The types of commercial products

Table 4. Micro-organisms that have been used as alternative hosts for cloning *B. thuringiensis* δ-endotoxin genes a

Bacillus cereus (Gonzalez, Brown and Carlton, 1982; Aronson and Beckman, 1987)
Bacillus subtilis (Shivakumar *et al.*, 1986; Earp and Ellar, 1988)
Bacillus sphaericus (Bar *et al.*, 1991)
Bacillus megaterium (Donovan *et al.*, 1988; Mettus and Macaluso, 1990)
Pseudomonas fluorescens (Obukowicz *et al.*, 1986a,b)
Pseudomonas cepacia (Stock *et al.*, 1990)
Caulobacter crescentus (Thanabalu *et al.*, 1992)
Agmellenum quadruplicatum (Angsuthanasombat and Panyim, 1989; Murphy and Stevens, 1992)
Rhizobium meliloti (Bezdicek, Quinn and Kahn, 1991)
Rhizobium leguminosarum (Bezdicek, Quinn and Kahn, 1991)
Clavibacter zyli subsp. *cynodontis* (Turner *et al.*, 1991)
Insect baculoviruses (Martens *et al.*, 1990; Merryweather *et al.*, 1990; Pang, Frutos and Federici, 1992)

envisioned to result from this type of work are pesticidal inocula that are applied to seeds or soil at the time of planting. Subsequent colonization of roots could then conceivably supply protection against soil-borne insect pests and minimize the need for multiple applications of pesticide during the growing season of the crop.

Since the practical application of this technology involved the release of a living recombinant bacteria into the field, many questions were raised as to its ecological impact. The general guidelines of the Federal Insecticide Fungicide Rodenticide Act were followed in assessing the ecological risk associated with releasing a living recombinant micro-organism. A number of studies were carried out to identify and monitor the engineered pseudomonad in the environment. Studies were designed to assess the genetic stability and likelihood of genetic exchange, in addition to the toxicity towards non-target species.

A native, genetically marked parental strain and its engineered derivative containing a chromosomal copy of the δ-endotoxin gene exhibited essentially identical survival profiles on the corn rhizoplane under growth chamber conditions. In those experiments, corn seeds had been treated with inocula and planted in non-sterile soil. The engineered and native isolates also showed virtually identical rapid decline survival profiles in sewage and surface water samples from various rivers and lakes. The only time persistence of the inocula was demonstrated was when water samples were filter sterilized prior to the addition of test inocula, suggesting that natural competitors, predators or parasites of microbes play an active role in limiting the persistence of introduced bacterial inocula. The case for genetic stability of the engineered pseudomonad and for lack of genetic exchange with other micro-organisms was based on the fact that the δ-endotoxin gene was inserted into the chromosome of *P. fluorescens* rather than being present on a plasmid. Chromosomal insertion of δ-endotoxin genes was preferred over plasmid insertion because chromosomal DNA is inherently more stable than plasmids and the *P. fluorescens* strain lacks indigenous plasmids and has demonstrated a limited capacity to accept plasmids from broad host-range incompatibility groups. The authors also postulated that if a plasmid could be introduced naturally into these isolates, the probability of transfer of the engineered trait from the chromosome to a second microbe would be very remote. Under optimal laboratory conditions, the estimated frequency of this transfer was 10^{-9}.

Toxicity assessments that were performed for engineered and native isolates of *P. fluorescens* included studies in non-target organisms such as mice, quail, fish, honeybees, earthworms and *Daphnia* spp. With the exception of target lepidopteran species such as *Manduca, Trichoplusia, Heliothis* and *Agrotis*, none of the studies demonstrated toxic effects resulting from the microbial treatments. The host range and relative pesticidal efficacy of the engineered *P. fluorescens* paralleled those of *B. thurinigiensis* var. *kurstaki*. Although initial results from this work appeared promising, anticipated field releases met with considerable opposition at the time, presumably due to the fact that the biocontrol product was a living recombinant organism.

Waalwijk, Dullemans and Maat (1991) developed another delivery system based on a grass root colonizing *Pseudomonas* to aid in the control of the root feeding dipteran *Tipula oleracea*. They cloned the cryIVB gene from *B. thuringiensis* var. *morrisoni* PG-14 and inserted it into *P. fluorescens* P1. To reduce the mobility of the introduced gene, Waalwijk *et al.* developed suicide vectors where integration is dependent upon recombination between homologous DNA sequences. These suicide vectors offer an improvement over the *Pseudomonas* constructs used by Obukowicz *et al.* (1986a,b) described above, because the Obukowicz strains would still allow mobilization of inserted genes by transposases provided by incoming plasmids. Waalwijk *et al.* tested 27 *P. fluorescens* isolates and found only one that hybridized with the recombinant DNA from strain P1. They suggested this indicated that the sequences flanking the *cryIVB* gene are relatively specific to the P1 strain and the probability that transgenic P1 cells would encounter these sequences in the field would be rather remote.

Stock *et al.* (1990) chose a strain of *Pseudomonas cepacia* (Pc) isolated from soil as a means of delivering a CryIA(c) protein to target insects. *Pseudomonas cepacia* 526 was found to colonize a variety of plant roots and leaves. The *cryIA(c)* gene from *B. thuringiensis* var. *kurstaki* HD-1 was cloned into a broad host range vector and the resulting plasmid conjugated into Pc526. Southern blot analysis confirmed the presence of the *cryIA(c)* gene in the transconjugants. The authors also noted that the Pc 526 colonies carrying the *cryIA(c)* gene were smaller than the wild-type Pc colonies and found that the Pc 526-Bt strain grew slower in liquid culture with a doubling time 2.2 to 2.5-fold longer than the wild-type Pc. They suggested that the introduction of the *cryIA(c)* gene had deleterious effects on Pc bacterial growth.

Western blot analysis demonstrated that the Pc 526-Bt strain produced a 78 kDa truncated version of the CryIA(c) protein that reacted weakly with HD-1 specific antiserum. The engineered strain was active against tobacco hornworm (*Manduca sexta*; Lepidoptera) at a concentration of 10^5–10^6 bacteria cm^{-2}. The establishment of Pc 526-Bt and its insecticidal activity were determined on axenically grown tobacco plants. The stability of the plasmid carrying the *cryIA(c)* gene was monitored because laboratory results suggested that the plasmid was unstable under non-selective conditions. Three days after tobacco leaves were inoculated with Pc 526-Bt and Pc 526, population estimates suggested that 10^8 bacteria per gram fresh weight remained and that the majority of that population was Pc 526 wild-type. The plasmid carrying the *cryIA(c)* gene was found in only 1.0% of the Pc 526-Bt population, confirming the observation that this plasmid is unstable under non-selective conditions. However, the proportion of Pc 526-Bt population expressing the *cryIA(c)* gene was large enough to protect the tobacco plants from *Manduca* feeding damage. Most larvae on plants treated with the Pc 526-Bt were killed, whereas most on the control plants survived.

These experiments estimate the number of leaf-colonizing Pc that would be required to protect a tobacco plant from *Manduca* feeding damage. However, since these data were generated on axenically grown tobacco, they may not reflect the true field environment, where competition with other bacteria would be present. Further experiments in greenhouse environments and actual outdoor field trials will help address the question.

The plant endophyte *Clavibacter xyli* subsp. *cynodontis (Cxc)* was engineered to express a δ-endotoxin and subsequently introduced into corn plants (Dimock, Beach and Carlton, 1989; Kostka, Reeser and Miller, 1988) to protect against European corn borer (*Ostrinia nubilalis*; Lepidoptera). This work focused on genetically altering *Cxc* so that it contains the new DNA inserted into its chromosome. The inserted DNA contains the *cryIA(c)* gene from *B. thuringiensis* var. *kurstaki* HD-73 fused to a gene that confers kanamycin resistance, a tetracycline resistance gene, and an *E. coli* replicon that is not functional in the new host. The wild-type strain of *Cxc* described by Davis *et al.* (1984) is frequently found in nature colonizing the vascular system of Bermuda grass. In addition to Bermuda grass, the wild-type and engineered *Cxc* have been shown to colonize several important crop plants including maize (Kostka, Reeser and Miller, 1988). This organism is one of the only alternative hosts for Bt δ-endotoxins which has been approved for field testing.

Another example of using an endophyte to deliver a δ-endotoxin to soil-inhabiting insects is the work of Bezdicek, Quinn and Kahn (1991). They introduced the *cryIIIA* gene from *B. thuringiensis* var. *tenebrionis* into *Rhizobium meliloti* and *R. leguminosarum* BV. *vicaea* to protect alfalfa and pea from the feeding damage of two coleopteran insects, the clover root curculio (*Sitona hispidulus*) and the pea leaf weevil (*Sitona lineatus*), respectively. Two *cryIIIA* constructs were made that have NH_2-terminus-truncated forms of CryIIIA in the broad host range vector pRK311 and differ only in the promoters that drive expression of the *cryIIIA* gene. In one, expression relies on the lacZ promoter, which is expressed by *Rhizobium in vitro* and *in planta*, whereas the other construct uses the nifH promoter, which is active only when the *Rhizobium* are fixing nitrogen inside nodules. Western blot analysis showed that CryIIIA protein was produced *in vitro* and in nodules. Bioassays indicated that the CryIIIA protein expressed in rhizobia and nodules was toxic to Colorado potato beetle when fed recombinant *Rhizobium*-treated potato leaves and to larvae of *S. hispidulus* and *S. lineatus* feeding on alfalfa and pea nodules, respectively, containing the recombinant *Rhizobia*.

Other workers (Barnes and Cummings, 1987) have developed a different approach to using a pseudomonad as an alternative δ-endotoxin host. In an attempt to improve the foliar persistence of Bt insecticidal activity, they developed a delivery system based on a recombinant micro-organism that expresses a δ-endotoxin, but has been killed using heat and chemical treatment prior to field release. Because the organisms are dead, this product has the additional advantage of relative freedom from the environmental and safety concerns associated with outdoor testing or the release of living recombinant organisms. Using this system, researchers have produced lepidopteran active and coleopteran active δ-endotoxins in a non-pathogenic strain of *Pseudomonas fluorescens*. When transformed cells are grown in submerged culture, the δ-endotoxin forms a typical δ-endotoxin-like crystal within the cell. However, unlike *Bacillus* cells, which normally burst at the end of the growth cycle releasing naked spores and crystals into the culture medium, the δ-endotoxin-producing *Pseudomonas* cells remain intact at the completion of the fermentation cycle. While still in the fermentation tank, the cells are chemically treated to kill the *Pseudomonas* cells and cause the bacterial cell wall to become more rigid through cross-linking of the cell wall components. The dead bacterial cell wall now serves as a protective 'microcapsule' for the enclosed δ-endotoxin. Because this bacterium produces no spores, it avoids the release of living bacteria when the product is applied to crop plants in the field. In part due to the lack of viable spores and DNA, this delivery system should be particularly useful for recombinant δ-endotoxins.

In 1985, the EPA approved testing of this encapsulated product on small fields with a lepidopteran active δ-endotoxin, making it the first recombinant product approved for outdoor testing. Field experiments conducted during 1988 suggested that a foliar application of this engineered microbe protected cabbage from lepidopteran pests for 7 days, whereas insecticidal activity from traditional Bt sprays dissipated by 96 h post-application. This increase in persistence resulted in better insect control, and therefore higher yields. Similar tests have been conducted with engineered microbes targeted against the Colorado potato beetle (CPB) on potatoes. Results from these tests confirm that when equivalent toxin rates of the engineered microbe and its naturally occurring counterpart, *B. thuringiensis* var. *san diego*, are applied to potato foliage, the persistence conferred by the microbially encapsulated product resulted in superior CPB control. It was also shown that the protected δ-endotoxin could be applied at two-thirds the toxin rate and still maintain higher levels of CPB control than the full rate delivered by the *B. thuringiensis* var. *San diego* product.

The microbial encapsulation delivery system is said to offer a number of advantages over traditional Bt products (Gaertner, 1990; Gelernter, 1990): the longevity of insect control may decrease the number of applications, and therefore the cost, required for season-long insect control; a two-fold increase in foliar persistence results in better insect control, when applied at the same rate and same application schedule as naturally occurring Bts; higher persistence levels may allow reductions in the recommended field rate, while still maintaining excellent levels of insecticidal activity; and product shelf life is enhanced by the absence of living organisms in these preparations.

ALTERNATE HOSTS AS CONTROL AGENTS FOR AQUATIC INSECTS

One of the major factors limiting the duration of mosquito control following the application of Bt strains is the rapid sedimentation of spores and insecticidal crystals. Since most mosquito larvae feed at or near the water surface, the effective larvicidal activity of the toxins is limited to the short period of time prior to sedimentation. A potential approach to circumvent these problems is to engineer micro-organisms living in the upper layers of aquatic habitats to synthesize toxic proteins. Several workers have recognized this and have transferred δ-endotoxin genes from *B. thuringiensis* var. *israelensis* into various aquatic insect food sources.

Thanabalu *et al.* (1992) cloned the 130 kDa (*cryIV*) toxin gene from *B. thuringiensis* var. *israelensis* into the broad host range plasmid pRK248 and expressed it in *Caulobacter crescentus* CB15. The recombinant *C. crescentus* cells were shown to be toxic to mosquito larvae. *Caulobacter* species are ubiquitous micro-organisms residing in the upper regions of aquatic environments. In the flagellate swarmer stage, *Caulobacter* species are motile, thus allowing distribution through the habitat. In the swarmer and stalked stages, they are capable of attachment to solid particles at or near the water surface. Thus *Caulobacter* species provide the potential for prolonged control by maintaining mosquitocidal toxins in larval feeding zones.

Angsuthanasombat and Panyim (1989) introduced the 130 kDa (*cryIV*) δ-endotoxin gene from *B. thuringiensis* var. *israelensis* into the cyanobacterium *Agmenellum quadruplicatum* (Aq) PR-6 by plasmid transformation. Lysates produced by sonication of recombinant Aq cells were shown to be toxic to mosquito larvae, but larvicidal activity of intact recombinant cyanobacteria was not demonstrated. The larvicidal activity of *E. coli* containing pAQPc388i was more effective than that of its Aq PR-6 counterpart, presumably due to the *E. coli* expressing higher amounts of undergraded δ-endotoxin.

INSECT BACULOVIRUSES AS ALTERNATE HOSTS

Baculoviruses are naturally occurring pathogens that infect some important lepidopteran and hymenopteran insects and are used in the control of these insects. A major drawback for a more widespread use of these viruses is their relatively slow insecticidal action. Baculoviruses produce a virion-occluding protein called polyhedrin late in the development of the virus. These viruses are capable of expressing foreign genes under the control of the polyhedrin promoter (Luckow and Summers, 1988). The polyhedrin gene is dispensable for virus replication, and the expression of foreign genes is based on the allelic replacement of the polyhedrin gene (Smith, Fraser and Summers, 1983). Genetically engineered variants of baculoviruses incorporating δ-endotoxin genes have been developed and are currently being evaluated to see if they are more effective control agents (Martens *et al.*, 1990; Merryweather *et al.*, 1990; Pang, Frutos and Federici, 1992).

Martens *et al.* (1990) introduced a *cryIA(b)* gene, isolated from *B. thuringiensis* var. *aizawai* 7.21, into the genome of the *Autographa californica* nuclear polyhedrosis virus (AcNPV). A transfer vector containing the β-galactosidase marker gene facilitated the screening of recombinants. All the β-galactosidase positive recombinant viruses analysed contained an insertion of the *cryIA(b)* gene. Recombinants were easily retrieved from infected *Spodoptera frugiperda* (Lepidoptera) cells, suggesting that the CryIA(b) δ-endotoxin is not toxic when produced within the cell. *Spodoptera* cells infected with recombinant ACNPV expressed the CryIA(b) protein at 5% of the total cell protein. CryIA(b) was found predominantly in the cytoplasm of infected cells as large, crystalline inclusions, sometimes bipyramidal in shape. The ultrastructure is similar to that of native Bt δ-endotoxins. Although these crystalline inclusions have been shown in organisms other than native Bt strains (Shivakumar *et al.*, 1986; Oeda *et al.*, 1989), the observation of Martens *et al.* (1990) is the first report of crystal formation occurring in eukaryotic cells. The toxicity of CryIA(b) δ-endotoxin expressed by AcNPV recombinants was comparable with that of CryIA(b) expressed by a corresponding *E. coli* recombinant when tested against the European cabbageworm (*Pieris brassicae*; Lepidoptera).

Merryweather *et al.* (1990) inserted the *cryIA(c)* gene from *B. thuringiensis* var. *kurstaki* HD-73 into the genome of the baculovirus *Autographa californica* using two transfer vector systems. In the first, the δ-endotoxin gene was placed under the control of the polyhedrin gene promoter, thus giving a polyhedrin-negative recombinant virus, Ac(PH^-)Bt. In the second, the δ-endotoxin gene was inserted under the control of a copy of the AcMNPV p10 promoter positioned upstream of the polyhedrin gene to produce a polyhedrin-positive recombinant virus, Ac(PH^+)Bt. An analysis of the cells infected with each recombinant virus demonstrated that a large polypeptide of about 130 kDa was synthesized, which reacted with antiserum specific for the Bt HD-73 δ-endotoxin. Two additional peptides of 62 kDa and 44 kDa were also detected. The authors speculate that these smaller peptides were degradation products of the larger 130 kDa protein. The peak time of synthesis for the 130 kDa protein in insect cells was about 18 h post-infection.

When larvae of the cabbage looper (*Trichoplusia ni*, Lepidoptera) were exposed to a diet treated with cell extracts or virus particles from recombinant virus infected cells, the larvae refused to eat. This was attributed to the presence of δ-endotoxin, since feeding inhibition is a well-known characteristic of insects that ingest natural Bt δ-endotoxins. The Ac(PH^+)Bt virus had an LD_{50} value about two-fold higher than that of the unmodified AcMNPV. The authors contend that this was within known experimental variation of AcMNPV viruses and that the recombinant virus and the wild type were essentially of equal toxicity. They also suggested that trace amounts of δ-endotoxin in the virus preparation used in the bioassays may have affected the infection of cells in the midgut by the virus. Clearly more work needs to be done before any conclusions regarding the utility of these constructs are made.

A recombinant baculovirus containing a mosquitocidal *cryIVD* gene from *B. thuringiensis* var. *morrisoni* (Pg-14) fused with the lacZ gene from *E. coli* under control of the polyhedrin promoter was constructed by Pang, Frutos and Federici (1992). Strong expression of the chimeric protein was obtained in infected cells and in *Trichoplushia* larvae. In both, crystalline cuboidal inclusions were formed in the cell cytoplasm during late stages of infection. These inclusions were purified and shown by Western blot analysis to contain a protein that reacted with antibody raised against the CryIVD δ-endotoxin from *B. thuringiensis* var. *israelensis*. This chimeric protein, however, was not active against mosquito larvae.

The large protein capsules (polyhedra) containing the virions are the infectious stages of baculoviruses as well as the most environmentally stable form of the virus. The application of polyhedra-negative viruses in the field is impractical, since naked virions are quickly inactivated (Bishop, 1989). Thus, recombinant viruses that have maintained the polyhedrin gene and therefore produce polyhedra are more desirable. Recombinants that have a δ-endotoxin gene inserted in another locus under the control of a promoter other than the polyhedrin promoter, for example the p10 gene, may be of more practical significance. The fact that δ-endotoxins produced in recombinant baculovirus-infected cells are biologically active provides a basis for a strategy in which Bt δ-endotoxins may be used to enhance the effectiveness of baculoviruses as microbial control agents.

Finally, an interesting approach to controlling Colorado potato beetle (CPB: *Leptinotarsa decemlineata*; Coleoptera) using an alternate host for δ-endotoxins is suggested by Bove and Hackett (1992). They propose engineering a δ-endotoxin into a spiroplasma (helical, wall-less bacteria in the Class Mollicutes) that is a commensal gut organism of CPB. Since no transformation system exists for introducing δ-endotoxin genes into the spiroplasma genome, their initial objectives are to develop such a system using either a native spiroplasma virus, such as SpV1 or *E. coli* plasmids. An *E. coli* plasmid containing a streptococcal transposon (Tn916) has already been used successfully as a vector system for *Mycoplasma* spp. (Dybvig and Cassell, 1987).

Other suitable hosts for expressing δ-endotoxins may be found in the groups of entomopathogenic fungi, protozoa and nematodes. Studies of the genetics and molecular biology of these groups, however, are in the very early stages of development and it is not likely that reports of inserting Bt genes into these organisms will appear in the near future.

Naturally occurring insect pathogens, including bacteria, fungi and viruses, occupy environmental niches in which they function as pest control agents; however, in most cases, this attribute cannot be readily adapted to crop protection. Clearly, there are major factors constraining the wider use of microbial pesticides. These include lack of environmental persistence, narrow host range, limited virulence and high production costs. Recombinant DNA technology will provide the tools for developing safe, efficient and cost-effective microbial control agents. It is now possible to combine the best traits of several different organisms into a single strain, precluding the need for complex mixtures of micro-organisms expressing δ-endotoxins which exhibit enhanced insecticidal activity and greater persistence in the environment.

Production of *Bacillus thuringiensis* insecticidal proteins in plants

The long history of efficacy and safety as microbial insecticides has placed the δ-endotoxins among the first genes of commercial interest to be engineered into plants. Since the δ-endotoxin is the product of a single gene, it is an ideal candidate for use in the rapidly progressing field of plant genetic engineering. Over the past decade, various Bt microbial sprays have been used to control insects on important crops such as corn, cotton, tomato and potato. Such crops have the most to benefit from acquiring resistance to insect attack through genetic engineering. Insecticidal sprays have inherent disadvantages that make production of δ-endotoxins in plants attractive: farmers must spray their crops several times a season; the insecticide may be washed off by rains; and short residual time on the leaf, believed to be due to various environmental factors such as temperature, leaf pH and UV inactivation, may lead to efficacy problems. The creation of transgenic plants which effectively utilize the insecticidal potential of Bt was initially encumbered by technical difficulties. Over the past few years, however, a combination of advances in plant transformation and gene expression has made the production of effective transgenic Bt plants a viable possibility. It is expected that cotton, potato and corn varieties containing a Bt δ-endotoxin gene will be marketed within the next few years and varieties of genetically transformed vegetable crops containing a δ-endotoxin gene will follow not long after.

At present, the majority of transgenic crops containing Bt genes have been transformed with *cryI* genes which have activity against the lepidopterans, including the tobacco hornworm, tobacco budworm, tomato pinworm, corn earworm and European corn borer. More recently, the *cryIIIA* gene has been engineered into plants for protection against the coleopterans, including the Colorado potato beetle (Perlak *et al.*, 1993). Bt δ-endotoxin genes with activity against insects of the genus *Diabrotica*, such as corn rootworm, a major corn pest, will likely be produced in the next generation of transgenic plants.

First-generation transgenic Bt plants

The first published accounts of transgenic plants containing the Bt δ-endotoxin gene include Vaeck *et al.* (1987), Barton, Whiteley and Yang (1987), Adang *et al.* (1987), and Fischhoff *et al.* (1987). The first three groups published experiments using transgenic tobacco, whereas Fischhoff *et al.* published results on transgenic tomato transformed using *Agrobacterium*-mediated T-DNA transfer. Vaeck *et al.* used the 2′ T-DNA mannopine synthetase promoter and 3′ polyadenylation region of T-DNA gene 7 to drive the expression of the *cryIA(b)* gene from *B. thuringiensis* var. *berliner* 1715. The constructs tested in tobacco included the full-length gene coding for a 1155 amino acid protein, a fragment of the *cryIA(b)* gene coding for 610 NH_2-terminal amino acids which had previously been shown to be fully insecticidal (Hofte *et al.*, 1986), and two constructs with translational fusions between NH_2-terminal fragments of CryIA(b) and the neomycin phosphotransferase (neo) gene from transposon Tn5, which confers kanamycin resistance to plants. The authors anticipated that δ-endotoxin-neo fusion transformants selected on high levels of kanamycin would aid in the selection of transformants containing high levels of δ-endotoxin. Transgenic tobacco plants were obtained by leaf disk transformation of *Nicotiana tabacum* var. Petit Havana SR1. Tobacco transformed with the δ-endotoxin-neo fusions and the NH_2-terminal truncated δ-endotoxin gene constructs produced 75–100% mortality against tobacco hornworm (*Manduca sexta*; Lepidoptera) larvae. The transformants containing the translational fusion constructs produced at a high frequency plants exhibiting 70–100% mortality against *Manduca*, whereas the unfused truncated endotoxin gene gave similar levels of toxicity but in a smaller percentage of transformants. This work also showed that plants transformed with the full-length endotoxin gene produced very little δ-endotoxin protein and were non-insecticidal. The levels of δ-endotoxin expressed in the highly insecticidal plants ranged from 2.6 to 190 ng CryIA(b) per mg soluble protein, or 0.0002–0.02% of total soluble protein. Further analysis of transformed tobacco showed that the plants expressing the highest levels of CryIA(b) contained five copies of the gene. Normal inheritance patterns of the Bt gene were confirmed in progeny plants (Vaeck, Reynaerts and Hofte, 1989).

Barton, Whiteley and Yang (1987) studied the expression of a NH_2-terminal fragment of the *cryIA(a)* gene from *B. thuringiensis* var. *kurstaki* HD-1 in transgenic tobacco. Transformants were obtained in *Nicotiana tabacum* cv. Havana 425 using an *Agrobacterium tumefaciens* binary vector system. They obtained insecticidal plants using constructs with either a full-length δ-endotoxin gene or a truncated δ-endotoxin gene encoding a 644 amino acid protein. Both δ-endotoxin genes were driven by a 425 bp cauliflower mosaic virus (CaMV) 35S promoter sequence, including the 5′ untranslated region of AMV RNA 4 and the nopaline synthetase 3′ polyadenlyation region. Tobacco calli expressing the full-length *cryIA(a)* gene at levels of 10–50 ng Bt per mg soluble protein became necrotic and died. None of the plants which were regenerated from the calli transformed with the full-length gene produced detectable levels of CryIA(a), mRNA or insecticidal activity, and many plants were found to contain deleted or broken *cryIA(a)* genes. Transgenic tobacco plants exhibiting the highest levels of toxicity contained 3–5 copies of the Bt gene. Northern blot analysis of transgenic plants showed hybridization of a δ-endotoxin-specific probe to a mRNA species shorter than the expected full-length transcript and also significant hybridization to distinct shorter fragments. The existence of fragmented Bt mRNA in transgenic plants was attributed to inefficient post-transcriptional processing or rapid turnover.

Adang *et al.* (1987) generated transgenic tobacco, *Nicotiana tabacum* cv. Xanthi, containing the full-length *cryIA(c)* gene from *B. thuringiensis* var. *kurstaki* HD-73 using a binary vector system and *Agrobacterium tumefaciens*. The 3.7 kb *cryIA(c)* gene was under control of the promoter and polyadenylation sequences of the T-DNA gene encoding the first step of mannopine synthesis (ORF 24). Northern blot analysis of transgenic plant RNA identified a 1.7 kb RNA corresponding to the 3′ end of the *cryIA* gene. At least one plant was identified with low levels of CryIA(c) protein (2 ng per mg soluble protein) and some degree of toxicity (53% mortality) towards tobacco hornworm (*Manduca sexta*; Lepidoptera).

Fischhoff *et al.* (1987) published the first transformation of tomato (*Lycopersicon esculentum* line VF36) via *Agrobacterium*-mediated T-DNA gene transfer with the *cryIA(b)* gene from *B. thuringiensis* var. *kurstaki* HD-1. Two truncated versions of the Bt gene were used; one encoded a protein of

646 amino acids and the other a protein of 725 amino acids. Each version was driven by the CaMV 35S promoter and the 3′ polyadenylation region from the nopaline synthetase gene. The findings of Vaeck *et al.* (1987) were strengthened by this study, which demonstrated that expression of the NH_2-terminal fragment of a δ-endotoxin produced plants with a high level of activity against tobacco hornworm (*Manduca sexta*), whereas plants containing a full-length δ-endotoxin gene did not confer 100% insecticidal activity. The study showed significant mortality and severe growth inhibition of *Heliocoverpa zea* (Lepidoptera) and *Heliothis virescens* (Lepidoptera) populations, which require higher levels of Bt δ-endotoxin for mortality. Progeny plants demonstrated the heritability patterns of a typical Mendelian dominant gene.

Transformation of dicots using *Agrobacterium* T-DNA was very successful in producing engineered plants containing Bt genes. There are published reports of δ-endotoxin expressing transgenic potato (Vaeck, Reynaerts and Hofte, 1989), cotton (Perlak *et al.*, 1990), tobacco and tomato (Vaeck, Reynaerts and Hofte, 1989; Honee, van der Salm and Visser, 1989). In 1989, Vaeck, Reynaerts and Hofte published results of studies on the transformation of tomato and potato plants with a portion of the *cryIA(b)* gene encoding the NH_2 terminal portion of the δ-endotoxin. Transgenic tomato plants exhibiting 100% mortality against tobacco hornworm (*Manduca sexta*) produced 60–80 ng CryIA(b) per gram of leaf tissue and potato plants with similar activity produced 90–150 ng CryIA(b) per gram of leaf tissue. In 1992, Jansens *et al.* demonstrated significant, although not complete, control of the American bollworm (*Helicoverpa armigera*; Lepidoptera) with transgenic tomatoes expressing a truncated CryIA(b) δ-endotoxin.

At least seven research groups had conducted transgenic field trials using plants expressing a δ-endotoxin gene by the end of 1990 (Goldberg and Tjaden, 1990), but the majority of results were not published. The first published report on the field performance of transgenic tomato plants expressing a δ-endotoxin gene was published in 1989 by Delannay *et al.* The authors demonstrated that transgenic tomato plants expressing the *cryIA(b)* gene were insecticidal against tobacco hornworm (*Manduca sexta*), but that higher levels of CryIA(b) expression would need to be achieved to fully control the agronomically important pests such as the tomato fruitworm (*Heliocoverpa zea*; Lepidoptera) and tomato pinworm (*Keiferia lycopersicella*; Lepidoptera). Their plants were homozygous progeny of the best T_0 plants in terms of expression of the transformed truncated *cryIA(b)* gene encoding a 725 amino acid protein derived from *B. thuringiensis* var. *kurstaki* HD-1.

The results of three Bt field trials were published in 1992 concerning transgenic tobacco (Warren *et al.*, 1992; Carozzi *et al.*, 1992; Hoffmann *et al.*, 1992) and cotton (Wilson *et al.*, 1992) using truncated δ-endotoxin genes. The test described by Warren *et al.* (1992) and Carozzi *et al.* (1992) consisted of six transgenic tobacco lines expressing a 645 amino acid truncated CryIA(b) protein from *B. thuringiensis* var. *kurstaki* HD-1 inserted at a single locus in the tobacco genome. The site of insertion contained a single copy of the truncated *cryIA(b)* gene. This gene was under control of the CaMV 35S promoter and 35S 3′ polyadenylation sequences. Transgenic plants were generated using *Agrobacterium*-mediated leaf disk transformation of *Nicotiana tabacum* var. Havana 38. Both homozygous and hemizygous plants of each line were tested. Five of the six CryIA(b) lines tested provided economic control against both tobacco hornworm (*Manduca sexta*; Lepidoptera) and tobacco budworm (*Heliothis virescens*; Lepidoptera), an insect that is 40-fold less sensitive than *Manduca* to the CryIA(b) protein. During the course of a 14 week field test under extremely high insect pressure, it was demonstrated that hemizygous plants can express sufficient levels of CryIA(b) δ-endotoxin to protect tobacco against economic damage from tobacco hornworm and tobacco budworm. The authors observed that Bt δ-endotoxin levels increased throughout the course of plant development, with a substantial increase at the time of flowering. Levels at flowering ranged from 400 to 1000 ng CryIA(b) per gram fresh weight or up to 0.01% total soluble protein. Bt δ-endotoxin mRNA of the expected size was readily detected, but there were also distinct truncated RNA forms, perhaps resulting from incomplete transcripts or cleavage products.

Hoffmann *et al.* (1992) also reported on the field performance of transgenic tobacco plants (*Nicotiana tabacum* var. Xanthi) containing a translational fusion of the *cryIA(c)* gene with the neo gene (kanamycin resistance) from transposon Tn5. Transgenic plants contained a truncated *cryIA(c)* gene derived from *B. thuringiensis* var. *Kurstaki* HD-73 encoding a 612 amino acid NH_2-terminal fragment, with the translational fusion protein under control of the CaMV 35S promoter and the 3′ region of the tomato protease inhibitor 1 gene. The data from this study demonstrated control against artificial infestations of tomato fruitworm (*Heliocoverpa zea*; Lepidoptera). This *cryIA(c)*-neo fusion has also been transformed in potato, *Solanum tuberosum*, using an *Agrobacterium tumefaciens* binary vector system (Cheng *et al.*, 1992). Transgenic potato plants showed some activity against *Manduca sexta*, primarily limited to a reduction in feeding.

The results of the field tests of the first-generation transgenic tobacco plants were encouraging, but the studies concurred in concluding that the expression levels of Bt in plants were relatively low and mRNA analysis demonstrated the presence of substantial amounts of truncated transcripts. Clearly, the expression levels of Bt in crops such as tomato, cotton and corn would not be high enough to protect totally plants from the major pests of agronomic importance. Modifications to the gene and/or regulatory sequences were required to enhance expression.

Murray *et al.* (1991) examined the expression of *cryIA(b)*, *cryIA(c)* and *cryIIIA* genes in both transgenic tobacco and electroporated carrot protoplasts to study mRNA stability. The study examined full-length and truncated *cryIA(b)* and *cryIA(c)* genes from *B. thuringiensis* var. *kurstaki* and a full-length *cryIIIA* gene from *B. thuringiensis* var. *tenebrionis*. The *cryIA(b)* gene was under control of the CaMV 35S promoter and 3′ polyadenylation sequences from ORF 26 of the TR-DNA and the *cryIA(c)* gene was under control of the mannopine synthetase promoter. Northern analysis of δ-endotoxin mRNA isolated from transgenic tobacco plants transformed with either full-length or truncated genes from both *cryIA(b)* and *cryIA(c)* showed hybridization only to truncated transcripts. Codon usage in the native δ-endotoxin genes is considerably different from that found in typical plant genes, which have a higher G + C content. Truncated δ-endotoxin gene transcripts in transgenic plants could result from a number of events relating to the high A + T content of these genes. These include premature transcriptional termination or polyadenylation in regions of high A + T content or inappropriate splicing or cleavage at sites recognized by the plant as splice sites. Instability of the mRNA could be the result of endonucleolytic or exonucleolytic degradation at specific sequences that destabilize the message during transcription or create pausing due to the formation of secondary structures. Instability of the mRNA could also be the result of inefficient translation due to codon usage differences between the Bt δ-endotoxin gene and plant genes. Northern blot analysis of *cryIA(b)* and *cryIA(c)* constructs electroporated into carrot protoplasts showed full-length non-degraded forms during the first 8 h after electroporation, but by 18 h the *cryIA* mRNA was degraded. The results of these transient expression experiments led the authors to conclude that the truncated transcripts observed in plants were the result of message instability rather than truncation due to premature transcriptional termination or polyadenylation. Using a series of 3′ deletion constructs in electroporated carrot cells, they demonstrated that deletion of sequences in the 3′ end of the *cryIA(b)* did not increase message stability, and the RNA instability was retained in the first 570 bases of the gene. The *cryIIIA* gene was also poorly expressed in electroporated carrot cells.

Second-generation transgenic Bt plants

Efforts to enhance Bt δ-endotoxin gene expression in second-generation transgenic plants include expressing δ-endotoxin genes with sequence modifications and/or new promoters, including tissue-specific promoters. Second-generation plants now include maize, which had, until recently, been particularly recalcitrant to genetic engineering. Recent advances in transformation have made it possible to introduce δ-endotoxin genes into maize.

To overcome problems presented by trying to express an A + T rich gene in an organism with a high G + C genomic content, several laboratories have made partially or completely modified *cryI* δ-endotoxin genes that have resulted in significant improvements in gene expression in cotton (Perlak *et al.*, 1990; Wilson *et al.*, 1992), tobacco, tomato (Barton and Miller, 1993) and maize (Koziel *et al.*, 1993). In 1990, Perlak *et al.* published reports on the performance of transgenic cotton plants, *Gossypium hirsutum* var. Coker 312, expressing modified forms of two δ-endotoxin genes. Constructs included genes expressing amino acids 1–612 of the CryIA(b) protein of *B. thuringiensis* var. *kurstaki* HD-1 and amino acids 1–640 or 1–615 of the *cryIA(c)* gene of *B. thuringiensis* var. *kurstaki* HD-73. Both genes were driven by a CaMV 35S promoter containing a duplicated enhancer region. To achieve these higher expression levels, the authors modified the DNA sequence without changing the amino acid sequence. The regions targeted for sequence modification included those with potential roles as regulatory sequences, or sequences with predicted mRNA secondary structure. These modifications increased the levels of both CryIA(b) and CryIA(c) δ-endotoxins to 0.05–0.1% of the total soluble protein, or what the authors estimated to be a 100-fold increase in expression compared to the truncated wild-type gene. Insect bioassays using leaves from cotton plants containing the wild-type *cryIA(b)* gene failed to demonstrate consistent insect control, while the plants containing the highly modified *cryIA(b)* gene were toxic to the cabbage looper (*Trichoplusia ni*; Lepidoptera) and the 100-fold less sensitive beet armyworm (*Spodoptera exigua*; Lepidoptera). Whole plant assays performed to determine the ability of transgenic cotton plants to control cotton bollworm (*Helicoverpa zea*; Lepidoptera) showed an 80% protection of squares and bolls with the *cryIA(c)* gene and 70–75% protection with the *cryIA(b)* gene, which is 3 to 5-fold less active against *Helicoverpa*.

In a subsequent publication, Perlak *et al.* (1991) examined several versions of the modified *cryIA(b)* and *cryIA(c)* genes in both transgenic tobacco and tomato to determine more closely the increased expression associated with various sequence modifications. All *cryIA* genes were under the control of a CaMV 35S promoter with a duplicated enhancer region. Two versions of modifications were made – one was partially modified, while the second was fully modified. In neither case did sequence modifications alter the amino acid sequence of the *cryIA* genes. The partially modified *cryIA(b)* gene had 62 of 1743 bases changed to remove regions with potential plant polyadenylation signals and A + T rich regions. The fully modified *cryIA(b)* gene had 390 of 1845 bases changed to remove all ATTTA sequences, most potential plant polyadenylation sites, and regions of potential mRNA secondary structure. The fully modified *cryIA(b)* gene also replaced bacterial codons with plant-preferred codons. The partially modified *cryIA(b)* gene had 97% homology to the wild-type gene and a G + C content of 41%, compared with 37% in the wild-type *cryIA(b)* gene. The fully modified *cryIA(b)* had 79% homology with the wild-type gene with a G + C content increased to 49%. The majority of the transgenic tomato and tobacco plants expressing the partially modified *cryIA(b)* gene produced δ-endotoxin at levels of 20–200 ng CryIA(b) per mg total protein and 1–200 ng CryIA(b) per mg total protein, respectively. Over 10% of the plants containing fully modified *cryIA(c)* and *cryIA(b)* genes expressed between 600–2000 ng CryIA per mg of total protein. The most highly expressing transgenic plants containing the partially modified gene and the fully modified gene increased expression 10- and 100-fold, respectively, compared with the wild-type truncated gene. Levels of expression with the modified *cryIA* genes were now up to 0.2% of the total protein in the best plants.

Constructs containing different combinations of the changes were tested in transgenic tobacco to study the effects of particular sequences (Perlak *et al.*, 1991). Modifications in the 5′ one-third of the *cryIA(b)* gene were sufficient to produce expression levels comparable to the partially modified gene, whereas modifications to the 3′ half of the gene had no effect on wild-type expression levels. The 5′ one-third of the *cryIA(b)* gene and nucleotides 246–283, which contain three potential polyadenylation signals, were identified as important, but no single region was identified as being critical to expression levels. Northern analysis of the modified *cryIA(b)* genes in transgenic plants indicated that, although the levels of mRNA were increased, the mRNA increase was not proportional to the level of protein increase. This observation led Perlak *et al.* to conclude that the problem associated with gene expression originated at the level of protein translation and not at the level of transcription.

Wilson *et al.* (1992) field tested three transgenic cotton lines containing a modified truncated *cryIA(b)* gene from *B. thuringiensis* var. *kurstaki* to examine resistance against the pink bollworm (*Pectinophora gossypiella*; Lepidoptera). Losses due to damage from the pink bollworm, as well as the expense of insecticides, make the pink bollworm one of the most important pests in cotton. Pink bollworm larvae entered the bolls of the transgenic lines as easily as the control lines. However, there was a 99.8% reduction in the number of live insects recovered from transgenic bolls and a 97% reduction in seed damage. Resistance to pink bollworm in the transgenic cotton lines was the same or greater than expected with insecticidal protection. The transgenic lines were highly resistant to cotton leaf perforator (*Bacculatrix thurberiella*; Lepidoptera), and sustained little damage due to the beet armyworm (*Spodoptera exigua*; Lepidoptera).

Barton and Miller (1993) constructed a synthetic *cryIA(a)* gene which resembled plant genes more closely, using a series of constructs with blocks of synthetic DNA representing 50 codons. Each of the constructs was studied to examine the effect of the modified block. Large increases (100-fold) in gene expression were seen with as little as 10% of the peptide coding region modified in the NH_2-terminal region.

Koziel *et al.* (1993) synthesized a truncated *cryIA(b)* gene encoding amino acids 1–648 of the *cryIA(b)* gene from *B. thuringiensis* var. *kurstaki* HD-1. This synthetic gene was completely redesigned to replace the bacterial codons with maize-preferred codons. The synthetic *cryIA(b)* gene had 65% homology with the wild-type gene and had a G + C content of 65%, compared with 37% for the native gene. The synthetic *cryIA(b)* gene, under control of various promoters, was introduced into immature embryos of an elite cultivar of maize using microprojectile bombardment. Two transgenic maize lines expressing this gene were field-tested in 1992. One line contained the synthetic *cryIA(b)* gene under control of the CaMV 35S promoter and 3′ polyadenylation sequences. The other line contained two versions of the synthetic *cryIA(b)* gene, one under control of the maize phosphoenolpyruvate carboxylase (PEPC) promoter, which is expressed in green tissues, and the other version under control of a maize pollen-specific promoter. Southern data and segregation ratios of progeny indicated a single site of insertion of the *cryIA(b)* gene with a few copies of the gene. Transgenic plants containing the two synthetic *cryIA(b)* genes under control of the PEPC and pollen-specific promoters produced over 1000 ng CryIA(b) per mg soluble protein at 7 weeks post-transplantation and up to 4000 ng CryIA(b) per mg soluble protein in certain plants later in the season. CryIA(b) protein levels in certain plants with the CaMV 35S promoter were as high at week 7 as the levels in the PEPC and pollen promoter plants. Overall, the levels of CryIA(b) protein in the CaMV 35S promoter line were lower and much more variable (25–2000 ng CryIA(b) per mg soluble protein), both within a particular cross and among crosses with different genotypes, whereas the PEPC and pollen promoter expression levels were much less variable. Tissue-specific expression patterns were examined in pith, root, pollen/anther, and kernel tissues of the transgenic plants. The CaMV line had high levels of CryIA(b) in pith and root (25–4000 ng CryIA(b) per mg soluble protein), moderate levels in the kernel (25–280 ng CryIA(b) per mg soluble protein) and no detectable CryIA(b) in the pollen/anther. The PEPC and pollen promoter line had moderate levels in the pollen/anther (50–400 ng CryIA(b) per mg soluble protein), lower levels in root and pith (25–120 ng CryIA(b) per mg soluble protein) and very low levels in kernels (1–18 ng CryIA(b) per mg soluble protein). The results of the field test showed that the transgenic maize lines provided season-long protection from repeated heavy infestations of European corn borer (*Ostrinia nubilalis*; Lepidoptera), which totally devastated control plants.

Concerns that constitutive season-long expression of a δ-endotoxin gene in transgenic crops might lead to the development of resistance to δ-endotoxins in insects has led to the engineering of plants with δ-endotoxin genes expressed by tissue-specific or chemically inducible promoters. Williams *et al.* (1992) used a chemically induced promoter from tobacco to drive the expression of a truncated 645 amino acid *cryIA(b)* gene from *B. thuringiensis* var. *kurstaki* HD-1. The inducible promoter, PR-1a, is from a family of tobacco genes involved in systemic acquired resistance to pathogen infection. Chemically induced transgenic tobacco plants expressing the PR-1a/*cryIA(b)* gene were shown to provide resistance to tobacco hornworm (*Manduca sexta*; Lepidoptera). It is also expected that at least some future transgenic plants will be transformed with multiple δ-endotoxin genes that are known to bind different midgut receptors within the target insect to help deter the development of resistance.

In addition to using tissue-specific promoters for resistance management, several laboratories are looking at second-generation promoters to enhance expression of δ-endotoxin genes. A modified truncated *cryIA(c)* gene from *B. thuringiensis* var. *kurstaki* HD-73 under the control of the *Arabidopsis thaliana* ribulose-1,5-bisphosphate carboxylase small subunit promoter was tested in transgenic tobacco. According to Wong, Hironaka and Fischhoff (1992), the small subunit promoter was tested in transgenic tobacco. According to Wong, Hironaka and Fischhoff (1992), the small subunit promoter with its own 5′ untranslated leader and chloroplast transit peptide provided a 10 to 20-fold increase in CryIA(c) expression levels compared to the CaMV 35S promoter with a double enhancer. It was determined that the increase in expression was the result of a combined effect due to both the 5′ untranslated leader and the transit peptide. A similar increase in expression of the

CryIA(c) protein was observed when the 5′ untranslated leader and transit peptide were fused behind the CaMV 35S promoter.

The difficulty that has been encountered in expressing a Bt δ-endotoxin gene in plants has caused scientists to examine closely the factors that are important for optimal transgene expression in plants. The success of the modified synthetic *cryIA* genes has opened the door for studies on transcriptional and translational regulation of the *cryIA* genes, as well as other genes, in plants. In the future, we are likely to see more tissue-specific and inducible promoters replacing the constitutive promoters like the CaMV 35S promoter. Further refinement will bring focus to the role of 5′ untranslated regions, enhancers and 3′ polyadenylation signals.

Transgenic plants with insecticidal activity against coleopteran pests

Bacillus thuringiensis var. *tenebrionis* produces a δ-endotoxin (CryIIIA) with activity against coleopterans such as Colorado potato beetle (CPB; *Leptinotarsa decemlineata*). An engineered *cryIIIA* gene starting at amino acid 48 was introduced into tomato and potato plants using *Agrobacterium*-mediated transformation (Perlak *et al.*, 1988). The expression levels in plants transformed with the *cryIIIA* gene were very low, less than 0.001% of total soluble protein. Expression levels have been improved by engineering synthetic *cryIIIA* genes using the same strategy used for the synthetic *cryIA* genes. A modified version of the *cryIIIA* gene was engineered to remove potential polyadenylation sites and A + T rich regions, with a final G + C content of 49% as compared with 37% in the wild-type gene (Perlak *et al.*, 1993). The modified *cryIIIA* gene was introduced into Russet Burbank potato plants via *Agrobacterium*-mediated transformation. Expression of the *cryIIIA* gene was under control of the CaMV 35S promoter with a duplicated enhancer region. Expression levels of the CryIIIA protein were between 0.002 and 0.3% of total soluble protein. Tests conducted under both greenhouse and field conditions demonstrated complete control of neonate CPB larvae as well as reduced feeding of adults accompanied by reduced fecundity.

Several field tests of potato lines transformed with the *cryIIIA* gene of *B. thuringiensis* var. *tenebrionis* were conducted in 1992 (Boylan-Pett *et al.*, 1992; Sewell and Storch, 1992; Stoltz and Matteson, 1992). The transgenic potato plants had significantly lower defoliation ratings by the CPB than non-transformed plants. Descriptions of the constructs and expression levels were not published.

Efforts in many laboratories continue to produce new results on the expression levels of the truncated and synthetic δ-endotoxin genes in plants. Bt δ-endotoxin genes other than *cryIA* and *cryIIIA* are also being introduced into plants. Visser *et al.* (1992) have transformed tobacco with a partially modified *cryIC* gene for control of *Spodoptera* spp. insects. In the near future, a number of transgenic crops expressing a variety of δ-endotoxin genes (probably modified for high expression levels) will be available as part of the overall pest management system. Within the next 2–5 years, the first commercial crops are expected to be cotton and maize.

Insect resistance to Bt δ-endotoxins

The growing use of Bt microbial spray-on products together with reports of insect resistance to Bt δ-endotoxins have led to an increasing awareness of, and concern for, the possible development of insect resistance to Bt δ-endotoxins on a scale of economic significance. The concern is shared, to a greater or lesser degree, by academic and industrial researchers, governmental regulatory officials and environmental organizations (National Audubon Society, 1991; USDA, 1992; Fitt, 1992). In recognition of the need for further research and an industry-wide consideration of management strategies, a group of 13 US and European companies formed the Bt Management Working Group in 1988 and has since funded seven research projects in universities on several highly relevant topics.

The attention given to the potential for insect resistance to Bt δ-endotoxins in advance of their widespread use results from several factors. First, the experience with chemical insecticides to which over 500 insect species have become resistant (Georgiou and Taylor, 1986) suggests strongly that insects have a high capacity to evolve resistance in the absence of careful management of the agents supplying selection pressure. Second, the δ-endotoxins of Bt are viewed as a 'natural resource'. The loss or reduction of their usefulness for insect control by the development of insect resistance is perceived, somewhat romantically, as a greater loss than that of a synthetic chemical pesticide. This sentiment must be tempered by consideration of the fact that the number of novel Bt strains and δ-endotoxins identified and the number of species affected is growing exponentially as, mostly, industrial researchers screen, with ever greater precision and breadth, the tens of thousands of strains collected from all over the world. More new 'active ingredients' are likely to become available, faster and at lower cost, from these collections than from the laboratories of synthetic chemists. In addition, as noted above, the engineering of new insecticidal proteins based upon δ-endotoxins is a reality with steadily increasing potential. Are man-made Bt-based insecticides a 'natural resource'? Third, the creation of transgenic plants constitutively producing a δ-endotoxin is seen by some as a prophylactic use of an insecticide and, as such, counter to the current trend towards the managed use of insecticides in integrated pest management programmes.

Bt δ-endotoxins, however, are insecticides. The history and facts of insect resistance to insecticides is relevant (Roush and McKenzie, 1987). In the majority of cases, field resistance has been based upon alleles with major effects at one or two loci. Selection of such alleles has been under conditions of continuous high-intensity use of single insecticides. Polygenic resistance based upon additive alleles with minor effects has been rare. Most high-level resistance genes produce altered targets for, or increased metabolism of, insecticides. Critical parameters in the development of resistance include frequency of resistance alleles in the population, mode of inheritance and fitness of those alleles, number of generations per year, and influx of susceptible individuals into the population (Georgiou and Taylor, 1986).

Field resistance has been based upon alleles that generally do not confer large fitness disadvantages (Denholm and Rowland, 1992).

FIELD AND LABORATORY OCCURRENCES

Most of the insects resistant to Bt δ-endotoxins have been derived from laboratory selection experiments. The majority of these laboratory selections have begun with individuals brought in from field populations that had been treated previously with microbial Bt δ-endotoxins. Laboratory selection of resistance can provide important information about how a species can adapt to the stress of δ-endotoxins in its diet, but certain caveats must be kept in mind when considering the relevance to the potential for field resistance.

First, the pool of genetic variability in a laboratory population is a subset of that in field populations. Rare alleles capable of conferring high levels of resistance in the field could be absent from a laboratory population. There is usually infrequent, or no, immigration into the laboratory population once it is established. Selection occurs in a special environment less stressful than most field environments. Partial resistance alleles have a greater probability of becoming homozygous and a reduced likelihood of conferring a negative fitness. Laboratory seledction regimes can be controlled to allow the accumulation of recessive or additive alleles that produce polygenic resistance. Because most field insecticide treatments, at least on initial application, are high level, only highly resistant homozygous or heterozygous individuals survive. The partially resistant progenitors of a polygenic high-level resistance do not survive such selection stringency.

Indian meal moth (Plodia interpunctella)

The first report of significant insect resistance to Bt was made by McGaughey (1985). After noticing less susceptibility in laboratory colonies derived from Indian meal moth populations from grain bins treated with Bt relative to those not treated with Bt, he began selecting in one more resistant colony for increased resistance to Dipel®, a commercial formulation of strain HD-1, which produces the three CryIA proteins and CryIIA. The increase in resistance was rapid, with a 27-fold increase in just two generations and 97-fold after 15 generations. Other colonies were also successfully selected for increased resistance, although considerable variation was seen in the level of resistance among colonies. In one colony, with higher selection pressure, a 250-fold increase in resistance was achieved. Resistance was lost when selection was removed in the early generations but was maintained after populations had 'plateaued'. If this lack of reversion to susceptibility in the absence of selection was translatable to a field situation, rotation of different Bt δ-endotoxins to manage resistance to any one would be unlikely to be very effective. Inheritance of resistance was partially recessive and was estimated to involve only one or a few major loci (McGaughey and Beeman, 1988).

One Dipel®-resistant Plodia colony was compared with a susceptible colony in two-dose and LD_{50} assays with 57 strains of Bt known to be toxic to the insect (McGaughey and Johnson, 1987). Twenty-one of the strains were active on the resistant colony and represented five of the eight serotypes tested. The toxicity was not a result of β-exotoxin. Examination of the midgut proteases from both Bt-susceptible and -resistant insects showed no difference in either the ability to activate δ-endotoxins or in total proteolytic activity on casein (Johnson *et al.*, 1990).

In further laboratory selection experiments involving a susceptible colony not previously exposed to Bt and a selected strain with a >250-fold decrease in susceptibility to Dipel®, McGaughey and Johnson (1992) were able to increase resistance of the susceptible colony 21-, 28- and 61-fold over 20 generations of selection to strains HD-198, HD-112 and HD-133, respectively. The rate of increase in resistance was slower than with Dipel®. Simultaneous selection of resistance to both Dipel® and HD-133 was effective (15×) but slower than to each strain alone. The Dipel® resistant colony developed an approximately 21-fold increased resistance to HD-133 in only 12 generations. These results are significant because they demonstrate the possibility of an insect population's relatively rapid acquisition of moderate resistance to several different endotoxin classes at once or sequentially. The HD-133 strain contains the CryIC and CryID proteins not present in the HD-1 (Dipel®) strain.

Van Rie *et al.* (1990a,b) determined the molecular basis for the most Dipel®-resistant colony. Brush border membrane vesicles from insects of the resistant colony were found to contain CryIA(b) binding proteins whose dissociation constants were approximately 50 times less than those of the unselected Dipel®-susceptible strain, while the number of binding proteins was unchanged. In contrast, the binding affinity for CryIC δ-endotoxin was about the same in the resistant and susceptible insects but the concentration of CryIC binding proteins was increased about three-fold in the resistant insects. This increase in CryIC binding proteins was reflected in an increased susceptibility of the Dipel®-resistant strain to CryIC. In the first case of a laboratory selected high level of Bt resistance, in a population of insects derived from one previously exposed to Bt in a 'field' situation, the mechanism of resistance seems to be an altered target site, one of the two most frequent mechanisms of high-level field resistance to chemical insecticides.

Tobacco budworm (Heliothis virescens)

Stone, Sims and Marrone (1989) reported selection of a laboratory strain of *Heliothis virescens* (Lepidoptera), resistant by about 24-fold versus the unselected population, *Pseudomonas fluorescens* expressing CryIA(b). The development of resistance to this level was rapid, just 7 generations, and then fluctuated between 13 and 20 × through generation 12. An interesting result was the marginal (4×) increase in resistance of the population to Dipel® and to the purified endotoxins from Dipel®. The difference was not the result of a differential resistance to the *Pseudomonas*. When the strain was further selected for five generations on Dipel® and, subsequently, seven generations again on the recombinant Bt-expressing *Pseudomonas*, 75-fold resistance to the *Pseudomonas*, 71-fold resistance to purified CryIA(b), 57-fold resistance to Dipel® and 16-fold resistance to CryIA(c) had been achieved (Macintosh *et al.*, 1991).

The mode of inheritance of the resistance (Sims and Stone, 1991) was found to be several incompletely dominant autosomal factors. The resistance was unstable in the absence of selection for five generations, declining from 69 to 13 times that of the susceptible population. Resistance to CryIA(c) was, however, stable in the absence of selection (Marrone and MacIntosh, 1993).

The mechanism of resistance underlying the complex phenotype of this strain of *Heliothis* is also complex. Proteolytic composition of the gut juices and proteolytic activation of the endotoxins were not altered in the resistant strain (Marrone and MacIntosh, 1993). However, for CryIA(c) (resistance ratio = 16) the midgut binding protein affinity was reduced four-fold but the concentration of binding proteins increased four-fold, whereas for CryIA(b) (resistance ratio = 70) the binding affinity was reduced two-fold while the binding protein concentration increased six-fold. Also, in the resistant insects CryIA(b) was less effective in competing with CryIA(c) for the same binding protein than in the susceptible insects (MacIntosh *et al.*, 1991).

Another laboratory-selected Bt-resistant population of *H. virescens* was reported by Gould *et al.* (1992). This strain was selected at an average larval mortality of 75% per generation on purified trypsin-activated CryIA(c) toxin. Its resistance factor showed a slow increase to about 22× through generation 16 and then increased to 50× in generation 17. What is novel about this strain is its cross-tolerance not only to related δ-endotoxins [13-fold increase in resistance to CryIA(b); increased relative weight gain on CryIA(a)-containing diet], but to δ-endotoxins of different classes (53-fold increase in resistance to CryIIA; increased relative weight gain on CryIC- and CryIB-containing diet). However, it should be noted that CryIA(a), CryIB and CryIC are essentially inactive on *H. virescens*; differences in weight gain could only be detected at concentrations of at least 0.1% of diet (w/v). The significance of these results to the potential for the development of economically relevant broad-spectrum resistance to Bt δ-endotoxins is difficult to determine.

Neither the affinity of midgut membrane proteins binding CryIA(b) or CryIA(c), nor their concentration, was significantly different between the selected and control strains. The F_1 progenies were intermediate in resistance relative to the parents. The number of genes underlying the resistance phenotype was not determined. Gould *et al.* (1992) point out that at doses of CryIA(c) sufficient to kill 80–90% of the susceptible insects, heterozygotes would survive. This, of course, would lead to rapid fixation of resistance genes in the population and increase the proportion of homozygous-resistant individuals. A balanced consideration of the dose–response curves for the three populations (susceptible, resistant and F_1) leads to the observation that, at current levels of δ-endotoxin expression in transgenic plants (about 5–50 μg g^{-1} fresh weight), the heterozygotes would not be likely to survive and many of the homozygotes, if formed, would perish. In fact, larvae from the inbred resistant strain (CP73-3) die after feeding on moderately expressing (Bt approximately 0.01% of total leaf protein) CryIA(b) transformed tobacco plants (Warren, unpublished results).

Diamondback moth (Plutella xylostella)

The diamondback moth (*Plutella xylostella*; Lepidoptera) presents the first case of field-selected resistance to Bt (Tabashnik *et al.*, 1990). LC_{50} and LC_{95} laboratory determinations were made using insects from fields minimally and heavily treated with commercial formulations of Bt collected at two different times. Compared to untreated laboratory strains, the treated strains showed increases of 25- and 33-fold, respectively. The LC_{50} resistance ratio for a heavily treated field population increased by about two over a 1 year period. Further selection for nine generations in the laboratory for Bt resistance among three strains established from resistant field populations resulted in 5- to 7-fold increases in LC_{50}, which was an increase in resistance of 150- to 190-fold over unselected laboratory strains (Tabashnik, Finson and Johnson, 1991). Another five generations of selection led to a resistance ratio of 820. The LC_{50}s of Bt-treated field populations did not differ significantly over three generations. The resistance of field populations did not decline within a period of about 4 months in the absence of Bt treatments. As the authors point out, these studies do not address how fast and to what degree resistance can arise in untreated susceptible populations, but do suggest that high levels of resistance can develop quickly with stringent selection among moderately resistant populations. It should be pointed out that the moderate levels of resistance arose in populations repeatedly treated with the same set of sprayed-on δ-endotoxins in a relatively geographically isolated area.

In another case of field resistance to intense use of Bt sprays (Dipel®), the LC_{50} for CryIA(b) in the resistant insects was >200 times that of a susceptible laboratory strain, while the LC_{50}s for CryIB and CryIC, to which *Plutella* is also susceptible, were not significantly different (Ferre *et al.*, 1991; Van Rie, Van Mellaert and Peferoen, 1992). When these three toxins were radiolabelled and independently bound to midgut brush border membrane vesicles (BBMVs), CryIB and CryIC saturably bound to both susceptible and resistant insects. However, CryIA(b) bound only to susceptible insect BBMV. Competitive heterologous binding experiments showed that CryIA(b), CryIB and CryIC bind to independent proteins in the midgut membrane. Because Dipel® does not contain CryIB or CryIC, the resistance can be rationalized on the basis of the complete lack of binding of CryA(b). No data were reported on the toxicity or binding of CryIA(c), a component of Dipel® highly active on *Plutella*.

Plutella populations resistant to Bt have also been found in Florida, USA (Shelton *et al.*, 1993). These populations varied in their degree of resistance, relative to a susceptible laboratory population, to several Bt microbial products. It should be noted that neither the specific endotoxins present nor their concentrations were determined for the products, which makes the interpretation of the variation of product performance, LC_{50}s and resistance ratios difficult. Nonetheless, a correlation was clearly established between a history of constant *Brassica* production, a high number of *Plutella* generations per year, repeated intense spraying of Bt and higher levels of resistance. Resistance ratios for the so-called 'kurstaki' products were from 321- to 461-fold, but for the 'aizawai' products these ratios varied from 3 to 4.1. The latter products have been in use for fewer years than the former. Neither the genetics nor mechanisms of resistance of these populations has been reported.

Almond moth (Cladra cautella)

McGaughey and Beeman (1988) reported only a seven-fold increase in resistance to strain HD-1 in a 10-year-old laboratory population subjected to selection for 21 generations.

Colorado potato beetle (Leptinotarsa decemlineata)

Another example of selection of increased resistance to Bt in a laboratory population derived from a field population previously treated with the same Bt strain was reported by Miller, Rahardja and Whalon (1990). After 10 generations of selection with a CryIIIA-containing product, 67-fold resistance was noted.

Significance of resistances identified to date

Several themes are evident in the cases of resistance cited above. First, most of the cases of field resistance, with the exception of three Florida populations of *Plutella*, involved moderate levels of resistance, which nontheless led to a reduction in the control of the treated insects. Second, the selection of high levels of resistance was rapid in the laboratory when starting with weakly to moderately resistant subpopulations from field populations previously treated with the same selective agent. Third, those previously treated populations were treated intensively over a period of time with a single Bt product, produced large numbers of generations per year and, in the cases of *Plodia* and Hawaiian and Philippine *Plutella*, saw little immigration. Fourth, laboratory selections initiated with previously untreated populations resulted in only moderate polygenic resistance. Fifth, high-level resistance derived from Bt-treated field populations was, in the only two cases known, *Plodia* and *Plutella*, correlated with target site modification, the elimination of gut membrane binding of at least one of the active δ-endotoxins. Finally, the only strain cross-resistant to more than one class of δ-endotoxin is only marginally so and incapable of survival on a constant diet of Bt or on a moderately expressing transgenic plant.

A couple of tentative conclusions from the studies on resistance to Bt so far are those already learned with chemical insecticides: (1) economically important field resistance develops under variable (the efficacy of Bt sprays declines with time) but continuous selection pressure with a single insecticide on many generations with little or no immigration; (2) levels of resistance in field situations leading to commercially important control failure most likely result from one or two target site changes. The assumptions behind these conclusions are that a lack of control of some *Plodia* populations in grain storage bins resulted from the altered Bt receptor genes later selected in the laboratory and that the inheritance of resistance in the Philippine *Plutella* population is simple and correlated with the altered CryIA(b) binding. It will be interesting to see if the molecular and genetic analysis of the resistant *Plutella* populations from Florida and Hawaii result from simply inherited recessive genes encoding altered Bt receptors that no longer bind one or more of the active δ-endotoxins of the selective strain or reduce the number of such receptors in the midgut epithelium membranes.

An obvious conclusion from the above studies is that economically significant insect resistance to multiple component microbial Bt sprays *can* occur in field populations. However, the significance of these occurrences cannot be fully appreciated in the absence of definitive analyses of the mechanism of resistance, at the molecular level, the mode of inheritance of the resistance trait and co-segregation of the resistance and the altered molecular phenotypes. The relevance of results with laboratory selections within populations not previously exposed to Bt in decision making about resistance management for Bt products *in the field* is not obvious, beyond providing an indication of the genetic potential of the species. Of course, none of the results to date bear much on the issue of insect resistance to δ-endotoxin expressing transgenic plants.

Population genetics and resistance modelling

Much of the concern about the risk for widespread insect resistance to Bt products has arisen from the coincidence of three elements: (1) the predictions of population genetics models; (2) the occurrence of Bt-resistant insect populations; and (3) the creation of transgenic Bt expressing plants that deliver δ-endotoxins to insects in novel ways (McGaughey and Whalon, 1992). And much of the alarm arising from population genetics modelling derives from models incorporating transgenic plants, which are seen to be unique Bt pesticides, primarily because, in principle, they can challenge an insect population uniformly and stringently at all of its susceptible stages for all of its generations. Whether this uniqueness is, in balance, positive or negative depends on the model and, critically, on its assumptions.

Although population genetics models can be useful as heuristic tools for estimating the consequences of different patterns of insecticide use on the development of resistance (Gould, 1986, 1988, 1991), they involve numerous assumptions, most of which lack experimental validation (Tabashnik, 1986). Many of the known or suggested factors influencing the selection of resistance in field populations (Georgiou and Taylor, 1986) are not varied in the models most frequently used to predict resistance to Bt. Most of the models accommodate only one or, at most, two resistance (R) loci with only one R allele at each locus, and make normally unvalidated assumptions about dominance and initial frequency of R alleles and about fitness of different genotypes. Tabashnik (1986) stated that resistance models needed experimental validation and, in particular, would benefit from baseline data on population susceptibility before selection and experimental estimates of biological parameters. With Bt, experimental validation of model parameters and assumptions has begun but results have not yet been published. The key parameters of resistance models that, together with experience with resistance to chemical insecticides, seem to be most in need of experimental attention are number of generations per year, intensity of selection, immigration of susceptible individuals (refugia), mode of inheritance of resistance alleles, number of selection agents and how they are applied in time and the particulars of species biology.

The priorities for future research in the nascent field of Bt resistance management seem obvious. First, for each insect species standardized assays must be adopted and used by all. Second, behavioural studies are needed on specific insect-crop combinations with +/–Bt transgenic versions of the crop where available. Third, baseline population susceptibilities must be established *before* Bt is used. These baselines should be determined by geographical location, insect species and selective agent used. The selective agent (transgenic plant or microbial spray) should be precisely characterized as to the nature and amount of endotoxins present. Domesticated laboratory cultures derived from geographically distant field populations are not relevant controls for monitoring the susceptibilities of field populations exposed to Bt. Fourth, a gut binding and toxicity (LC_{50}) database should be developed for each δ-endotoxin–insect combination. Many current Bt resistance data are uninterpretable because the actual active ingredient in a Bt product for a given insect species is unknown. Fifth, studies are needed to determine the effect on resistance development of multiple Bt δ-endotoxins, with different target sites used sequentially or simultaneously. The efficacy of multiple insecticide tactics is unresolved in general (Tabashnik, 1989). Finally, the most critical assumptions (e.g. initial R allele frequency) are in need of experimental validation.

Potential strategies for management of resistance to Bt

The body of experimental and modelling data suggests one seemingly obvious basic principle that can be applied to the use of Bt products as well as chemical insecticides: *Avoidance of fixation of resistance alleles*. The fixation of resistance alleles in homozygotes must be prevented. This can be done by adjusting the dose well above the LC_{50} of treated susceptible insects and maintaining it there. Of course, 'well above' cannot be determined precisely in advance of knowing what resistance alleles are in the populations to be treated and how they are inherited. So far, high-level Bt resistance genes from field populations have been recessive and, even in polygenic laboratory selected populations, additive. Dominant resistance genes are quite rare and have not yet been identified for Bt δ-endotoxins. In most cases, the 'high-dose' approach would eliminate heterozygotes. Bt is most effective on early stages of susceptible insects. Consistent high doses applied to early stage instars are most easily achieved via transgenic plants. In practice, the 'high-dose' approach is probably not possible with spray-on products because applications are not continuous, persistence is relatively short and insects can escape treatment.

Another and much touted approach to avoiding homozygous resistant insects is to provide a source of susceptible insects for mating and maintaining resistance alleles in the homozygous state. Refugia of various sorts can supply susceptible insects. Refuges can occur outside the crop for polyphagous insects with alternate hosts or within the crop, either by design, through the planting of seed mixtures or susceptible border rows, or as a result of the choices made by individual growers. Another form of refuge, in the case of transgenic Bt crops, is the organ or tissue in which the Bt gene is not expressed but upon which the insect can feed without causing economic damage. Whether the tissue-specific or chemically controlled promoters that could condition such expression will be used will depend less on the availability of the technology than on the specifics of insect–crop interaction and the nature of the marketable product. Refugia used together with a consistent high dose might be a particularly powerful approach to delaying resistance for a large number of insect generations.

Fixation of resistance could also be potentially avoided through the use of more than one insecticide (>1 Bt δ-endotoxin or a Bt δ-endotoxin + a different, but equally effective, active principle). The efficacy of multiple active principles depends on the nature of the principles and how they are used. The effectiveness of simultaneous use derives from the multiplication of rare events, two rare events, where two loci and dominant resistance alleles are involved and four rare events are needed for resistant insect phenotype (where two loci and recessive resistance alleles are involved). Individuals resistant to two independently but simultaneously acting insecticides will have a very low probability of arising unless resistance alleles are present at a high frequency and the dose of the insecticide is such that it allows heterozygous individuals to survive. If multiple active principles are used in sequence, one at a time, their efficacy in delaying resistance will depend on the fitness of the resistance alleles and rapidity with which the population returns to susceptibility versus the frequency of alternating insecticides.

A potentially risky scenario might arise with microbial sprays applied to the same insect on the same crop in the same geographical area where transgenic plants are grown that express the same endotoxin delivered in the spray. In this situation, heterozygous individuals might arise in response to sprays, mate, emigrate and lay homozygous resistant eggs on transgenic crop plants that select stringently for high levels of resistance. Resistance would then be expected to develop relatively rapidly, depending on the frequency of resistance alleles in the population (F. Gould, pers. comm.). In this situation, multiple endotoxins with different sites of action (binding proteins) could be used, between microbe and plant, to manage resistance.

Conclusions regarding resistance

1. Field resistance to Bt has been very limited in occurrence and, in the two cases studied, was based upon altered δ-endotoxin binding. Insects resistant to one δ-endotoxin could be killed by another.
2. The significance of laboratory selections for what can happen in the field is a matter of speculation.
3. There is limited knowledge of baseline susceptibilities of various species from various geographical locations. Therefore, it is not easy to distinguish natural variation among susceptible populations from low to moderate resistance.
4. Knowledge of insect behaviour relevant to the population genetics models that raise concerns about the likelihood of insect resistance to Bt is limited.
5. Each crop–species–geographical location combination will present different challenges and opportunities for resistance management.
6. The management of resistance to a particular insect species will require consideration of the total picture of that species on all of its hosts and for all of the available control measures.
7. Bt δ-endotoxins remain a class of insecticides with many of the properties desired of modern insect control agents: relatively narrow spectrum of activity; non-toxic to mammals, birds, fish; usually inactive on arthropod species that prey on the pests susceptible to Bt; little environmental persistence; no bio-accumulation; stomach poisons consistent with IPM practices. They are large in number. Knowledge of their mode of action, with the increased potential for rational management that it brings, is preceding their widespread use. This is unprecedented for an insecticide.

Summary

Since the first use of *Bacilus thuringiensis* as a biological control agent for insects, a great deal of progress has been made. This progress continues with increasing speed. New Bt strains with improved insecticidal activity have been identified. New strains with new activities have also been identified, thereby increasing the spectrum of activity for microbial sprays as a whole. The current tools of molecular biology in concert with techniques of classical genetics, is making it possible to design and construct improved Bt strains for use as microbial sprays. As more novel activities are identified and more improved strains are constructed, the use of such microbial sprays in agriculture will increase. As the proteins responsible for the insecticidal activity, and their respective genes, become better characterized, more options will become available for their use. The ability to precisely introduce and express foreign genes in a variety of organisms presents the opportunity to deliver these proteins in a controlled manner. These new host organisms are also able to solve some of the problems inherent in spray formulations such as uneven application and the necessity for precise timing. New organisms, however, will not completely replace microbial sprays, since these have a number of uses for which it is not feasible to engineer alternative hosts. The use of alternative hosts to express the insecticidal proteins from *B. thuringiensis* merely presents more options for insect control using a safe active ingredient.

As the mode of action of the δ-endotoxins is better understood and the structure of the proteins and their interactions with the insect gut cells is elucidated, the opportunity for engineering new activities will increase. This may allow the broadening of the spectrum of activity of a given δ-endotoxin or the engineering of completely novel activities. While such protein engineering possibilities are now in their infancy, the tools are in place to design and construct Bt strains with a desired collection of δ-endotoxins. This ability to design and engineer Bt strains with a desired spectrum of activity and a desired pattern of receptor recognition provides valuable tools both for insect control and for the management of resistance. Presently, the greatest concern about the rapidly increasing use of the Bt δ-endotoxins is that their widespread utilization will create a widespread insect resistance problem which will render them ineffective as a class of insect control compounds. Understanding the basis of resistance development, as well as its cost to the insect, will help develop strategies that prevent, or at least significantly delay, the development of resistance to a given δ-endotoxin. As more δ-endotoxins are discovered, and as more activities are likely to be engineered in the future, we will have even more tools for resistance management as well as for pest management.

Acknowledgements

The authors are grateful to Dr F. Perlak for providing a preprint of his research prior to its publication and Dr F. Gould for sharing unpublished data on resistance modelling.

References

ADANG, M.J., STAVER, M.J., ROCKELEAU, T.A., LEIGHTON, J., BARKER, R.F. AND THOMPSON, D.V. (1985). Characterized full-length and truncated plasmid clones of the crystal protein of *Bacillus thuringiensis* subsp. *kurstaki* HD-73 and their toxicity to *Manduca sexta*. *Gene* **36**, 289–300.

ADANG, M.J., FIROOZABADY, E., KLEIN, J., DEBOER, D., SEKAR, V., KEMP, J.D., MURRAY, E., ROCHELEAU, T.A., RASHKA, K., STAFFIELD, G., STOCK, C., SUTTON, D. AND MERL, D.J. (1987). Application of a *Bacillus thuringiensis* crystal protein for insect control. In *Molecular Strategies for Crop Protection* (C.J. Arntzen and C. Ryan, Eds), pp. 345–353. Alan R. Liss, New York.

ANDREWS, G.L. AND SIKOROWSKI, P.P. (1973). Effects of cotton leaf surfaces on the nuclear polyhedrosis virus of *Heliothis zea* and *Heliothis virescens* (Lepidoptera: Noctuidae). *Journal of Invertebrate Pathology* **22**, 290–291.

ANGSUTHANASOMBAT, C. AND PANYIM, S. (1989). Biosynthesis of 130-kilodalton mosquito larvicide in the cyanobacterium *Agmenellum quadruplicatum* PR-6. *Applied and Environmental Microbiology* **55**, 2428–2430.

ARANTES, O. AND LERECLUS, D. (1991). Construction of cloning vectors for *Bacillus thuringiensis*. *Gene* **108**, 115–119.

ARONSON, A.I. AND BECKMAN, W. (1987). Transfer of chromosomal genes and plasmids in *Bacillus thuringiensis*. *Applied and Environmental Microbiology* **53**, 1525–1530.

BAR, E., LIEMAN, J., RAHAMIN, E., KEYNAN, A. AND SANDLER, N. (1991). Cloning and expression of *Bacillus thuringiensis israelensis* δ-endotoxin DNA in *B. sphaericus*. *Journal of Invertebrate Pathology* **57**, 149–158.

BARNES, A.C. AND CUMMINGS, S.G. (1987). Cellular encapsulation of pesticides produced by expression of heterologous genes. United States Patent No. 4,695,455.

BARTON, K.A. AND MILLER, M.J. (1993). Production of *Bacillus thuringiensis* insecticidal proteins. In *Plants: Transgenic Plants, Vol. 1: Engineering and Utilization*, pp. 297–315. Academic Press, New York.

BARTON, K.A., WHITELEY, H.R. AND YANG, N. (1987). *Bacillus thuringiensis* delta-endotoxin expressed in transgenic *Nicotiana tabacum* provides resistance to lepidopteran insects. *Plant Physiology* **85**, 1103–1109.

BEZDICEK, D.F., QUINN, M.A. AND KAHN, M. (1991). Genetically engineering *Bacillus thuringiensis* genes into rhizobia to control nodule-feeding insects. *USDA CRIS Report* No. 9131200.

BISHOP, D.H.L. (1989). Genetic engineered viral insecticides: A progress report. 1986–1989. *Pesticide Science* **27**, 173–189.

BOSSE, M., MASSON, L. AND BROUSSEAU, R. (1990). Nucleotide sequence of a novel crystal protein gene isolated from *Bacillus thuringiensis* subspecies *Kenyae. Nucleic Acids Research* **18**, 7443.

BOVE, J.M. AND HACKETT, K.J. (1992). Development of a vector for transferring genes to spiroplasmas. *USDA CRIS Report* No. 1275–2400–008–015.

BOYLAN-PETT, W., GRAFIUS, E., BISHOP, B., PYLE, A., PARKER, G. AND STONE, T. (1992). Effectiveness of *Bacillus thuringiensis* transgenic potato plants for control of Colorado potato beetles, 1991. In *Insecticide and Acaricide Tests* (A.K. Burditt, Jr, Ed), Vol. 17, pp. 124–125. Entomological Society of America, Lanham, MD.

BRADFISCH, G.A., MICHAELS, T. AND PAYNE, J.M. (1993). Biologically active *Bacillus thuringiensis* isolates. United States Patent No. 5,208,017.

BRADLEY, D., BIEVER, D., BAUER, L.S. AND HARKEY, M.A. (1992). The insecticidal CryIB crystal protein of *Bacillus thuringiensis* subsp. *thuringiensis* has dual specificity to coleopteran and lepidopteran larvae. In *Abstracts of the XXV Annual Meeting of the Society for Invertebrate Pathology*, Heidelberg, August, p. 234.

BRAVO, A., JANSENS, S. AND PEFEROEN, M. (1992). Immunocytochemical localization of *Bacillus thuringiensis* insecticidal crystal proteins in intoxicated insects. *Journal of Invertebrate Pathology* **60**, 237–246.

BRAVO, A., HENDRICKX, K., JANSENS, S. AND PEFEROEN, M. (1992). Immunocytochemical analysis of specific binding of *Bacillus thuringiensis* insecticidal crystal proteins to lepidopteran and coleopteran midgut membranes. *Journal of Invertebrate Pathology* **60**, 247–253.

BRIZZARD, B.L. AND WHITELEY, H.R. (1988). Nucleotide sequence of an additional crystal protein cloned from *Bacillus thuringiensis* subsp. *thuringiensis. Nucleic Acids Research* **16**, 2723–2724.

BROWN, K.L. AND WHITELEY, H.R. (1992). Molecular characterization of two novel crystal protein genes from *Bacillus thuringiensis* subsp. *thompsoni. Journal of Bacteriology* **174**, 549–557.

BURGESS, D.R. AND JARRETT, P. (1985). Preparation of strains of *Bacillus thuringiensis* having an improved activity against certain lepidopterous pests and novel strain produced thereby. United States Patent Number 4,935,353.

CAROZZI, N.B., KRAMER, V.C., WARREN, G.W., EVOLA, S. AND KOZIEL, M.G. (1991). Prediction of insecticidal activity of *Bacillus thuringiensis* strains by polymerase chain reaction product profiles. *Applied and Environmental Microbiology* **57**, 3057–3061.

CAROZZI, N.B., WARREN, G.W., DESAI, N., JAYNE, S.M., LOTSTEIN, R., RICE, D.A., EVOLA, S. AND KOZIEL, M.G. (1992). Expression of a chimeric CaMV 35S *Bacillus thuringiensis* insecticidal protein gene in transgenic tobacco. *Plant Molecular Biology* **20**, 539–548.

CAROZZI, N.B., KRAMER, V.C., WARREN, G.W. AND KOZIEL, M.G. (1993). *Bacillus thuringiensis* strains active against coleopteran insects. United States Patent Number 5,204,100.

CHAMBERS, J.A., JELEN, A., GILBERT, M.P., JANY, C.S., JOHNSON, T.B. AND GAWRON-BURKE, C. (1991). Isolation and characterization of a novel insecticidal crystal protein gene from *Bacillus thuringiensis* subsp. *zizawai*. *Journal of Bacteriology* **173**, 3966–3976.

CHENG, J., BOLYARD, M.G., SAXENA, R.C. AND STICKLEN, M.B. (1992). Production of insect resistant potato by genetic transformation with a delta-endotoxin gene from *Bacillus thuringiensis* var. *kurstaki*. *Plant Science* **81**, 83–91.

CHUNJATPORNCHAI, W., HOFTE, H., SEURNICK, J., ANGUSUTHANASOMBAT, C. AND VAECK, M. (1988). Common features of *Bacillus thuringiensis* toxins specific for Diptera and Lepidoptera. *European Journal of Biochemistry* **173**, 9–16.

CIDARIA, D., CAPPAI, A., VALLESI, A. AND PIRALI, G. (1991). A novel strain of *Bacillus thuringiensis* (NCIMB 40152) active against coleopteran insects. *FEMS Microbiology Letters* **81**, 129–134.

DARDENNE, F., SEURINCK, J. AND PEFEROEN, M. (1990). Nucleotide sequence and deduced amino acid sequence of a *cryIA(c)* gene variant from *Bacillus thuringiensis*. *Nucleic Acids Research* **18**, 5546.

DART, P.J. (1990). Agricultural microbiology: Introduction. In *Agricultural Biotechnology: Opportunities for International Development* (G.J. Persley, Ed), pp. 53–77. Cambridge University Press, Cambridge.

DAVIS, M.J., GILLASPIE, A.J., VIDAVER, A.K. AND HARRIS, R.W. (1984). *Clavibacter*: A new genus containing some phytopathogenic coryneform bacteria including *Clavibacter xyli* sp. nov. and *Clavibacter xyli* subsp. nov., pathogens that cause ratoon stunting disease of sugarcane and bermuda-grass stunting disease. *International Journal of Systematic Bacteriology* **34**, 107–117.

DELANNAY, X., LAVALLEE, B.J., PROKSCH, R.K., FUCHS, R.L., SIMS, S.R., GREENPLATE, J.T., MARRONE, P.G., DODSON, R.B., AUGUSTINE, J.J., LAYTON, J.G. AND FISCHHOFF, D.A. (1989). Field performance of transgenic tomato plants expressing the *Bacillus thuringiensis* var. *kurstaki* insect control protein. *Bio/Technology* **7**, 1265–1269.

DENHOLM, I. AND ROWLAND, M.W. (1992). Tactics for managing pesticide resistance in arthropods: Theory and practice. *Annual Review of Entomology* **37**, 91–112.

DIMOCK, M.B., BEACH, R.M. AND CARLTON, P.S. (1989). Endophytic bacteria for delivery of crop protection agents. In *Biotechnology, Biopesticides and Novel Plant Pest Resistance Management* (D.W. Roberts and R.R. Granados, Eds), pp. 88–92. Boyce Thompson Institute for Plant Research, Ithaca, NY.

DONOVAN, W.P. (1991). CryIIB crystal protein gene from *Bacillus thuringiensis*. United States Patent Number 5,073,632.

DONOVAN, W.P., DANKOCSIK, C.C., GILBERT, M.P., GAWRON, M.C., GROAT, R.G. AND CARLTON, B.C. (1988). Amino acid sequence and entomocidal activity of the P2 crystal protein. *Journal of Biological Chemistry* **263**, 561–567.

DONOVAN, W.P., DANKOCSIK, C.C. AND GILBERT, M.P. (1988). Molecular characterization of a gene encoding a 72-kilodalton mosquito-toxic crystal protein from *Bacillus thuringiensis* subsp. *israelensis*. *Journal of Bacteriology* **170**, 4732–4738.

DONOVAN, W.P., RUPAR, M.J. AND SLANEY, A.C. (1992). *Bacillus thuringiensis* CryIIIC(b) toxin gene and protein toxic to coleopteran insects. PCT International Patent Application Number WO 92/13954.

DONOVAN, W.P., GONZALEZ, J.M., JR, LEVINSON, B.L. AND MACALUSO, A. (1991). Coleopteran active microorganism, related insecticide compositions and methods for their production and use. United States Patent Number 5,024,837.

DONOVAN, W.P., RUPAR, M.J., SLANEY, A.C., MALVAR, T., GAWRON-BURKE, M.C. AND JOHNSON, T. (1992). Characterization of two genes encoding *Bacillus thuringiensis* insecticidal crystal proteins toxic to Coleoptera species. *Applied and Environmental Microbiology* **58**, 3921–3927.

DONOVAN, W.P., RUPAR, M.J., SLANEY, A.C. AND JOHNSON, T.J. (1993). *Bacillus thuringiensis cryIIIC* gene encoding toxic to coleopteran insects. United States Patent Number 5,187,091.

DULMAGE, H. AND AIZAWA, K. (1982). Distribution of *Bacillus thuringiensis* in nature. In *Microbial and Viral Pesticides* (E. Kurstak, Ed.), pp. 209–237. Marcel Dekker, New York.

DYBVIG, K. AND CASSELL, G.H. (1987). Transposition of gram-positive transposon Tn916 in *Acholeplasma laidlawii* and *Mycoplasma pulmonis. Science* **235**, 1392–1394.

EARP, D.J. AND ELLAR, D.J. (1988). Assessment of the toxicity of the cloned 27 kDa toxin of *Bacillus thuringiensis* var. *morrisoni* strain PG14. *FEMS Microbiology Letters* **52**, 29–32.

ENGLISH, L. AND SLATIN, S.L. (1992). Mode of action of delta-endotoxins from *Bacillus thuringiensis*: A comparison with other bacterial toxins. *Insect Biochemistry and Molecular Biology* **22**, 1–7.

FEITELSON, J.S., PAYNE, J. AND KIM, L. (1992). *Bacillus thuringiensis*: Insects and beyond. *Bio/Technology* **10**, 271–275.

FERRE, J., REAL, M.D., VAN RIE, J., JANSENS, S. AND PEFEROEN, M. (1991). Resistance to the *Bacillus thuringiensis* bioinsecticide in a field population of *Plutella xylostella* is due to a change in a midgut membrane receptor. *Proceedings of the National Academy of Sciences, USA* **88**, 5119–5123.

FISCHHOFF, D.A., BOWDISH, K.S., PERLAK, F.J., MARRONE, P.G., MCCORMICK, S.M., NIEDERMEYER, J.G., DEAN, D.A., KUSANO-KRETZMER, K., MAYER, E.J., ROCHESTER, D.E., ROGERS, S.G. AND FRALEY, R.T. (1987). Insect tolerant transgenic tomato plants. *Bio/Technology* **5**, 807–813.

FITT, G.P. (1992). *Workshop on the Management of Bt Toxins in Transgenic Plants and Conventional Sprays*. Narrabri Agricultural Research Station, August. Cotton Research and Development Corporation. CSIRO Cotton Research Unit. Final Report.

FITZJAMES, P.C. (1984). Production of spore-free insecticides for Diptera larvae by cultivation of *Bacillus thuringiensis* mutant esp. ATCC59152. European Patent Application, Publication Number 0 099 301.

FONCERRADA, L., SICK, A.J. AND PAYNE, J.M. (1992). Novel coleopteran-active *Bacillus thuringiensis* isolate and a novel gene encoding a coleopteran-active toxin. European Patent Application, Publication Number 0 498 537 A2.

GAERTNER, F. (1990). Cellular delivery systems for insecticidal proteins: Living and non-living microorganisms. In *Controlled Delivery of Crop Protection Agents* (R.M. Wilkins, Ed.), pp. 245–257. Taylor and Frances, London.

GAERTNER, F.H., SICK, A.J. AND SCHWAB, G.E. (1993). Universal and specific probes for *Bacillus thuringiensis* endotoxin genes. United States Patent Number 5,204,237.

GAMEL, P.H. AND PIOT, J.-C. (1992). Characterization and properties of a novel plasmid vector for *Bacillus thuringiensis* displaying compatibility with hosts plasmids. *Gene* **120**, 17–26.

GAWRON-BURKE, M.C., CHAMBERS, J.A. AND GONZALEZ, J.M. (1991). *Bacillus thuringiensis* cryIF and cryIX genes and proteins toxic to lepidopteran insects. PCT International Patent Application Number WO 91/16434.

GE, A.Z., SHIVAROVA, N.I. AND DEAN, D.H. (1989). Location of the *Bombyx mori* specificity domain on a *Bacillus thuringiensis* delta-endotoxin protein. *Proceedings of the National Academy of Sciences, USA* **86**, 4037–4041.

GE, A.Z., RIVERS, D., MILNE, R. AND DEAN, D.H. (1991). Functional domains of *Bacillus thuringiensis* insecticidal crystal proteins: Refinement of *Heliothis vires-*

cens and *Trichoplusia ni* specificity domains on CryIA(c). *Journal of Biological Chemistry* **266**, 17954–17958.

GEISER, M., HARTMANN, H. AND ODDOU, S.P. (1993). New antibody against *Bacillus thuringiensis* endotoxin binding protein – used e.g. in assays, determination of cross-reactivity and for identification or isolation of new toxins. European Patent Application, Publication Number 526 397 A1.

GELERNTER, W.D. (1990). Targeting insecticide-resistant markets: New developments in microbial-based products. In *Managing Resistance to Agrochemicals: From Fundamental Research to Practical Strategies* (M.B. Green, H.M. LeBaron and W.K. Moberg, Eds), pp. 105–117. American Chemical Society, Washington, DC.

GEORGIOU, G.P. AND TAYLOR, C.E. (1986). Factors influencing the evolution of resistance. In *Pesticide Resistance: Strategies and Tactics for Management*, pp. 157–169. Committee on Strategies for the Management of Pesticide Resistant Pest Populations, Board of Agriculture, National Research Council. National Academy Press, Washington, DC.

GILL, S.S., COWLES, E.A. AND PIETRANTONIO, P.V. (1992). The mode of action of *Bacillus thuringiensis* endotoxins. *Annual Review of Entomology* **37**, 615–636.

GLEAVE, A.P., HEDGES, R.J. AND BROADWELL, A.H. (1992). Identification of an insecticidal protein from *Bacillus thuringiensis* DSIR517 with significant sequence differences from previously described toxins. *Journal of General Microbiology* **138**, 55–62.

GOLDBERG, R.J. AND TJADEN, G. (1990). Are B.T.K. plants really safe to eat? *Bio/Technology* **8**, 1011–1015.

GONZALEZ, J.M., JR AND MACALUSO, A. (1992). Novel *Bacillus thuringiensis* strains, and related insecticidal compositions. United States Patent Number 5,080,897.

GONZALEZ, J.M., BROWN, B. AND CARLTON, B.C. (1982). Transfer of *Bacillus thuringiensis* plasmids coding for δ-endotoxin among strains of *B. thuringiensis* and *B. cereus*. *Proceedings of the National Academy of Sciences, USA* **79**, 6951–6955.

GOULD, F. (1986). Simulation models for predicting durability of insect resistant germplasm: A deterministic diploid, tow-locus model. *Environmental Entomology* **15**, 1–10.

GOULD, F. (1988). Evolutionary biology and genetically engineered crops. *BioScience* **38**, 26–32.

GOULD, F. (1991). The evolutionary potential of crop pests. *American Scientist* **79**, 496–507.

GOULD, F., MARTINEZ-RAMIREZ, A., ANDERSON, A., FERRE, J., SILVA, F.J. AND MOAR, W.J. (1992). Broad-spectrum resistance to *Bacillus thuringiensis* toxins in *Heliothis virescens*. *Proceedings of the National Academy of Sciences, USA* **89**, 7986–7990.

GURTLER, H. AND PETERSEN, A.S. (1991). Mutants or variants of *Bacillus thuringiensis* producing high yields of delta endotoxin. PCT International Patent Application Number WO 91/07481.

HAIDER, M.Z., KNOWLES, B.H. AND ELLAR, D.J. (1986). Specificity of a *Bacillus thuringiensis* var. *colmeri* insecticidal delta-endotoxin by differential processing of the protoxin by larval gut proteases. *European Journal of Biochemistry* **156**, 531–540.

HEIMPEL, A.M. AND ANGUS, T.A. (1959). The site of action of crystalliferous bacteria in Lepidoptera larvae. *Journal of Insect Pathology* **1**, 152–170.

HERRNSTADT, C. AND GAERTNER, F.H. (1987). Mutants of *Bacillus thuringiensis*. European Patent Application, Publication Number 0 228 228.

HERRNSTADT, C., GILROY, T.E., SOBIESKI, D.A., BENNETT, B.D. AND GAERTNER, F. H. (1987). Nucleotide sequence and deduced amino acid sequence of a coleopteran-active delta-endotoxin gene from *Bacillus thuringiensis* subsp. *san diego*. *Gene* **57**, 37–46.

HICKLE, L.A., BRADFISCH, G.A. AND PAYNE, J.M. (1992). Use of a *Bacillus thuringiensis* microbe for controlling lesser mealworm, *Alphitobius diaperinus*. United States Patent Number 5,098,705.

HOFMANN, C., VANDERBRUGGEN, H., HOFTE, H., VAN RIE, J., JANSENS, S. AND VAN MELLAERT, H. (1988a). Specificity of *Bacillus thuringiensis* delta-endotoxins is correlated with the presence of high-affinity binding sites in the brush border membrane of target insect midguts. *Proceedings of the National Academy of Sciences, USA* **85**, 7844–7848.

HOFMANN, C., LUTHY, P., HUTTER, R. AND PLISKA, V. (1988b). Binding of the delta endotoxin from *Bacillus thuringiensis* to brush-border membrane vesicles of the cabbage butterfly, *Pieris brassicae*. *European Journal of Biochemistry* **173**, 85–91.

HOFFMANN, M.P., ZALOM, F.G., WILSON, L.T., SMILANICK, J.M., MALYJ, L.D., KISER, J., HILDER, J.A. AND BARNES, W.M. (1992). Field evaluation of transgenic tobacco containing genes encoding *Bacillus thuringiensis* delta-endotoxin or cowpea trypsin inhibitor: Efficacy against *Heliocoverpa zea* (Lepidopteran: Noctuidae). *Journal of Economic Entomology* **6**, 2516–2522.

HOFTE, H. AND WHITELEY, H.R. (1989). Insecticidal crystal proteins of *Bacillus thuringiensis*. *Microbiological Reviews* **53**, 242–255.

HOFTE, H., DE GREVE, H., SEURINCK, J., JANSENS, S., MAHILLON, J., AMPE, C., VANDEKERCKHOVE, J., VANDERBRUGGEN, H., VAN MONTAGU, M., ZABEAU, M. AND VAECK, M. (1986). Structural and functional analysis of a cloned delta endotoxin of *Bacillus thuringiensis* Berliner 1715. *European Journal of Biochemistry* **161**, 273–280.

HOFTE, H., VAN RIE, J., JANSENS, S., VAN HOUTVEN, A., VANDERBRUGGEN, H. AND VAECK, M. (1988). Monoclonal antibody analysis and insecticidal spectrum of three types of lepidopteran-specific insecticidal crystal proteins of *Bacillus thuringiensis*. *Applied and Environmental Microbiology* **54**, 2010–2017.

HOFTE, H., SOETAERT, P., JANSENS, S. AND PEFEROEN, M. (1990). Nucleotide sequence and deduced amino acid sequence of a new Lepidoptera-specific crystal protein gene from *Bacillus thuringiensis*. *Nucleic Acids Research* **18**, 5545.

HONEE, G., VAN DER SALM, T. AND VISSER, B. (1988). Nucleotide sequence of a crystal protein gene isolated from *Bacillus thuringiensis* subspecies *entomocidus* 60.5 coding for a toxin highly active against *Spodoptera* species. *Nucleic Acids Research* **16**, 6240.

HONEE, G., VAN DER SALM, T. AND VISSER, B. (1989). Insect resistant tomato plants. Symposium on Plant Gene Transfer held at the 18th Annual UCLA Symposia on Molecular and Cellular Biology, Park City, Utah, April. *Journal of Cellular Biochemistry* **13**, 337 (suppl.).

HONEE, G., CONVENTS, D., VAN RIE, J., JANSENS, S., PEFEROEN, M. AND VISSER, B. (1991). The carboxyl-terminal domain of the toxic fragment of a *Bacillus thuringiensis* crystal protein determines receptor binding. *Molecular Microbiology* **5**, 2799–2806.

ISHIWATA, S. (1901). On a kind of severe flacherie (sotto disease) (I). *Dainihin Sanshi Keiho* **9**, 1–5.

JANSENS, S., DE CLERCQ, R., REYNAERTS, A. AND PEFEROEN, M. (1992). Greenhouse evaluation of transgenic tomato plants expressing a *Bacillus thuringiensis* insecticidal crystal protein, for control of *Helicoverpa armigera*. Lepidoptera: Noctuidae. International Symposium on Crop Protection, Gent, Belgium. *Meded Faclandbouwwet Rijksuniv Gent* **57** (2–3 Part A–B), 515–522.

JOHNSON, D.E. (1981). Preparing entomocidal products with oligosporogenic mutants of *Bacillus thuringiensis*. United States Patent Number 4,277,564.

JOHNSON, D.E., BROOKHART, G., KRAMER, K.J., BARNETT, B.D. AND MCGAUGHEY, W.H. (1990). Resistance to *Bacillus thuringiensis* by the Indian meal moth, *Plodia interpunctella*: Comparison of midgut proteinases from susceptible and resistant larvae. *Journal of Invertebrate Pathology* **55**, 235–244.

KINGAN, S.L. AND ENSIGN, J.C. (1968). Isolation and characterization of three autolytic enzymes associated with sporulation of *Bacillus thuringiensis* var. *thuringiensis*. *Journal of Bacteriology* **96**, 629–638.

KOSTKA, S.J., REESER, P.W. AND MILLER, D.P. (1988). Experimental host range of *Clavibacter xyli* ssp. *cynodontis, Cxc,* and a *Cxc* and *Bacillus thuringiensis* recombinant, *CxcBt*. Annual Meeting of the American Phytopathology Society and the Pacific Division, San Diego, CA, November. *Phytopathology* **78**, 1540.

KOZIEL, M.G., BELAND, G.L., BOWMAN, C., CAROZZI, N.B., CRENSHAW, R., CROSSLAND, L., DAWSON, J., DESAI, N., HILL, M., KADWELL, S., LAUNIS, K., LEWIS, K., MADDOX, D., MCPHERSON, K., MEGHJI, M.R., MERLIN, E., RGODES, R., WARREN, G.W., WRIGHT, M. AND EVOLA, S.V. (1993). Field performance of elite transgenic maize plants expressing an insecticidal protein derived from *Bacillus thuringiensis*. *Bio/Technology* **11**, 194–200.

KRIEG, A., HUGER, A.M., LANGENBRUCH, G.A. AND SCHNETTER, W. (1983). *Bacillus thuringiensis* var. *tenebrionis*, new variety, a new pathotype effective against larvae of Coleoptera. *Journal of Applied Entomology* **96**, 500–508.

LAMBERT, B., HOFTE, H., ANNYS, K., JANSENS, J., SOETAERT, P. AND PEFEROEN, M. (1992a). Novel *Bacillus thuringiensis* insecticidal crystal protein with a silent activity against coleopteran larvae. *Applied and Environmental Microbiology* **58**, 2536–2542.

LAMBERT, B., THEUNIS, W., AGUDA, R., VAN AUDENHOVE, K., DECOCK, C., JANSENS, S., SEURINCK, J. AND PEFEROEN, M. (1992b). Nucleotide sequence of gene *cryIIID* encoding a novel coleopteran-active crystal protein from strain BT109P of *Bacillus thuringiensis* subsp. *kurstaki*. *Gene* **110**, 131–132.

LEE, M.K., MILNE, R.E., GE, A.Z. AND DEAN, D.H. (1992). Location of a *Bombyx mori* receptor binding region on a *Bacillus thuringiensis* delta-endotoxin. *Journal of Biological Chemistry* **267**, 3115–3121.

LI, J., CARROLL, J. AND ELLAR, D.J. (1991). Crystal structure of insecticidal delta-endotoxin from *Bacillus thuringiensis* at 2.5 Å resolution. *Nature (London)* **353**, 815–821.

LILLEY, M., RUFFELL, R.N. AND SOMERVILLE, H.J. (1980). Purification of the insecticidal toxin in crystals of *Bacillus thuringiensis*. *Journal of General Microbiology* **118**, 1–11.

LUCKOW, V.L. AND SUMMERS, M.D. (1988). Trends in the development of baculovirus expression vectors. *Bio/Technology* **6**, 47–55.

MACINTOSH, S.C., STONE, T.B., JOKERST, R.S. AND FUCHS, R.L. (1991). Binding of *Bacillus thuringiensis* proteins to a laboratory selected line of *Heliothis virescens*. *Proceedings of the National Academy of Sciences, USA* **88**, 8930–8933.

MARRONE, P.G. AND MACINTOSH, S.C. (1993). Resistance to *Bacillus thuringiensis* and resistance management. In *Bacillus thuringiensis, an Environmental Biopesticide: Theory and Practice* (P.F. Entwistle, J.S. Cory, M.J. Bailey and S. Higgs, Eds), pp. 221–235. John Wiley, New York.

MARTENS, J.W.M., HONEE, G., ZUIDEMA, D., VAN LENT, J.W.M., VISSER, B. AND VLAK, J.M. (1990). Insecticidal activity of a bacterial crystal protein expressed by recombinant baculovirus in insect cells. *Applied and Environmental Microbiology* **56**, 2764–2770.

MASSON, L., MOAR, W.J., VAN FRANKENHUYZEN, K., BOSSE, M. AND BROUSSEAU, R. (1992). Insecticidal properties of a crystal protein gene product isolated from *Bacillus thuringiensis* subsp. *kenyae*. *Applied and Environmental Microbiology* **58**, 642–646.

MCGAUGHEY, W.H. (1985). Insect resistance to the biological insecticide *Bacillus thuringiensis*. *Science* **229**, 193–195.

MCGAUGHEY, W.H. AND BEEMAN, R.W. (1988). Resistance to *Bacillus thuringiensis* in colonies of Indian meal moth and almond moth (Lepidoptera: Pyralidae). *Journal of Economic Entomology* **81**, 28–33.

MCGAUGHEY, W.H. AND JOHNSON, D.E. (1987). Toxicity of different serotypes and toxins of *Bacillus thuringiensis* to resistant and susceptible Indian meal moths

(Lepidoptera: Pyralidae). *Journal of Economic Entomology* **80**, 1122–1126.

McGaughey, W.H. and Johnson, D.E. (1992). Indian meal moth (Lepidoptera: Pyralidae) resistance to different strains and mixtures of *Bacillus thuringiensis*. *Journal of Economic Entomology* **85**, 1595–1600.

McGaughey, W.H. and Whalon, M.E. (1992). Managing insect resistance to *Bacillus thuringiensis* toxins. *Science* **258**, 1451–1455.

McGuire, M.R. and Shasha, B.S. (1992). Adherent starch granules for encapsulation of insect control agents. *Journal of Economic Entomology* **85**, 1425–1433.

Merryweather, A.T., Weyer, U., Harris, M.P.G., Hirst, M., Booth, T. and Possee, R.D. (1990). Construction of genetically engineered baculovirus insecticides containing the *Bacillus thuringiensis* ssp. *kurstaki* HD-73 delta endotoxin. *Journal of General Virology* **71**, 1535–1544.

Mettus, A.-M. and Macaluso, A. (1990). Expression of *Bacillus thuringiensis* delta-endotoxin genes during vegetative growth. *Applied and Environmental Microbiology* **56**, 1128–1134.

Miller, D.L., Rahardja, U. and Whalon, M.E. (1990). Development of a strain of Colorado potato beetle resistant to the delta-endotoxin of B.t. *Pest Resistance Management, WRCC Newsletter* **2**, 25.

Murphy, R.C. and Stevens, S.E. (1992). Cloning and expression of the CryIVD gene of *Bacillus thuringiensis* subsp. *israelensis* in the cyanobacterium *Agmenellum quadruplicatum* PR-6 and its resulting larvicidal activity. *Applied and Environmental Microbiology* **58**, 1650–1655.

Murray, E.E., Rocheleau, T., Eberle, M., Stock, C., Sekar, V. and Adang, M. (1991). Analysis of unstable RNA transcripts of insecticidal crystal protein genes of *Bacillus thuringiensis* in transgenic plants and electroporated protoplasts. *Plant Molecular Biology* **16**, 1035–1050.

National Audubon Society (1991). Insect resistance to Bt delta-endotoxin: What it means for farming practices and the environment. Report of a Workshop for University Scientists, Pesticide Managers, Environmentalists and Policy Makers Convened by the National Audubon Society, October.

Obukowicz, M.G., Peralk, F.J., Kusano-Kretzmer, K., Mayer, E.J., Bolten, S.L. and Watrud, L.S. (1986a). Tn5-mediated integration of the delta-endotoxin gene from *Bacillus thuringiensis* into the chromosome of root-colonizing Pseudomonads. *Journal of Bacteriology* **168**, 982–989.

Obukowicz, M.G., Perlak, F.J., Kusano, K., Mayer, E.J. and Watrud, L.S. (1986b). Integration of the delta-endotoxin gene of *Bacillus thuringiensis* into the chromosome of root-colonizing strains of Pseudomonads using Tn5. *Gene* **45**, 327–331.

Oddou, P., Hartmann, H. and Geiser, M. (1991). Identification and characterization of *Heliothis virescens* midgut membrane proteins binding *Bacillus thuringiensis* of delta-endotoxins. *European Journal of Biochemistry* **202**, 673–680.

Oeda, K., Inouye, K., Ibuchi, Y., Oschie, K., Shimizu, M., Nakamura, K., Nishioka, R., Takada, Y. and Ohkawa, H. (1989). Formation of crystals of the insecticidal proteins of *Bacillus thuringiensis* subsp. *aizawai* IPL7 in *Escherichia coli*. *Journal of Bacteriology* **171**, 3568–3571.

Pang, Y., Frutos, R. and Federici, B.A. (1992). Synthesis and toxicity of full-length and truncated bacterial cryIVD mosquitocidal proteins expressed in lepidopteran cells using a baculovirus vector. *Journal of General Virology* **73**, 89–102.

Payne, C.C. (1986). Insect pathogenic viruses as pest control agents. *Fortschritte der Zoologie* **32**, 183–200.

Payne, J. and Sick, A. (1990). Novel *Bacillus thuringiensis* isolates active against lepidopteran pests, and genes encoding novel lepidopteran-active toxins. European Patent Application, Publication Number 0 401 979.

Payne, J. and Sick, A.J. (1991a). Novel *Bacillus thuringiensis* isolate active against lepidopteran pests, and genes encoding novel lepidopteran-active toxins. European Patent Application, Publication Number 0 405 810 A2.

PAYNE, J. AND SICK, A.J. (1991b). Novel *Bacillus thuringiensis* isolate denoted B.T. PS81F, active against lepidopteran pests, and a gene encoding a lepidopteran-active toxin. United States Patent Number 5,039,523.

PAYNE, J., SICK, A.J. AND THOMPSON, M. (1992). Gene encoding a lepidopteran-active toxin from *Bacillus thuringiensis* isolate denoted B.T. PS81GG active against lepidopteran pests. United States Patent Number 5,135,867.

PAYNE, J.M., CUMMINGS, D.A. AND CANNON, R.J.C. (1993). Novel *Bacillus thuringiensis* isolate denoted B.t. PS158C2, active against lepidopteran insects, and genes encoding lepidopteran-active toxins. PCT International Patent Application Number WO 93/05657.

PAYNE, J.M., KENNEDY, M.K., RANDALL, J.B., MEIER, H. AND UICK, H.J. (1992). Novel *Bacillus thuringiensis* isolates active against hymenopteran pests and gene(s) encoding hymenopteran-active toxins. European Patent Application, Publication Number 0 516 306 A2.

PEFEROEN, M. (1992). Engineering of insect-resistant plants with *Bacillus thuringiensis* crystal protein genes. In *Biotechnology in Agriculture, No. 7. Plant Genetic Manipulation for Crop Protection* (A.M.R. Gatehouse, V.A. Hilder and D. Boulter, Eds), pp. 135–153. CAB International, Wallingford.

PERLAK, F.J., MCPHERSON, S.A., FUCHS, R.L., MACINTOSH, S.C., DEAN, D.A. AND FISCHHOFF, D.A. (1988). Expression of *Bacillus thuringiensis* protein in transgenic plants. In *Proceedings of a Conference on Biotechnology, Biological Pesticides and Novel Plant–Pest Resistance for Insect Pest Management* (D.W. Roberts and R.R. Granados, Eds), pp. 77–81. Boyce Thompson Institute for Plant Research, Ithaca, NY.

PERLAK, F.J., DEATON, R.W., ARMSTRONG, T.A., FUCHS, R.L., SIMS, S.R., GREENPLATE, J.T. AND FISCHHOFF, D.A. (1990). Insect resistant cotton plants. *Bio/Technology* **8**, 939–943.

PERLAK, F.J., FUCHS, R.L., DEAN, D.A., MCPHERSON, S.L. AND FISCHHOFF, D.A. (1991). Modification of the coding sequence enhances plant expression of insect control protein genes. *Proceedings of the National Academy of Sciences, USA* **88**, 3324–3328.

PERLAK, F.J., STONE, T.B., MUSKOPF, Y.M., PETERSON, L.J., PARKER, G.B., MCPHERSON, S.A., WYMAN, J., LOVE, S., REED, G., BIEVER, D. AND FISCHHOFF, D.A. (1993). Genetically improved potatoes: Protection from damage by Colorado potato beetles. *Plant Molecular Biology* **22**, 313–321.

POZSGAY, M., FAST, P., KAPLAN, H. AND CAREY, P.R. (1987). The effect of sunlight on the protein crystals from *Bacillus thuringiensis* var. *kurstaki* HD1 and NRD12: A Raman spectroscopic study. *Journal of Invertebrate Pathology* **50**, 246–253.

PREFONTAINE, G., FAST, P., LAU, P.C.K., HEFFORD, H.A., HANNA, Z. AND BROUSSEAU, R. (1987). Use of oligonucleotide probes to study the relatedness of delta-endotoxin genes among *Bacillus thuringiensis* subspecies and strains. *Applied and Environmental Microbiology* **53**, 2808–2814.

ROUSH, R. AND MCKENZIE, J.A. (1987). Ecological genetics of insecticide and acaricide resistance. *Annual Review of Entomology* **32**, 361–380.

ROWE, G.E. AND MARGARATIS, A. (1987). Bioprocess developments in the production of bioinsecticides by *Bacillus thuringiensis. CRC Critical Review of Biotechnology* **6**, 87–127.

SCHNEPF, H.E. AND WHITELEY, H.R. (1981). Cloning and expression of the *Bacillus thuringiensis* crystal protein gene in *Escherichia coli. Proceedings of the National Academy of Sciences, USA* **78**, 2893–2897.

SCHNEPF, H.E., WONG, H.C. AND WHITELEY, H.R. (1985). The amino acid sequence of a crystal protein from *Bacillus thuringiensis* deduced from the DNA base sequence. *Journal of Biological Chemistry* **260**, 6264–6272.

SCHNEPF, H.E., SCHWAB, G.E., PAYNE, J.M., NARVA, K.E. AND FONCERRADA, L.

(1992). Novel nematode-active toxins and genes which code therefor. European Patent Application, Publication Number 0 517 367 A1.

SCHWAB, G.E., NARVE, K.E. AND KIM, L. (1992). Isolation characterization and cloning of toxins from *Bacillus thuringiensis* active against plant parasitic nematodes. Keystone Symposium on Crop Improvement Via Biotechnology: An International Perspective, Keystone, Colorado, USA. *Journal of Cellular Biochemistry* suppl. **16F**, 238.

SCOTT, J.K. AND SMITH, G.P. (1990). Searching for peptide ligands with an epitope library. *Science* **249**, 386–390.

SEKAR, V., THOMPSON, D.V., MARONEY, M.J., BOOKLAND, R.G. AND ADANG, M.J. (1987). Molecular cloning and characterization of the insecticidal crystal protein gene of *Bacillus thuringiensis* var. *tenebrionis. Proceedings of the National Academy of Sciences, USA* **84**, 7036–7040.

SEWELL, G.H. AND STORCH, R.H. (1992). Transgenic potato plants, control of Colorado potato beetle, 1991. In *Insecticide and Acaricide Tests* (A.K. Burditt Jr, Ed.), Vol. 17, pp. 138–139. Entomological Society of America, Lanham, MD.

SHELTON, A.M., ROBERTSON, J.L., TANG, J.D., PEREZ, C., EIGENBRODE, S.D., PREISLER, H.K., WILSEY, W.T. AND COOLEY, R.J. (1993). Resistance of diamondback moth (Lepidoptera: Plutellidae) to *Bacillus thuringiensis* subspecies in the field. *Journal of Economic Entomology* **86**, 697–705.

SHIVAKUMAR, A.G., GUNDLING, GJ., BENSEN, T.A., CASUTO, D., MILLER, M.F. AND SPEAR, B.B. (1986). Vegetative expression of the delta-endotoxin genes from *Bacillus thuringiensis* subsp. *kurstaki* in *Bacillus subtilis. Journal of Bacteriology* **166**, 194–204.

SICK, A.J. (1991). *Bacillus thuringiensis* genes encoding dipteran-active toxins. European Patent Application, Publication Number 0 457 498 A2.

SICK, A. AND GILROY, T.E. (1991). *Bacillus thuringiensis* gene encoding a coleopteran-active toxin. United States Patent Numbers 4,996,115 and 5,133,962.

SICK, A., GAERTNER, F. AND WONG, A. (1990). Nucleotide sequence of a coleopteran-active toxin gene from a new isolate of *Bacillus thuringiensis* subsp. *tolworthi. Nucleic Acids Research* **18**, 1305.

SIMS, S.R. AND STONE, T.B. (1991). Genetic basis of tobacco budworm resistance to an engineered *Pseudomonas fluorescens* expressing the δ-endotoxin of *Bacillus thuringiensis kurstaki. Journal of Invertebrate Pathology* **57**, 206–210.

SIVASUBRAMANIAN, N. AND FEDERICI, B. (1991). Method and means for extending the host range of insecticidal proteins. PCT International Patent Application, Publication Number WO 91/17254.

SMITH, G.E., FRASER, M.J. AND SUMMERS, M.D. (1983). Molecular engineering of *Autographa californica* nuclear polyhedrosis virus genome: Deletion mutants with the polyhedrin gene. *Journal of Virology* **46**, 584–593.

SMULEVITCH, S.V., OSTERMAN, A.L., SHEVELEV, A.B., KALUGER, S.V., KARASIN, A.I., KADYROV, R.M., ZAGNITKO, O.P., CHESTUKHINE, G.G. AND STEPANOV, V.M. (1991). Nucleotide sequence of a novel δ-endotoxin gene *cryIG* of *Bacillus thuringiensis* subsp. *galleriae. FEBS Letters* **293**, 25–28.

STOCK, C.A., MCLOUGHLIN, T.J., KLEIN, J.A. AND ADANG, M.J. (1990). Expression of a *Bacillus thuringiensis* crystal protein gene in *Pseudomonas cepacia* 526. *Canadian Journal of Microbiology* **36**, 879–884.

STOLTZ, R.L. AND MATTESON, N.A. (1992). Colorado potato beetle control with genetically engineered, transgenic, potato plants, 1991. In *Insecticide and Acaricide Tests* (A.K. Burditt Jr, Ed.), Vol. 17. pp. 143–144. Entomological Society of America, Lanham, MD.

STONE, T.B., SIMS, S.R. AND MARRONE, P.G. (1989). Selection of tobacco budworm for resistance to a genetically engineered *Pseudomonas fluorescens* containing the δ-endotoxin of *Bacillus thuringiensis* subsp. *kurstaki. Journal of Invertebrate Pathology* **53**, 228–234.

TABASHNIK, B.E. (1986). Computer simulation as a tool for pesticide resistance management. In *Pesticide Resistance: Strategies and Tactics for Management*, pp. 94–206. Committee on Strategies for the Management of Pesticide Resistant Pest Populations, Board of Agriculture, National Research Council. National Academy Press, Washington, DC.

TABASHNIK, B.E. (1989). Managing resistance with multiple pesticide tactics: Theory, evidence and recommendations. *Journal of Economic Entomology* **82**, 1263–1269.

TABASHNIK, B.E., FINSON, N. AND JOHNSON, M.W. (1991). Managing resistance to *Bacillus thuringiensis*: Lessons from the diamondback moth (Lepidoptera: Plutellidae). *Journal of Economic Entomology* **84**, 49–55.

TABASHNIK, B.E., CUSHING, N.L., FINSON, N. AND JOHNSON, M.W. (1990). Field development of resistance to *Bacillus thuringiensis* in diamondback moth (Lepidoptera: Plutellidae). *Journal of Economic Entomology* **83**, 1671–1676.

TAILOR, R., TIPPETT, J., GIBB, G., PELLS, S., PIKE, D., JORDAN, L. AND ELY, S. (1992). Identification and characterization of a novel *Bacillus thuringiensis* δ-endotoxin entomocidal to coleopteran and lepidopteran larvae. *Molecular Microbiology* **6**, 1211–1217.

TALEKAR, N.S. AND SHELTON, A.M. (1993). Biology, ecology and management of the diamondback moth. *Annual Review of Entomology* **38**, 275–301.

THANABALU, T., HINDLEY, J., BRENNER, S., OEI, C. AND BERRY, C. (1992). Expression of the mosquitocidal toxins of *Bacillus sphaericus* and *Bacillus thuringiensis* subsp. *israelensis* by recombinant *Caulobacter crescentus*, a vehicle for biological control of aquatic insect larvae. *Applied and Environmental Microbiology* **58**, 905–910.

THOMPSON, M. AND GAERTNER, F.H. (1991). Novel *Bacillus thuringiensis* isolate having anti-protozoan activity. European Patent Application, Publication Number 0 461 799 A2.

THORNE, L., GARDUNO, F., THOMPSON, T., DECKER, D., ZOUNES, M., WILD, M., WALFIELD, A.M. AND POLLOCK, T.J. (1986). Structural similarity between Lepidoptera- and Diptera-specific insecticidal endotoxin genes of *Bacillus thuringiensis* subsp. *kurstaki* and *israelensis*. *Journal of Bacteriology* **166**, 801–811.

TURNER, J.T., LAMPEL, J.S., STEARMAN, R.S., SUNDIN, G.W., GUNYUZLU, P. AND ANDERSON, J.J. (1991). Stability of the δ-endotoxin gene from *Bacillus thuringiensis* subsp. *kurstaki* in a recombinant strain of *Clavibacter xyli* subsp. *cynodontis*. *Applied and Environmental Microbiology* **57**, 3522–3528.

US DEPARTMENT OF AGRICULTURE, COOPERATIVE STATE RESEARCH SERVICE, AGRICULTURAL RESEARCH SERVICE (1992). Scientific evaluation of the potential for pest resistance to the *Bacillus thuringiensis* (Bt) delta-endotoxins: A conference to explore resistance management strategies, January.

VAECK, M., REYNAERTS, A. AND HOFTE, H. (1989). Protein engineering in plants: Expression of *Bacillus thuringiensis* insecticidal protein genes. In *Cell Culture and Somatic Cell Genetics of Plants* (J. Schell and I.K. Vasil, Eds), Vol. 6, pp. 425–440. Academic Press, Troy, MI.

VAECK, M., REYNAERTS, A., HOFTE, H., JANSENS, S., DE BEUCKELEER, M., DEAN, C., ZABEAU, M., VAN MONTAGU, M. AND LEEMANS, J. (1987). Transgenic plants protected from insect attack. *Nature (London)* **328**, 33–37.

VAN RIE, J., VAN MELLAERT, H. AND PEFEROEN, M. (1992). Mechanism of insect resistance to *Bacillus thuringiensis* in *Plodia interpunctella* and *Plutella xylostella*. In *Molecular Mechanisms of Insecticide Resistance: Diversity Among Insects* (C.A. Mullin and J.G. Scott, Eds), pp. 191–199. American Chemical Society Symposium Series 505. American Chemical Society, Washington, DC.

VAN RIE, J., JANSENS, S., HOFTE, H., DEGHEELE, D. AND VAN MELLAERT, H. (1989). Specificity of *Bacillus thuringiensis* δ-endotoxins: Importance of specific receptors on the brush border membrane of the mid-gut of target insects. *European Journal of Biochemistry* **186**, 239–247.

VAN RIE, J., JANSENS, S., HOFTE, H., DEGHEELE, D. AND VAN MELLAERT, H.

(1990a). Receptors on the brush border membrane of the insect midgut as determinants of the specificity of *Bacillus thuringiensis* delta-endotoxins. *Applied and Environmental Microbiology* **56**, 1378–1385.

VAN RIE, J., MCGAUGHEY, W.H., JOHNSON, D.E., BARNETT, B.D. AND VAN MELLAERT, H. (1990b). Mechanism of insect resistance to the microbial insecticide *Bacillus thuringiensis*. *Science* **247**, 72–74.

VISSER, B. (1989). A screening for the presence of four different crystal protein gene types in 25 *Bacillus thuringiensis* strains. *FEMS Microbiology Letters* **58**, 121–124.

VISSER, B., MUNSTERMAN, E., STOKER, A. AND DIRKSE, W.G. (1990). A novel *Bacillus thuringiensis* gene encoding a *Spodoptera exigua*-specific crystal protein. *Journal of Bacteriology* **172**, 6783–6788.

VISSER, B., MUNSTERMAN, E., HONEE, G. AND VAN DER SALM, T. (1992). Transgenic tobacco plants expressing a modified *Bacillus thuringiensis cryIC* gene. Keystone Symposium on Crop Improvement Via Biotechnology: An International Perspective, Keystone, Colorado, USA. *Journal of Cellular Biochemistry* suppl. **16F**, 213.

VON TERSCH, M.A., ROBBINS, H.L., JANY, C.S. AND JOHNSON, T. (1991). Insecticidal toxins from *Bacillus thuringiensis* subsp. *kenyae*: Gene cloning and characterization and comparison with *B. thuringiensis* CryIA(c) toxins. *Applied and Environmental Microbiology* **57**, 349–358.

WAALWIJK, C., DULLEMANS, A. AND MAAT, C. (1991). Construction of a bioinsecticidal rhizosphere isolate of *Pseudomonas fluorescens*. *FEMS Microbiology Letters* **77**, 257–264.

WABIKO, H., RAYMOND, K.C. AND BULLA, L.A. (1986). *Bacillus thuringiensis* entomocidal protoxin gene sequence and gene product analysis. *DNA* **5**, 305–314.

WAKISAKA, Y., MASAKI, E., KOIZUMI, K., NISHIMOTO, U., ENDO, Y., NISHIMURA, M.S. AND NISHIITSUTSUJI-UWO, J. (1982). Asporogenous *Bacillus thuringiensis* mutant producing high yields of delta-endotoxin. *Applied and Environmental Microbiology* **43**, 1473–1480.

WARD, E.S. AND ELLAR, D.J. (1987). Nucleotide sequence of a *Bacillus thuringiensis* var. *israelensis* gene encoding a 130 kDa delta-endotoxin. *Nucleic Acids Research* **15**, 7175.

WARREN, G.W., CAROZZI, N.B., DESAI, N. AND KOZIEL, M.G. (1992). Field evaluation of transgenic tobacco containing a *Bacillus thuringiensis* insecticidal protein gene. *Journal of Economic Entomology* **5**, 1651–1659.

WATRUD, L.S., PERALK, F.J., TRAN, M., KUSANO, K., MAYER, E.J., MILLER-WIDEMAN, M.A., OBUKOWICZ, M.G., NELSON, D.R., KREITINGER, J.P. AND KAUFMAN, R.J. (1985). Cloning of the *Bacillus thuringiensis* subsp. *kurstaki* delta-endotoxin gene into *Pseudomonas fluorescens*: Molecular biology and ecology of an engineered microbial pesticide. In *Engineered Organisms in the Environment: Scientific Issues* (H.O. Halvorson, D. Pramer and M. Rogul, Eds), pp. 40–44. American Society for Microbiology, Washington, DC.

WIDNER, W.R. AND WHITELEY, H.R. (1989). Two highly related insecticidal crystal proteins of *Bacillus thuringiensis* subsp. *kurstaki* possess different host range specificities. *Journal of Bacteriology* **171**, 965–974.

WIDNER, W.R. AND WHITELEY, H.R. (1990). Location of the dipteran specificity region in a lepidopteran–dipteran crystal protein from *Bacillus thuringiensis*. *Journal of Bacteriology* **172**, 2826–2832.

WILLIAMS, S., FRIEDRICH, L., DINCHER, S., CAROZZI, N., KESSMANN, H., WARD, E. AND RYALS, J. (1992). Chemical regulation of *Bacillus thuringiensis* delta-endotoxin expression in transgenic plants. *Bio/Technology* **10**, 540–543.

WILSON, F.D., FLINT, H.M., DEATON, W.R., FISCHHOFF, D.A., PERLAK, F.J., ARMSTRONG, T.A., FUCHS, R.L., BERBERICH, S.A., PARKS, N.J. AND STAPP, B.R. (1992). Resistance of cotton lines containing a *Bacillus thuringiensis* toxin to pink bollworm (Lepidopteran: Gelechiidae) and other insects. *Journal of Economic Entomology* **4**, 1516–1521.

WONG, E.Y., HIRONAKA, C.M. AND FISCHHOFF, D.A. (1992). *Arabidopsis thaliana* small subunit leader and transit peptide enhance the expression of *Bacillus thuringiensis* proteins in transgenic plants. *Plant Molecular Biology* **20**, 81–93.

WU, D., CAO, X.L., BAI, Y.Y. AND ARONSON, A.I. (1991). Sequence of an operon containing a novel δ-endotoxin gene from *Bacillus thuringiensis*. *FEMS Microbiology Letters* **81**, 31–36.

7
Biotechnological Inventions and the Patent Law: Outstanding Issues

R.S. CRESPI

Blackbird Cottage, Walton Avenue, Bognor Regis, Sussex, UK

Introduction

Although nearly 4 years have elapsed since the last review on patent law in this series (Crespi, 1989), many of the issues previously discussed have not yet been properly resolved. Delays in the settlement of law suits can arise just as much from the tenacity of the contestants, who will go on appealing from adverse decisions in the hope of obtaining a favourable one, as from the delays inherent in the legal process itself.

Some of the legal disputes on specific patents outlined in the last review seem to have come to their natural conclusion, or at least to the point beyond which one or other party decides not to proceed, and these will be discussed later. The other kind of delay is that which comes from the political process, that is, questions of public policy as to what the law should allow or refuse to be patented in the sphere of biotechnology. This latter topic was addressed in the last review in the light of earlier initiatives of public bodies including the World Intellectual Property Organization in Geneva, which is at present inactive on this subject, and more especially the European Commission in Brussels, which is continuing its effort to forge a Directive to EC Member States to improve and clarify their national laws in this respect. The industry most affected by these public initiatives is that of agrobiotechnology, and the contentious issues can perhaps be seen as part of the wider ongoing debate over the General Agreement on Tariffs and Trade (GATT) and other matters involving the special problems of the agricultural industry and its dependant communities, which are unlikely to be solved in the short term.

This review will broadly follow the structure of the previous one and discuss some of the more crucial legal questions arising in specific patent cases,

---

Abbreviations: bGH, bovine growth hormone; CAFC, Court of Appeals for the Federal Circuit; cDNA, complementary DNA; EC, European Commission; EPC, European Patent Convention; epo, erythropoietin; EPO, European Patent Office; EST, expressed sequence tag; GATT, General Agreement on Tariffs and Trade; MRC, Medical Research Council; NIH, National Institutes of Health; *N*-met, *N*-terminal methionine; PVR, plant variety rights; t-PA, tissue plasminogen activator; UPOV, International Union for the Protection of New Varieties of Plant; USPTO, US Patent and Trade Mark Office.

Biotechnology and Genetic Engineering Reviews — Vol. 11, December 1993
0264–8725/93/11/229–261 $20.00 + $0.00 © Intercept Ltd, P.O. Box 716, Andover, Hampshire SP10 1YG, UK

including law suits between industrial competitors. To place these in context, however, it will sometimes be necessary to sketch in the legal background against which current issues must be viewed. The most important features of this background are the concepts of novelty and inventiveness which underpin the structure of the patent law. A further topic concerns the question of patent scope, that is, how much of the technology can be covered by the claims of the patent based on what the inventor has actually demonstrated experimentally. Reference to the very first review (Crespi, 1985) may also be helpful for an understanding of some of these basic concepts. The second part of this review will look at the political dimension, as a part of which it will be important to discuss developments in the controversial subject of plant patents and animal patents and to address ethical issues which have come to the fore in recent times.

Case law developments: Novelty and inventiveness (obviousness)

To obtain a valid patent and to enforce it successfully in the courts, the patent application and resulting patent must survive a number of legal tests. The first and most basic test is that the subject matter of the claims must be novel in the light of previous public knowledge. As a close second comes the requirement that what is claimed must also be inventive when judged against the same background. At the date an applicant files for a patent, therefore, the invention must not already have been disclosed to the public by others or even by the person who invented it. Previous public disclosure of an invention by an inventor has often ruined the chance of subsequently obtaining a patent for it, and there are some notorious examples of this in the field of biological chemistry.

Novelty in patent law

Novelty has always seemed to be a simple concept to grasp. However, the question has arisen in recent years as to what sort of disclosure will torpedo a patent application which has been filed *after* the disclosure has been made public. Following a number of cases on this point in the European Patent Office (EPO), the matter has also recently been decided by the House of Lords, the highest legal authority on the interpretation of law in the UK, in a particular case which we shall refer to as the Asahi case.

THE ASAHI CASE

Confirming the position taken by the EPO, the House of Lords has ruled (Asahi, 1991) that for a publication to be fatal to a subsequently filed patent application, it must be more than a *bare* disclosure of what is claimed in the application. The disclosure must be an *enabling* disclosure if it is to be an effective anticipation of the invention, that is, it must contain sufficient information to enable a person of ordinary skill and knowledge in the art successfully to repeat the experiment or process described. At first sight, this ruling may seem to be of some help to those research workers who are anxious to obtain earliest possible recognition of their work through scientific publication but who are also reluctant to jettison completely the chance, which a patent gives, of benefiting from possible commercial application of their work. A full literature publication which issues before the author seeks patent protection will usually give sufficient information to be enabling in the above sense. However, there arc exceptions, especially where repetition of the published work is impossible without access to novel strains of a micro-organism or other biological material and the publication does not provide this. But literature publication is not the only type of disclosure which is prejudicial to patent prospects. Any form of prior public disclosure may possibly be damaging in patent terms, including oral disclosures at scientific meetings, non-confidential information provided to visitors on conducted tours of the laboratory, and information volunteered to the press or other media. Posters at scientific meetings which effectively disclose the information in an enabling manner are a regular trap for the unwary (some readers may think that this point is being rammed home unduly, but others may not be fully aware of these pitfalls!).

The points made above apply primarily to the patent law in Europe. US law is different from European law in important respects and these differences will be mentioned later. In the language of European patent law, all such information which is made available in one way or another to the public (it does not have to be the general public, the interested public will do) is said to be part of the 'state of the art'. The law in European countries extends this principle further by also including in the state of the art pending patent applications of an earlier date, even though these have not reached the stage of actual publication. This is a sort of special case, a legal fiction designed to ensure that competitive applicants claiming the same invention cannot both have a patent but are treated on a 'first come, first served' basis.

THE FIRST-TO-FILE PRINCIPLE

The case before the House of Lords involved the patent applications of two Japanese companies, Asahi and Dainippon, the question being which of the two applications was entitled to priority in regard to claims to certain DNA sequences. Priority in patent law is decided not only by the date of filing an application but also by the content of what is filed. This means that the application must be a proper disclosure of what is claimed. The legal detail of the case is complex, involving a number of applications filed in the UK, Europe, USA and Japan and the chronology of these is shown in *Figure 1*. But the matter boiled down essentially to the question whether the earliest application to be filed, that of Dainippon in Japan, was entitled to its date for priority purposes. The Dainippon Japanese application had merely disclosed the DNA sequence without giving any experimental detail of the preparation of this material. Asahi therefore claimed that Dainippon were not entitled to this earlier date and that Asahi's application, which was filed later and which did contain an enabling disclosure of the preparation of the DNA, should take precedence over Dainippon's.

One of the Law Lords stated: 'if published information does no more than disclose the existence of a product which is not physically obtainable by the public that product cannot be said to have been "made available" unless and until the public has been told how it can be produced. The product is not available in the absence of an enabling disclosure.' According to another of their Lordships, to accord priority on the basis of a patent application which lacked such a disclosure 'would seem to encourage the filing of applications for "inventions" which may or may not be capable of commercial exploitation in the hope that in the ensuing 12 months something will turn up to render them patentable, such applications having the effect of preventing anyone else who may be better informed from obtaining a patent therefor.'

It might be argued that if a nucleotide sequence is specified, it would be obvious how to synthesize the DNA and therefore unnecessary to describe a method of preparation. The decision specifically noted that there may be cases in which the method of producing a compound may be self-evident from a mere recital of the formula, but their Lordships decided that this was not such a case. The message to applicants not to rely on such an argument is therefore rather clear – there is no substitute for doing the work and properly describing how it was done. Asahi secured their patent because they had the fuller patent disclosure.

CONSEQUENCES OF THE ASAHI CASE

Legal aspects

Although the decision seems fair, it could possibly give rise to some legal uncertainty. There may often be considerable room for doubt as to whether a prior document is enabling, since this sort of judgment involves the question 'enabling to whom?' and this in turn means assessing the degree of skill and knowledge possessed by the notional 'person skilled in the art'. The uncertainty could even be felt by the author of the publication, who might want to know whether by publishing the information the chance of obtaining a patent has been lost, and it could apply also to third parties who might want to assess their freedom to exploit the published information.

Figure 1. Chronology of Asahi/Dainippon applications.

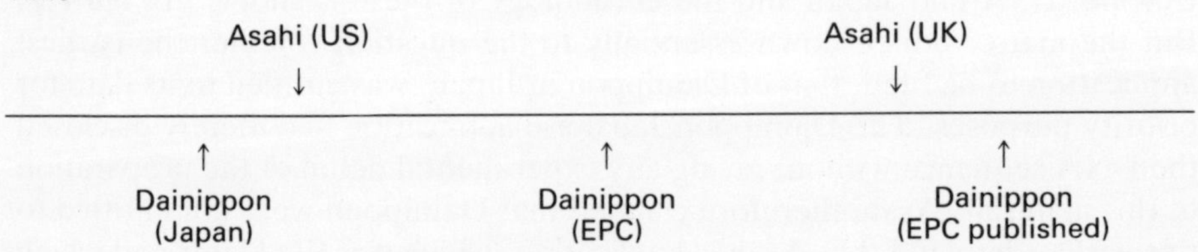

Practical effect on inventors

Will this decision encourage research workers to be even less cautious than before about damaging their potential patent rights in their enthusiasm for rapid disclosure of information? It is not uncommon for papers read at scientific meetings to disclose results without providing an enabling disclosure of how these results could be duplicated. The Asahi case would seem to make this a respectable practice from the patent viewpoint, but it does involve skating on thin ice. Even if such disclosures do not sabotage a subsequent patent application from the point of view of strict novelty, they may undermine it in relation to the requirement of inventiveness (see later). Most patent attorneys tend to be cautious and prefer the old maxim 'file first and publish second'.

THE FIRST-TO-INVENT PRINCIPLE OF US LAW

The patent law of the USA also recognizes the distinction between a bare disclosure and an enabling one and is in line with the European view. However, in relation to an inventor's own enabling publication, the US law is more generous than the European (see Crespi, 1989, for detail). In deciding the question of priority, the actual date of invention is taken into account and the US resident can prove this date from laboratory notebook records (however, the foreign inventor does not have this privilege). Incidentally, this is a complex area of law and the present simplified explanation is not a substitute for proper professional advice. An inventor's own prior publication will not ruin the chances of obtaining a US patent provided the US application is filed within a 'grace period' of 1 year from the publication date (this applies to all inventors who file a US application and not only to US nationals).

INTERNATIONAL HARMONIZATION

In recent international moves to harmonize patent laws on the above issues, outlined briefly in Crespi (1989), it appeared that the USA would consider changing to the first-to-file system if other countries were willing to adopt the grace period in their laws. This proposal, along with many others, reached the stage of a Diplomatic Conference in June 1991, but unfortunately it transpired that the US Government had not received enough support at home for making the proffered change. The conclusion of a treaty was therefore postponed and is unlikely to be resumed before 1994.

A worldwide acceptance of the grace period would no doubt be welcome to the academic institutional inventors who, as a group, are prominent in biotechnology. In industry opinion is divided on this question. From experience, US industry is comfortable with the grace period, but in other countries industry is more dubious about change. Both sides must consider whether the trade-off between the first-to-file principle and the grace period is worthwhile.

Before leaving the question of novelty, special comment must be made on the patentability of those substances which exist in Nature.

Natural product patents

Many patents have been granted for substances produced in Nature. These substances can range from relatively simple to highly complex chemical structures, the extreme form of this complexity being that of living organisms themselves, which must also be classified under this heading. Whether natural products are living or inanimate, the same considerations apply as far as patent law is concerned. Naturally occurring substances might appear to pose a problem for the concept of novelty (and inventiveness), since it may be asked how something can be patented that already exists in Nature, whether it be an inanimate substance, an enzyme, a micro-organism or higher life form, or a gene.

The patent laws of most countries make a distinction between 'discovery' and 'invention' and exclude *mere* discoveries from patentability. Some might say that the isolation of substances from Nature is mere discovery and not invention. Some might even raise the objection that because the substance already existed it cannot be new, let alone be an invention.

The product of Nature problem has a long history in patent law, but in most industrially developed countries the extraction of valuable substances from natural sources in a way which makes them available for the first time in a form in which their properties can be utilized in medicine, agriculture or any other useful art is recognized as worthy of patent protection. In terms of what belongs to the state of the art – that is, what is already *available* to the public (the true test of novelty) – these materials can be just as new and inventive as the synthetic materials created by the chemist and biochemist.

THE US APPROACH

Most of the instructive precedents in this field have come from the US case law on purified natural products. The early court decisions pointed to the conclusion that a known substance cannot be patented as a pure product *solely* on the basis of its purity. However, where the substance was known previously only in the form of crude extracts which were of no practical use, an inventor who had devised a practically useful method of making pure product in quantity could patent not only the specific method but could claim the pure product *per se*.

The classical example of this (Merck & Co. v Chase Chemical Co., discussed briefly in Crespi, 1985) was the production of vitamin B_{12} by fermentation and its isolation in bulk as the pure substance. Previously, the substance had been available only as crude liver extracts which were of no use for therapeutic application. The US District Court upheld a product *per se* claim to pure crystalline B_{12}. The patentee had devised a fermentation process which enabled the substance to be produced on a commercial scale and this was an unquestionably meritorious advance. Considering such a product, the court stated that 'until . . . [the inventors] . . . had made it available to the world [it] . . . did not exist'. To object to a natural product patent in these circumstances seems to this writer to be legalistic in the extreme.

The US Patent and Trade Mark Office (USPTO) recognizes purity as a change of form from the natural material which is sufficient to justify the claim. For the discovery of micro-organisms and their isolation from natural sources, the USPTO applies the same logic and allows these to be patented if claimed as 'pure cultures'.

Two important modern examples of patents for purified natural products (Scripps Clinic and Research Foundation v Genentech, and Amgen Inc. v Genetics Institute) were discussed in Crespi (1989). The first of these involved Factor VIII, the blood-clotting factor important in the treatment of haemophilia, and the second erythropoietin, the hormone which stimulates the formation of red blood cells. These patents disclose particular chromatographic methods of purifying or isolating the protein. However, the claims are not limited to products produced by these methods; they also cover the proteins defined solely in terms of units of activity, a parameter connected with the degree of purity which these methods have made possible.

FACTOR VIII LITIGATION

In the Scripps case, the crucial claim of their US Re-issue patent 32,011 was: 'A human VIII:C preparation having a potency in the range of 134 to 1172 units per ml and being substantially free of VIII:RP'. Scripps asserted that this claim was infringed by the defendant's manufacture of recombinant Factor VIII, since the latter would fall within its purity criteria. The defendant argued that this claim should be interpreted in the context of the Scripps method of purification and therefore limited to material purified from natural sources. The case went before the District Court of Northern California on a motion for Summary Judgment (a procedure for obtaining a speedy decision preliminary to a full trial), and in 1987 the court decided in favour of Scripps on this issue. However, on appeal to the Court of Appeals for the Federal Circuit (CAFC), it was decided in 1991 that this and the other issues in the case were too complex to be resolved in preliminary proceedings. The CAFC, therefore, remanded the lower court's findings for full trial. At the present time, the trial is still pending.

ERYTHROPOIETIN LITIGATION

In the Amgen case, Genetics Institute Inc. had obtained US patent 4,677,195 for a method of purification of epo from human urine by reverse phase high-performance liquid chromatography. But the patent also has broad product claims, including the following: 'Homogeneous erythropoietin characterised by a molecular weight of about 34,000 daltons on SDS PAGE, movement as a single peak on reverse phase high performance liquid chromatography and a specific activity of at least 160,000 IU per absorbance unit at 280 nanometers.'

The dispute in this case involves issues similar to those in the Factor VIII case. For similar reasons, the US court of first instance held that Amgen's recombinant product infringed GI's claim to the purified protein. However, this holding has been suspended on appeal.

The two instances outlined above are examples of conflict between the respective proprietors of a natural product patent and a recombinant DNA patent in which it has to be decided who is free to operate commercially with the recombinant protein. In these two examples, one question for the court is whether a patent based entirely on the purification of the natural product can be permitted to cover the protein as made by recombinant methods. Unfortunately, single issues are rare in patent lawsuits and the answers are often complicated by other factors. It seems possible that the US courts could still affirm in principle the dominance of the natural product patent.

THE EUROPEAN APPROACH TO NATURAL PRODUCTS

The European Patent Office analyses this question in its Official Guidelines (EPO 1992) as follows:

> To find a substance freely occurring in nature is also mere discovery and therefore unpatentable. However, if a substance found in nature has first to be isolated from its surroundings and a process for obtaining it is developed, that process is patentable. Moreover, if the substance can be properly characterised either by its structure, or by the process by which it is obtained or by other parameters and it is 'new' in the absolute sense of having no previously recognised existence, then the substance *per se* may be patentable. An example of such a case is that of a new substance which is discovered as being produced by a micro-organism.

The Guideline speaks of products having no previously recognized existence and it is therefore rather limited. Much research in the natural product field is concerned with active substances which are known to exist, the object being to isolate the substance in highly pure form or to devise synthetic methods of producing the substance. The most topical example of this is the preparation of pure proteins by recombinant DNA methods.

Inventiveness in patent law

Of all the concepts of patent law, inventiveness is perhaps the most difficult to apply in practice. The subject matter of a patent is said to involve an inventive step if it is not obvious to a worker of ordinary skill in the art in the light of what is publicly known at the relevant date. This formulation is almost universally accepted and intended to be of assistance, although it does little more than transfer the difficulty to the term 'obvious', which itself then calls for interpretation.

WHAT IS OBVIOUSNESS?

The notion of obviousness is usually explained as one of simple or logical progression from what has gone before. In the research laboratories of academic institutions and of the science-based industries, it is common to pursue particular lines of investigation because they are the obvious paths to take in the search for new understanding or the solution of practical problems. This is one purpose of all scientific training. Patent law reflects this assumption in the concept of the 'ordinary person skilled in the art' who is supposed to be capable of pursuing such investigations from a common stock of skill and knowledge which can be exercised without drawing upon any inventive faculty. Such persons are expected to possess a certain level of scientific deductive power and to make practical use of it in their daily work.

Inventiveness calls for something more than the application of the foregoing qualities, but what that extra amounts to cannot be readily defined. Although the exercise of scientific deduction is often involved in the innovative process, it is not to be equated with inventiveness as understood in patent law. Patents have not hitherto been intended as rewards for scientific ability, nor are they given to research workers in recognition of their being gifted experimentalists.

The degree of inspiration required to support a patent on the ground of inventiveness cannot be easily stated in a general way, but must be assessed in every specific case on its own merits. An element of subjectivity must therefore enter into the process of judging inventiveness, but this ought to be minimal. The patent system would lose credibility if the courts of different countries were frequently to reach conflicting conclusions on equivalent patents litigated in their national jurisdictions.

RELEVANCE OF THE CLAIMS

In determining whether an inventive step has been taken, the claims of the patent are crucial. No matter how ingenious the practical details of the invention described in the inventor's examples of the process or the specific illustrations in the patent drawings, what has usually counted is the way in which the invention is claimed. The claims can easily go beyond the inventive step and be unduly broad. Over-claiming may be a result of faulty professional judgement or may be due to the demands of a covetous applicant. If this is not forestalled at the application stage by the critical examination of the Patent Office, it may sooner or later be subjected to kill or cure treatment in litigation.

Where a patent is for a new chemical substance, the usual form of claim recites the chemical structural formula of the compound. It may be difficult to see how a chemical structure *per se* can be said to be inventive, especially where it differs only slightly from a known structure. Even where the difference is more striking, one might ask what this has to do with inventiveness. The answer lies in the properties which flow from the structure in an unpredictable manner. Inventiveness is therefore a combination of a structure and an unexpected utility or degree of utility. For this reason, the patent is granted for the product *per se*. The inventor must describe a way of making the product (one will suffice), but the claim is not restricted to that particular method of manufacture.

A more restricted form of claim, the 'product-by-process' claim, is used when the product can only be defined by the way it has been made. In the UK and a number of other countries, this form of claim is still available when the product is not new – that is, it has been made before by older methods – and the invention resides primarily in a new process for making it. European and US patent case law have virtually ruled out the product-by-process claim for known compounds, but under both systems the protection given by a process claim extends to the product of the process. These considerations of claims to chemical substances apply equally to large molecules including proteins and nucleic acids.

OBVIOUSNESS IN UK LAW

In English law, obviousness is considered to be a question of fact to be determined against the technological background or 'state of the art' at the time the patent application is filed. This determination usually requires taking the testimony of experts as to the state of knowledge which the competent person in the particular field is expected to have at the application date or priority date of the patent application. Expert evidence can go almost to the point of clinching the matter, but the court reserves to itself the right to decide the issue. Considering the demands made on the judges in grasping the deep technicalities of the many 'high-tech' patents that come before them, it is remarkable that the system works as well as it does.

UK case law shows that the question of obviousness can arise in different ways. First, the subject of a patent may be plainly obvious in the sense that the prior art falls just short of but nevertheless points towards what is claimed either directly or by strong implication. Secondly, even in the absence of specific pointers in the prior art, it may have been 'obvious to try' to make the claimed product or to carry out the claimed process. A third category is that of the 'obvious desideratum', where the claim defines something that everyone would like to have if only they could actually achieve it. This third category includes the case of the 'problem claim', in which the claim states only a problem to be solved but not the solution itself.

In the field of chemical inventions, the classical approach to the question of obviousness in UK law was first formulated as the celebrated 'Cripps test' (Sharpe & Dohme v Boots Drug 1927). The case involved the question whether, in order to manufacture certain higher alkyl resorcinols, it was obvious to use a process which was already known for making the lower alkyl homologues. As counsel for the patentee, Stafford Cripps put the issue as follows:

> Was it for all practical purposes obvious to any skilled chemist, in the state of chemical knowledge at the date of the patent (which consists of the

literature available and general chemical knowledge), that he could manufacture valuable therapeutic agents by making the higher alkyl resorcinols by the condensation and reduction processes described?

The court answered this question affirmatively because to establish antibacterial properties in the higher members of the series and to use a method of preparation known for the lower members was no more than *mere verification* of what was reasonably predictable.

An important later case in the *obvious to try* category (Johns Manville Corp 1967) involved the new use of a known material. Here the claim was to the use of polyacrylamides as filtration aids in the making of asbestos cement articles. The same materials had previously been used in a similar way for filtering minerals, industrial waste and the processing of clay materials. The court concluded that in order to show the obviousness of a claimed process or use, 'it is enough that the person versed in the art would assess the likelihood of success as sufficient to warrant actual trial'. The claim was for this reason held invalid in this instance. The argument that something was obvious to try will therefore fail unless there would have been a strong presupposition that the trial was likely to succeed.

The Cripps test was later adapted by the High Court in a pharmaceutical case (Olin Mathieson v Biorex 1970) involving the replacement of the Cl substituent in the structure of a known drug (chlorpromazine) by the CF_3 group (leading to trifluoroperazine). The question was modified to include the notion of obvious to try and re-stated as: 'Would the notional research group . . . be directly led as a matter of course to try the CF_3 group in place of Cl . . . in the expectation that it might well produce a useful alternative or a better drug.' This time a negative answer was given.

The test was elaborated further in a microbiological case (American Cyanamid v Berk Pharmaceuticals 1976), in which the claim was to the use of *Streptomyces* strains which would produce tetracycline without co-producing the unwanted chlortetracycline. The question (paraphrased here) was whether the notional research group would, from prior knowledge, be directly led to try to find or induce a novel strain descended from the type strain A-377 and which would have the desired properties. On this occasion, the court gave an affirmative answer, even though the search for the strain was not certain to succeed.

OBVIOUSNESS IN US LAW

In US legal thought, obviousness is said to be a question of law, although based on certain factual enquiries. There is an extensive case law on this subject, but it will suffice here to reproduce the oft-quoted passage from the leading case (Graham v John Deere Co 1966):

> . . . the scope and content of the prior art are to be determined; differences between the prior art and the claims at issue are to be ascertained; and the level of ordinary skill in the pertinent art resolved. Against this background, the obviousness or non-obviousness of the

> subject matter is determined. Such secondary considerations as commercial success, long felt but unsolved needs, failure of others, etc may be utilised to give light to the circumstances surrounding the origin of the subject matter sought to be patented. As indicia of obviousness or non-obviousness, these enquiries may have relevancy.

The US courts have stressed repeatedly that *obvious to try* is not the standard test, since many things may be obvious to try in a general sense and yet the prior art may give no guidance as to the specific path along which success will be found. There may be obviousness even though absolute predictability of success was lacking. Unexpected results can show that what was apparently obvious was not obvious in law after all. What is required to show obviousness is that there was a reasonable expectation of success.

OBVIOUSNESS IN EUROPEAN LAW

The European Patent Office was the first and is still perhaps the only Patent Office to offer guidance on this difficult topic (EPO 1992):

> The term 'obvious' means that which does not go beyond the normal progress of technology but merely follows plainly or logically from the prior art i.e. something which does not involve the exercise of any skill or ability beyond that to be expected of the person skilled in the art.
>
> It should be remembered that an invention which at first sight appears obvious might in fact involve an inventive step. Once a new idea has been formulated it can often be shown theoretically how it might be arrived at, starting from something known, by a series of apparently easy steps. The Examiner should be wary of *ex post facto* analysis of this kind. He should always bear in mind that the documents produced in the search have, of necessity, been obtained with foreknowledge of what matter constitutes the alleged invention. In all cases he should attempt to visualise the overall state of the art confronting the skilled man before the applicant's contribution and he should seek to make a 'real life' assessment of this and other relevant factors.

Inventiveness in biotechnology

In classical microbiology, the prototype invention is the isolation or development of living organisms from natural sources. Here, the question of inventive step ought to be relatively uncontroversial. Assuming the inventor has for the first time isolated or produced, say, a new strain of micro-organism which is superior to known strains in some commercially significant respect, there cannot be any real doubt that he or she is justified in obtaining a patent for it. One might say that it is obvious to search for new strains, but it can hardly be said that the particular new strain for which protection is sought would have been obvious. This type of invention is one in the 'lucky find' category and so long as the strain performs as stated such a patent ought to be water-tight. Patents of this kind are fairly easy to read and understand and the nature of the inventive step is easy to perceive.

RECOMBINANT DNA PATENTS

To move from patents in the field of classical microbiology to those in the recombinant field is to enter a different world, populated by a new species of patent literature. For many of these patents dealing with the production of known proteins ('first-generation products'), such as the therapeutically important serum proteins, one often finds an introductory presentation of the following type:

1. Protein X is a substance of therapeutic importance.
2. Protein X is obtained from natural sources in minute amounts.
3. rDNA technology is now in place and offers the prospect of producing protein X on a commercial scale.
4. It was difficult to do it but we have succeeded in doing it.
5. Our invention provides 'a DNA sequence coding for protein X' (and sometimes 'recombinant protein X' as well).

A recombinant DNA patent for products of this type will typically describe the isolation or construction of the DNA which codes for the protein, followed by the insertion of the DNA into the vector and transformation of the host micro-organism. The patent disclosure will typically give a dictionary of technical terms, flow-sheet drawings of cloning strategy, and an impressive but densely detailed description of the experimental protocol. It is in the experimental section of the disclosure that this new kind of patent drafting is most characteristic, being so full of references to methods taken from the scientific literature as to justify a long reference list. Such lists are of course standard for scientific papers, but have hitherto been most rare in patents. Indeed, many patents of this type are 'mini-textbooks' which even include a table of contents! As noted in Crespi (1989), the patent claims will be typically selected from the following types:

(a) Recombinant protein products (and alleles, variants, derivatives).
(b) DNA sequences coding for the products of (a).
(c) Vectors containing the DNA sequences of (b).
(d) Micro-organisms, cell lines and other organisms transformed with vectors (c).
(e) Processes for constructing the micro-organisms, etc., of (d).
(f) Processes of producing products as in (a) by expression of DNA sequences (b) in a recombinant host organism.

What signposts have the courts given us to evaluate levels of inventiveness in technology of this kind?

APPROACH OF UK COURTS

The British case on tissue plasminogen activator (t-PA) (Genentech v Wellcome 1989, UK patent 2,119, 804 discussed in Crespi, 1989) is a prime example of reliance on a claim of type (a) listed above. The patent contained a number of such final product claims of which the following is representative for purposes of discussion: 'Human tissue plasminogen activator as produced by recombinant DNA technology'.

For commercial purposes, this type of claim was exactly what was wanted, but it failed on legal grounds, being held obvious, first, because a product of this kind was a known desirable objective and, second, because the product was obtained by using known methods of gene cloning. This case is therefore a modern prototype of the *obvious desideratum* category and it allows the conclusion that, for the British courts, being first in the race to clone and express genes coding for known proteins is not enough to support so broad a legal monopoly in such circumstances.

Genentech had relied heavily on the fact that they were the first to discover the DNA sequence coding for human t-PA and to deduce the amino acid sequence of the protein. Prior to their efforts, the absence of sequence information was 'the one log holding up the log jam' in the recombinant route to this product. The Court of Appeal was not persuaded by this argument. In its view, the determination of a sequence was an act of acquiring knowledge and therefore a discovery rather than an invention. Since the recombinant methods used to apply this information were assessed as conventional for the skilled person at the priority date (May 1982), they did not add anything of inventive character to the process as a whole. One can only speculate on the judgment that might have been delivered had the methods used by the patentee been considered more original by the court.

Some commentators have considered the decisions of the UK Patents Court and Court of Appeal to have been unjustifiably severe in overturning a patent on a product of great potential medical importance. The importance of the legal issues would certainly have merited an appeal to the highest UK court (House of Lords), but this was not made. However, the lower court decisions lie squarely in the tradition of the British courts in approaching the question of obviousness in biological chemistry. The t-PA product was clearly a good candidate for production by these techniques judging by the fact that the subject of the claims was formulated conceptually as a worthwhile project long before any laboratory work was started. The patent itself made it clear that recombinant methods were perceived as an effective way of making large quantities of t-PA, even though the path might be tortuous, and that the invention was based on the discovery (some might say verification) that the technology could be successfully applied to it. Recombinant t-PA, as such, and viewed as a product, fell into the obvious desideratum category.

It should not be thought that the decision of the UK court was conditioned solely by the fact that the claims of main interest were of the final product type. The legal analysis would have been no different for claims of the type obtained in the corresponding US patent 4,766,075. The latter has no claims to the recombinant t-PA protein but relies on claims of types (b), (c) and (d) noted above, for example, 'A DNA isolate consisting essentially of a DNA sequence encoding human tissue plasminogen activator'.

However, it must be stressed that as a precedent the UK decision applies only to first-generation products. As regards next-generation products of the t-PA type, numerous patent applications are now in the pipeline for modified or mutant forms of natural t-PA. It is clear that patent prospects are reasonable for these compounds.

APPROACH OF US COURTS

The US Court of Appeals for the Federal Circuit have given a clear signal on the question of obvious to try (In re O'Farrell 1988) in the biotechnology field. The crucial prior art against the applicants was one of their own prior publications in the scientific literature. Had it succeeded, the application could have given rise to a patent of basic importance. It would have protected the 'fusion protein' method in genetic engineering, according to which a heterologous (foreign) gene is inserted into a micro-organism and tacked on to one of the micro-organism's homologous (indigenous) genes so that the protein-synthesizing machinery of the cell expresses the indigenous protein linked with the desired foreign protein.

The applicants' scientific paper had described the essentials of this fusion method except that the tacked-on frog DNA gene used in the paper coded for ribosomal RNA, which is not normally translated into protein. However, the authors had speculated as to the possibility of applying the method to the case of an RNA which was translatable into protein. The applicants argued that there was significant unpredictability in molecular biology at the date of the publication and it was not yet certain whether a specifically desired protein could be produced in this way. They asserted that the Patent Office rejection was based on the impermissible 'obvious to try' argument. But the court refused to accept this submission. From their own 'pioneering' paper, especially when combined with another item of prior art, there was a reasonable expectation that the method would be successful when applied to the making of proteins. By their publication, therefore, the applicants had foreclosed themselves from obtaining a patent.

In a further phase of the US litigation on erythropoietin (Amgen Inc. v Chugai Pharmaceutical Co. 1989), the validity of the Amgen patent was attacked. The patent contained a variety of claims to DNA sequences but no process claims for making the final recombinant product and no final product claims. One of the DNA claims was to 'A purified and isolated DNA sequence consisting essentially of a DNA sequence encoding human erythropoietin'.

As to inventiveness, the Massachusetts District Court reviewed the course of the research by Amgen and others from 1981 onward into the cloning of the erythropoietin (epo) gene. This was not a large gene and for this reason was seen in 1981 as a good candidate for cloning. But it took 2 years of painstaking work by the Amgen inventor before the successful strategy for achieving the result was worked out. This entailed determining the correct amino acid sequence of sufficient portions of the protein structure to enable oligonucleotide probes to be designed which would extract the gene from a genomic library. A large number of probes were constructed and it was necessary to use a combination of probes which would hybridize to different parts of the gene. The great magnitude of the task cannot be denied. Genetics Institute, Cal-Tech and Biogen had also embarked on the quest for this gene.

Here there seem to be some similarities to the t-PA case. The gene was a good candidate for cloning and a number of competitive groups were pursuing it along their own trails. But apparently there were important differences from the t-PA development. First, the epo protein was an exceptionally scarce material when the Amgen inventor started work, a fact which made sequence information that much more difficult to obtain. Second, and of greater significance, was the fact that by 1981 and even 1982 there had been little or no precedent in the screening of a genomic library using two sets of probes. The court therefore concluded that it was not within the power of one of ordinary skill in the art to reduce this invention to practice at that time.

Although the court did not say so in so many words, it might be fair to suggest that recombinant erythropoietin, as a desirable product, might have been *conceptually obvious* but that its reduction to practice was far from obvious. In approaching the question of the inventive step, the US court did not put the partial sequencing of epo into a separate category from the steps which followed. The correct sequencing and the consequential design of two sets of fully degenerate probes from different regions of the sequence to explore a genomic library amounted to an inspirational combination in the eyes of the court. The court concluded that 'the unique probing and screening method employed in isolating the EPO gene was what distinguished the invention from the prior art'.

In the decision on appeal the CAFC noted that it was the product that was claimed and not the process, but decided to follow the approach of the lower court and the parties to the litigation. The point was therefore bypassed.

APPROACH OF EPO BOARDS OF APPEAL

In the European case law of the EPC in the biotechnology field, the most prominent of the controversial issues to have arisen to date have involved questions of sufficiency of disclosure, adequacy of support for broad claims, and clarity in claim language. There has been no decided case on inventive step of the calibre of the t-PA case in the UK and the erythropoietin case in the USA. Incidentally, the European patent (EP 148,605) equivalent to the US epo patent discussed above has so far survived a formal Opposition filed by six industrial Opponents in the EPO.

The EPO Boards of Appeal have given highly favourable consideration to the early applications filed in this field and in which these difficulties have been raised, either by the EPO Examiners or in Opposition proceedings. The first example was one of the earliest and broadest scope patents in the field of genetic engineering (Genentech I, T 292/85). The invention was typified by the main claim directed to a recombinant plasmid, the most important features of which were the presence of a homologous regulon and a heterologous DNA coding for a desired polypeptide and in proper reading frame. The principal item of prior art was the paper which figured prominently in the O'Farrell case mentioned above. This had shown the use of the claimed homologous/heterologous combination to produce a transcript which would not have been translated into protein, coupled with a speculation as to the possibility of expressing eukaryotic DNA in bacteria. Despite this, the Board found generously in favour of the applicant on the issue of inventive step, and by somewhat tortuous reasoning decided that the prior document actually pointed away from the claimed invention.

The most important point of this case as a valuable precedent concerned the question of scope of claim. The Examining Division had objected to the use, in the claims, of the broad functional terms 'bacteria, regulon, plasmid etc', arguing that these should be limited to specified known materials in order to meet the requirement of sufficiency and to avoid covering future inventions or discoveries of new materials falling within these broad descriptions. The Board rejected this view and held that for this particular invention only the use of such functional terminology could give fair protection.

In one recombinant DNA application which failed on the ground of 'obvious to try' (Monsanto, T 249/88), the application claimed a method of increasing the yield of cow's milk by administering a recombinant pituitary bovine growth hormone (bGH) having an *N*-terminal methionine group (*N*-met bGH) instead of the natural bGH as in the prior art. One prior document contained a suggestion to use a recombinant bGH for this purpose and another, dealing with the human analogue, expressed lack of surprise that the terminal *N*-met group did not affect the activity of bGH (for growth promotion in rats).

Declarations from three scientific experts were filed to show lack of predictability of the activity of *N*-met bGH in the cow from the prior art. The Board held that 'inventive step is not assessed from the viewpoint of the outstanding but cautious scientist reluctant to make a prediction but from that of the notional skilled person of average ability and knowledge'. The inability to predict, in the strict scientific sense, or the necessity to confirm experimentally that *N*-met bGH would increase milk yield, did not make the claimed method un-obvious. There were clear hints from the prior art and the results could reasonably have been expected even though experimental confirmation was necessary. The patent was therefore refused. In latching on to the idea of 'reasonable expectation' of obtaining useful results, the EPO Board is clearly in line with UK and US precedents.

THE BIOGEN ALPHA INTERFERON CASE

Crespi (1989, p. 232) gave an account of Biogen's success in defending their European patent for recombinant alpha interferon against formal Opposition by nine industrial Opponents. Among the important points in this decision one must be singled out for further comment in view of its significance to the relationship between patents and scientific publication. As will be clear from the above discussion of the Asahi case, and as also explained in Crespi (1985, pp. 19–20), one of the crucial concepts of patent law is that of *priority date*. The filing of a patent application creates a priority date for what it discloses. This date is obviously significant in disputes between rival applicants for the same invention. But its main significance is that it defines the line between what is prior art and what is not. Consequently, an inventor who has filed an application will normally be told that it is then safe to publish the work in the journals, the reason being that any publication coming out after the priority date cannot torpedo the patent application. This rule applies to the 100 or more countries (members of the Paris Convention) which have agreed that a patent application filed in any one country establishes a priority date which will apply also to corresponding applications filed within the following year in the other member countries. Consequently, if soon after filing the original (priority) application the inventor chooses to publish in the journals during the year before filing abroad, the publication will not harm the foreign applications.

THE PRIORITY DATE QUESTION IN THE BIOGEN CASE

Biogen had claimed the priority under the Paris Convention of three earlier applications (I, II, III) covering DNA sequences, each of which represented a stage in the development of the invention during the year preceding the European filing. The inventors had followed the policy of publishing the subject matter of each of the three applications in the scientific literature soon after it had been filed, rather than suspending publication until the final European application had been lodged at the EPO.

The Opposition Division had held that scientific publication of the subject matter of application I prior to the filing of application II had made that subject matter part of the prior art citable against any of the final claims which were only entitled to the priority of application II or subsequent applications. As a result, certain claims which were entitled only to the priority of application II were held lacking in inventive step over the scientific publication of the subject matter of application I. However, the EPO Technical Appeal Board rejected this view and decided that so long as the publication of the subject matter of application I did not go beyond the content of the application, it could not be used against any claim in the latter application.

In effect, therefore, the filing of a patent application before publication of whatever is disclosed in the application not only protects the subject matter of that disclosure in itself, but also removes it from the realm of material which can be cited against subsequent developments of that disclosure which are protected by filing a further patent application before they are themselves published. In enunciating this principle so clearly for the first time, the Technical Board of Appeals has laid a milestone of major significance. Being of general application, this must be assessed as one of the more progressive judgments of recent years. It is highly favourable to that group of inventors who accept the value, even the necessity, of filing patent applications, but for whom early scientific publication of research results is equally necessary for career advancement or for reasons of scientific or public interest. Unfortunately, however, the Biogen decision is not popular with those who hold a strict 'purist' view of the law and the President of the EPO has referred the decision for review by the Enlarged Board of Appeal of the EPO. At present, therefore, the point of law cannot be regarded as entirely certain; it might also be raised before national courts. If international harmonization were to result in the widespread adoption of the US grace period, the problem would disappear.

HYBRIDOMA TECHNOLOGY

Crespi (1989, p. 234) discussed the Hybritech case, one of the earliest cases on monoclonal antibodies to go before the US CAFC. This court is the highest authority for patent cases below the Supreme Court itself. In brief, the issue was whether at the relevant date it had been inventive or obvious to substitute monoclonals for polyclonals in a known type of sandwich assay. After a thorough examination of the inventors' notebooks and the various articles published by others predicting the widespread applications of monoclonals following their first appearance in 1975, the CAFC decided that at the date of conception of this particular invention (early 1979), it was not obvious to make the substitution. Many of the published articles were dated after the conception date and in the year preceding the filing date of the patent application (therefore not damaging prior art), while the others were mere 'invitations to try' monoclonal antibodies in immunoassays without saying how this would be achieved (thus the court rejected the 'obvious to try' approach).

The Hybritech case had been preceded by a decision of the US Patent Office (Ex parte Old 1985) allowing a claim to monoclonal antibodies to malignant human renal cell antigens. The Examiner had argued that polyclonals to surface antigens of malignant melanoma cells being known, and polyclonals to human renal cell antigens being also known, it was obvious to make monoclonals to malignant human renal cell antigens by the Kohler Milstein technique. The Board of Appeal said: 'We cannot subscribe to this rationale . . . hybridoma technology is an empirical art . . . no expected results can thus be said to be present'. If making monoclonals to cancer antigens was routine and predictable, the Board considered that 'cancer immunology would have produced cancer cures'.

Against the above background, the unsuccessful 10 year struggle of Cetus to obtain a US patent for hybridomas producing monoclonals to human fibroblast interferon stands in sharp contrast (Ex parte Erlich 1992). This time the US Board of Appeals supported the Examiner's rejection based on obviousness. By 1980, the Cetus invention date, the Kohler and Milstein method had been successfully applied to the construction of hybridomas secreting monoclonals to a wide variety of antigens including human leukocyte interferon. The Board rejected the Cetus argument of 'unpredictability', brushed aside the Hybritech and Old decisions as based on different factual situations, and concluded:

> Therefore one of ordinary skill in the art at the time of the present invention, recognising the antigenicity of the human fibroblast interferon and its potential therapeutic value, would have had ample motivation to make monoclonal antibodies to this protein using the classical Kohler and Milstein hybridoma technology. Such a person would have entered this venture with a reasonable expectation of success given the large number of successes other researchers had at that point in time in adapting hybridoma technology to other antigens.

US commentators (McGough and Burke, 1992) have asked whether the Erlich decision signals the end of monoclonal patents. While concluding 'probably not', McGough and Burke point out that Examiners will undoubtedly start to rely on Erlich as a precedent for rejecting applications. To overcome such rejections, it will be necessary to be able to point to some inventive feature of the technique used or to some novel specificity in the product.

Also in contrast to the US Hybritech approach, in the EPO case of Unilever PLC v Boehringer Mannheim (T 499/88, immunoglobulins), the Technical Board of Appeals held that the replacement of polyclonal antibodies by monoclonal antibodies in an immunopurification process required no inventive step at the priority date of the opposed patent (February 1981). The Unilever claims were to a process for the recovery of immunoglobulins of high purity and potency from milk by selective binding to a low-affinity monoclonal antibody specific to one or more of the immunoglobulins and itself bound to an insoluble carrier material. It was clear that the novelty lay only in the use of monoclonal antibodies in substitution for the prior use of monospecific polyclonal antibodies in the same basic process. This substitution was held to be the 'next logical step' and the 'desired logical and obvious step' to improve the purification process of the nearest prior art.

POSSIBLE CONCLUSIONS REGARDING INVENTIVENESS

The litigated applications and patents discussed above represent only a fraction of those being filed and granted across a wide field, ranging over pharmaceutical proteins of many kinds, antigenic DNA sequences for use in diagnosis, and in synthetic and recombinant vaccine manufacture. In the agricultural field, patents for plant gene sequences and genetically manipulated plants are also being routinely obtained. In many such patents, extremely broad claims are being granted in response to strong commercial pressures to secure dominance over selected areas of the technology. These legal prizes awarded to those 'first past the post' are highly pre-emptive and have the effect that others in second or third place are forced to seek licences or leave the field altogether.

The response of the patent law to this situation will be expressed largely through the application of the test of inventiveness to these all-embracing claims. It would therefore be highly desirable to be able to derive a common pattern from the cases discussed above. In pursuing such an ambition, a British commentator must bear in mind the cautionary words of the UK Court of Appeal in the Johns-Manville case (*supra*):

> I have endeavoured to refrain from coining a definition of obviousness which counsel may be tempted to cite in subsequent cases relating to different types of claims. Patent law can too easily be bedevilled by linguistics, and the citation of a plethora of cases about other inventions of different kinds . . .
> I doubt whether there is any verbal formula which is appropriate to all classes of claims.

Nevertheless, the following suggestion has been made (Crespi, 1991):

> Even though the patent application is the first to show the preparation of a DNA sequence coding for a particular known peptide or of a hybridoma which produces monoclonal antibodies to a particular known antigen, there will be a presumption of obviousness if it can be shown that, at the relevant date,
> (a) to apply the known methods of recombinant DNA or hybridoma technology to derive the particular product was obviously desirable, and
> (b) the methods available to the skilled person were considered likely to succeed, and
> (c) the available methods did in fact succeed without the need for significant modification.
> Where these three conditions are met the presumption of obviousness can of course be displaced in any particular case involving special factors. The only firm general conclusion must therefore be that no such test formula will ever avoid the need, case by case, for advice from patent lawyers.

Patents for higher organisms (plants and animals)

As outlined previously (Crespi, 1985, 1989), the patentability of living organisms, from bacteria to plants to animals, has shown a linear development over the past decade. This may be summarized as follows. After a period of some uncertainty and (in the USA) strong opposition, micro-organism patents are now routinely granted by the US, European and Japanese Patent Offices. The US Supreme Court had previously held that the 'discovery of some of the handiwork of Nature' was unpatentable, but in the celebrated Chakrabarty case in 1980, the US Supreme Court decided that a micro-organism was not precluded from patentability solely because it was *alive*. Thus a *Pseudomonas* manipulated to contain more than one plasmid controlling the breakdown of hydrocarbons (therefore allegedly more useful in dispersing oil slicks than the natural organism containing only one such plasmid) was 'a new bacterium with markedly different characteristics from any found in Nature' and hence not Nature's handiwork but that of the inventor. The 'product of Nature' objection therefore failed and the modified organism was patentable. This decision was influential in most other industrially developed countries and the issue is now settled in law.

Plant patents are also obtainable in the USA, Europe and Japan. The US Plant Patent Act of 1930 is restricted to asexually propagated plants and over 6500 of such plant patents have been granted (mostly for rose and fruit trees). In the Hibberd case (Ex parte Hibberd 1985), following the principle established in the Chakrabarty case, it was decided that normal US 'utility' patents could be granted for other types of plant, for example genetically modified plants.

In Europe, the patent law was originally considered unsuitable for protecting new plant varieties developed by traditional breeding methods. Special national laws of plant breeders' rights (also called plant variety rights) were therefore established in some countries in the 1960s, as well as the International Union for the Protection of New Varieties of Plant (UPOV). To avoid legal confusion, patent law in Europe subsequently excluded plant varieties from patentability, for example in the prototype provision of EPC Article 53(b), which excludes patents for 'plant and animal varieties' and 'essentially biological processes for the production of plants and animals'.

Plant breeders' rights have been highly successful in their own sphere. However, it is now generally recognized that patent law is the better suited to the protection of recombinant methods for producing transgenic plants and the resulting products. Patents of this type, claiming methods and products *per se*, are now granted by the EPO.

Animal breeds produced by traditional methods have no legal system for their protection comparable to plant breeders' rights. Based on the microorganism and plant patent precedents, in 1987 the US Commissioner of Patents declared that US patents would be granted for 'non-naturally occurring non-human multicellular living organisms including animals'. The first transgenic animal patent was US 4,736,866, issued in 1988 to Harvard University with claims covering the 'onco-mouse', one in which an onco-gene has been introduced to make the animal more susceptible to cancer and therefore more sensitive to testing possible carcinogens. The grant of further US transgenic mouse patents was held up until 1992 when three more issued. After initial reluctance by the EPO to grant the corresponding European patent (and a successful appeal to the Appeal Board), the European patent issued. This is now under formal Opposition by anti-vivisection and animal rights groups.

By those who oppose such patents, certain freedoms for breeders and farmers are said to be threatened by patents on transgenic plants and animals. Under plant variety rights (PVR), breeders previously enjoyed the so-called 'breeder's privilege' or 'research exemption', which gave them the freedom not only to use protected plant varieties in their breeding programmes but also to commercialize the further varieties developed therefrom (often only 'cosmetically' different from the original) without any royalty payment to the owner of the initial variety. The UPOV Convention as revised in 1991 now expands the scope of the breeder's right to include what are termed 'essentially derived' varieties (this term is defined). This revision awaits ratification by Member States.

The *freedom to research* is safeguarded equally under both patent law and PVR law. But the freedom to *commercialize* the resulting products of research depends on whether or not they infringe the patent claims or are 'essentially derived' under PVR law. A strengthened UPOV-type protection would therefore go part of the way towards the strong protection given by patents. Neither system is a threat to the free use of existing germ-plasm, since these rights can in no sense monopolize known material as such. Again, until the UPOV revision is taken up in national laws, farmers legitimately sowing seed of a protected variety are legally free to save part of the seed from the first crop of plants for sowing on their own farms to produce a second and subsequent crops (the 'farmer's privilege'). Recognizing that the current scale of use of farm-saved seed thus deprives the breeder of significant royalty income, the strengthened right under the 1991 version of UPOV would now make this subject to authorization of the breeder. However, contracting states can 're-introduce' this freedom under their national legislation 'within reasonable limits and subject to the safeguarding of the legitimate interests of the breeder'.

Now that the 1991 UPOV no longer prohibits the availability of both types of legal right (patent and PVR), plant breeders who are themselves using the techniques of biotechnology alongside traditional breeding methods can obtain both types of protection as appropriate.

Gene patents and current issues

The tissue plasminogen activator and erythropoietin patents discussed above are examples of gene patents which have had to stand the test of litigation. In the course of this ordeal, the principle of granting gene patents was never in question before the court. But outside the legal forum, the concept of such patents has been seen as a controversial issue. It may be asked whether the biological function of DNA is so fundamentally different from that of other bioactive substances (e.g. antibiotics) that different legal considerations should apply when considering the question of legal rights. Patent law itself cannot make such a distinction. The intimate connection between DNA and life processes is not a legal reason for excluding it from patent protection or even for treating it according to criteria different from those that apply to other large molecules, such as the products of synthetic polymer chemistry.

Genes are a special example of the broad class of naturally occurring materials which in appropriate circumstances can be patented, as discussed above. Where it is necessary to isolate and characterize a natural product and to devise a process for producing it in quantity before it can be utilized by human beings for any practical purpose, the patent law offers scope for protection. Mere pre-existence of the substance, in admixture with vast quantities of other materials, is insufficient to contradict this view. This is the declared position of the EPO quoted above and of the European Commission (see later). It is also the implicit view of the US patent authorities in relation, for example, to naturally occurring micro-organisms which can be patented as 'biologically pure' cultures. So the feature of isolation or purification of the natural material comes into play to allow such materials to be patented, not in the form in which they exist in their natural condition but in the form resulting from human intervention.

A highly instructive example of a gene patent is US patent 4,885,25112 (Eli Lilly Co.). This is based on the isolation of a gene from an antibiotic-producing micro-organism and the determination of its nucleotide sequence. The gene codes for an enzyme which catalyses the cyclization of a certain tripeptide into the iso-penicillin N cyclic structure in the biosynthesis of pencillin and cephalosporin antibiotics by the *Pencillium* and *Cephalosporium* moulds. The isolation of the gene and its incorporation into an expression vector is said to offer the prospect of driving the natural fermentation process with greater efficiency. Also, it may be used to adapt the natural process so as to convert other tripeptides into products which are different from the natural antibiotics, leading to further antibiotic diversity. The patent describes the construction of plasmids containing the gene and their insertion into a wide range of host micro-organisms. One of the claims is: 'An isolated DNA compound that encodes isopenicillin N synthetase from *Cephalosporium*'. It will be noted that by specifying the term 'isolated', this claim does not cover the gene in its natural environment.

Another claim of the patent specifies the full sequence of 339 codons in the coding strand of the gene, and yet another claim is directed to the enzyme expression product in terms of its amino acid sequence.

INTERMEDIATES AND cDNA FRAGMENTS

The patentability of gene/enzyme patents of the type exemplified in the last section is clear-cut. One of the patentability criteria of US law is that the invention must be 'useful' and it is evident that the isopenicillin N synthetase invention qualifies in this respect. Patent utility is not to be equated with commerical utility. The patent must identify the utility and describe how the skilled worker is to use the invention, but the Patent Office is not entitled to demand proof of the commercial feasibility of the claimed process or product. A patent may be perfectly valid on the grounds of novelty, inventiveness and utility, even though the invention fails to meet commercial criteria.

UTILITY IN US LAW

A substantial US case law exists on the subject of utility in relation to intermediates useful in synthetic chemistry, especially in the field of steroids (Maebius, 1992; Daus, 1992). The US Supreme Court has stated that 'a patent is not a hunting license. It is not a reward for the search, but compensation for its successful conclusion' (Brenner v Manson 1966). The court held that chemical compounds are not 'useful' simply because they are used in research. Patent utility arises 'where specific benefit exists in currently available form'. Consequently, certain steroid intermediates were held unpatentable in the absence of evidence that the end-products into which they could be converted were themselves biologically active and useful.

The Brenner v Manson ruling must surely be one of the factors relevant to the current debate over attempts to patent DNA sequences, the biological function of which is as yet unknown. This debate has arisen over the attempt first by the US National Institutes of Health (NIH) and then by the UK Medical Research Council (MRC) to patent 'expressed sequence tags' (ESTs), fragments of DNA isolated from cDNA libraries prepared from human brain. This debate has been unusual in many respects. First, the NIH patent application filed in the US Patent Office (and later in other countries) has been the subject of commendable open disclosure by NIH and extensive comment by individual eminent scientists and both public and private sector organizations. The second remarkable feature of the situation is that the scientists have dominated the controversy leaving the patent attorneys to concentrate on the legal question whether these sequences are patentable. Few are yet prepared to pronounce dogmatically upon this latter point, this being yet one more instance of the tendency of biotechnology to put the patent lawyers under strain to make predictions. Various potential utilities are asserted for these sequences: use for isolation of full coding regions of the corresponding genes, forensic uses, and use for gene mapping, tissue typing and antisense technology. One independent commentator has assessed this catalogue of uses as 'more of an invitation to experiment than concrete utilities' and has described the first quoted use as a 'use of compounds as tools for the researcher in search of valuable end-products' (Maebius, 1992).

The scientists and the biotechnology companies have argued over the merits of allowing patents of this sort and even more strenuously over whether patenting may conflict with the principle of international scientific collaboration in the human genome project. This third aspect of the EST debate is notable in that the applicants for the patents have themselves presented an ambivalent attitude to the question whether or not they are entitled to a patent. Thus they claim primarily to be testing the point of law, coupled with a readiness to abide by a decision from the proper authority, a situation which must be extremely rare if not totally unprecedented in patent history!

The initial rejection of the NIH application(s) by the US Patent Office has been contested by NIH and the issues may have to go to appeal unless, for policy reasons unconnected with patent law, the application is abandoned by NIH. If these sequences eventually turn out to be patentable, none of the parties wishes at this stage to forego the possibility of protection and to lose out to the others by being over-zealous to maintain scientific purity. Precisely how relevant patents on such cDNA fragments will be to eventual attempts to patent the whole genes to which they relate is not immediately clear.

INDUSTRIAL APPLICABILITY IN EUROPEAN LAW

In the regional law of the European Patent Convention (EPC) and in the corresponding European national laws, there is no requirement for 'utility'.

The counterpart provision is that the invention must be 'susceptible of industrial application' as stated in EPC Article 52(1). A particular group of inventions which are expressly classed as not susceptible to industrial application is set out in Article 52(4). These are 'Methods for the treatment of the human or animal body by surgery or therapy and diagnostic methods practised on the human or animal body'. This exclusion applies only to *method* claims, the philosophy being that a patent should not impede the activity of medical practitioners. Of course, patents for drugs, devices and other products for use in any of these methods are not excluded from patentability, since there is no philosophical problem involved in legally controlling the manufacturers of such products. There is some EPC case law on inventions refused under the specific Article 52(4), but little or none under the general Article 52(1).

By way of speculation, consider the possible case of a novel and inventive method of *in vitro* fertilization, which is effective for both human and animal propagation. Such a method would not be clearly ruled out by the wording of Article 52(4) above. But it is possible that the human application of the method might be objected to on the ground that it is normally carried out in a hospital or clinic rather than in an industrial context. The animal application would not be open to this criticism. However, bearing in mind the liberality of the European patent system and the rule that all exclusionary provisions must be interpreted as narrowly as possible, it would be unwise to attempt a conclusion on the point.

A good illustration of the rule of interpreting exclusions narrowly is provided by a series of decisions of the EPO Boards of Appeal on the question of diagnostic methods. In one remarkable case (Bruker T385/86), the claim was to a particular 'Method for the non-invasive determination of chemical and/or physical conditions inside a whole intact living animal or human body' using magnetic resonance spectral measurements to determine the pH or temperature of local body areas. The purpose of obtaining this data was clearly in the context of diagnosis and the EPO had therefore rejected the application under Article 52(4). But because the claim did not include the final step of reaching any diagnostic conclusion from the data, the Technical Board of Appeal held that it could not be excluded on this ground. The claimed method was one of data acquisition or processing that could be used in a diagnostic method but did not itself immediately indicate a diagnosis. The fineness of this distinction is clearly very helpful if the claims can be written in this indirect way.

Industrial applicability is in one sense a more focused test for patentability than that of US utility, which need not be industrial. On the other hand, the notion of mere 'susceptibility', that is, capability rather than proven actuality, does seem to open the door to the admission of inventions whose industrial applicability is potential and perhaps entirely speculative. On a more practical note, the EPO insist that an invention be presented wherever possible as a solution of a problem and their Appeal Boards regularly adopt this sort of analysis when deciding on disputes over inventive step. Thus the problem is first identified, the nearest prior art is selected for comparison, the patent description is examined to see if the invention actually solves the problem, and a conclusion is drawn as to whether inventiveness was necessary to arrive at the solution. It is hard to see how an entirely speculative piece of research, which might or might not eventually lead to something of practical importance, would stand up to this kind of legal analysis. It will be some time before these issues are settled in the official examination of the EST patent applications and others which may already be in the pipeline. In the meantime, the EPO Examiners are no doubt following the public debate with considerable interest.

Opposition to biotechnology patents

The industries that utilize biotechnology are convinced that intellectual property should be obtainable for the inventions that stem from research and which have commercial potential. Increasingly, this view is shared by biotechnology research workers in universities and other academic research institutions in need of research funding. A serious challenge to this assumption has come from a number of special interest groups concerned variously with matters of ecology, animal welfare and rights, moral issues and the interests of small farmers and the developing countries. Some of these groups have formally opposed specific European patents and demanded their revocation. By many such groups, 'patenting life' is considered unethical in principle. The opposition extends also to possible structural change in the agricultural industry which might stem from biotechnology, and especially from the acquisition in the hands of the larger corporations of monopoly rights on the advances that are being made.

LEGAL AND MORAL OBJECTIONS

Many of the arguments used by opponents of biotechnology patents are outside the permissible grounds of objection provided in the patent law. The authorities are therefore not empowered to revoke patents on the strength of such arguments. However, one legally permissible ground of objection is that genes are naturally occurring entities and that the methods for transferring them to plants or animals are well known and straightforward. This is a challenge to the inventiveness content of the particular patent at issue: it is an argument that industrial competitors will sometimes use against each other's patents, but so far it has not achieved a high success rate.

Some people claim it to be wrong for a patent to confer on anyone the 'ownership' of living things. This is particularly sensitive as regards animals, which patents are said to reduce to the status of mere objects which can be treated according to the whim of their owner. This emotive view seems to ignore the fact that plants and animals are owned by the farmers who produce them and use them as agricultural commodities. All such owners are bound to respect animal welfare legislation.

ETHICAL ISSUES

The question which immediately prompts itself is whether the patent law is an appropriate vehicle for ethical ideas and moral judgements. Some would suggest that these issues are more properly in the province of laws dealing with individual and corporate behaviour and not at all the concern of the patent law. The alternative view is that patents are granted by public bodies, which by their nature cannot ignore issues of concern to the population at large, especially those having ethical relevance.

European patent law opens the door to this debate by prohibiting the grant of patents on inventions 'the publication or exploitation of which would be contrary to "ordre public" or morality' [EPC Article 53(a)]. In addition, as discussed above in connection with industrial applicability under EPC Article 52(4), certain methods of treatment and of diagnosis are also prohibited from patentability. This latter exclusion is primarily based on the view that these methods lack industrial character, but perhaps it also contains an undercurrent of the idea that preventing medical operations from taking place is unethical. National patent laws in Europe also contain these exclusions. Since these provisions exist, it must be expected that on occasion patent offices will have to make judgements on moral questions.

According to the discussion of Article 53 (a) in the EPO Guidelines (EPO, 1992), it is stated that 'This provision is likely to be invoked only in rare and extreme cases. A fair test to apply is to consider whether it is probable that the public in general would regard the invention as so abhorrent that the grant of patent rights would be inconceivable'. The only example given in the guideline is the rather unrealistic one of a letter-bomb. It is not easy to suggest a range of more likely examples.

THE ANIMAL PATENT ISSUE

In the onco-mouse case, the EPO allowed the claim to the transgenic mammal but reserved an open position on the generality of transgenic animal patents in view of ethical issues. The EPO had originally stated that it was not equipped to deal with the morality of animal welfare. Nevertheless, the Appeal Board instructed the EPO to consider the question and suggested a line of approach for the task.The EPO were to balance the potential of the invention for human good against the possible risk of harm to the environment and the undoubted harm to the test animal. The EPO then decided that the benefit to cancer research outweighed the other factors and granted the patent. As this patent has been formally opposed on ethical grounds, the argument continues. In another pending application involving a transgenic mouse in the screening of substances as hair-growth stimulants, the EPO took the view that the benefit did not justify the use of such an animal model.

PATENTING HUMAN MATERIAL

The patenting of human cellular material has been accepted for many years by the authorities as a natural extension of the patentability of lower life-forms such as bacteria and yeasts. Indeed, there are precedents for the patenting of cell lines derived from human tissue which are useful in research and in industrial processes. Although of human origin, these cells remain as individual cells incapable of being organized into any higher level of life-form. The patenting of biological material of this kind should remain non-controversial.

At a lower level still, genetic material – DNA sequences – isolated from human sources and biochemically modified for transfer to industrially usable simpler organisms, are the basis of recombinant DNA patents granted in many countries. These DNA sequences code for the production of therapeutically important proteins useful in human medicine, and there is no doubt that such inventions are morally acceptable and desirable.

However, one European patent for a DNA sequence derived from human source material has been the subject of a formal Opposition from a political party. The Green Party have opposed European patent 112,149 granted to the Howard Florey Institute of Experimental Physiology and Medicine for a gene sequence coding for human relaxin, a hormone involved in reproduction. The gene was isolated from ovarian tissue removed in the treatment of an ectopic pregnancy. One of the arguments on which this Opposition is based is that it is 'an offence against morals to exploit the pregnancy condition of a woman by removing tissue from her ovary and using it as the basis of a profit-oriented technical process'. The Opposers seem to have overlooked the fact that the termination of an ectopic pregnancy is frequently essential to preserve the life of the mother and is morally permissible in both Judaistic and Christian moral thought.

But at the higher level of human cellular organization there is at least a theoretical question whether any limit should be put upon the kinds of biological material that can be patented. It is perhaps far-fetched to contemplate the possibility of reconstructing human organs by means of biotechnology, although the genetic modification of animal organs for use in human spare-part surgery is already in prospect. Whether a human being, as such, could be patented is a related question which keeps haunting this debate, even though one can scarcely imagine the circumstances in which it could become a real and practical question. Legislators have not hitherto dreamed of the necessity to discountenance such a bizarre notion. Nowadays, humans cannot be objects of property rights and it is arguable that provisions such as EPC Article 53(a) would be effective to dispose of any attempt to pursue this absurdity. Similarly, as humans are not 'products' in an industrial sense, it could be held that inventions to 'produce' them are not susceptible of industrial application and are therefore ruled out by EPC Article 52(1).

DIFFICULTIES IN APPLYING THE MORAL TEST

The whole question of morality is beset with difficulty. Is the morality in question to be based on institutional religious belief in the predominantly Judaeo-Christian culture of Western Europe, or is it the prevailing morality of a secular modern society? Whatever moral backcloth is relevant, there are deep-rooted divisions of public opinion on major issues such as pacifism, abortion, human *in vitro* fertilization and animal rights. To intervene in these matters involves a burden which patent officials are not trained to bear.

It ought to be possible to side-step this problem by appeal to the principle, mentioned above in connection with diagnostic methods, that all exclusionary provisions should be narrowly construed. Applying this principle to the morality issue would suggest that a patent should be refused on this ground only where there is an overwhelming consensus that a particular invention is immoral (the EPO has in fact expressed this view in a more recent case). Thus the objection should not be invoked by patent authorities merely on the basis that some section of society condemns the activity as immoral. This would be consistent with the original EPO guideline, but it does not seem to have been invoked by the Appeal Board in the Harvard case. The test devised by the Appeal Board, which is evidently to be used as a standard for transgenic animal cases, is one which is highly subjective. It will often be very difficult if not impossible for patent officials to assess the benefit against the harm of particular inventions, since these may only be properly revealed in the longer term and not at the early stage at which patent applications come before the authorities for adjudication. The ethical questions discussed above have recently assumed importance in connection with the proposals of the European Commission concerning the legal protection of biotechnological inventions. This topic was also discussed in Crespi (1989), but deserves recapitulating in the light of subsequent events in the political arena.

The proposed EC directive

In November 1988, the European Commission proposed a Directive on the legal protection of biotechnological inventions from the Council of Ministers to EC Member States. The Directive sought to secure harmony in the expanding community between national laws and the EPC, uniform interpretation in the courts, and to upgrade national patent laws in Europe to US and Japanese standards as far as possible without clashing with the EPC. The draft Directive was discussed sufficiently in Crespi (1989) and therefore only the recently proposed changes to it need be outlined.

Article 2 provides that no invention is to be refused patent protection for the *sole* reason that biological material is involved. This includes any matter capable of being replicated through a biological system and it therefore covers living matter, viruses, genes and other types of DNA and RNA. This principle has been accepted for many years in the major industrial countries. Article 3 provides specifically for the patentability of plants and animals and parts of these except for ‘plant and animal varieties’.

At the instigation of the European Parliament, Article 2(3) of the current Draft excludes from the above any invention 'the publication or exploitation of which would be contrary to public policy or morality'. As indicated above, the EPC and laws of most Member States already contain a similar exclusion [e.g. in EPC Article 53(a)]. However, Article 2(3) goes on to specify particular examples which for this reason cannot be patented. These include 'the human body or parts of the human body *per se*' and processes for modifying the genetic constitution of humans and animals which, in the case of humans, are 'for a non-therapeutic purpose contrary to the dignity of man' and, in the case of animals, 'are likely to inflict suffering or physical handicap upon them without any benefit to man or animals'.

As far as parts of the human body are concerned, it is perfectly permissible to patent artificial hip joints and other prostheses. But the notion of patenting real arms and limbs seems so far removed from reality that one wonders how the European Parliament came on to this track. It is to be hoped that 'parts' of the body cannot be interpreted to extend down as far as cell lines and sub-cellular particles, including genetic material, but the explanations given by the Commission in the accompanying commentary on this exclusion do not altogether rule out such an interpretation.

Natural products which have biological utility can qualify for patent protection in certain circumstances (usually as the purified material). Article 7 of the Directive confirms that patents for these products should not be ruled out in principle as mere 'discoveries'. Thus the presence of a product as part of a pre-existing material is not *alone* a sufficient ground for refusing a patent for it.

Article 10 confirms that a patent on a biological material (or a process for producing it) covers the first and all subsequent generations of material obtained by multiplication or propagation provided the crucial characteristics of the original are retained. Patent rights in a product normally become exhausted when the product is marketed by the patent owner or a licensee. However, for a product which can be multiplied biologically, the purchaser can obviously propagate the purchased product for the purpose implied in the sale, but is not legally free to multiply it further for use as propagation material. The European Parliament called for an important exception to this rule by the demand for a 'farmer's privilege' (see above). Article 13 now gives freedom to farmers to re-sow seed saved from the first crop and freedom to breed from the patented animal for renewal of farm stock.

Article 14 covers the situation in which a third party has bred a new plant variety from a patented transgenic plant and has obtained a plant breeder's right for it. If it is deemed 'in the public interest', a compulsory licence to the third party must be granted from the owner of the plant patent 'upon payment of an appropriate royalty'. Articles 13 and 14 are objected to by the agrobiotechnology industry because they detract from the patent right in an unprecedented way.

The final form of the Directive has yet to be determined. From the aspect of intellectual property, the Commission's original proposals were rather modest and fully consistent with what is already the law in major industrial

European countries. As far as agricultural biotechnology is concerned, the combined effect of the compulsory licence and farmer's privilege proposals will be the reverse of what was intended by the Commission's original initiative. Consequently, this industry must seriously question whether a Directive of the kind now envisaged will help or hinder it in the difficult markets for agriculture that are likely to prevail for some time to come.

References

CRESPI, R.S. (1985). Microbiological inventions and the patent law: The international dimension. In *Biotechnology and Genetic Engineering Reviews* (G.E. Russell, Ed.), Vol. 3, pp. 1–37. Intercept, Andover.

CRESPI, R.S. (1989). Microbiological inventions and the patent law: International developments. In *Biotechnology and Genetic Engineering Reviews* (G.E. Russell, Ed.), Vol. 7, pp. 221–258. Intercept, Andover.

CRESPI, R.S. (1991). Inventiveness in biological chemistry: An international perspective. *Journal of the Patent Office Society (USA)* **73**, 351–370.

DAUS, D.G. (1992). The patent law of (bio)chemical precursor/intermediate compounds. *Journal of the Patent Office Society (USA)* **74**, 659–668.

EUROPEAN PATENT OFFICE (1992). *European Patent Office Guidelines*, Part C, Ch. IV: *Natural Products* (p. 33); *Obviousness* (pp. 46, 73); *Morality* (p. 36). EPO, Munich.

MAEBIUS, S.B. (1992). Novel DNA sequences and the utility requirement: The human genome initiative. *Journal of the Patent Office Society (USA)* **74**, 651–658.

MCGOUGH, K.J. AND BURKE, D.B. (1992). The end of monoclonal patents? *Biotechnology* **10**, 1081–1083.

Patent cases

Citations of UK patent cases are to Reports of Patent Cases (RPC); citations of US patent cases are to US Patent Quarterly; citations of European cases are to Official Journal of the European Patent Office (OJEPO).

UK CASES

American Cyanamid v Berk Pharmaceuticals 1976 RPC 231.
Asahi case 1991 RPC 21, pp. 485–552.
Johns Manville Corp. 1967 RPC 479.
Olin Mathieson v Biorex 1970 RPC 157.
Sharpe & Dohme v Boots Drug 1927 45 RPC 367.

US CASES

Amgen Inc. v Chugai Pharmaceutical Co. 1989 13 USPQ 2d 1737.
Amgen Inc. v Genetics Institute 10 USPQ 2d 1187.
Brenner v Manson, US Supreme Court, 148 USPQ (1966) 689, 696.
Ex parte Erlich 1992 22 USPQ 2d 1463.
Ex parte Old 1985 229 USPQ 196.
Graham v John Deere Co. 1966 148 USPQ 459.
Hybritech Inc. v Monoclonal Antibodies Inc. 231 USPQ 81.

In re O'Farrell 1988 7 USPQ 2d 1673.
Merck & Co. v Chase Chemical Co. 155 USPQ 155 (US patent 2,563,794).
Scripps Clinic and Research Foundation v Genentech 11 USPQ 2d 1187.

EDC CASES

Genentech I, T 292/85 OJEPO 1989/7, 275–296.
Monsanto, T 249/88, unpublished.
Unilever, T 499/88, unpublished.
Biogen, T 301/87, OJEPO 1990/8, 335–358.
Bruker, T 385/86, OJEPO 1988/8, 308–316.

8

The Role of Trehalose and Other Carbohydrates in Biopreservation

Y.M. NEWMAN,[1] S.G. RING[1] AND C. COLACO[2]

[1] *AFRC Institute of Food Research, Norwich and* [2] *Quadrant, Trumpington, Cambridge, UK*

General introduction

Trehalose is a non-reducing disaccharide of D-glucose (α-D-glucopyranosyl-1,1-α-D-glucopyranoside) which is found in fungi, insects and some lower plants, such as the resurrection plant *Selaginella lepidophylla*. The biological role of trehalose in these organisms is still discussed. In some situations it can function as a reserve carbohydrate, in others as a protectant. For example, it has been noted that trehalose accumulates under conditions of environmental stress, such as desiccation. As the trehalose is rapidly metabolized on rehydration, a reserve function could be inferred. Trehalose can also accumulate in response to other stresses, such as heat or osmotic shock. This response to environmental stress has led to the suggestion that the accumulation of trehalose protects biological structures. Currently, the exact mechanisms of protection are debated. In an effort to identify protective mechanisms, a range of *in vitro* studies have been carried out, examining the effectiveness of trehalose in the preservation of the structure and biological activity of proteins and membrane structures. At the same time, processes for the utilization of trehalose for effective biopreservation have been developed. In part, this has developed as a result of the current need to preserve proteins and other biological structures in the dry state, and a potential increasing requirement to preserve genetically engineered proteins.

Although trehalose is a major disaccharide found in fungi, in other organisms, other carbohydrates and polyols can accumulate when exposed to environmental stress. In the examination of the role of trehalose in biopreservation, it is useful to compare the behaviour of trehalose with these other carbohydrates. Using this approach, it should be possible to identify the specific molecular mechanisms and any special properties or effectiveness of trehalose.

Abbreviations: AMP, adenosine 5′ phosphate; FTIR, Fourier transform infra-red; NMR, nuclear magnetic resonance.

Biotechnology and Genetic Engineering Reviews — Vol. 11, December 1993
0264–8725/93/11/263–294 $20.00 + $0.00 © Intercept Ltd, P.O. Box 716, Andover, Hampshire SP10 1YG, UK

In this chapter, we will examine the biosynthesis and occurrence of trehalose, and its ability to protect membranes and proteins *in vitro*.

Biosynthesis and occurrence of trehalose

FUNGI AND BACTERIA

A first step in the biosynthesis of trehalose is the formation of trehalose-6-P from UDP-glucose and glucose-6-P with the enzyme UDP-glucose: D-glucose-6-P-1-glucosyl transferase (EC 2.4.1.15). The phosphate group is then removed with a specific phosphatase, trehalose-6-P phosphohydrolase (EC 3.2.1.28), to form trehalose. In its catabolism, trehalose is hydrolysed by trehalase. In yeast there are two main forms of the enzyme, one of which is regulated by a cyclic AMP-dependent phosphorylation.

The levels of trehalose obtained are dependent on the balance between synthesis and catabolism. This balance changes with life-cycle, nutrient availability and environmental stress (Thevelein, 1984). Both gluconeogenesis and glycogenolysis can provide the glucosyl units for synthesis (François, Neves and Hers, 1991). In yeast growing on a glucose medium, the levels of trehalose are generally low during logarithmic growth of the cells, typical values being a few milligrams per gram of cells. As the carbon source becomes limiting and growth slows, there is a rapid accumulation of trehalose, to a level in the region of 80 mg of trehalose per g of cells. At the same time, there is an accumulation of glycogen. In this instance, glycogenolysis is not the source of the glycosyl units of trehalose. This general pattern of accumulation shows some genetic dependency with a few strains not showing this accumulation (Panek *et al.*, 1980).

Mono- and disaccharides are utilized to different extents by different species of yeast (Barnett, 1981), and changing the carbohydrate in the growth medium – for example, substituting maltose for glucose – can change the pattern of accumulation of trehalose (Panek and Mattoon, 1977). In addition to the dependence of trehalose levels on the availability and type of carbon source, the levels of trehalose found are also dependent on the presence of other nutrients (Thevelein, 1984). When growing yeast cells were transferred to media deprived of nitrogen, sulphur or phosphorus, an accumulation of both trehalose and glycogen was observed (Lillie and Pringle, 1980), with levels of trehalose increasing from 0.3% to 16.0% of dry weight. The comparable change in glycogen level was from 2.5 to 14.0 dry weight. The level of these carbohydrates is also affected by oxygen availability, with anaerobic conditions favouring accumulation (Chester, 1963).

While nutrient limitation is a common cause of trehalose accumulation in yeast and other fungi (Thevelein, 1984), resumption of growth, either through the further provision of a carbon source or by adding a deficient nutrient, brings rapid breakdown of the accumulated trehalose. In an NMR study on the mobilization of trehalose during the germination of yeast spores, it was concluded that its usage only accounted for 9% of the energy consumed during germination and that the primary source was exogenous glucose

(Barton *et al.*, 1982). While trehalose may not be an important source of metabolic energy during germination, it was suggested that it could function as a reserve during dormancy (Barton *et al.*, 1982). There have been a number of studies of the regulation of trehalose metabolism in fungi and the reader is referred to Thevelein (1984) for a review. Trehalose is also found in bacteria such as *Streptomyces* (McBride and Ensign, 1987b), *Rhizobium* (Streeter, 1985) and *Micrococcus* (Ahmad, Alden and Montague, 1980). In some bacterial spores (e.g. *Streptomyces griseus*), accumulation of trehalose can be substantial and up to 25% of the dry weight of the spore. Transfer of the spores to conditions which favour germination leads to the metabolism of the accumulated trehalose (McBride and Ensign, 1987b). While the trehalose present could function as a reserve carbohydrate, its presence is also associated with preservation, in that the resistance of the spores to damage by heat or desiccation increases with increasing content of trehalose (McBride and Ensign, 1987a). For example, the spores showed a five-fold improvement in resistance to exposure to 60°C for 10 min, as the trehalose level increased from 0.2 to 0.8 mg/mg protein (McBride and Ensign, 1987a). These spores contain both trehalose and the enzyme responsible for its breakdown, trehalase. For fungal spores, it has been suggested that both the different compartmentation of trehalose and trehalase, and the regulation of trehalase activity through a cyclic AMP-dependent phosphorylation of the enzyme, are responsible for the simultaneous occurrence of trehalose and trehalase (Thevelein, 1984). A study on yeast suggested that the trehalase was located in the vacuole and the trehalose in the cytosol (Keller, Schellenberg and Wiemken, 1982). In a study on *Streptomyces griseus* spores, it was also suggested (McBride and Ensign, 1990) that the level of hydration of the spores was important and that trehalase activity was sensitive to the extent of hydration, increasing as the water content of the spore increased.

PLANTS

Selaginella lepidophylla is a desert xerophyte of North America. It is well known for its ability to survive desiccation. On rehydration, the brown curled-up mass of the plant rapidly opens and becomes green again, and has the general appearance of a healthy plant, hence the popular name for the species is the 'resurrection plant'. There have been a few studies on the carbohydrate metabolism of this species (Adams, Kendall and Kartha, 1990). In the growing plant, two main disaccharides were present – sucrose and trehalose – representing approximately 14% of the dry weight, with trehalose being the major component (90%). On desiccation, there was a small decrease in the total amount of carbohydrate. Sucrose levels increased to 23% of the carbohydrate present, while the amount of trehalose fell slightly (75%). Although the levels of the disaccharides actually fell slightly during desiccation, it was argued that this desert plant must at all times be prepared for drought, hence the high levels of the disaccharides found.

INVERTEBRATES

Trehalose is also found in insects (Wyatt and Kalf, 1957); for example, it is the major carbohydrate found in the blood of the blowfly *Phormia terraenovae* (Wilps and Gade, 1990), where it is a metabolic source of energy for flight. The levels of trehalose found in the blood of insects are influenced by diet (Friedman *et al.*, 1991) and environmental stress, particularly as a response to cold temperatures during wintering (Zachariassen, 1985). Trehalose is also found in the cysts of the brine shrimp *Artemia salina*; these cysts are considered to be dormant embryos. Trehalose accumulation only occurs in embryos entering dormancy, probably as a result of the conversion of glycogen. The amount of trehalose accumulated is approximately 15% (w/w) of the dry weight; glycogen (1.5% w/w) and glycerol (3.3% w/w) are also present (Clegg, 1964). At the end of dormancy, there is a dramatic fall in the level of trehalose and an increase in the levels of glycogen and glycerol. Although some of the trehalose is catabolized and could be considered to be a reserve carbohydrate, the interconversion of glycogen and trehalose at different stages in the life-cycle has prompted the suggestion that trehalose has another function, associated with the successful dormancy of the cysts (Clegg, 1965). As it is not metabolized during the dormant period, it is thought that the trehalose functions as a 'preservative'.

Trehalose is also found in nematodes, and accumulates at the onset of desiccation. It was observed that species which showed good survival after desiccation accumulated higher levels of trehalose than the poor survivors (Womersley and Smith, 1981).

CHANGES IN LEVELS OF TREHALOSE IN RESPONSE TO ENVIRONMENT STRESS

Heat shock

The accumulation of trehalose in *Saccharomyces cerevisiae* is dependent upon the temperature of growth. At ambient temperatures, only traces of the disaccharide are found in yeast cells growing exponentially, with glucose as the carbon source and with no deficiency of other nutrients. If these cells are then transferred to a higher, but non-lethal temperature, there is a rapid accumulation of trehalose, with levels approaching 1 g/g protein (Hottiger, Schmutz and Wiemken, 1987). Following the initial rapid synthesis, there is a gradual reduction in trehalose levels to a new 'steady-state' value, which, in the temperature range 27–40°C, increases with increasing temperature. The activities of both trehalose-6-P synthase and trehalase increase with increasing temperature in this range, the dominant effect being on the synthetic enzyme leading to an accumulation of trehalose (Hottiger, Schmutz and Wiemken, 1987). It is also suggested that the primary cause of increased trehalose levels is the increase in the levels of substrate for trehalose synthase. The heat shock response on trehalose levels is rapidly reversed if the cells are transferred back to a lower temperature (Attfield, 1987).

Changes in trehalose levels are associated with improved tolerance to desiccation and heat. Cells containing approximately 1 g trehalose per g protein showed a six-fold improvement in desiccation tolerance, and a similar correlation was noted between trehalose content and the ability to survive heating to 40°C (Hottiger, Boller and Wiemken, 1987). Transfer of the cells back to the temperature of normal growth, with the consequent fall in trehalose levels, resulted in the disappearance of desiccation and thermotolerance. These observations led to the suggestion that trehalose functions as a thermoprotectant, (Hottiger, Boller and Wiemken, 1989) rather than a reserve carbohydrate (Wiemken, 1990). Similar observations were reported on the heat resistance of spores of *Dictyostelium discoideum* with an association between the level of trehalose and resistance (Emianitoff and Wright, 1979).

In addition to the increase in trehalose levels as a response to heat shock, heat shock proteins are also synthesized (Sanchez and Lindquist, 1990). These proteins may be classified as chaperonins, necessary for the successful folding and assembly of proteins (Cheng, 1989). The observation by Winkler *et al.* (1991) that yeast mutants lacking the heat shock-induced protein HSP104 were killed by exposure to 50°C despite accumulating trehalose, has led to debate on the possible protection conferred by trehalose. It has been suggested that the molecular chaperone HSP104 is involved in the repair of damaged structures, while the trehalose confers protection over a limited temperature range (De Virgino *et al.*, 1991). The heat shock protein is required after severe heat stress, which has caused damage to biological assemblies, and is involved in the refolding of the damaged protein.

Osmotic stress

When microbial cells are subjected to osmotic stress, achieved for example by the addition of salt to the growing medium, a common response is the accumulation of low molecular weight solutes, which protect against loss of water from the cell through an osmotic effect. Often these are relatively simple carbohydrates or polyols such as sucrose and glycerol. It has been suggested that the structurally more complex cyclic glucans of *Agrobacterium* and *Rhizobium* could function in this way (Miller, Kennedy and Reinhold, 1986; Breedveld, Zevenhuizen and Zehnder, 1990). It has also been suggested that trehalose could function as an osmoprotectant in rhizobia and bradyrhizobia species (Elsheikh and Wood, 1990). In a survey of the osmoregulatory solutes of cyanobacteria, a number of solutes were identified, including sucrose, trehalose, O-α-D-glucopyranosyl-(1-2)-glycerol, glycine betaine and L-glutamate betaine (Mackay, Norton and Borowitzka, 1984). Trehalose accumulation in response to osmotic stress has also been noted in *Escherichia coli* (Welsh, Reed and Herbert, 1991). In this case, the accumulation occurs when the organism is grown on a glucose/salt medium without added betaine (Strom, Falkenberg and Landfald, 1986; Larsen *et al.*, 1987). In the case of a halophilic bacterium *Ectothiorhodospira halochloris*, betaine was the primary osmoprotectant, although trehalose was synthesized under nitrogen-limiting conditions and could have a subsidiary role as an osmoprotectant (Galinski and Herzog, 1990).

The main conclusion to be drawn from these studies is that although trehalose is accumulated in response to osmotic stress in some organisms, a variety of other low molecular weight solutes apparently fulfil the same role in other organisms.

Low temperature stress

The association between the accumulation of trehalose and survival at low temperatures has mainly been studied in insects. Generally, this biochemical adaptation is of two forms. In freeze-tolerant species, ice formation in the extracellular fluid is tolerated. In some species, the extent of ice formation can approach 65% of the available water before injury occurs (Zachariassen, 1985). In freeze-avoiding species, a supercooled liquid state is maintained, partly through the avoidance of ice nucleation (Storey, 1990). It has been found that some freeze-tolerant species can survive cooling to –85°C, while the lowest temperature recorded for survival of freeze-avoiding species is in the region of –50°C (Zachariassen, 1985). In both cases, the biosynthesis of cryoprotectants is common. One property of the cryoprotectant – indeed, of all solutes – is to depress the equilibrium melting point of ice through a colligative action. A further property of a cryoprotectant is to inhibit the crystallization, on practical time-scales, of a fraction of the water present (see below). Common cryoprotectants found in insects are carbohydrates and other polyols, including glycerol, sorbitol, mannitol, xylitol, inositol, rhamnitol, fucitol, ribitol, threitol, erythritol and ethylene glycol (Zachariassen, 1985; Storey, 1990). In the freeze-tolerant gall fly, *Eurosta solidaginis*, typical levels of glycerol and sorbitol accumulated are 25 and 18 mg g wet weight^{-1}, respectively. *Eurosta scudderiana* can accumulate up to 19% of its wet weight of glycerol (Storey, 1990).

The disaccharides trehalose and sucrose are also found. In the poplar sawfly *Trichiopcampus populi* – a freeze-tolerant insect – trehalose increases from 10 to 50 mg g wet weight^{-1} during winter. Comparable levels (40 mg g^{-1}) have been found in the overwintering larvae of the red sunflower seed weevil, *Smicronyx fulvus*, which is freeze-avoiding (Rojas, Charlet and Leopold, 1991). It is also suggested that free amino acids and inorganic ions could also play a role. Survival at these low temperatures implies a very substantial accumulation of the cryoprotectant.

Desiccation tolerance

Desiccation of biological cells might involve different environmental stresses. During drying under ambient conditions, the cells might experience a form of osmotic stress. Drying at higher temperatures could involve both heat and osmotic stresses. Freezing of water external to the cell could also lead to desiccation through two effects; first, from freeze-concentration of external solutes with consequent osmotic stress and, second, from the freezing of water vapour escaping from the supercooled cell. Biochemical and physiological aspects of anhydrobiosis have been reviewed (Womersley, 1981).

In yeast cells and the dormant embryos of the brine shrimp, the accumulation of trehalose is associated with improved desiccation tolerance (see above). Trehalose is found in many fungal spores; for example, the spores of *Phycomyces blakesleeanus* contain up to 35% of their dry weight as trehalose. The spores can support 'complete desiccation' without loss of viability. Germination occurs on rehydration and is associated with a depletion in trehalose levels. Since the breakdown exceeds the respiratory requirement, this strengthens the argument in favour of the proposed role of trehalose as a protective factor against desiccation (Barton *et al.*, 1982; Laere *et al.*, 1987).

In some micro-organisms and invertebrates, trehalose is the major disaccharide present, which increases in response to some environmental stresses, largely as a result of the conversion of glycogen. In higher plants, there is a starch-to-sucrose interconversion, and it has been observed that sucrose accumulates when plants are exposed to cold and desiccation stress. It is thought that sucrose, alone and together with non-reducing oligosaccharides such as raffinose, could play a part in the desiccation tolerance of seeds (Koster and Leopold, 1988) and pollen (Hoekstra and van Roekel, 1988).

Concluding remarks

In vivo studies have shown that trehalose, and other low molecular weight carbohydrates and polyols, might have a number of biological roles. There is a strong association between the accumulation of these substrates on environmental stress, and a subsequent tolerance of that stress. This has led to the suggestion that these molecules have a 'preservative' action *in vivo*, and could be used *in vitro* to preserve delicate biological structures. The preservative action *in vitro* could be improved with a better understanding of the mechanisms involved. *In vivo* studies have largely provided evidence for an association between the accumulation of solutes and preservation action. These have been followed by *in vitro* studies examining the preservation of isolated biological structures such as proteins and membranes. In this chapter, we would like to compare *in vitro* and *in vivo* behaviour.

To aid comparison we have compiled a table of the levels of trehalose, and other carbohydrates and polyols, in various micro-organisms and invertebrates, and the changes in these levels that occur under certain conditions of environmental stress (*Table 1*). We will review current understanding on the role of trehalose and other low molecular weight carbohydrates in conferring protection to biological structures on exposure to sub-ambient temperatures, freezing, elevated temperatures and desiccation.

Solid and solution conformations of trehalose

Trehalose is the name commonly given to α, α-trehalose, although two other possible isomers, α,β-trehalose (neotrehalose) and β,β-trehalose (isotrehalose), are known. Trehalose consists of two D-glucopyranosyl residues which

Table 1. Levels of trehalose found in different organisms and their change with environmental stress

Organism	Condition	% Trehalose (w/w)	Reference
Nematodes			
Anguina tritici	Desiccated	9.1	Womersley and Smith (1981)
Ditylenchus dipsaci	Desiccated	4.6	Womersley and Smith (1981)
Brine shrimp			
Artemia salina	Dormant embryo	15.0	Clegg (1965)
Artemia francisana	Dormant embryo	16.9	Clegg and Jackson (1992)
Seed weevil			
Smicronyx fulvus		7.3	Rojas, Charlet and Leopold (1991)
Blowfly			
Phormia terraenovae	Blood	30 mg ml^{-1}	Wilps and Gade (1990)
Yeast			
Saccharomyces cerevisiae	Heat shock	1.12 g g $protein^{-1}$	Hottiger, Boller and Wiemken (1987a)
	Nutrient limit	16	Lillie and Pringle (1980)
Bacterial spore			
Streptomyces griseus		25	McBride and Ensign (1987a)

are linked by a glycosidic oxygen bridge between carbons 1 and 1′.

Trehalose can exist in several solid forms. The common form is that of the crystalline dihydrate which crystallizes from 80% ethanol (Birch, 1963) and has a melting point of 97°C. If the dihydrate is heated to 130°C, the trehalose loses water of crystallization and resolidifies to form anhydrous crystals which melt at 214–216°C (Shafizadeh and Susott, 1973). A new form of anhydrous trehalose which has different vibrational and NMR characteristics has recently been reported (Belton and Gil, in press). Additionally, an amorphous glassy solid may be obtained by supercooling the molten crystalline dihydrate (Green and Angell, 1989).

The crystal and molecular structures of the dihydrate have been confirmed by several independent X-ray studies (Brown *et al.*, 1972; Taga, Senma and Osaki, 1972), which show four $C_{12}H_{22}O_{11}.2H_2O$ units arranged in an orthorhombic cell. The two glucopyranose residues exist in the chair $^4C_1$ form, and the trehalose molecule has an approximate two-fold symmetry. Departures from symmetry are found in the torsion angles about the glycosidic linkages. In most of the common disaccharides (e.g. sucrose, lactose and maltose) the conformation is influenced by intramolecular hydrogen bonding between monosaccharide residues. For trehalose dihydrate, indirect inter-residue hydrogen bonds through water affect the conformation. The average length of these hydrogen bonds is relatively short (1.825 Å) and therefore unusually strong (Jeffrey and Saenger, 1991). For this reason, the crystalline dihydrate readily crystallizes from concentrated aqueous solutions.

The crystal structure of the anhydrous form of trehalose has also been studied (Jeffrey and Nanni, 1985). This was carried out using X-ray analysis at low temperature (–150°C). The molecular structure was similar to that found for the dihydrate crystals.

From ^{1}H NMR studies and optical rotation data (Duda and Stevens, 1990) it has been shown that in aqueous solution the D-glucopyranosyl residues of the trehalose are physically equivalent and the expected $^4C_1$ conformation is adopted. There have been a few studies on the hydration of trehalose in aqueous solution, using different experimental approaches, including measurement of the enthalpy of solution (Gaffney *et al.*, 1988), partial molar volumes and isoentropic compressibilities (Galema and Hoiland, 1991) and, finally, determination of the unfrozen water content associated with trehalose in frozen aqueous solutions (Kawai *et al.*, 1992). These data were used to estimate or evaluate a hydration number for trehalose which was then compared to that of other disaccharides. The values obtained for trehalose were at the upper end of the range of values found for the disaccharides. The significance of this observation for the preservative action of trehalose is difficult to assess at the current time.

The effect of carbohydrates on the thermal transitions of proteins and lipids

PROTEINS

The stability of the native conformation of proteins is a balance of stabilizing and destabilizing forces and effects. Intermolecular forces stabilizing the structure include van der Waals interactions and hydrogen bonding. As the temperature is raised, these interactions weaken, thereby contributing to the change in the protein conformation from the native to the denatured state. A further interaction to be considered is the interaction of water with the native and denatured forms of the protein and should include consideration of the intermolecular forces between water and both polar and non-polar groups on the protein, and the effect of these forces on the configurational entropy of the system. For example, a major stabilizing contribution is the unfavourable entropic interaction between water and non-polar residues which would occur if these residues are exposed at the protein surface, as in the denatured form. Often the interaction between water and the non-polar residues of the protein is separately identified and described under a general hydrophobic interaction or hydrophobic effect (Tanford, 1980; Privalov and Gill, 1988).

As the stabilizing entropic contribution of the hydrophobic interaction decreases with decreasing temperature, the stability of the protein can also decrease (Jaenicke, 1990), and a phenomenon known as cold denaturation has been predicted and observed to occur (Brandts, Fu and Nordin, 1970; Franks, Mathias and Hatley, 1990). A further influence arises as a result of the temperature dependence of the pK of charged low molecular weight solutes, water and charged groups on macromolecules. This dependence affects the solvent qualities of the aqueous medium, solvent–macromolecule interactions, and the interaction between charged groups on the macromolecule. The net result is temperature-dependent changes in conformation and biological activity of the macromolecule (Franks, Mathias and Hatley, 1990). A further observed effect on multi-subunit proteins, is the phenomenon of cold dissociation of subunits (Jaenicke, 1990).

The denaturation of proteins is conveniently studied by differential scanning calorimetry (Privalov, 1979, 1982), where the denaturation process on heating is observed as an endothermic transition, from which both the denaturation temperature T_d and the enthalpy change on denaturation $\triangle H_d$ can be determined. In some cases, the denaturation process can be reversed on cooling and, as a result, denaturation has been likened to, and analysed as, a melting process. For the ordered crystalline structures of synthetic polymers, the observed melting temperature T_m increases with decreasing volume fraction of solvent. Similar behaviour has been observed for the denaturation of proteins (Rupley and Careri, 1991). For example, in a study on the denaturation behaviour of β-lactoglobulin, T_d increased from 80°C to 99°C as the water content decreased from 4.6 to 0.18 g g protein $^{-1}$. Similarly, in a study of the denaturation of ovalbumin, the T_d of ovalbumin increased sharply from 77°C as the water content decreased below 0.4 g g^{-1} and approached 115°C at a water content of 0.1 g g^{-1}. The drying of a protein therefore increases its thermal stability, although whether after drying the protein can be rehydrated and regain its biological activity is another matter.

As it has been observed *in vivo* that carbohydrates and polyols are biosynthesized in response to heat shock or cold, it is useful to examine *in vitro* behaviour, and to see if the magnitude of the stabilizing effect observed is consistent with the proposed protective role *in vivo*. To do this, we need to examine the way in which carbohydrates interact with, and affect, the denaturation process.

The interaction of carbohydrates with proteins in aqueous solution has been conveniently studied through the measurement of solution densities (Arakawa and Timasheff, 1982). In the usual form of this experiment (Lee and Timasheff, 1981), the density of an aqueous carbohydrate solution of known molality is determined as a function of added protein. The apparent partial specific volume of the protein, $\phi_2{}^0$, in a carbohydrate solution of constant molality, may then be obtained by extrapolation of the data to zero concentration. For the second measurement, the aqueous carbohydrate solution containing protein is dialysed against the carbohydrate solution alone. During the course of the dialysis, solvent components redistribute across the dialysis membrane. The redistribution depends on the relative affinity of the solvent components for the macromolecule. From measurement of the density of the protein solution, it is possible to determine the apparent partial specific volume of the protein at constant chemical potential of the carbohydrate, $\phi_2{}^0$. This term contains information on the redistribution of solvent components. For example, if the carbohydrate had a preferred interaction with the protein, the apparent partial specific volume of the protein at constant chemical potential would be smaller than that at constant molality, reflecting an increase in density in the domain of the protein arising from its interaction with the carbohydrate. A more formal way of expressing this interaction is through a preferential interaction parameter of the carbohydrate (component 3) with the protein which is calculated (Casassa and Eisenberg, 1964; Lee, Gekko and Timasheff, 1979):

$$\xi_3 = \rho_0 (\phi_2{}^0 - \phi'_2{}^0) / (1 - \rho_0 v_3) \qquad (1)$$

where ρ_0 is the density of the solvent medium and v_3 is the partial specific volume of the carbohydrate.

The preferential interaction parameter obtained for carbohydrates and polyols mixed with proteins in aqueous solution is negative (Gekko and Morikawa, 1981; Arakawa and Timasheff, 1982), indicating that the protein has a preferred interaction with water, with the carbohydrate or polyol being excluded from the domain of the protein. The interaction parameter increases with increasing concentration of carbohydrate (Arakawa and Timasheff, 1982). This means that the aqueous solution of the carbohydrate is a less effective solvent than water alone. A consequence of this is that the ordered or native conformation of the protein is stabilized and the denaturation is shifted to higher temperatures. In contrast, reagents which promote the denaturation of proteins such as guanidine hydrochloride (Lee and Timasheff, 1974) have a preferred interaction with the protein.

The preferential interaction parameter shows a small dependence on carbohydrate/polyol structure. On a molar basis, it increases with molecular weight on going from a low molecular weight polyol (glycerol) to a monosaccharide and disaccharide (Lee and Timasheff, 1981; Gekko and Timasheff, 1981; Arakawa and Timasheff, 1982). The thermal stability of α-chymotrypsin in aqueous solution increased by approximately 8°C on the addition of sucrose to produce a 1 M solution (Lee and Timasheff, 1981). In comparisons (Back, Oakenfall and Smith, 1979) of the stabilizing effect of different carbohydrates and proteins on the denaturation of ovalbumin (Back, Oakenfall and Smith, 1979) and lysozyme (Gekko, 1982), it was found that polyols and carbohydrates generally increased the denaturation temperature (an exception was ethylene glycol). Typical values of the increase were in the region of 4–6°C for a 1.5 M solution of the solute (Back, Oakenfall and Smith, 1979).

In a recent study on the interaction of trehalose with bovine serum albumin in aqueous solution, it was found that, as for the other carbohydrates, the protein was preferentially hydrated and the magnitude of the preferential interaction parameter was comparable to that of other disaccharides. The stabilizing effect of trehalose in aqueous solution on protein denaturation was similar to that of the other carbohydrates. Trehalose at a concentration of 500 g l^{-1} (~1.5 M) would generally be expected to raise the denaturation temperature by about 5°C. In the heat shock response in yeast, the maximum concentration of trehalose obtained is in the region of 1.1 g trehalose per g protein. For a yeast cell containing 80% (w/w) water, this level of trehalose would represent an average concentration of approximately 0.25 M. At this concentration, the protective effect on protein denaturation would be relatively small, perhaps less than a degree. The idea that trehalose is protective against heat shock through its effect on protein denaturation is not, at present, convincing. *In vitro* experiments can give indications of the *in vivo* effect of the accumulation of these solutes, although the *in vivo* situation differs in that the protein content is appreciably greater. This in itself will have an effect both on the denaturation temperature and the aggregation of partially unfolded proteins. Additionally, the distribution of trehalose within the cell might lead to high local concentrations, which could be effective in conferring some protection. As far as we are aware, the effect of carbohydrates and polyols on cold denaturation have not been studied to date, although increased stabilization would be expected.

When considering the effect of carbohydrate on the drying of protein solutions, it might be predicted – by analogy with the behaviour of synthetic polymers in mixed solvent systems (Flory, 1953) – that as drying proceeds and the concentration of the macromolecule increases, a phase separation would occur in which a macromolecule/water phase separates from a carbohydrate/water phase. In this case, the function of the carbohydrate might be to maintain some hydration of the protein in the 'dry' state.

LIPIDS AND MEMBRANES

Carbohydrates and polyols are also thought to play a role in the stabilization of biological membranes. The stabilizing action is thought to arise as a result of the ability of the carbohydrate species to maintain the lipid in a particular phase (Crowe *et al.*, 1988). The structure of biological membranes is very complex, containing proteins, a variety of phospholipids and glycolipids and other lipid species. Although the phase behaviour of individual classes of lipids may be well defined (Small, 1986), the phase behaviour of the mixture of compounds found in the membrane is less easy to measure. As a consequence of the structural complexity, it is extremely difficult to predict the temperature dependence of the phase behaviour of the membrane, the effect of changing the level of hydration and, finally, the effect of solutes such as polyols and carbohydrates on the observed behaviour.

One approach is to study the interaction of lipids of defined structure with carbohydrates. In one experimental study, phospholipids such as 1,2-dipalmitoyl-L-phosphatidylcholine were spread as a monolayer onto water, and the surface tension measured as a function of the extent of compression of the film (Johnston *et al.*, 1984). From the surface tension, it is possible to calculate the surface pressure. It was found that if carbohydrates were present in the aqueous phase, then at a given surface pressure the film was more expanded than a similar film spread over pure water. The extent of expansion was dependent on the type of carbohydrate present. An interpretation of the observed behaviour was that the carbohydrate had a preferred interaction with the lipid head group, the relative increase in concentration of the carbohydrate at the film surface leading to the observed expansion of the monolayer (Johnston *et al.*, 1984). A similar approach was used in an earlier study which compared the effect of glycerol and oligosaccharides such as trehalose and sucrose (J.H. Crowe *et al.*, 1984). The effectiveness of the carbohydrates in preserving membrane structures was correlated with their ability to expand the monolayer. Since that time, a further study has demonstrated that the observed effects could be due to very potent surface active impurities, which were present in the 'pure' preparations of the carbohydrate (Arnett *et al.*, 1986).

In other experimental studies, the effect of carbohydrates on the 'phase behaviour' of lipids has been determined using techniques such as differential scanning calorimetry. Amphiphilic lysophosphatidylcholines swell when placed in water, the extent of water uptake depending on the acyl chain, ranging from 20 to 40 water molecules per lipid molecule (Small, 1986). The swollen lipid forms a bilayer structure with the head group on the external surface. There are several polymorphic forms of this structure in which the arrangement and packing of the acyl chains differ. In the Lα form of the lamellar structure, the acyl chains retain fluidity. On cooling the fluidity of the chains is reduced, and at defined transition temperatures the acyl chains can adopt more ordered conformations, such as a lamellar Lβ form. The extent of polymorphism of the lamellar form is dependent on the structure of the phospholipid. At higher temperatures, the lamellae may be disrupted to form a structure based on the hexagonal packing of cylinders, hexagonal 11, in which the head group and associated water occupy the centre.

The transition from a liquid crystalline lamellar structure to a more ordered lamellar structure such as Lβ, is dependent on water content. For example, for a 1,3-dipalmitoylphospatidyl choline, this transition decreases from 70°C to 40°C as the water content is increased from 2 to 20% (w/w) (Small, 1986). In this temperature range, the addition of water encourages the stabilization of a more fluid lamellar structure. A decrease in temperature, or drying, promotes the occurrence of the lamellar gel structure Lβ. This transition to the gel form is associated with a reduction in integrity of lamellar structures containing a single lipid, and also membranes containing a range of lipids and proteins. In the former case, it can be envisaged that this occurs as a result of the reduction in flexibility of the lamella as the acyl chains adopt an ordered conformation. For the membrane, the phase behaviour of the lipid mixture, the potential phase separation of different types of lipids and the aggregation of the proteins present must also be considered. The protein aggregates may lead to defects in the membrane. Damage to the membrane may also occur both during the initial cooling, inducing a particular species of lipid to undergo a transition from liquid crystalline to gel, and also during heating through the same transition (Quinn, 1985).

The phase behaviour of mixtures of 1,2-dipalmitoyl phosphatidylcholine mixed with sucrose (Hentschel, Miethe and Meyer, 1989) and trehalose (Crowe, Crowe and Chapman, 1984) was investigated by differential scanning calorimetry. The events observed in the calorimeter were associated with transitions from a fluid lamellar form to a more ordered form. It was found that the addition of sucrose and other carbohydrates to the 'dry' lipid depressed the transition. For example, the addition of sucrose to a mole fraction of 0.7 (2 g sucrose per g lipid: Hentsche, Miethe and Meyer, 1989) depressed the transition at 70°C by 10°C. The same depression in transition temperature can be achieved through the addition of 10% (w/w) water. In a study on the effect of trehalose on this transition, it was found that a maximal depression of the transition temperature to 24°C was observed at concentrations of trehalose in excess of 1.25 g g lipid^{-1}. The extent of depression observed was dependent on the method of preparation of the mixed system (Crowe and Crowe, 1988).

As a result of these and similar studies on dry phospholipids, it is argued that carbohydrates could stabilize membranes in the dry state through their effect on depressing the transition from lamellar liquid crystalline phase to a lamellar gel phase, thus ensuring the retention of fluidity in the dry state (Crowe *et al.*, 1987). There have been a number of structural studies on the interaction of trehalose with phosphatidylcholines. NMR experiments (Lee, Waugh and Griffin, 1986; Lee *et al.*, 1989) have established that in mixtures of trehalose with dry phospholipid, the trehalose hinders the motion of the head group, suggesting a direct interaction. X-ray diffraction studies have confirmed that the lipid was in a lamellar structure. From experiments on bilayer periodicity (Nakagaki, Nabase and Ueda, 1992), it was deduced that the trehalose was intercalated between the lamellae either as a monolayer, or at higher trehalose concentrations as a bilayer. This intercalation affected the extractability of the trehalose on subsequent rehydration (Viera *et al.*, 1993).

The effect of carbohydrates on the polymorphism of hydrated phospholipids has also been examined. It was found that the addition of trehalose and sucrose lead to a small increase in the temperature of the transition from lamellar gel to lamellar liquid crystalline phase (Ram, Mazzola and Prestegard, 1989). Additionally, it was shown that carbohydrates promote the formation of a hexagonal 11 phase at the expense of the lamellar liquid crystalline phase (Bryszewska and Epand, 1988; Aurell Wistrom *et al.*, 1989; Koyonova, Tenchov and Quinn, 1989). This effect would cast some doubt on the function of trehalose in the preservation of membrane structures during heat shock.

Biological membranes are considerably more complex, since they consist of mixtures of lipids and proteins. In one example (Williams, 1990), the lipid extract of a chloroplast membrane when dispersed in aqueous media did not reform a bilayer structure, and it was suggested that the proteins in the membrane help stabilize the bilayer. An attempt has been made to define the general phase behaviour of membranes (Williams, 1990), the stability of the membrane being associated with the preservation of a fluid bilayer structure. Environmental changes which potentially destabilize the membrane, include heating, cooling and drying.

A biological response to a change in the ambient temperature could be a change in the chemical composition and hence phase behaviour of the membrane. Alternatively, it is suggested that protectants such as polyols could function by helping the retention of a suitable membrane structure under adverse conditions.

Drying of biological molecules

If an aqueous glucose solution is dried at room temperature, the viscosity of the solution will increase progressively as it becomes more concentrated. At sufficiently high concentrations, the liquid glucose–water mixture will vitrify if crystallization of the carbohydrate does not intervene.

Relaxation is the time-dependent response in a property following a perturbation. The shear stress relaxation time, τ, is related to the shear viscosity, η, by the relationship:

$$\tau = \eta / G_{\infty}$$

where G_{∞} is the high-frequency limit of the shear modulus which is 10^{10} N m^{-2} for many materials. A commonly accepted value for the shear viscosity of a glass is 10^{12} Pa s, giving a shear stress relaxation time of 100 s. On the time-scale of a few seconds the material behaves as a solid, whereas if observations are conducted over longer periods of time fluid-like behaviour is observed. Once vitrified, the carbohydrate glass is relatively stable. A consequence of the enormous slowing of translational diffusion in the very viscous liquid is that even if nuclei of the crystalline solid are present, the rate of growth at these very high viscosities is slow. In fact, carbohydrate glasses have stabilities to crystallization of several years, crystallization of the carbohydrate generally commencing at a defect in the glassy structure (Parks, Huffman and Cattoir, 1928). A further effect of the very high viscosity of the glass and the predicted retardation in diffusion, is that the rate of drying of these materials is very slow. One consequence of the vitrification of biological structures may be the entrapment of sufficient water so that the material never really drys to completion, and subsequent rehydration is facilitated.

The glass transition temperature, T_g, of carbohydrates and polyols shows a dependence on molecular weight, and a weak dependence on molecular structure (Slade and Levine, 1988). Some data (Green and Angell, 1989; Finegold, Franks and Hatley, 1989; Roos, 1993; Orford, Parker and Ring, 1990) on carbohydrates and polyols are shown in *Table 2*. The T_g of the alditol is less than that of the corresponding aldose; for example, D-glucitol has a T_g of 0°C, whereas that of D-glucose is 38°C. For hexoses, pentoses and their alditols, there are relatively small differences in observed T_g within any group. The T_g of a series of oligomers increases with increasing degree of polymerization. For a series of dry malto-oligomers, the T_g increases from 38°C to 95°C to 175°C on going from monomer to dimer to hexamer.

For carbohydrates and polyols, water is a very effective plasticizing agent (Orford *et al.*, 1989; Roos, 1993), and the addition of even small amounts of

Table 2. Glass transition temperatures, T_g, of low molecular weight carbohydrates[a]

Carbohydrate	T_g (°C)
Fructose	5–13
Glucose	31–39
Glucitol	–9 to 0
Trehalose	78
Maltose	87–95
Sucrose	57–70

[a] From Finegold, Franks and Hatley (1989), Green and Angell (1989), Orford, Parker and Ring (1990) and Roos (1993).

water to the dry carbohydrate leads to a marked depression in T_g. The glass transition behaviour of trehalose and its plasticization by water is comparable to that of other disaccharides (Green and Angell, 1989). It is also useful to examine the molecular weight dependence. For the malto-oligomer series, the water content of the malto-oligomer/water mixture with a T_g of 25°C is 7.5%, 10% and 15% for the dimer, trimer and hexamer, respectively. For the high molecular weight polymer, the corresponding value is 18%. If it is accepted that glassy matrices might have a protective role through the trapping and encapsulation of water, then we predict that the high molecular weight polymer should be the most effective.

Structural proteins and enzymes can also show glassy behaviour. The glass transition behaviour of several structural proteins has been described, including elastin (Kakivaya and Hoeve, 1975) and cereal proteins (Hoseney, Zeleznak and Lai, 1986; Kalichevsky and Blanshard, 1992; Kalichevsky, Jaroszkiewicz and Blanshard, 1992). The glass transition behaviour of these proteins are broadly similar. Water is an effective plasticizing agent and its addition depresses the T_g of a protein/water mixture. For an elastin/water mixture, the water content of the mixture with a glass transition at ambient temperature is in the region of 24% (w/w). Although it is likely that amorphous biopolymers, including proteins and polysaccharides, show a glass transition, it is less likely for proteins which have an ordered conformation in aqueous solution. Even proteins with a defined three-dimensional structure (Iben *et al.*, 1989) can show structural fluctuations, ranging from the local dynamics of a few amino acid residues, to global structural changes as in the hinge bending motion of lysozyme and L-arabinose-binding protein (McCammon and Harvey, 1987). The ability of the structure to fluctuate on a practical time-scale is temperature-dependent (Frauenfelder, Sligar and Wolynes, 1991), and in calorimetric experiments sharp changes in heat capacity are observed, indicative of a change from a rigid to a mobile structure. These transitions are analogous to the glass transitions of amorphous polymers and low molecular weight solutes. A further indication of these dynamic transitions is the disappearance of enzyme activity or the ability to bind ligands over a narrow temperature range (Steinbach *et al.*, 1991; Rasmussen, Stock and Petsko, 1992). For example, in a study of the heat capacity of lysozyme/water mixtures (Yang and Rupley, 1979), a relatively sharp increase in heat capacity, indicating a change from a rigid to a mobile structure, was observed on increasing the water content from 0.1 to 0.27 g water per g protein. The onset of detectable enzymic activity coincided with the transition to a mobile structure (Careri *et al.*, 1980).

The concept that the vitrification of biological structures is a useful preservative mechanism is attractive. For example, a 'glassy' enzyme would not show enzymic activity on practical time-scales. If the translational diffusion of components in the glassy matrix was determined by the macroscopic shear viscosity, then the rate of diffusional encounter of an enzyme and its substrate would be greatly retarded. In a matrix with the shear viscosity of a glass of 10^{12} Pa s, a water molecule would take several hours to diffuse its molecular diameter. The enzyme and substrate could be mixed and there would be no significant reaction on practical time-scales until the mixture was plasticized, for example by the addition of water, leading to a sharp reduction in the observed viscosity.

Drying as a means of preservation could have several practical advantages. It is probable that the drying of biological structures results in the formation of a glassy matrix containing biopolymers and low molecular weight solutes. Bacterial and fungal spores, and plant seeds (Koster and Leopold, 1988), might be examples of this phenomenon. The high viscosity of the glass would be protective through its effect on diffusive processes. Enzymic action is effectively arrested, and the glassy matrix would prevent further loss of water. It is difficult from the current literature to estimate the magnitude of the latter effect. In many experiments on the drying of an organism or a protein preparation, the material is dried to a constant weight under a defined set of conditions for a specified length of time, often overnight. The achievement of an apparent constant weight could simply mean that the rate of water loss from the sample is very low. Drying a protein could have further advantages in improving its thermal stability. Additionally, biopolymer glasses could potentially trap more water at ambient temperatures than glasses prepared from low molecular weight solutes and this may benefit subsequent rehydration. A remaining question concerns the usefulness and effectiveness of polyols and carbohydrates for the protection of biological structures on drying. A disadvantage of drying proteins or lipid structures alone is that the drying encourages protein aggregation and fusion of the lipid structures. The carbohydrate could therefore function as a diluent, preventing contact and aggregation of the active species. As the carbohydrate is a polyhydroxy compound, it is sufficiently like water in its solution properties, that the aqueous solution conformation of a protein, or the organization of a lipid structure in aqueous dispersion, is retained. This has led to the proposal that the carbohydrate or polyol replaces water in these systems (Clegg *et al.*, 1982; Clegg, 1986).

Freezing of biological materials

The biophysics and biochemistry of biological materials at low temperatures were considered in a recent textbook (Franks, 1985). Pure water may be supercooled below the equilibrium melting temperature of the crystalline solid. Bulk samples of the order of a gram in weight can, with care, usually be supercooled to –12°C. If smaller samples are used, contained in capillaries of 0.5 mm diameter, then it is possible to supercool to –34°C. The extent of supercooling that can be achieved depends on the initial elimination of nucleating impurities. One approach is to use very small sample volumes (e.g. μm, such as in emulsion droplets). The lower limit of supercooling obtained using this approach is –42°C (Angell, 1983).

It has been suggested that insects and other organisms utilize two approaches to prevent ice crystallization. The first is the 'removal' of nucleating impurities. Second, polyols are synthesized, which both depresses the equilibrium melting point of water through a colligative action, and more importantly extends the limit of undercooling (Franks, Mathias and Hatley, 1990; Storey, 1990).

On cooling below the limit of supercooling, crystallization of the water is inevitable. Through the condensation of water vapour on a liquid helium cooled cryoplate, an amorphous solid form of water can be prepared. On heating, the material undergoes a glass transition and then rapidly crystallizes. Recently, through the employment of extremely fast cooling rates in the region of 10^7 °C per second, it has been possible to vitrify liquid forms of water. Calorimetric experiments have permitted the determination of a glass transition in the region of 136 K (Hallbrucker and Mayer, 1987; Hallbrucker, Mayer and Johari, 1989).

Compared to water, supercooled liquid carbohydrates are much more readily vitrified, and crystallization is avoided even using comparatively slow cooling rates of the order of a few degrees per minute. The behaviour of aqueous solutions of polyols and carbohydrates has also been examined (Slade and Levine, 1988; Levine and Slade, 1988; Franks, 1990). If an aqueous solution of the carbohydrate is cooled below the melting temperature of ice, and ice growth is initiated through nucleation, ice growth will proceed and the remaining aqueous solution of the carbohydrate will become progressively more concentrated. Providing the crystallization of the carbohydrate does not intervene, the viscosity of the solution will also increase progressively until ice crystal growth is not observed on experimental time-scales, and a vitreous carbohydrate/water mixture is obtained. This behaviour is usually described using a solid/liquid state diagram (Levine and Slade, 1986, 1988), which incorporates information on the effect of the solute on the depression of the freezing point, and the dependence of the glass transition temperature of the amorphous carbohydrate/water mixture on composition.

The freezing behaviour of biological materials is relevant to the behaviour of these materials during the process of freeze-drying.

Preservation of proteins by carbohydrates

Investigations have shown that some protection of proteins during preparation and storage, both in aqueous solution and the dry state, can be attained by the addition of stabilizing solutes. Amino acids, salts, polyols and sugars have been found to stabilize proteins in aqueous solution. Carbohydrates and polyols might protect proteins in a number of ways. We have already pointed out that when sugar and protein are mixed in aqueous solution, there is a deficiency in the stabilizing solute in the immediate vicinity of the protein, and the protein is preferentially hydrated. If this effect persists during drying, the added carbohydrate may help the protein retain sufficient water to maintain its conformation. In the dry state, the carbohydrate may also be sufficiently like water in its solvent properties to stabilize protein conformations, similar to those found in aqueous solution. The carbohydrate may also prevent protein aggregation through its effect as a diluent. Another suggestion is that a specific interaction occurs between the sugar and the protein.

Studies of Fourier transform infra-red (FTIR) data (Carpenter and Crowe, 1989) have led to the proposal that intermolecular hydrogen bonds are formed between trehalose and a protein when dried.

There have been a number of studies on the preservation of phosphofructokinase, both during and after freeze-drying (Crowe *et al.*, 1987; Carpenter, Crowe and Crowe, 1987; Carpenter *et al.*, 1988; Carpenter and Crowe, 1988). It was found that the activity of this enzyme was easily lost on freeze-drying, and therefore that it was a good protein on which to test the preservative action of different solutes. The preservative action immediately after drying a number of mono- and disaccharides was tested but the two monosaccharides examined, glucose and galactose, showed no significant preservative effect (Crowe *et al.*, 1987). The disaccharides were effective, their effectiveness increasing as the weight ratio of protein to carbohydrate increased from 400 : 1 to 3500 : 1. The protein activity was monitored soon after freeze-drying, and it was found that at high relative concentrations of the carbohydrate, it was possible to recover between 60 and 80% of the original enzyme activity. In this study, trehalose and maltose were marginally more effective than sucrose. The origin of the increased effectiveness of the disaccharides relative to the monosaccharides is unclear. It cannot simply be that the carbohydrate is functioning as a diluent, since the monosaccharides were found to be ineffective (Crowe *et al.*, 1987; Carpenter, Crowe and Crowe, 1987). In a subsequent study, it was demonstrated that glucose, a monosaccharide, was partially effective (Carpenter *et al.*, 1988). A protective mechanism based solely on a specific molecular interaction is also ruled out, since the disaccharides tested showed comparable protective ability.

At this stage, it is useful to examine some of the potential interactions and reactions that could occur during the freeze-drying process. The protein could denature, or multi-subunit proteins dissociate as a result of cooling prior to freezing. On freezing, the crystallization of water freeze-concentrates the remaining solutes. Depending on the freezing rate, and the rate of freeze concentration, the remaining solutes could either crystallize, phase-separate or form an amorphous glass. A consequence of phase separation or crystallization is the separation of a protein-rich phase, in which the aggregation of protein molecules would be encouraged. The observation by Izutsu, Yoshioka and Takeda (1991) that carbohydrates which crystallize either during the freeze-drying process or on storage are less effective preservatives, would support this suggestion and the general conclusion that for successful freeze-drying, the crystallization of additives should be avoided (Franks, 1990).

Although trehalose readily crystallizes from aqueous solution, it has been shown that during its drying with proteins an amorphous form is retained. FTIR spectra of freeze-dried preparations of trehalose (Carpenter and Crowe, 1989) clearly show the presence of the crystalline dihydrate, whereas on freeze-drying with protein the trehalose gives a different spectrum. Although this spectral change has been interpreted as showing a direct interaction between trehalose and the protein, the spectrum obtained is that of an amorphous form of trehalose (Newman *et al.*, submitted). The presence of the protein has therefore interfered with the crystallization of the trehalose.

If it is assumed that an amorphous matrix has been obtained, then it is worthwhile to examine other factors which might influence the ability of carbohydrates to preserve proteins. A potentially important factor is the reactions which could occur between reducing sugars and proteins. It is well known that on heating a protein with a reducing sugar, a brown product is obtained. If these reactions can occur in the dry state at ambient temperature, then non-reducing sugars would be predicted to be better preservatives, and would therefore be preferred. Nevertheless, it is still useful to examine this reaction for a number of reasons. First, reducing sugars are invariably present with non-reducing sugars *in vivo*, and it is useful to consider whether the reducing sugars have the potential to react with the protein and alter its activity. Second, since non-reducing sugar preparations often contain reducing sugar impurities, it is therefore useful to be able to define the required degree of purity. Finally, one must also consider the potential cleavage of the glycosidic bond of non-reducing disaccharides to produce reactive products. Although the glycosidic linkage of trehalose is relatively resistant to acid hydrolysis, that of sucrose is much more labile, and for this reason trehalose may be more suitable.

THE MAILLARD REACTION

The Maillard reaction (for a review, see Ellis, 1959) is defined as the reaction that occurs initially between the amino groups of amino acids, peptides or proteins with the glycosidic hydroxyl group of sugars. More complex reactions then occur after this initial reaction, which lead to the formation of brown pigments and polymers. The reaction has also been termed the ‘browning reaction’ and ‘non-enzymic browning’. The formation of Maillard polymers is the result of a number of chemical reactions which are dependent on reaction parameters such as pH, reactant concentration, temperature and type of reactant. A general reaction scheme for the initial stages of this reaction is shown in *Figure 1*. The initial reaction occurs between the sugar and the amino group. This leads to the formation of a Schiff base, which then undergoes cyclization to the corresponding *N*-substituted glycosylamine. The glycosylamine is then converted into the 1-amino-2-ketose via the Amadori rearrangement.

The Maillard reaction has been studied because of its relevance to the colour, flavour and nutritional quality of foods (Mauron, 1981). These studies generally examine the reaction products formed on incubation of a protein and a reducing sugar at relatively high temperatures (*c*. 100°C or higher). At these temperatures, degradation products of the Amadori rearrangement are formed. It is also recognized that the initial Maillard reactions are clinically important (Ahmed, Baynes and Thorpe, 1986; Ahmed *et al.*, 1988; Negia *et al.*, 1985; Baynes, 1991; Baynes *et al.*, 1989, 1991; Bunn, 1979; Higgins and Bunn, 1981; Shapiro *et al.*, 1980) and contribute to the cross-linking and thus ageing of tissue proteins (Baynes *et al.*, 1991). Furthermore, acceleration of

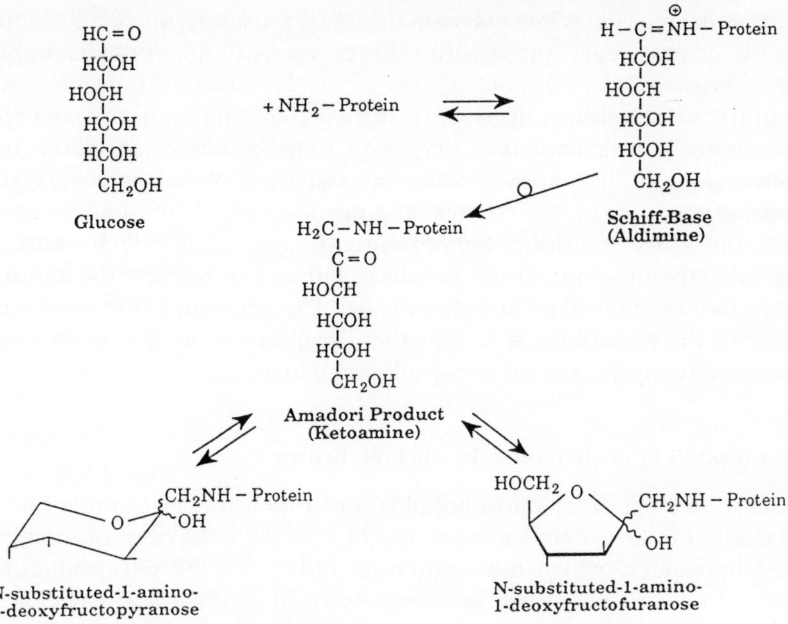

Figure 1. General scheme for the reaction of reducing sugars with proteins.

these reactions through elevated blood glucose levels, is relevant to the pathogenic effects associated with diabetes. The non-enzymatic glycosylation of proteins can alter their conformation and function (Bunn, Garlick and Shaklai, 1984), and lead to progressive enzyme inactivation (Watkins, Thorpe and Baynes, 1985). Further reaction can lead to the cross-linking of proteins (Grandhee and Monnier, 1991) and further affect their biological activity. Recent studies have concentrated on the initial stages of the Maillard reaction, and have shown that even at 37°C the extent of reaction obtained can significantly affect biological function and activity. The reaction is therefore relevant to the use of carbohydrates in the preservation of proteins in the dry state.

In considering potential reaction rates between proteins and reducing sugars in the dry state, it is useful to examine the dependence of the viscosity of the supercooled carbohydrate on both the type of carbohydrate and water content. The viscosity of the liquid influences the predicted rate of diffusional encounter of reacting species, and hence the rate of reaction. *Figure 2* shows the variation in viscosity with temperature for a hexitol, a hexose and a disaccharide. The temperature dependence of viscosity is of the same general form. As the liquid is cooled, its viscosity increases progressively, until a vitreous solid is obtained with a viscosity in excess of 10^{12} Pa s. As this viscosity is approached, very marked changes in viscosity are observed with relatively small changes in temperature. At ambient temperature, the viscosity of the hexitol is in the region of 10^4 Pa s, whereas that of the disaccharide is effectively infinite. At this temperature, the disaccharide matrix is expected to be much more effective in retarding the rate of reaction. In addition, the stability of the matrix with respect to crystallization is much reduced as the viscosity is decreased. While glassy matrices of a viscosity of 10^{12} Pa s may be stable for several years, those with a lower viscosity may only be stable for hours or days.

In practical situations, it is very difficult to maintain the amorphous carbohydrate in a completely dry state. The addition of water to the carbohydrate will dramatically affect its viscosity. Water is a very potent plasticizing agent for carbohydrates, and the addition of 10% (w/w) water can depress the glass transition temperature of a disaccharide by 60°C. For storage at ambient temperature, the disaccharide can tolerate the addition of 7% (w/w) water and still be in a glassy state. The addition of the same amount of water to the hexitol leads to a further, dramatic reduction in its viscosity and hence effectiveness as an encapsulating matrix.

Preservation of lipid structures by carbohydrates

The phase behaviour of phospholipids and other triacylglycerols has been much studied in an attempt to gain insight into the behaviour of membranes such as liposomes, which have potential utility for the encapsulation and delivery of active agents such as pharmaceutical products. It is necessary to preserve and stabilize these products prior to use. One strategy is to use carbohydrates as protecting and stabilizing matrices. The usual approach is to freeze-dry a liposome preparation which is dispersed in an aqueous solution of the carbohydrate. The freeze-drying process usually results in the formation of an amorphous matrix of the carbohydrate. The effectiveness of the carbohydrate in preserving the liposome is judged from its ability to encapsulate and retain an active agent.

The mechanism of action of the carbohydrate in preserving the liposome structure is not entirely clear, and one effect of the carbohydrate might be the preservation of the appropriate polymorph of the lipid (e.g. a particular liquid crystalline lamellar structure) during both drying and cooling to freeze-drying

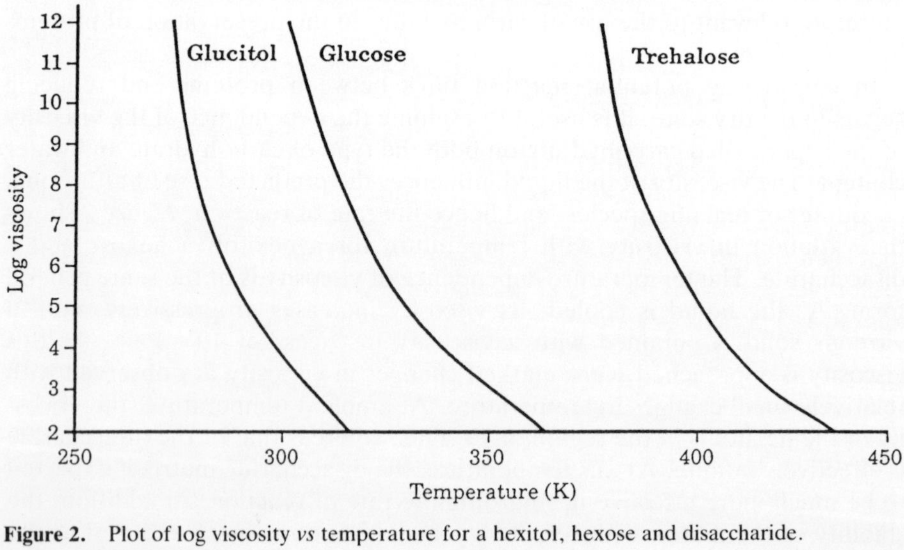

Figure 2. Plot of log viscosity *vs* temperature for a hexitol, hexose and disaccharide.

temperatures. In the freeze-drying operation, the sample may be cooled in liquid nitrogen prior to freeze-drying at, for example, –40°C. From model studies on the effect of carbohydrates on the phase behaviour of phospholipids, it is difficult to imagine how the carbohydrate could prevent the appearance of different polymorphs of the lipid, such as a lamellar gel or lamellar crystalline phase, when the sample is quenched to such an extent. It it more likely that the carbohydrate is more important in preserving the desired structure in the dry state at ambient temperature and on initial rehydration.

The carbohydrate may also prevent fusion of the lipid vesicles, which would result in loss of the encapsulated contents, and it may achieve this in several ways. First, it has been suggested that carbohydrate has a preferred interaction with the head group of the phospholipid. This 'layer' of carbohydrate on the liposome surface could inhibit contact between droplets and subsequent fusion. The greater the matrix viscosity of the undercooled liquid carbohydrate, the slower the rate of diffusional encounter of vesicles, and the greater the stability. From this consideration it would be predicted that the effectiveness of preservation would be in the order disaccharide>monosaccharide>hexitol>glycerol. Second, carbohydrate could reduce the rate of encounter of vesicles simply through its action as a diluent.

The effectiveness of different carbohydrates in preventing fusion and leakage in freeze-dried liposomes has been explored in a number of studies (Crowe *et al.*, 1984, 1986; Strauss and Hauser, 1986; Strauss, Schurtenberger and Hauser, 1986; Madden *et al.*, 1985). The extent of preservation increases with increasing carbohydrate content and approaches a plateau at around 0.5 g carbohydrate per g lipid. Trehalose is generally found to be among the most effective of the disaccharides, which are more effective than monosaccharides and low molecular weight polyols.

The possible mechanisms through which trehalose and other carbohydrates could exert a preservative effect on liposomes were evaluated in a recent review by Baltes and Bochman (1986). A continuing theme of many studies is that the interaction of the phospholipid head group with the carbohydrate, particularly trehalose, is an important factor. In studies of this nature, it is extremely difficult to identify the dominant protective mechanism, and the preservation of these structures probably depends on a range of equilibrium and non-equilibrium factors. In accounts of the biopreservation of liposomes, there has been a tendency to emphasize specific molecular interactions between the carbohydrate and the lipid, but it is also necessary to consider other factors such as the role of the matrix viscosity in influencing aggregation and fusion in these systems.

Current biotechnological use

Air-drying and freeze-drying methods have been used by the food industry for many years to preserve food. These methods of preservation increase the shelf-life of product and there is the advantage of reducing the weight and volume of the product. There are disadvantages though. By using these methods, the appearance of food may be altered, and the taste and aroma changed. Roser (1991) has claimed that foods dried in the presence of trehalose retain their original texture, colour, taste and cooking properties. Pureed fresh fruit has been dried in the presence of trehalose at 25–50°C and after prolonged storage showed no detectable changes in colour, but it had no aroma. On rehydration, the colour, viscosity and texture of the fresh purees were restored, and after a few minutes the aroma of the fresh fruit was noticeable. Roser (1989) has patented the use of trehalose as a method of protection against denaturation in proteinaceous foodstuffs during drying at temperatures above ambient. Examples of foods that have been treated with trehalose include milk, coffee, orange juice and egg. Trehalose was claimed to be more efficient than both sucrose and glucose in retaining the freshness and structure of the original food when the food was dried at elevated temperatures and then reconstituted.

Restriction enzymes are important in genetic engineering, cutting the DNA into precise and reproducible fragments. Some of the members of this class of enzyme are stable at elevated temperatures, but others are extremely unstable. As a result, these reagents usually have to be stored in solid CO_2, making them expensive to store, difficult to handle and the shipping costs are high. A technique has been developed (Colaco *et al.*, 1992) which enables even the most unstable restriction enzymes to be stored at up to 55°C for prolonged periods with no detectable loss of activity. This patent describes how trehalose and other polyhydroxy compounds have been used to preserve delicate biological substances such as enzymes, antibodies and vaccines, in the dry state and/or at elevated temperatures and/or under irradiated conditions.

Trehalose is an important storage compound for yeasts and fungi. It has also been found that high levels of trehalose in fungal spores enhances tolerance to stress conditions such as increased osmotic pressure, elevated temperature and desiccation. Yeast strains have been developed that provide a yeast with a higher content of trehalose, and as a result the strain has improved stress resistance (Driessen, Osinga and Herweifer, 1991).

Trehalose and other polyhydroxy compounds have been considered for use in medicine as cryoprotectants (Borelli, Semino and Hernandez, 1987; Fahy *et al.*, 1984; Blakeley *et al.*, 1990). It is common practice for organs, etc, to be frozen and then thawed, but there are problems involved with this method. It has been reported (Borelli, Semino and Hernandez, 1987) that isolated islets of Langerhans (used in the treatment of clinical and experimental diabetes) can be cryopreserved using trehalose.

Conclusions

At the beginning of this chapter, we examined the occurrence of trehalose and the association between its accumulation and protection against environmental stress. From *in vitro* studies, it appears that trehalose is best suited to the protection of biological materials in the dry state. Other molecular species are better osmoprotectants and protectants against chill damage. As for the mechanisms of protection, these have been clarified to a certain extent. There is good evidence that trehalose interacts with the head group of charged phospholipids and can intercalate between bilayers and prevent their fusion. Evidence of a specific molecular interaction with proteins has not yet been obtained. In the preservation of both proteins and lipids *in vitro*, the matrix properties of the carbohydrate, such as its reactivity, stability, concentration and viscosity, are important and need to be recognized so that experiments may be designed to test for specific interactions. The current use of trehalose as a preservative is increasing, and should expand further. The elucidation of the mechanisms of action should improve technological use.

Acknowledgements

The authors would like to thank the Department of Trade and Industry for financial support through the LINK scheme.

References

ADAMS, R.P., KENDALL, E. AND KARTHA, K.K. (1990). Comparison of free sugars in growing and desiccated plants of *Selaginella lepidophylla*. *Biochemical Systematics and Ecology* **18**, 107–110.

AHMAD, Z.I., ALDEN, J.R. AND MONTANGUE, M.D. (1980). The occurence of trehalose in *Micrococcus* species. *Journal of General Microbiology* **121**, 483–486.

AHMED, M.U., BAYNES, J.W. AND THORPE, S.R. (1986). Identification of *N*-carboxymethyllysine as a degradation product of fructoselysine in glycated protein. *Journal of Biological Chemistry* **261**, 4889–4894.

AHMED, M.U., BAYNES, J.W., DUNN, J.A., WALLA, M.D. AND THORPE, S.R. (1988). Oxidative degradation of glucose adducts to proteins. *Journal of Biological Chemistry* **263**, 8816–8821.

ANGELL, C.A. (1983). Supercooled water. *Annual Review of Physics and Chemistry* **84**, 593–630.

ARAKAWA, T. AND TIMASHEFF, S.N. (1982). Stabilization of protein structure by sugars. *Biochemistry* **21**, 6536–6544.

ARNETT, E.M., HARVEY, N., JOHNSON, E.A., JOHNSTON, D.S. AND CHAPMAN, D. (1986). No phospholipid monolayer-sugar interactions. *Biochemistry* **25**, 5239–5242.

ATTFIELD, P.V. (1987). Trehalose accumulates in *Saccharomyces cerevisiae* during exposure to agents that induce heat shock response. *FEBS Letters* **225**, 259–263.

AURELL, WISTROM, C., RAND, R.P., CROWE, L.M., SPARGO, B.J. AND CROWE, J.H. (1989). Direct transition of dioleoylphosphatidylethanolamine from lamellar gel to inverted hexagonal phase caused by trehalose. *Biochimica Biophysica Acta* **984**, 238–242.

BACK, J.F., OAKENFALL, D. AND SMITH, M.B. (1979). Increased thermal stability of proteins in the presence of sugars and polyols. *Biochemistry* **18**, 5191–5196.

BALTES, W. AND BOCHMANN, G. (1986). Model reactions on roast aroma formation: The reaction of serine and threonine with sucrose. *Developments in Food Science* **13**, 245–255.

BARNETT, J.A. (1981). The utilization of disaccharides and some other sugars by yeasts. *Advances in Carbohydrate Chemistry and Biochemistry* **39**, 347.

BARTON, J.K., DEN HOLLANDER, J.A., HOPFIELD, J.J. AND SHULMAN, R.G. (1982). ^{13}C-nmr study of trehalose mobilization in yeast spores. *Journal of Bacteriology* **151**, 177–185.

BAYNES, J.W. (1991). Age-dependent accumulation of *N*-(carboxymethyl)lysine and *N*-(carboxymethyl)hydroxylysine in human skin collagen. *Biochemistry* **30**, 1205–1210.

BAYNES, J.W., DUNN, J.A., PATRICK, J.S. AND THORPE, S.R. (1989). Oxidation of glycated proteins: Age-dependent accumulation of *N*-(carboxymethyl)lysine in lens protein. *Biochemistry* **28**, 9464–9468.

BAYNES, J.W., BLACKLEDGE, J.A., DYER, D.G. AND THORPE, S.R. (1991). Formation of pentosidine during nonenzymatic browning of proteins by glucose. *Journal of Biological Chemistry* **266**, 11654–11660.

BELTON, P.S. AND GIL, A. (in press). Spectroscopic studies of three forms of solid α, α-trehalose. *Carbohydrates Research*.

BIRCH, G.G. (1963). Trehaloses. *Advances in Carbohydrate Chemistry* **18**, 201–225.

BLAKELEY, D., TOLLIDAY, B., COLACO, C. AND ROSER, B. (1990). Dry instant blood typing for bedside use. *Lancet* **336**, 854–855.

BORELLI, M.I., SEMINO, M.C. AND HERNANDEZ, R.E. (1987). Cryopreservation of islets of Langerhans: The use of trehalose as a cryoprotective agent. *Medical Science Research* **15**, 299–300.

BRANDTS, J.F., FU, J. AND NORDIN, J.H. (1970). The low temperature denaturation of chymotrypsinogen in aqueous solution and in frozen aqueous solution. In *The Frozen Cell* (Wolstenholme and O'Connor, Eds), pp. 189–212. Churchill, London.

BREEDVELD, M.W., ZEVENHUIZEN, L.P.T.M. AND ZEHNDER, A.J.B. (1990). Osmotically induced oligo- and poly-saccharide synthesis by *Rhizobium meliloti* SU-47. *Journal of General Microbiology* **136**, 2511–2519.

BROWN, G.M., ROHRER, D.C., BERKING, B., BEEVERS, C.A., GOULD, R.O. AND SIMPSON, R. (1972). The crystal structure of trehalose dihydrate from three independent X-ray determinations. *Acta Crystallographica* **B28**, 3145.

BRYSZEWSKA, M. AND EPAND, R.M. (1988). Effects of sugar alcohols and disaccharides in inducing the hexagonal phase and altering membrane properties: Implications for diabetes mellitus. *Biochimica Biophysica Acta* **943**, 485–492.

BUNN, H.F. (1979). Structural heterogeneity of human haemoglobin A due to nonenzymatic glycosylation. *Journal of Biological Chemistry* **254**, 3892–3898.

BUNN, H.F., GARLICK, R.L. AND SHAKLAI, N. (1984). Nonenzymatic glycosylation of human serum albumin alters its conformation and function. *Journal of Biological Chemistry* **259**, 3812–3817.

CARERI, G., GRATTON, E., YANG, P.-H. AND RUPLEY, J.A. (1980). Correlation of IR spectroscopic, heat capacity, diamagnetic susceptibility and enzymatic measurements on lysozyme powder. *Nature* **284**, 572–573.

CARPENTER, J.F. AND CROWE, J.H. (1988). Modes of stabilization of a protein by organic solutes during desiccation. *Cryobiology* **25**, 459–470.

CARPENTER, J.F. AND CROWE, J.H. (1989). An infrared spectroscopic study of the interactions of carbohydrates with dried proteins. *Biochemistry* **28**, 3916–3922.

CARPENTER, J.F., CROWE, L.M. AND CROWE, J.H. (1987). Stabilization of phosphofructokinase with sugars during freeze-drying: Characterisation of enhanced protection in the presence of divalent cations. *Biochimica Biophysica Acta* **923**, 109–115.

CARPENTER, J.F., MARTIN, B., LOOMIS, S.H. AND CROWE, J.H. (1988). Long-term preservation of dried phosphofructokinase by sugars and sugar/zinc mixtures. *Cryobiology* **25**, 372–376.

CASASSA, E.F. AND EISENBERG, H. (1964). Thermodynamic analysis of multicomponent systems. *Advances in Protein Chemistry* **19**, 287–395.

CHENG, M.Y. (1989). Mitochondrial heat-shock protein hsp 60 is essential for assembly of proteins imported into yeast mitochondria. *Nature* **337**, 620–625.

CHESTER, V.E. (1963). The dissimilation of the carbohydrate reserves of a strain of *Saccharomyces cerevisiae*. *Biochemical Journal* **86**, 153.

CLEGG, J.S. (1964). The control of emergence and metabolism by external osmotic pressure and the role of free glycerol in developing cysts of *Artemia salina. Journal of Experimental Biology* **41**, 879–892.

CLEGG, J.S. (1965). The origin of trehalose and its significance during the formation of encysted dormant embryos of *Artemia salina. Comparative Biochemistry and Physiology* **14**, 135–143.

CLEGG, J.S. (1986). The physical properties and metabolic status of *Artemia* cysts at low water contents: The water replacement hypothesis. In *Membranes, Metabolism and Dry Organisms* (J.S. Clegg and A.C. Leopold, Eds), pp. 169–187. Cornell University Press, Ithaca, NY.

CLEGG, J.S. AND JACKSON, S.A. (1992). Aerobic heat shock activates trehalose synthesis in embryos of *Artemia franciscana. FEBS Letters* **303**, 45–47.

CLEGG, J.S., SEITZ, P., SEITZ, W. AND HAZLEWOOD, C.F. (1982). Cellular responses to extreme water loss: The water-replacement hypothesis. *Cryobiology* **19**, 306–316.

COLACO, C., SENS, S., THANGAVELU, M., PINDER, S. AND ROSER, B. (1992). Extraordinary stability of enzymes dried in trehalose: Simplified molecular biology. *Biotechnology* **10**, 1007–1011.

CROWE, J.H., CROWE, L.M. AND CHAPMAN, D. (1984). Preservation of membranes in anhydrobiotic organisms: The role of trehalose. *Science* **223**, 701.

CROWE, J.H., WHITTAM, M.A., CHAPMAN, D. AND CROWE, L.M. (1984). Interactions of phospholipid monolayers with carbohydrates. *Biochimica Biophysica Acta* **769**, 151–159.

CROWE, J.H., CROWE, L.M., CARPENTER, J.F. AND AURELL WISTROM, C. (1987). Stabilization of dry phospholipid bilayers and proteins by sugars. *Biochemical Journal* **242**, 1–10.

CROWE, J.H., CROWE, L.M., CARPENTER, J.F., RUDOLPH, A.S., AURELL WISTROM, C., SPARGO, B.J. AND ANCHORDOGUY, T.J. (1988). Interactions of sugars with membranes. *Biochimica Biophysica Acta* **947**, 367–384.

CROWE, J.H. AND CROWE, J.H. (1988). Trehalose and dry dipalmitoylphosphatidylcholine revisited. *Biochimica Biophysica Acta* **946**, 193–201.

CROWE, J.H., MOURADIAN, R., CROWE, J.H., JACKSON, S.A. AND WOMERSLEY, C. (1984). Effects of carbohydrates on membrane stability at low water activities. *Biochimica Biophysica Acta* **769**, 141–150.

CROWE, J.H., WOMERSLEY, C., CROWE, J.H., REID, D., APPEL, L. AND RUDOLPH, A. (1986). Prevention of fusion and leakage in freeze-dried liposomes by carbohydrates. *Biochimica Biophysica Acta* **861**, 131–140.

DE VIRGILIO, C., PIPER, P., BELLER, T. AND WIEMKEN, A. (1991). Acquisition of thermotolerance in *Saccharomyces cerevisiae* without heat shock protein hsp 104 and in the absence of protein synthesis. *FEBS Letters* **288**, 86–90.

DRIESSEN, M., OSINGA, K.A. AND HERWEIJER, M.A. (1991). New yeast strains with enhanced trehalose content, process to obtain such yeasts and the use of these yeasts. European Patent Application No. 91200686.3.

DUDA, C.A. AND STEVENS, E.S. (1990). Trehalose conformation in aqueous solution from optical rotation. *Journal of the American Chemistry Society* **112**, 7406–7407.

ELLIS, G.P. (1959). The Maillard reaction. *Advances in Carbohydrate Chemistry* **14**, 63–134.

ELSHEIKH, E.A.E. AND WOOD, M. (1990). Rhizobia and bradyrhizobia under salt stress: Possible role of trehalose in osmoregulation. *Letters in Applied Microbiology* **10**, 127–129.

EMYANITOFF, R.G. AND WRIGHT, B.E. (1979). Effect of intracellular carbohydrates on heat resistance of *Dictyostelium discoideum* spores. *Journal of Bacteriology* **140**, 1008–1012.

FAHY, G.M., MACFARLANE, D.R., ANGELL, C.A. AND MERYMAN, H.T. (1984). Vitrification as an approach to cryopreservation. *Cryobiology* **21**, 407–426.

FINEGOLD, L., FRANKS, F. AND HATLEY, R.H.M. (1989). Glass/rubber transitions and heat capacities of binary sugar blends. *Journal of the Chemical Society, Faraday Transactions I* **85**, 2945–2951.

FLORY, P.J. (1953). *Principles of Polymer Chemistry*. Cornell University Press, Ithaca, NY.

FRANCOIS, J., NEVES, M. AND HERS, H. (1991). The control of trehalose biosynthesis in *Saccharomyces cerevisiae*: Evidence for a catabolite inactivation and repression of trehalose-6-phosphate synthase and trehalose-6-phosphate phosphatase. *Yeast* **7**, 575–587.

FRANKS, F. (1985). *Biophysics and Biochemistry at Low Temperatures*. Cambridge University Press, Cambridge.

FRANKS, F. (1990). Freeze-drying: From empiricism to predictability. *Cryo-letters* **11**, 93–110.

FRANKS, F., MATHIAS, S.F. AND HATLEY, R.H.M. (1990). Water, temperature and life. *Philosophical Transactions of the Royal Society, London* **B326**, 517–533.

FRAUENFELDER, H., SLIGAR, S.G. AND WOLYNES, P.G. (1991). The energy landscape and motions of proteins. *Science* **254**, 1598–1603.

FRIEDMAN, S., WALDBAUER, G.P., EERTMOED, J.E., NAEEM, M. AND GHENT, A.W. (1991). Blood trehalose levels have a role in the control of dietary self-selection by *Heliothis zea* larvae. *Journal of Insect Physiology* **37**, 919–928.

GAFFNEY, S.H., HASLAM, E., LILLEY, T.H. AND WARD, T.R. (1988). Homotactic and heterotactic interactions in aqueous solutions containing some saccharides. *Journal of the Chemical Society, Faraday Transactions I* **84**, 2545–2552.

GALEMA, S.A. AND HOILAND, H. (1991). Stereochemical aspects of hydration of carbohydrates in aqueous solutions. 3. Density and ultrasound measurements. *Journal of Physical Chemistry* **95**, 5321–5326.

GALINSKI, E.A. AND HERZOG, R.M. (1990). The role of trehalose as a substitute for nitrogen-containing compatible solutes (*Ectothiorhodospira halochloris*). *Archives of Microbiology* **153**, 607–613.

GEKKO, K. (1982). Calorimetric study on thermal denaturation of lysozyme in polyol–water mixtures. *Journal of Biochemistry* **91**, 1197–1204.

GEKKO, K. AND MORIKAWA, T. (1981). Preferential hydration of bovine serum albumin in polyhydric alcohol–water mixtures. *Journal of Biochemistry* **90**, 39–50.

GEKKO, K. AND TIMASHEFF, S.N. (1981). Mechanism of protein stabilization by glycerol: Preferential hydration in glycerol–water mixtures. *Biochemistry* **20**, 4667–4676.

GRANDHEE, S.K. AND MONNIER, V.M. (1991). Mechanism of the formation of the Maillard protein cross-link pentosidine. *Journal of Biological Chemistry* **266**, 11649–11653.

GREEN, J.L. AND ANGELL, C.A. (1989). Phase relations and vitrification in saccharide–water solutions and the trehalose anomoly. *Journal of Physical Chemistry* **93**, 2880–2882.

HALLBRUCKER, A. AND MAYER, E. (1987). Calorimetric study of the vitrified liquid water to cubic ice phase transition. *Journal of Physical Chemistry* **91**, 503–505.

HALLBRUCKER, A., MAYER, E. AND JOHARI, G.P. (1989). The heat capacity and glass transition of hyperquenched glassy water. *Philoshophical Magazine* **B60**, 179–187.

HENTSCHEL, M., MIETHE, P. AND MEYER, H.W. (1989). The phase diagram of 1,2-diapalmitoyl-sn-glycero-3-phosphocholine/sucrose in the dry state: Sucrose substitution for water in lamellar mesophases. *Biochimica Biophysica Acta* **980**, 169–174.

HIGGINS, P.J. AND BUNN, H.F. (1981). Kinetic analysis of the nonenzymatic glycosylation of haemoglobin. *Journal of Biological Chemistry* **256**, 5204–5208.

HOEKSTRA, F.A. AND VAN ROEKEL, T. (1988). Desiccation tolerance of *Papayer dubium* L. pollen during its development in the anther. *Plant Physiology* **88**, 626–632.

HOSENEY, R.C., ZELEZNAK, K. AND LAI, C.S. (1986). Wheat gluten: A glassy polymer. *Cereal Chemistry* **63**, 285–286.

HOTTIGER, T., BOLLER, T. AND WIEMKEN, A. (1987). Rapid changes of heat and desiccation tolerance correlated with changes of trehalose content in *Saccharomyces cerevisiae* cells subjected to temperature shifts. *FEBS Letters* **220**, 113–115.

HOTTIGER, T., BOLLER, T. AND WIEMKEN, A. (1989). Correlation of trehalose content and heat resistance in yeast mutants altered in the RAS/adenylate cyclase pathway: Is trehalose a thermoprotectant? *FEBS Letters* **255**, 431–434.

HOTTIGER, T., SCHMUTZ, P. AND WIEMKEN, A. (1987). Heat-induced accumulation and futile cycling of trehalose in *Saccharomyces cerevisiae. Journal of Bacteriology* **169**, 5518–5522.

IBEN, I.E.T., BRAUNSTEIN, D., DOSTER, W., FRAUENFELDER, H., HONG, M.K., JOHNSON, J.B., LUCK, S., ORMOS, P., SCHULTE, A., STEINBACH, P.J., XIE, A.H. AND YOUNG, R.D. (1989). Glassy behaviour of a protein. *Physical Review Letters* **62**, 1916–1919.

IZUTSU, K., YOSHIOKA, S. AND TAKEDA, Y. (1991). The effect of additives on the stability of freeze-dried β-galactosidase stored at elevated temperatures. *International Journal of Pharmacy* **71**, 137–146.

JAENICKE, R. (1990). Protein structure and function at low temperatures. *Philosophical Transactions of the Royal Society, London* **B326**, 535–553.

JEFFREY, G.A. AND NANNI, R. (1985). The crystal structure of anhydrous α, α-trehalose at –150°C. *Carbohydrate Research* **137**, 21–30.

JEFFREY, G.A. AND SAENGER, W. (1991). Hydration of proteins. In *Hydrogen Bonding in Biological Structures* (G.A. Jeffrey and W. Saenger, Eds), pp. 459–486. Springer, Berlin.

JOHNSTON, D.S., COPPARD, E., PARERA, G.V. AND CHAPMAN, D. (1984). Langmuir film balance study of the interactions between carbohydrates and phospholipid monolayers. *Biochemistry* **23**, 6912–6919.

KAKIVAYA, S.R. AND HOEVE, C.A.J. (1975). The glass point of elastin. *Proceedings of the National Academy of Sciences, USA* **72**, 3505–3507.

KALICHEVSKY, M.T. AND BLANSHARD, J.M.V. (1992). A study of the effect of water on the glass transition of 1:1 mixtures of amylopectin, casein and gluten using DSC and DMTA. *Carbohydrate Polymers* **19**, 271–278.

KALICHEVSKY, M.T., JAROSZKIEWICZ, E.M. AND BLANSHARD, J.M.V. (1992). Glass transition of gluten. 2. The effects of lipids and emuslifiers. *International Journal of Biological Macromolecules* **14**, 267–273.

KAWAI, H., SAKURAI, M., INOUE, Y., CHUJO, R. AND KOBAYASHI, S. (1992). Hydration of oligosaccharides: Anomalous hydration ability of trehalose. *Cryobiology* **29**, 599–606.

KELLER, F., SCHELLENBERG, M. AND WIEMKEN, A. (1982). Localization of trehalose in vacuoles and of trehalose in the crystal of yeast (*Saccharomyces cerevisiae*). *Archives of Microbiology* **131**, 298–301.

KOSTER, K.L. AND LEOPOLD, A.C. (1988). Sugars and desiccation tolerance in seeds. *Plant Physiology* **88**, 829–832.

KOYONOVA, R.D., TENCHOV, B.G. AND QUINN, P.J. (1989). Sugars favour formation of hexagonal phase at the expense of lamellar liquid-crystalline phase in hydrated phosphatidylethanolamines. *Biochimica Biophysica Acta* **980**, 377–380.

LAERE, A.V., FRANCOIS, A., OVERLOOP, K., VERBEKE, M. AND VAN GERVEN, L. (1987). Relation between germination, trehalose and the status of water in *Phycomyces blakeleeanus* spores as measured by proton NMR. *Journal of General Microbiology* **133**, 239–245.

LARSEN, P.I., STROM, A.R., SYNDES, L.K. AND LANDFALD, B. (1987). Osmoregulation in *Escherichia coli* by accumulation of organic osmolytes: Betaines, glutamic acid and trehalose. *Archives of Microbiology* **147**, 1–7.

LEE, C.W.B., WAUGH, J.S. AND GRIFFIN, R.G. (1986). Solid-state NMR of trehalose/1,2-Dipalmitoyl-sn-phosphatidylcholine interactions. *Biochemistry* **25**, 3737–3742.

LEE, C.W.B., DAS GUPTA, S.K., MATTAI, J., SHIPLEY, G.G., ABDEL-MAGEED, O.H., MAKRIYANNIS, A. AND GRIFFIN, R.G. (1989). Characterization of the Lλ phase in trehalose-stabilized dry membranes by solid-state NMR and X-ray diffraction. *Biochemistry* **28**, 5000–5009.

LEE, J.C. AND TIMASHEFF, S.N. (1974). Partial specific volumes and interactions with solvent components of proteins in guanidine hydrochloride. *Biochemistry* **13**, 257–265.

LEE, J.C. AND TIMASHEFF, S.N. (1981). The stabilization of proteins by sucrose. *Journal of Biological Chemistry* **256**, 7193–7201.

LEE, J.C., GEKKO, K. AND TIMASHEFF, S.N. (1979). Measurements of preferential solvent interactions by densimetric techniques. *Methods in Enzymology* **61**, 26–49.

LEVINE, H. AND SLADE, L. (1986). A polymer physico-chemical approach to the study of commercial starch hydrolysis products (SHPs). *Carbohydrate Polymers* **6**, 213–244.

LEVINE, H. AND SLADE, L. (1988). Principles of cryostabilisation technology from structure/property relationships of carbohydrate/water systems: A review. *Cryoletters* **9**, 21–63.

LILLIE, S.H. AND PRINGLE, J.R. (1980). Reserve carbohydrate metabolism in *Saccharomyces cerevisiae*: Responses to nutrient limitation. *Journal of Bacteriology* **143**, 1384–1394.

MACKAY, M.A., NORTON, R.S. AND BOROWITZKA, L.J. (1984). Organic osmoregulatory solutes in cyanobacteria. *Journal of General Microbiology* **130**, 2177–2191.

MADDEN, T.D., BALLY, M.B., HOPE, M.J., CULLIS, P.R., SCHIEREN, H.P. AND JANOFF, A.S. (1985). Protection of large unilamellar vesicles by trehalose during dehydration: Retention of vesicle contents. *Biochimica Biophysica Acta* **817**, 67–74.

MAURON, J. (1981). The Maillard reaction in food: A critical review from the nutritional standpoint. *Progress in Food Nutrition Science* **5**, 5–35.

MCBRIDE, M.J. AND ENSIGN, J.C. (1987a). Effects of intracellular trehalose content on *Streptomyces griseus* spores. *Journal of Bacteriology* **169**, 4995–5001.

MCBRIDE, M.J. AND ENSIGN, J.C. (1987b). Metabolism of endogenous trehalose by *Streptomyces griseus* spores and by spores or cells of other actinomycetes. *Journal of Bacteriology* **169**, 5002–5007.

MCBRIDE, M.J. AND ENSIGN, J.C. (1990). Regulation of trehalose by *Streptomyces griseus* spores. *Journal of Bacteriology* **172**, 3637–3643.

MCCAMMON, J.A. AND HARVEY, S.C. (1987). *Dynamics of Proteins and Nucleic Acids*. Cambridge University Press, Cambridge.

MILLER, K.J., KENNEDY, E.P. AND REINHOLD, V.N. (1986). Osmotic adaptation by Gram-negative bacteria: Possible role for periplasmic oligosaccharides. *Science* **231**, 48–51.

NAKAGAKI, M., NABASE, H. AND UEDA, H. (1992). Stabilization of the lamellar structure of phosphatidylcholine by complex formation with trehalose. *Journal of Membrane Science* **73**, 173–180.

NEGIA, C.I., COHEN, H.J., GARBER, A.R., THORPE, S.R. AND BAYNES, J.W. (1985). Characterization of glycated proteins by ^{13}C n.m.r spectroscopy. *Journal of Biological Chemistry* **260**, 5406–5410.

NEWMAN, Y.M., RING, S.G., BELTON, P.S., COLACO, C. AND ROSER, B. (submitted). Trehalose/protein interactions in aqueous mixtures. *European Journal of Biochemistry*.

ORFORD, P.D., PARKER, R. AND RING, S.G. (1990). Aspects of the glass transition behaviour of carbohydrates of low molecular weight. *Carbohydrate Research* **196**, 11–18.

ORFORD, P.D., PARKER, R., RING, S.G. AND SMITH, A.C. (1989). The effect of water as a diluent on the glass transition behaviour of malto-oligosaccharides, amylose and amylopectin. *International Journal of Biological Macromolecules* **11**, 91–96.

PANEK, A.D. AND MATTOON, J.R. (1977). Regulation of energy metabolism in *Saccharomyces cerevisiae*. *Archives of Biochemistry and Biophysics* **183**, 306–316.

PANEK, A.D., SAMPAIO, A.L., BRAZ, G.C., BAKER, S.J. AND MATTOON, J.R. (1980). Genetic and metabolic control of trehalose and glycogen synthesis: New relationships between energy reserves, catabolite and maltose utilization. *Cellular and Molecular Biology* **25**, 345–354.

PARKS, G.S., HUFFMAN, H.M. AND CATTOIR, F.R. (1928). Studies on glasses. *Journal of Physical Chemistry* **32**, 1366–1379.

PRIVALOV, P.L. (1979). Stability of proteins. In *Advances in Protein Chemistry* (C.B. Anfinsen, J.T. Edsall and F.M. Richards, Eds), Vol. 33, pp. 167–242. Academic Press, London.

PRIVALOV, P.L. (1982). Stability of proteins: Proteins which do not present a single cooperative system. In *Advances in Protein Chemistry* (C.B. Anfinsen, J.T. Edsall and F.M. Richards, Eds), Vol. 35, pp. 1–104. Academic Press, London.

PRIVALOV, P.L. AND GILL, S.J. (1988). Stability of protein structure and hydrophobic interaction. In *Advances in Protein Chemistry* (C.B. Anfinsen, J.T. Edsall and D.S. Eisenberg, Eds), Vol. 39, pp. 193–234. Academic Press, London.

QUINN, P.J. (1985). A lipid-phase separation model of low temperature damage to biological membranes. *Cryobiology* **22**, 128–146.

RAM, P., MAZZOLA, L. AND PRESTEGARD, J.H. (1989). Orientational analysis of micelle-associated trehalose using an NMR pseudoenergy approach. *Journal of the American Chemical Society* **111**, 3176–3182.

RASMUSSEN, B.F., STOCK, A.M. AND PETSKO, G.A. (1992). Crystalline ribonuclease A loses function below the dynamical transition at 220K. *Nature* **357**, 423–424.

ROJAS, R.R., CHARLET, L.D. AND LEOPOLD, R.A. (1991). Biochemistry and physiology of overwintering in the mature larva of the red sunflower seed weevil, *Smicronyx fulvus* Leconte (Coleoptera: Curculionidae). *Journal of Insect Physiology* **37**, 489–496.

ROOS, Y. (1993). Melting and glass transition behaviour of low molecular weight carbohydrates. *Carbohydrate Research* **238**, 39–48.

ROSER, B. (1989). Food process. International Patent Application No. PCT/GB88/00511.

ROSER, B. (1991). Trehalose, a new approach to premium dried foods. *Trends in Food Science and Technology* **2**, 166–169.

RUPLEY, J.A. AND CARERI, G. (1991). Protein hydration and function. In *Advances in Protein Chemistry* (J. Edsall, C.B. Anfinsen, F.M. Richards and D.S. Eisenberg, Eds), Vol. 41, pp. 38–173. Academic Press, London.

SANCHEZ, Y. AND LINDQUIST, S.L. (1990). HSP104 required for induced thermotolerance. *Science* **248**, 1112–1115.

SHAFIZADEH, F. AND SUSOTT, R.A. (1973). Crystalline transitions of carbohydrates. *Journal of Organic Chemistry* **38**, 3710.

SHAPIRO, R., MCMANUS, M.J., ZALUT, C. AND BUNN, H.F. (1980). Sites of nonenzymatic glycosylation of human haemoglobin A. *Journal of Biological Chemistry* **255**, 3120–3127.

SLADE, L. AND LEVINE, H. (1988). Non-equilibrium behaviour of small carbohydrate–water systems. *Pure and Applied Chemistry* **60**, 1841–1864.

SMALL, D.M. (1986). *Handbook of Lipid Research 4. The Physical Chemistry of Lipids: From Alkanes to Phospholipids*. Plenum Press, New York.

STEINBACH, P.J., ANSARI, A., BERENDZEN, J., BRAUNSTEIN, D., CHU, K., COWEN, B.R., EHRENSTEIN, D., FRAUENFELDER, H., JOHNSON, J.B., LAMB, D.C., LUCK, S., MOURANT, J.R., NIENHAUS, G.U., ORMOS, P., PHILIPP, R., XIE, A. AND YOUNG, R.D. (1991). Ligand binding to heme proteins: Connection between dynamics and function. *Biochemistry* **30**, 3988–4001.

STOREY, K.B. (1990). Biochemical adaptation for cold hardiness in insects. *Philosophical Transactions of the Royal Society, London* **B326**, 635–654.

STRAUSS, G. AND HAUSER, H. (1986). Stabilization of lipid bilayer vesicles by sucrose during freezing. *Proceedings of the National Academy of Sciences, USA* **83**, 2422–2426.

STRAUSS, G., SCHURTENBERGER, P. AND HAUSER, H. (1986). The interaction of saccharides with lipid bilayer vesicles: Stabilization during freeze-thawing and freeze-drying. *Biochimica Biophysica Acta* **858**, 169–180.

STREETER, J.G. (1985). Accumulation of α, α-trehalose by Rhizobium bacteria and bacteriods. *Journal of Bacteriology* **164**, 78–84.

STROM, A.R., FALKENBERG, P. AND LANDFALD, B. (1986). Genetics of osmoregulation in *Escherichia coli*: Uptake and biosynthesis of organic osmolytes. *FEMS Microbiology Review* **39**, 79–86.

TAGA, T., SENMA, M. AND OSAKI, K. (1972). The crystal and molecular structure of trehalose dihydrate. *Acta Crystallographica* **B28**, 3258.

TANFORD, C. (1980). *The Hydrophobic Effect: Formation of Micelles and Biological Membranes*, 2nd Edn. Wiley Interscience, New York.

THEVELEIN, J.M. (1984). Regulation of trehalose mobilization in fungi. *Microbiological Reviews* **48**, 42–59.

VIERA, L.I., ALONSO-ROMANOWSKI, S., BOROVYAGIN, V., FELIZ, M.R. AND DISALVO, E.A. (1993). Properties of gel phase lipid-trehalose bilayers upon rehydration. *Biochimica Biophysica Acta* **1145**, 157–167.

WATKINS, N.G., THORPE, S.R. AND BAYNES, J.W. (1985). Glycation of amino groups in protein. *Journal of Biological Chemistry* **260**, 10629–10636.

WELSH, D.T., REED, R.H. AND HERBERT, R.A. (1991). The role of trehalose in the osmadaptation of *Escherichia coli* NCIB 9484: Interaction of trehalose, K+ and glutamate during osmadaptation in continuous culture. *Journal of General Microbiology* **137**, 745–750.

WIEMKEN, A. (1990). Trehalose in yeast, stress protectant rather than reserve carbohydrate. *Antonie van Leeuwenhoek* **58**, 209–217.

WILLIAMS, W.P. (1990). Cold-induced lipid phase transitions. *Philosophical Transactions of the Royal Society, London* **B326**, 555–570.

WILPS, H. AND GADE, G. (1990). Hormonal regulation of carbohydrate metabolism in the blowfly *Phormia terraenovae*. *Journal of Insect Physiology* **36**, 441–449.

WINKLER, K., KIENLE, I., BURGERT, M., WAGNER, J. AND HOLZER, H. (1991). Metabolic regulation of the trehalose content of vegetative yeast. *FEBS Letters* **291**, 269–272.

WOMERSLEY, C. (1981). Biochemical and physiological aspects of anhydrobiosis. *Comparative Biochemistry and Physiology* **70B**, 669–678.

WOMERSLEY, C. AND SMITH, L. (1981). Anhydrobiosis in nematodes I. The role of glycerol, myo-inositol and trehalose during desiccation. *Comparative Biochemistry and Physiology* **70B**, 579–586.

WYATT, G.R. AND KALF, G.F. (1957). The chemistry of insect hemolymph. II. Trehalose and other carbohydrates. *Journal of General Physiology* **40**, 833–847.

YANG, P.H. AND RUPLEY, J.A. (1979). Protein–water interactions: Heat capacity of the lysozyme–water system. *Biochemistry* **18**, 2654–2661.

ZACHARIASSEN, K.E. (1985). Physiology of cold tolerance in insects. *Physiology Review* **65**, 799–832.

9

Modulation of Structure and Function of Enzymes and Other Biologically Active Proteins by Their Specific Antibodies

EZEKIEL Y. SHAMI

Hybrisens Ltd and Department of Biology, York University, Toronto, Ontario, Canada

Introduction

In nature, the primary role of antibodies is limited to tagging foreign antigens and thereby initiating their ultimate destruction by other components of the immune system. In some cases, the interacting antibodies have the capacity to neutralize the biological activity associated with the foreign antigens. To ensure efficient tagging and neutralization, the immune system generates antibodies to multitudes of different epitopes on the surface of the antigen – a polyclonal response. Each of these antibodies is specific for a single epitope and is originated in a pre-programmed clone of B-cell lymphocytes which, upon exposure to 'their' antigen and a rather complex chain of events associated with the immune system, produce and secrete their unique antibodies. Once the clonal origin of antibodies was revealed, attempts to clone B cells were initiated as it was recognized that the *in vitro* utilization of antibodies could be greatly increased, since a uniform antibody preparation could be generated from the cloned B cells. White cell cloning procedures were well established at the time, the major obstacle being the inability of B cells to survive, propagate and produce antibodies *in vitro*. This obstacle was finally removed in the late 1970s when Kohler and Milstein (1975) developed a new technology to 'immortalize' B cells. Briefly, a mouse is immunized with the appropriate antigen to elicit a standard polyclonal response. Once a positive polyclonal response is verified, the stimulated and differentiated B cells are harvested from either the mouse spleen or lymph nodes and fused with mouse myeloma cells to create stable hybrid cell lines that can be

Abbreviations: APAAP, alkaline phosphatase–anti-alkaline phosphatase; ELISA, enzyme-linked immunosorbent assay; Hex A and Hex B, heat labile and heat stable isoenzymes of *N*-acetyl-hexosaminidase; hGH, human growth hormone; IgG, immunoglobulin G; K_d, dissociation constant; MAb, monoclonal antibody; MW, molecular weight; NMR, nuclear magnetic resonance; PAGE, polyacrylamide gel electrophoresis; PAP, peroxidase–anti-peroxidase; SDS, sodium dodecyl sulphate; t-PA, tissue plasminogen activator.

Biotechnology and Genetic Engineering Reviews — Vol. 11, December 1993
0264–8725/93/11/295–316 $20.00 + $0.00 © Intercept Ltd, P.O. Box 716, Andover, Hampshire SP10 1YG, UK

propagated *in vitro* (a capacity that is absent in original B cells). This new feature (stable hybrids) enables one to clone and selct individual cell lines originated in a single B cell. As the new hybrid cell lines retain the capacity of the original parent B cell to produce antibodies, the antibodies are monoclonal (i.e. they are specific for a single epitope and can be produced in large quantities and with consistency *in vitro*).

The unique capacity of antibodies to react selectively and with relatively high affinity (an extended exposure to the antigen eventually yields high-affinity antibodies: $K_d < 10^{-9}$ M) with their antigen, has inspired scientists over the last three decades to formulate innovative uses for antibodies not always consistent with their role in nature. Some of the most important include the use of antibodies for the determination of the antigen concentration in a mixture with other proteins (as is the case for hormones in plasma), or for the purification or elimination of the antigen from such a mixture.

Current immunological techniques allow the generation of antibodies to practically any peptide or protein, including autologous protein. Most if not all of the naturally occurring peptides and proteins could be broadly defined as biologically active. The common characteristic of all biologically active proteins is that they possess a specific structure that enables them to interact specifically with other molecules. For enzymes, these specific interactions result in direct catalysis of specific chemical reactions. The interactions of other classes of biologically active proteins and peptides such as hormones, cytokines, etc, result in either stimulation or inhibition of a complex cascade of biological events that involve enzymes.

This chapter will review applications where antibodies are utilized (out of context of the immune system) to form interactions between antibodies and biologically active proteins for the purpose of modulating their activity. It will also address a related phenomenon, enzyme–protein (inhibitor) interactions. But first, the nature of protein–protein interactions will be described using antibody–antigen interaction as a model system.

Antibody–antigen interactions

Enzymes, due to their direct catalytic activity (which can be measured easily), motivated scientists to use them as models to study protein structure–function relations. Hence, some of the best described and studied proteins are enzymes. When advances in X-ray crystallography technology made it possible to study the interaction between antibodies and their antigens, most of the 'first-wave' studies selected an enzyme as the model antigen. Davies, Sheriff and Paldan (1988) recently reviewed X-ray crystallography studies concerning the interaction between hen egg lysozyme and the Fabs of three monoclonal antibodies (MAbs), as well as the interaction of neuraminidase of influenza virus with a Fab of one of its monoclonal antibodies. It was concluded that the interactions are tight, and all water molecules are excluded from the area of contact, which measures around 700 Å^2 (Sheriff *et al.*, 1987). If this is typical of the size of the contact area in antibody–antigen interactions, it will dictate an epitope comprising more than one continuous oligopeptide (Barlow, Edwards and Thornton, 1986), so that several discontinuous oligopeptides or residues (three for lysozyme) which serve as an epitope would be cross-linked by the antibody reaction. As many as 3 salt links, 10 hydrogen bonds and 74 van der Waals interactions are involved in the case of lysozyme (Sheriff *et al.*, 1987). While the specificity of the antibody will be determined primarily by the individual side-chain reactions, hydrogen bonding and van der Waals interactions, the affinity of the association and the reduction in free energy of the antigen is primarily due to hydrophobic interactions (Rees *et al.*, 1988).

Although these studies employed enzymes as proteins and ignored the consequences of this interaction on enzyme activity and structural integrity, they provide information which allows us to postulate the outcome of this interaction. For example, the extensive cross-linking driven by hydrophobic interactions should result in stabilization of the folded structure of the antigen. Also, it is interesting to note that the thermodynamic parameters determined for the antibody–antigen interaction in these studies are consistent with stabilization of the folded state of the antigen. As the average affinity of monoclonal antibody–antigen binding is around (10^8 M^{-1}), this interaction could lead to a reduction in free energy of the antigen of about 10 kcal mol^{-1} (40 kJ mol^{-1}) (Rees *et al.*, 1988). In a general thermodynamic sense, this is sufficient to confer increased stability, since the difference in free energy between the folded and unfolded states of active globular proteins is in the range 5–15 kcal mol^{-1} (20–60 kJ mol^{-1}) (Tanford, 1970).

Enzyme–protein interactions and their effect on enzyme structural and functional integrity

Antibody–enzyme interactions belong to the larger class of interactions which are described as protein–protein non-covalent associations. In nature, enzymes are involved in protein–protein interactions that in a broad thermodynamic sense resemble antibody–antigen interactions. These interactions can confer stability by excluding water from the interaction area on the enzyme surface, thereby reducing the free energy and driving the reaction towards the folded state (Chothia and Janin, 1975). Certain thermophilic organisms confer stability on their enzymes by elaborating protective macromolecules such as peptides and polyamines (Nakamura *et al.*, 1978; Prasad and Maheshwari, 1978; Oshima, 1982). While multimers or enzyme aggregates are often more stable and active than the constituent monomers (Mozahev and Martinek, 1984). Although protein–enzyme interactions can increase the resistance of the enzyme to thermal unfolding, it could also result in the total inhibition of enzyme activity, as is the case for subtilisin–subtilisin inhibitor complex (Takahashi and Sturtevant, 1981). Subtilisin inhibitor is a protein which binds specifically and stoichoimetrically with subtilisin resulting in the total inhibition of subtilisin. The transition temperature (folded to unfolded state) for subtilisin-inhibitor complex was 20°C higher than for the free enzyme, but this high degree of protection was not beneficial as the enzyme–inhibitor complex was not biologically active. More recently, Sweadner (1991) described another stabilizing aspect of the enzyme–inhibitor system, using trypsin and its soybean inhibitors. The interaction of soybean trypsin inhibitors with trypsin and the mechanism that is responsible for the efficient inhibition of trypsin is well documented. The interaction is tight, excluding all water from the interface (Rühlmann *et al.*, 1973; Sweet *et al.*, 1974), with very low dissociation constants (10^{-10}–10^{-14} M). Again, as for subtilisin, this association results in the total inhibition of trypsin enzyme activity, but confers a significant increase in resistance to thermal unfolding (Donovan and Beardslee, 1975). Moreover, Sweadner (1991) showed that the inactive trypsin in this complex was capable of withstanding the irreversible denaturing effect that exposure to sodium dodecyl sulphate (SDS) has on the free trypsin and 'paradoxically' regained its activity in the presence of two very powerful inhibitors. In this case, the extreme structural rigidity imposed by the inhibitor is attenuated by SDS to a certain extent, so that the enzyme regains its activity in an environment in which free enzyme would lose its activity in a matter of seconds.

Antibody–enzyme interactions

There is good reason to believe that the interaction of an enzyme with its specific polyclonal antibodies will, in most cases, result in a significant increase in structural integrity and resistance to thermal unfolding of the enzyme molecules. The effect of this interaction on functional integrity (the activity of the enzyme) is dependent on the location and the role of the epitopes (at which the antibodies bind) in the enzyme activity. It is impossible to predict the extent of the inhibition, or if it will occur at all – every enzyme has to be tested individually.

Most studies in the late 1960s and early 1970s dealing with the effect of specific antibodies on enzyme activity were concerned with enzyme characterization and with differences between wild-type and mutant enzymes, as well as isoenzymes. In probably the earliest study, Burnett and Schmidt (1921) reported that catalase in a complex with anti-catalase (antibodies to catalase) retained full activity, although the complex precipitated. Almost 20 years later, Tria (1939) and Campbell and Fourt (1939), unaware of the report by Burnett and Schmidt, reproduced their results. In the following years, similar studies with other enzymes were published; for example, Cohn and Torriani (1952) made the same observation for β-galactosidase, Stenberger *et al.* (1970) for peroxidase and Cordell *et al.* (1984) for alkaline phosphatase.

For many years, antibody–enzyme complexes have been used very effectively in immunocytochemistry applications. The peroxidase–anti-peroxoidase (PAP) and the alkaline phosphatase–anti-alkaline phosphatase (APAAP) systems have distinct advantages with regard to the reactivity of the antibodies and the enzyme activity when compared with their equivalent covalently conjugated immuno-reagents.

This phenomenon (i.e. that enzymes can retain full enzymatic activity even while in a precipitated complex with their antibodies) was considered an aberration. General statements such as 'the interaction of enzymes with their specific antibodies generally leads to a reduction in their enzymatic activity' (Ben-Yoseph, Geiger and Arnon, 1975; Solomon *et al.*, 1984) are standard in papers dealing with antibody–enzyme interactions, and they are based on a review by Arnon (1973). The author of the current review believes that in most cases there is only a limited initial loss of activity (10–20%), as described by Melchers and Messers (1970), Ben-Yoseph, Geiger and Arnon (1975), Zyk (1973) and Shami, Rothstein and Ramjeesingh (1989), and that total loss of activity due to an interaction with specific polyclonal antibodies is the exception rather than the rule. The absence of significant inhibition in some (if not most) cases could be explained by a combination of the following: (1) the active site of the enzymes used is a 'blind' spot to the immune system; (2) the inhibitory antibodies are of lower affinity and/or quantity and their binding to the enzyme active site is sterically hindered by adjacent high-affinity antibodies; (3) the low molecular weight synthetic substrates used in the tests are less susceptible to the steric interference imposed by inhibitory antibodies.

However, as soon as the generation of monoclonal antibodies to enzymes became possible in the early 1980s, reports describing total loss of activity by utilizing selected monoclonal antibodies became common. Early reports inclue those by Cotton *et al.* (1980), Frackelton and Rotman (1980), Park *et al.* (1980), Mather *et al.* (1980), Ross, Reis and Joh (1981), Webster, Hinshaw and Laver (1982) and Fambrough, Engel and Rosenberry (1982). Strong indications regarding the powerful utility of monoclonals in enzyme research were found by Solomon *et al.* (1984), who generated a battery of monoclonal antibodies to carboxipeptidase A (an enzyme exhibiting peptidase activity as well as esterase activity). Of the 25 clones stabilized, 2 produced antibodies that selectively inhibited peptidase activity by more than 80%, 2 selectively inhibited the esterase activity by 50%, 10 had no effect or only a slight effect (less than 20%) on both activities, and 11 had a partial effect (20–40% inhibition) on both activities. It is interesting to note that polyclonal antibodies to carboxypeptidase generated in rabbits were capable of only partial inhibition of the enzymatic activity (Amiraian and Plummer, 1971), although in all probability the polyclonal mixture contained industrial antibodies capable of total inhibition when employed alone. Some of the possible reasons for the reduced effectiveness of polyclonal antibodies are listed above.

Mounting evidence has shown that in many cases enzymes in complex with their specific polyclonal antibodies retain their full catalytic activity. This has set the stage for a more comprehensive use of antibodies in enzyme research dealing with the effects on the enzyme activity profile. The activity profile of an enzyme is a composite of activities obtained by gradually changing physical, chemical and biological parameters (such as temperature, pH, substrate, ions, proteolytic enzymes, etc).

Mutated enzymes or isoenzymes may exhibit reduced reaction rate when compared with their respective wild-type enzymes. 'New' calibration of the optimal conditions is generally ineffective in 'reviving' their activity. The degree of mutation has been monitored by interacting the mutant enzymes with antibodies generated to the wild-type enzymes and comparing it to the interaction of wild-type enzymes with the same antibodies. It was shown that in some cases the mere interaction of a mutant enzyme with antibodies to wild-type enzyme was sufficient to restore a large proportion of the lost enzyme activity, as reported by Rotman and Celada (1968) and Messer and Melchers (1969). They attributed this effect to the enforcement of proper (wild-type) folding by the interacting antibodies. Heat-inactivated acetylcholinesterase was partially reactivated following the addition of its specific antibodies (Michaeli *et al.*, 1969); again, the same general mechanism was implicated.

The next, logical, step taken in these early studies was to determine the effect of pre-treatment (of wild-type enzymes, mutant enzymes and isoenzymes with their respective specific polyclonal antibodies) on the capacity to resist increasingly 'denaturing' conditions such as heat, pH and proteolytic enzymes. These studies revealed that:

1. Mutant catalases sensitive to heat and mild alkalinity were stabilized by interacting them with antibodies generated against the wild-type catalase (Feinstein *et al.*, 1971).
2. Wild-type β-galactosidase and 10 of its 11 mutants, which were complexed with the antibodies generated against the wild-type enzyme, exhibited increased resistance to heat inactivation. The greatest shift (12°C) and overall highest value (66.5°C) of the transition temperature was recorded for the wild-type enzyme (Melchers and Messer, 1970).
3. The heat labile and heat stable isoenzymes of *N*-acetyl-hexosaminidase (Hex A and Hex B, respectively) were both stabilized against heat inactivation by their respective antibodies (Ben-Yoseph, Geiger and Arnon, 1975).

As these early studies were not concerned with the stabilization of enzymes for practical purposes, this short-lived and restricted burst of activity in the field of antibody–enzyme interaction dissipated without follow-up. Even the emergence of monoclonal antibodies in the late 1970s and early 1980s did not stimulate new activity in this field.

In the mid and late 1980s, advances in X-ray crystallography and protein engineering provided new tools for studying the structure–function relations of proteins, and antibody–antigen interactions became a popular model system, with most models utilizing enzymes as model antigens. The ability to modify almost at will the primary structure of antigens and antibodies by standard recombinant DNA techniques, coupled with the ability to achieve high-resolution images of protein tertiary structure by X-ray crystallography and, more recently, by nuclear magnetic resonance (NMR), has revived interest in this field.

In 1987, we embarked on the development of practical applications utilizing the unique properties of antibody–antigen interactions to confer resistance to inactivation and stabilize commercially valuable proteinaceous antigens. To test our hypothesis, we used a number of model enzymes and treated them with their specific antibodies (polyclonal antibodies for α-amylase, glucoamylase and subtilisin, and selected protective, non-inhibitory, monoclonal antibodies for L-asparaginase, singly or in mixtures) and subjected them to physical, chemical and biological inactivation. The results presented here are drawn from Shami, Rothstein and Ramjeesingh (1988, 1989), Shami *et al.* (1991) and Ramjeesingh *et al.* (1992).

Briefly, enzymes in their pure form were pre-incubated overnight at 4°C, with increasing concentrations of either their pure monoclonal antibodies or with the IgG fraction for their polyclonal antibodies, the control enzyme samples were 'spiked' with equivalent amounts of non-immune human IgG and pre-incubated under the same conditions. both enzyme samples were then exposed for varying lengths of time to various inactivation procedures.

Enzyme functional integrity can be irreversibly disrupted by physical, chemical and biological influences. We selected heating, freezing and lyophilization as physical methods, oxidation and the effects of ethanol as chemical methods, and proteolytic degradation by trypsin as a biological method. Protection was assessed by determining the residual activity of the enzyme, at the end of the exposure period. The optimal antibody–enzyme concentration ratio for protection was then determined for each model enzyme–AB system. This ratio was then used for all further protection experiments.

THERMAL INACTIVATION

This parameter was assayed with the enzymes α-amylase (*Figure 1A, B*), glucoamylase and subtilisin. In all three cases, the retained activity of the antibody complexed enzymes was substantially higher than that of the free enzyme. We recorded increases in short-term (5 min) heat resistance (i.e. the temperature for 50% inhibition) of 6.0°C for subtilisin, 16.0°C for glucoamylase and 20.5°C for α-amylase (*Table 1*). Long-term heat resistance was defined in terms of the half-life at a selected temperature. The temperature was selected so as to maximize the protective effect for each enzyme. Generally at this temperature, the short-term heat resistance for the protected enzyme was still close to 100% of the control, whereas for the unprotected enzyme it was well below 50%. The selected temperatures for long-term thermal inactivation ranged between 65°C and 70°C. As much as a 240-fold increase in the half-life of α-amylase was noted, and 60-fold and 65-fold increases were recorded for glucoamylase and subtilisin, respectively (*Table 1*). While the dramatic increase in the long-term heat resistance of α-amylase was associated with an impressive shift of 20.5°C in short-term heat resistance, a 16.0°C shift for glucoamylase resulted in less of an increase in long-term heat resistance than that recorded for subtilisin with only a 6.0°C shift.

Economically speaking, long-term heat resistance is the more important feature for industrial enzymes. However, for practical reasons, in most reports 'the most promising stabilized enzyme preparations' are designated as such, based on screening for short-term temperature-dependent heat resistance. Our results show clearly that care should be taken to analyse all preparations for time-dependent heat resistance as well, since larger tempera-

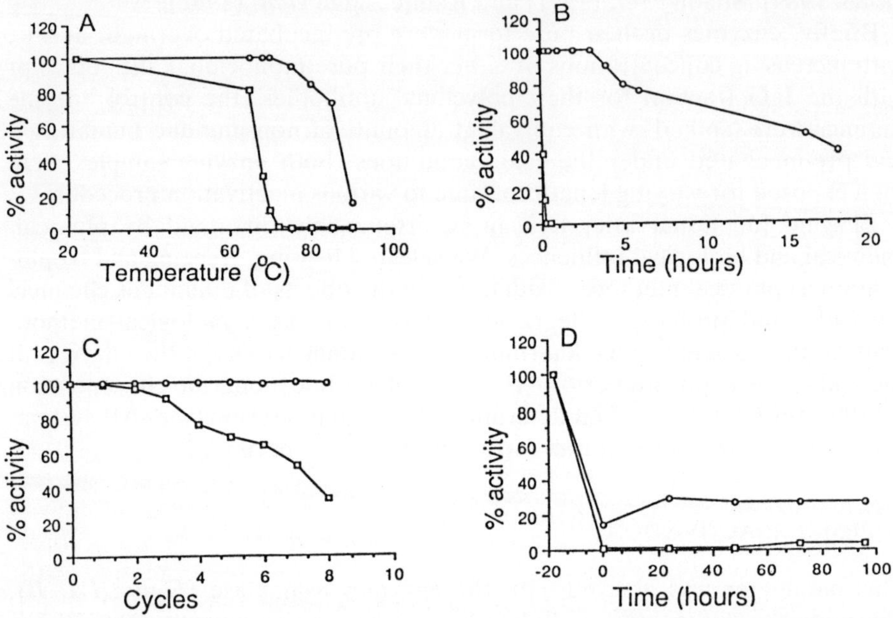

Figure 1. Protection of α-amylase by specific antibodies against physical disruption. From Shami *et al.* (1991).
(A) Heat inactivation of α-amylase. The results of residual activity (100% activity = activity at 23°C) of protected and unprotected α-amylase samples as a function of temperature. The same symbols for protected and unprotected enzymes are used throughout.
(B) Time-dependent inactivation of α-amylase at 70°C. The results of residual activity of protected and unprotected enzyme samples as a function of incubation time at 70°C.
(C) Protection of α-amylase against inactivation by freezing and thawing. The residual enzyme activities (expressed as percent of pre-treatment enzyme activity) for protected and unprotected samples as a function of number of freezing and thawing cycles.
(D) Lyophilization effect on protected and unprotected α-amylase. The results are expressed as percent of pre-treatment enzyme activity.

ture shifts are not always associated with a proportionally longer half-life.

FREEZING AND THAWING

After eight cycles of freezing and thawing, there was no significant loss of activity for antibody-protected α-amylase, whereas the unprotected enzyme lost over 65% of its initial activity (*Figure 1C*). This protective capacity might provide greater flexibility in exploiting industrial enzymes, providing an additional option for storage.

Table 1. Thermostability of enzyme–antibody complexes[a]

		Half-life			T(°C) for 50% inhibition		
Enzyme	T(°C)	A Free enzyme	B Enz–AB complex	B/A	C Free enzyme	D Enz–AB complex	D–C T(°C) shift
α-amylase[b]	70	4 min	16.0h	240	66.5	87.0	20.5
Glucoamylase[c]	66	3 min	3.0h	60	57.5	73.5	16.0
Subtilisin[c]	65	4 min	4.3h	65	63.0	69.0	6.0

[a] From Shami *et al.* (1991).
[b] Data obtained from *Figure 1A, B*.
[c] Data for glucoamylase and subtilisin were obtained using experimental protocols similar to those described for α-amylase.

LYOPHILIZATION

The unprotected α-amylase lost almost all of its enzyme activity after lyophilization, while the antibody-protected enzyme retained approximately 15% of its initial activity (*Figure 1D*). Twenty-four hours after reconstitution, the protected sample had recovered 25% of its initial activity, an activity that was maintained for a period of 96 h. The unprotected α-amylase did not exhibit significant recovery of activity. Of all the physical disruption forces tested, the protection provided by antibodies against lyophilization was the least impressive. However, as we tested only one enzyme for this effect, it is possible that other enzymes may benefit more from such protection, again allowing an additional storage option.

OXIDATION

The short-term (5 min) exposure of subtilisin to increasing concentrations of NaOCl revealed that the antibody protection factor was largest (81 *vs* 60%) at 0.05% (w/v) NaOCl. However, upon longer-term exposure (30 min) of subtilisin to 0.05% (w/v) NaOCl, only 25% of its original activity was retained, while the antibody-protected enzyme retained close to 80% of its activity (*Figure 2*). It would appear that following the initial damage due to oxidation, the protected enzyme remained unchanged at around 80%, whereas the unprotected enzyme kept on losing its activity. Oxidation is a major inactivation force in nature. *In vitro*, enzymes are rather vulnerable to oxidation even under normal atmospheric conditions, let alone if extra oxidant is added to the system, as is the case with the detergent enzyme subtilisin, which has to function in the presence of the bleach-oxidant, NaOCl. The action of subtilisin could be enhanced greatly by an increase in resistance to both high temperature and oxidation. Antibodies are capable of providing this dual protection.

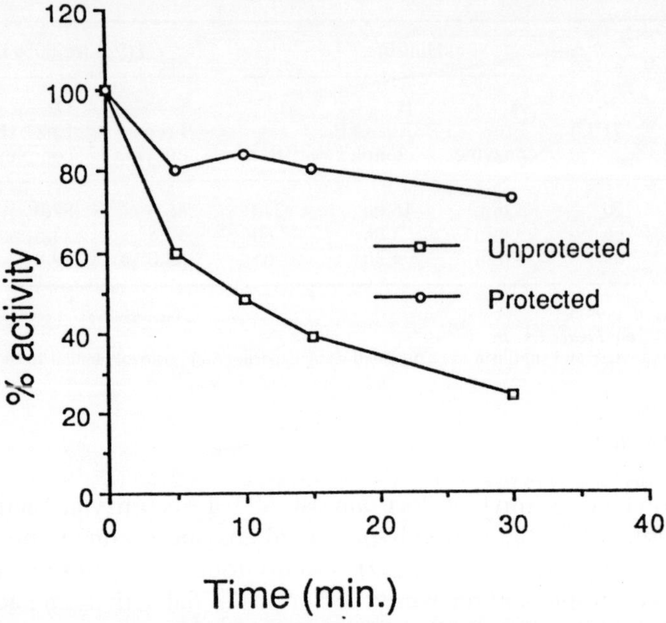

Figure 2. Time-dependent effect of oxidant (sodium hypochlorite: NaOCl). The residual enzyme activity, expressed as percent of the rate obtained for the corresponding untreated enzyme preparations, is plotted as a function of incubation time for the protected and unprotected enzyme. From Shami *et al.* (1991).

THE EFFECTS OF ETHANOL

Glucoamylase pre-incubated with 2.5% ethanol for 25 h retained only 10% of its original activity, whereas the antibody-protected enzyme retained 98% of its activity (*Figure 3*). The enzymatic conversion of sugars to ethanol is a well-known process. As the alcohol concentration rises in the system, some of the enzymes involved are subject to product inhibition by ethanol. In the brewing of beer, this phenomenon is actually exploited to terminate the reaction so that the level of ethanol does not rise above the required level. However, the production of industrial alcohol is geared towards the maximization of ethanol levels, so resistance to ethanol could be beneficial.

PROTECTION WITH ANTIBODIES AGAINST PROTEOLYTIC INACTIVATION

The clinical application of protein-based pharmaceuticals is rapidly increasing. For many, however, the prolonged maintenance of appropriate blood levels is a problem because of their relatively short *in vivo* half-life. Since the most common form of *in vivo* inactivation for many proteins is enzymatic proteolytic degradation (Tombs, 1985), any method of reducing proteolytic susceptibility might lead to reduced dosage, improved clinical efficacy and management, and reduced costs.

To a certain degree, proteolytic inactivation is related to inactivation by unfolding. In 'folded' proteins, the cleavages are limited in number because

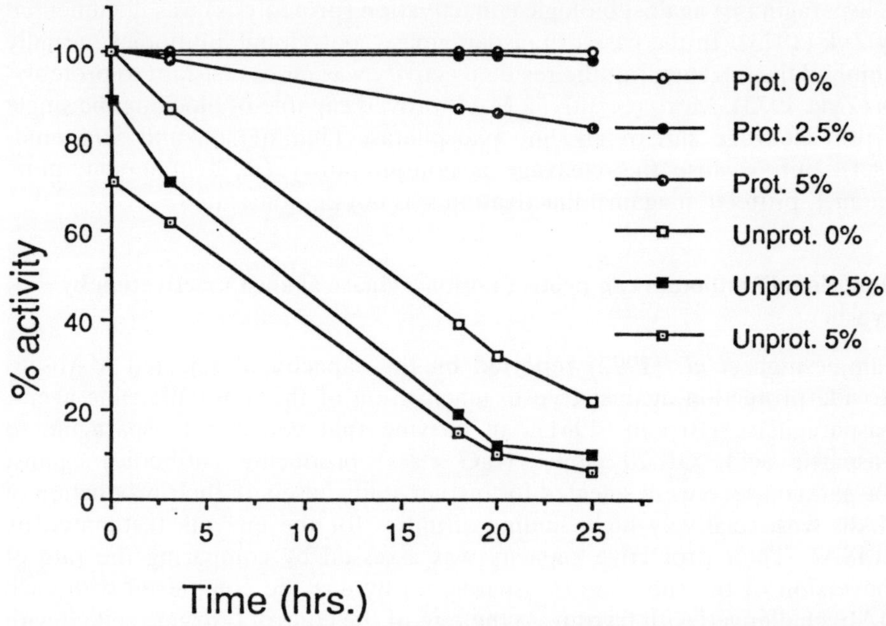

Figure 3. Time-dependent effect of ethanol on glucoamylase enzyme activity. The absorbance of the untreated protected (Prot.) and the untreated unprotected (Unprot.) samples at time zero were taken as 100% activity. The results for the time-dependent effect at 37°C of 0%, 2.5% and 5% ethanol on protected and unprotected glucoamylase are shown. From Shami (1991).

few or no proteolytic sites are exposed and such cleavages do not necessarily result in inactivation (Stoops *et al.*, 1978), due to the fact that catalytic domains are highly organized and more stable, whereas the peptide loops that link the domains are relatively unstructured and, therefore, more susceptible to digestion. In unfolded proteins, on the other hand, all potential cleavage sites are exposed, so that extensive degradation and loss of biological activity occurs (Mclendon and Radany, 1978). As to be expected from such considerations, a significant correlation between thermal stability and resistance to proteolytic inactivation has been noted. Thermostable enzymes from thermophilic bacteria are, for example, more resistant to proteolysis than are similar enzymes from mesophilic organisms (Daniel *et al.*, 1982). Enzymes with higher melting temperatures (T_m) are more resistant than those with a lower T_m (Mclendon and Radany, 1978; Daniel *et al.*, 1982). It can be inferred, therefore, that any process that confers thermal stability – as is the case with antibodies – might also confer some degree of protection against proteolytic inactivation.

One approach to stability which has achieved some success involves covalent conjugation of other macromolecules, such as polyethylene glycol (Abuchowski and Davis, 1981) or albumin (Poznansky *et al.*, 1982), to target proteins. The limitations, however, are that random covalent coupling in many cases results in reduced activity and that the overall stabilization is considerably lower than that achieved with the aid of specific antibodies.

The first use of antibodies to protect biologically active protein

(L-asparaginase) against biological inactivation (proteolysis) was documented by Zyk (1973). In the case of L-asparaginase, polyclonal antibodies partially inhibited the enzyme, but the residual activity was more resistant to proteolysis (Zyk, 1973). More recently, a MAB proved capable of blocking the single trypsin cleavage site of alkaline phosphatase (Jemmerson and Stigbrand, 1984), but because this cleavage is non-inhibitory (R. Jemmerson, pers. comm.), protection against inactivation was not demonstrated.

Monoclonal antibodies can protect L-asparaginase against inactivation by trypsin

Ramjeesingh *et al.* (1992) reported on the capacity of selected MAbs to provide protection against trypsin inactivation of the anti-leukaemic agent, L-asparaginase (Broom, 1961), an enzyme that converts L-asparagine to L-aspartic acid. Of 20 clones (IgG class) producing antibodies against L-asparaginase, 6 were selected for further study based on their production of MAbs with relatively high binding affinities for the enzyme (estimated by ELISA). Their protective capacity was assessed by comparing the rate of conversion of the substrate (L-asparagine) by enzyme complexed with each MAb, challenged with trypsin, to the rate of the control (enzyme spiked with equivalent amounts of non-immune human IgG, challenged with trypsin). The substrate consumption curves are shown in *Figure 4*. Neither non-immune human IgG (used as a control) nor any of the MAbs were inhibitory in the absence of trypsin. Trypsin treatment of the free enzyme reduced the substrate conversion rate to a minimal level (6% of the control), equivalent to 94% inactivation. Little or no protection against trypsin inactivation was afforded by the non-specific IgG or by three of the MAbs (Nos 19, 33 and 35). On the other hand, one MAb (No 12) was highly effective, preserving 72% of the control rate, and two others (Nos 29 and 34) provided intermediate protection (33 and 20%, respectively). In the experiment, substrate was depleted within 40 min (MAb No 12). During two additional cycles of substrate addition and depletion, MAb No 12 preserved the 72% level of activity, indicating that protection had been maintained for at least 120 min, despite the continuous presence of trypsin. The most effective MAb (No 12) was selected for further study. Its subclass type was determined as IgG2b with κ light chain.

TRYPSIN AND MAb CONCENTRATION

The protective action of MAb No 12 was influenced by the trypsin concentration and by the ratio of MAb to enzyme. Protection is defined as either the difference between, or the ratio of, the activity of MAb protected enzyme and the unprotected enzyme when challenged with trypsin. The highest protection (around 72% and a protection ratio of 12) was obtained with 37.5 units of trypsin per ml (*Figure 5*), using a ratio of MAb to enzyme tetramer of 10 : 1 (*Figure 6*). Increasing the ratio to 20 : 1 raised the overall activity of the protected sample, but provided no additional protection (as defined above)

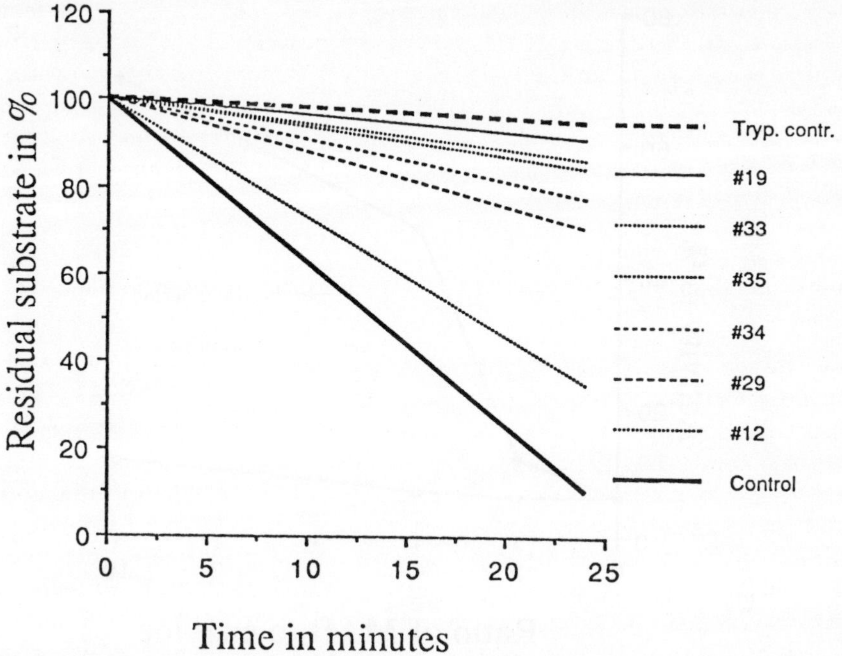

Figure 4. Screening for protective monoclonal antibodies. L-Asparaginase was protected with MAbs (Nos 19, 33, 35, 34, 29 and 12) and challenged with trypsin. Tryp. contr., free L-asparaginase spiked with equivalent amounts of non-immune human IgG and challenged with trypsin; Control, free L-asparaginase. Enzyme activity was determined as described in the experimental protocol. Residual substrate curves were generated by continuously monitoring the preparations at 197 nm (three readings per min per sample) and using the apparent initial rates obtained for residual substrate concentrations above 10% (100 μM) that corresponds in this case to 10 times the K_m value. From Ranjeesingh *et al.* (1992).

due to a similar increase in activity in the control sample (resulting from non-specific protection from high protein concentration).

STOICHIOMETRY OF THE MAb-ENZYME COMPLEX

Protection of enzyme activity implies that a MAb-enzyme complex has been formed. Direct evidence for complex formation was obtained by use of goat anti-mouse antibodies. Over 90% of asparaginase activity was precipitated, indicating that most of the enzyme tetramers were complexed with at least one MAb (MIgG). At the optimal 2.5 : 1 molar ratio of MAb: asparaginase monomer (*Figure 5*), 95% occupancy of binding sites would be predicted from the 2×10^{-8} M dissociation constant, K_d (determined by immunoprecipitation). Maximal protection, however, was only about 72%, which leads us to speculate that one monomer per tetramer was unprotected due to steric hindrance.

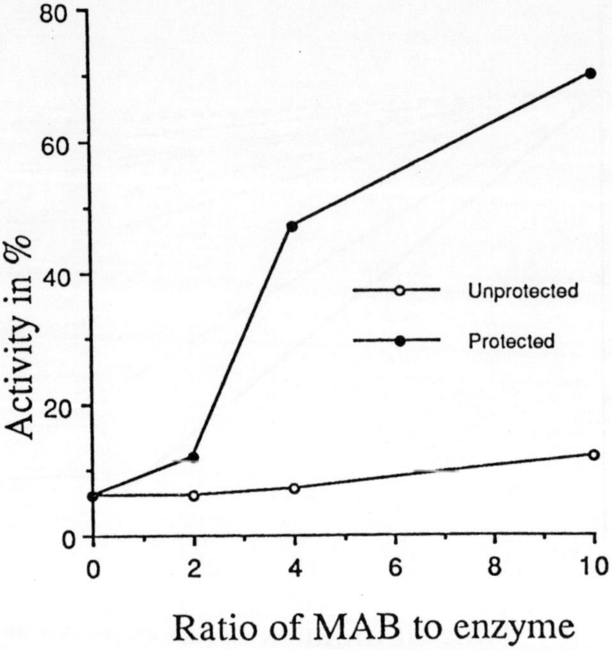

Figure 5. The protective effect of MAb No 12 as a function of trypsin concentration. Protected, L-asparaginase protected with MAb No 12; unprotected, L-asparaginase spiked with non-immune human IgG. Both preparations were challenged with increasing amounts of trypsin. Activity is expressed as percent of activity of the corresponding zero trypsin samples. From Ramjeesingh *et al*. (1992).

Mode of trypsin inactivation and location of its cleavage site

Based on SDS polyacrylamide gel electrophoresis, trypsin reduced the apparent molecular mass of asparaginase monomers by 3–4 kDa to about 31 kDa (*Figure* 7). On the other hand, enzyme protected by MAb No 12 was resistant to this cleavage. According to Maita, Morokuma and Matsuda (1974), *Escherichia coli* L-asparaginase II contains 321 amino acid residues per monomer (MW 34 080) in a sequence that was deduced from sequencing 27 non-overlapping peptides produced by trypsinization, and whose arrangement was deduced from cyanogen bromide peptides. A more recent sequence (Jenning and Beacham, 1990) based on molecular analysis of the isolated gene suggests 326 residues per monomer (MW 34 549).

To locate the inhibitory cleavage site, the major tryptic product (about 31 kDa) was recovered from the gel column and subjected to analysis. Its N-terminal sequence was Val-Gly-Val-Glu-Asn-Leu-. Only a single matching location was found, according to both published sequences (Maita, Morokuma and Matsuda, 1974; Jenning and Beacham, 1990) and the tryptic map. As expected, it is adjacent to lysine 29, since trypsin cleavages result in lysine or arginine C-terminal peptides. Thus the tryptic peptide must be cleaved from the N-terminal region and has a calculated molecular mass of 2647 Da rather than the 3–4 kDa estimated from the difference in mobility in SDS-gel (an inaccurate procedure) noted above. The role of the N-terminal domain of

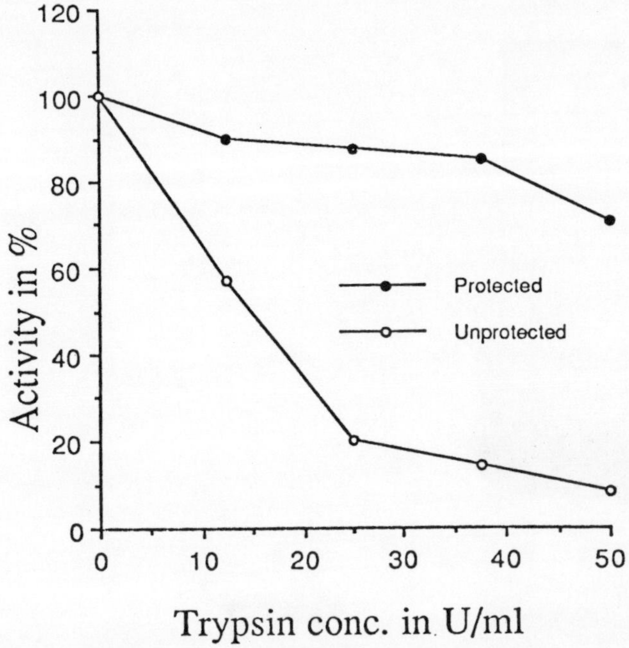

Figure 6. Protective effect of MAb no 12 against trypsin as a function of its concentration ratio with the enzyme. Protected and unprotected preparations were prepared, challenged with 37.5 U ml^{-1} trypsin and assayed for activity. Protected: L-asparaginase protected with increaasing amounts of MAb No 12; the ratio of MAb to enzyme is expressed as a molar ratio of MAb to tetramer enzyme. Unprotected: L-asparaginase spiked with equivalent amounts of human IgG. Activity is expressed as percent of rate of substrate consumption for the free enzyme, unchallenged with trypsin (shown in Fig. 4). From Ranjeesingh *et al.* (1992).

L-asparaginase, has been revealed by a recent study that determined its crystal structure (Swain *et al.*, 1993). This study proposed that this tetrameric enzyme is a dimer of identical intimate of dimers and the active sites, are located between residues of primarily the N-terminal and to a lesser extent residues of the C-terminal domains of monomers belonging to an intimate dimer. Thus, cleavage of the 2647 Da N-terminal peptide leads to deletion of some of the residues comprising the active site, resulting in inactivation of the enzyme and MAb No 12 protects activity by preventing the cleavage (*Figure* 7).

MAbs Nos 12 and 29, which both afford some protection, appear to bind to non-overlapping epitopes. MAb No 12 binds equally well to the intact enzyme monomer and the 31.9 kDa trypsin product (*Figure* 5). MAb No 29, on the other hand, binds to the intact enzyme monomer but not to the 31.9 kDa trypsin product. This could be interpreted as meaning that its epitope is largely located on the cleaved 2647 Da fragment. However, it is also possible that its epitope is actually on the 31.9 kDa fragment and the cleavage of 2647 Da by trypsin has disrupted the epitope so that it is no longer recognizable by the MAb.

This study demonstrates that a non-inhibitory monoclonal antibody can be selected which provides substantial and sustained protection against proteo-

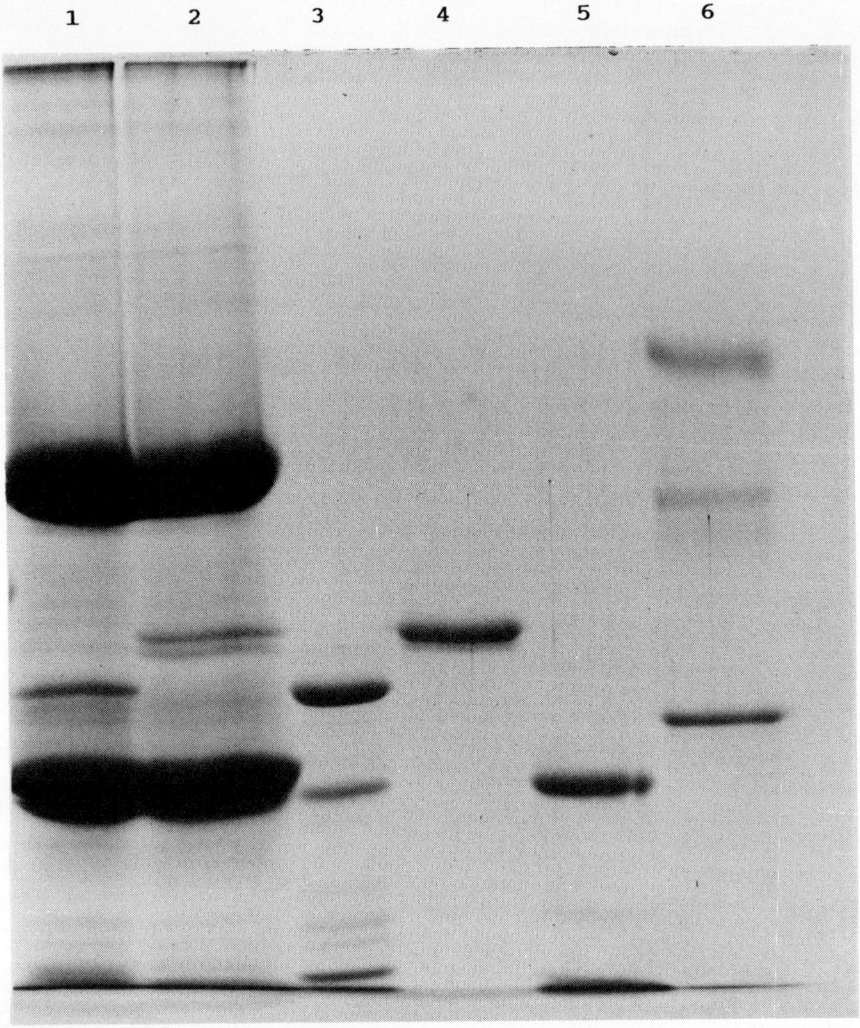

Figure 7. Comparison of protected and unprotected L-asparaginase by SDS-PAGE (10%). Effect of trypsin. Lane 1, L-asparaginase (5 μg) spiked with 10 : 1 molar ratio of non-immune human IgG, plus trypsin. Lane 2, L-asparaginase (5 μg) protected with 10 : 1 molar ration of anti-asparaginase MAB No 12, plus trypsin. Lane 3, free L-asparaginase (10 μg) treated with trypsin. Lane 4, free L-asparaginase (10 μg). Lane 5, trypsin (10 μg). Lane 6, molecular weight standards: BSA 66 kDa, ovalbumin 45 kDa, carbonic anhydrase 29 kDa. Trypsin-treated samples for lanes 1–3 were prepared as follows: 10 μg of L-asparaginase and 113 μg of non-immune human IgG (lane 1) or anti-asparaginase MAB No 12 (lane 2) were combined in 25 mM NaCl, 10 mM Tris pH 7.8 (final volume 100 μl; lane 3 sample just in buffer) and incubated over night at 4°C. The following day, samples for lanes 1–3 were treated with trypsin (20 μl, 20 μg, 200 U) for 5 min at 37°C, followed by addition of 20 μl of 'laemmli sample solubilizing solution', boiled for 5 min and concentrated by evaporation under nitrogen to a final volume of 40 μl, 20 μl (per lane) of which was used for SDS-PAGE (10%) according to Laemmli (1970). From Ramjeesingh *et al.* (1992).

lytic inactivation of L-asparaginase by trypsin. Of the six MAbs that were tested, none was inhibitory to the enzyme reaction. The most simple explanation is that none bind in the region of the active site of the enzyme in a fashion that blocks the access of the small size substrate (L-asparagine) to the active site. Polyclonal antibodies, which were reported by Zyk (1973) to partially inhibit, might hinder substrate interactions by binding to several epitopes near the active site.

Maximal protection was about 72% with a protection ratio of 12. On the other hand, at the optimal 2.5 : 1 ratio of MAb : asparaginase monomer, 95% occupancy of binding sites would be predicted from the 2×10^{-8} M dissociation constant (K_d). This discrepancy leads us to speculate that full occupancy (4 MAbs per tetramer) may be prevented by steric hindrance.

Other MAbs with similr binding affinity do not afford protection, indicating that binding *per se* is not sufficient. Presumably, in order to afford protection, a MAb must bind to a particular region of the protein's surface. However, more than one epitope may be associated with protection. For example, the most protective MAb (No 12) binds to the 31.9 kDa trypsin product, but No 29, which provides more modest protection, does not. At the most, only a small overlapping fraction of each MAb's binding area (estimated to be 26 × 19 Å: Sheriff *et al.*, 1987) can be involved in the protection action, or perhaps each MAb protects independently with no overlapping epitopes.

In vivo use of complexes comprising antibodies and biologically active proteins

Immunoassay is the most common application that utilizes monoclonal antibodies to form complexes with active proteins. While practically all the current 'approved' immunoassay applications are for *in vitro* diagnostics, there have been a number of studies concerning as yet unapproved *in vivo* diagnostic uses. However, since the purpose of this interaction (in immunoassay) is not the modulation of the biological activity of the target protein, this application will not be addressed here.

The ultimate utility of hormones, cytokines and clinically important enzymes is *in vivo*, in therapeutic applications. As mentioned earlier, their short half-life in circulation is a major drawback which hinders effective utilization. It is evident that, as proteins, their stability and activity can be modulated by complexing them with either specific polyclonal or monoclonal antibodies. The employment of these complexes *in vivo* has been studied by several groups.

Due to its antiviral, antiproliferative and immunomodulatory properties, interferon-αA was one of the first cytokines to be produced by recombinant DNA techniques. Pharmacokinetic studies conducted by Rosenblum *et al.* (1985) revealed that non-inhibitory anti-interferon monoclonal antibodies, when complexed with interferon prior to intravenous administration to rats, increased the half-life of interferon-αA three-fold (from 60 to 188 min). In this case, the monoclonal antibody was selected based on its capacity to interact with interferon, without neutralizing its activity. It is likely that reduced urinary excretion is due to the increased size of the complex (× 150 000) and is the main reason for the increase in half-life.

In another instance, 'Potentiation of the somatogenic and lactogenic activity of human growth hormone (hGH) monoclonal antibodies' was reported by Aston *et al.* (1986). They attributed this potentiation effect to the possibility that hGH in complex with a certain antibody may be capable of binding only to a particular class of hGH receptors, thereby enhancing the biological activity associated with these receptors. They carried out the experiments *in vivo*, both for the somatogenic assay (in Snell dwarf mice) and the lactogenic assay (Pigeon crop sac bioassay). They reported that only direct injection into the crop sac resulted in enhancement; injections at other locations were ineffective. This led them to conclude that systemic mechanisms such as prolonged half-life are not involved in the enhancement phenomenon.

Animal models for the therapeutic use of enzyme–antibody complexes have been used to test the effectiveness of both polyclonal and monoclonal antibodies. Surprisingly, Hadely, Lindmann and Sato (1987) found that L-gulunolactone oxidase (an enzyme that catalyses the final step in ascorbic acid – vitamin C biosynthesis), administered by the intraperitoneal route as a complex with either homologous or heterologous polyclonal antibodies to guinea pigs, doubled their survival time when maintained on an ascorbic acid deficient diet. The use of a polyclonal enzyme–antibody complex had no adverse reaction (with regard to allergic or toxic response), although in this study the enzyme–antibody complex was further fortified by covalently cross-linking it with gluteraldehyde. Other enzymes, including serum cholinesterase, L-asparaginase and histidase, were also tested in this study.

A more recent study by Dunn (1991) utilized a t-PA (tissue plasminogen activator)–anti-t-PA monoclonal antibody covalently linked complex to extend the half-life of t-PA in rabbits. However, as this study only monitored the presence of t-PA by ELISA (rather than t-PA enzymic activity), the reported extension in half-life may be irrelevant, as inactivated t-PA molecules are recognized as well by ELISA.

General features

Substantial protection of enzymes and other biologically active proteins by specific antibodies has been demonstrated against physical inactivation (high temperatures, freezing and thawing, and lyophilization), chemical inactivation (low, pH, oxidation with NaOCl and ethanol and biological inactivation (proteolysis). Considering the number of proteins that have been successfully protected by either polyclonal or monoclonal antibodies, it would appear likely that, in principle, protection can also be afforded to most other proteins. However, practical considerations restrict the use of antibodies to mainly non-industrial *in vitro* applications, such as stabilization of research and diagnostic reagents and stabilization of growth factors used in either research or large-scale production involving cell culture.

Barriers to the use of antibodies in industrial applications are mainly economic in nature. For example, industrial enzymes are inexpensive reagents and protecting them with antibodies that are four orders of magnitude more expensive is not practical, even if one can increase the half-life by two to three orders of magnitude. To exploit this protective phenomenon exhibited by antibodies, one would have to integrate via recombinant DNA techniques the antibody features that are responsible for the protection effect of the enzyme molecule, and to create a new modified enzyme that can be produced at the same cost as the native enzyme. This might be accomplished by using existing methodologies. For example, protective non-inhibitory monoclonal antibodies could be converted into single chain antibodies (see review by Wetzel, 1988). In this form, the DNA sequence encoding for thc antibody (variable regions) could be fused to the DNA encoding its antigen (the enzyme to be protected) for the construction of a fusion gene, similar to the one described by Chaudhary *et al.* (1989). This fusion gene, upon expression in *Escherichia coli* for example, might produce a chimeric protein comprised of two domains – the first an enzyme and the second a protective single chain antibody – that might provide stabilization by interacting with a vulnerable site on the enzyme.

In clinical applications, there is no economic barrier, as protein drugs are quite often more expensive to produce than antibodies. In clinical applications, it is not the utilization of murine monoclonal antibodies *per se* that is the problem (there are approved MAb drugs on the market), but the fact that in each protected preparation there are three different protein species (i.e. MAb-protein complex, free protein and free MAb), which is a major tumbling block in the process of drug approval. Again to overcome this barrier, the protective features of the antibody would have to integrate into the protein-drug via recombinant DNA, so that new more resistant drugs can be produced.

In principle, non-inhibitory MAbs should be selectable, which would provide protection against proteolysis of any protein-drug, potentially improving its efficacy. The protective MAb could be 'humanized' (Reichmann *et al.*, 1988), converted to a single chain antibody (Bird *et al.*, 1988) and its gene fused with that of the target protein. This would allow the production of a fusion protein containing, for example, a cytokine and a single chain antibody (Chaudhary *et al.*, 1989), connected via an appropriate spacer. The product might be biologically active, proteolysis-resistant, with low immunogenicity, economically producible by a single organism and suitable for clinical use.

References

ABUCHOWSKI, A. AND DAVIS, F.F. (1981). Soluble polymer–enzyme adducts. In *Enzymes as Drugs* (J.S. Holcenberg and J. Robert, Eds), pp. 367–383. John Wiley, New York.

AMIRAIAN, K. AND PLUMMER, T.H. JR (1971). Immunochemical studies on bovine carboxipeptidase A. *Journal of Immunology* **107**, 547–553.

ARNON, R. (1973). Immunochemistry of enzymes. In *The Antigens* (M. Sela, Ed.), Vol. 1, pp. 88–159 Academic Press, New York.

ASTON, R., HOLDER, A.T., PREECE, M.A. AND IVANYI, J. (1986). Potentiation of the somatogenic and lactogenic activity of human growth hormone with monoclonal antibodies. *Journal of Endocrinology* **110**, 381–388.

BARLOW, D.J., EDWARDS, M.S. AND THORNTON, J.M. (1986). Continuous and discontinuous protein antigenic determinants. *Nature* **332**, 747–748.

BEN-YOSEPH, Y., GEIGER, B. AND ARNON, R. (1975). Antibody-mediated thermal stabilization of human hexosaminidases. *Immunochemistry* **12**, 221–226.

BIRD, R.E., HARDMAN, K.D., JACOBSON, J.W., JOHNSON, S., KAUFMAN, B.N., LEE, S.M., LEE, T., POPE, S.H., RIORDAN, G.S. AND WHITLOW, M. (1988). Single-chain antigen binding proteins. *Science* **242**, 423–426.

BROOM, J.D. (1961). Evidence that the L-asparaginase activity of guinea pig serum is responsible for its antilymphoma effedts. *Nature* **191**, 1114–1115.

BURNETT, T.C. AND SCHIMDT, C.L.A. (1921). Immunological experiments with catalase. *Journal of Immunology* **6**, 255–262.

CAMPBELL, D.H. AND FOURT, L. (1939). Immunochemistry of catalase. *Journal of Biological Chemistry* **129**, 385–392.

CHAUDHARY, V.J., QUEEN, C., JUNGHANS, R.P., WALDMANN, T.A., FITZGERALD, D.J. AND PASTAN, I. (1989). A recombinant immunotoxin consisting of low antibody variable domains fused to *Pseudomonas* exotoxin. *Nature* **339**, 394–397.

CHOTHIA, C. AND JANIN, J. (1975). Principles of protein–protein recognition. *Nature* **256**, 705–708.

COHN, M. AND TORRIANI, A.M. (1952). Immunochemical studies with the β-galactosidase and structurally related proteins of *Escherichia coli. Journal of Immunology* **69**, 471–491.

COTTON, R.G., JENNINGS, I.G., CHOO, K.H. AND FOWLER, K. (1980). Isolation and characterization of a myeloma-spleen cell line hybrid producing antibody to phenylalanine hydrolase. *Biochemical Journal* **191**, 777–783.

CORDELL, J.L., FALINI, B., ERBER, W.N., GHOSH, A.K., ABDULAZIZ, Z., MACDONALD, S., FULFORD, K.A.F., STEIN, H. AND MASON, D.Y. (1984). Immunoenzymatic labeling of monoclonal antibodies complexes of alkaline phosphatase and monoclonal anti-alkaline phosphatase (APAAP). *Journal of Histochemistry and Cytochemistry* **32**, 219–229.

CREIGHTON, T.D. (1978). Experimental studies of protein folding and unfolding. *Progress in Biophysics and Molecular Biology* **33**, 231–297.

DANIEL, R.M., COWAN, D.R., MORGAN, H.W. AND CURRAN, M.P. (1982). A correlation between protein thermostability and resistance to proteolysis. *Biochemical Journal* **207**, 641–644.

DAVIES, D.R., SHERIFF, S. AND PALDAN, E.A. (1988). Antibody–antigen complexes. *Journal of Biological Chemistry* **263**, 10541–10544.

DONOVAN, J.W. AND BEARDSLEE, R.A. (1975). Heat stabilization by protein–protein association. *Journal of Biological Chemistry* **250**, 1966–1971.

DUNN, J.T. (1991). Therapeutic enzyme–antibody complexes. US Patent No 5,009,888.

FAMBROUGH, D.M., ENGEL, A.G. AND ROSENBERRY, T.L. (1982). Acetylcholinesterase of human erythrocytes and neuromuscular junctions: Homologies revealed by monoclonal antibodies. *Proceedings of the National Academy of Sciences, USA* **79**, 1078–1082.

FEINSTEIN, R.N., JAROSLOW, B.N., HOWARD, J.B. AND FAULHABER, J.T. (1971). Stabilization of mutant catalase by complex formation with antibody to normal catalase. *Journal of Immunology* **106**, 1316–1322.

FRACKELTON, A.R. JR AND ROTMAN, B. (1980). Functional diversity of antibodies elicited by bacterial β-D-galactosidase. *Journal of Biological Chemistry* **255**, 5286–5290.

HADELY, K.B., LINDMANN, D.A. AND SATO, P.H. (1987). Heterologous immunoprecipitates also have potential for therapeutic use. *Biotechnology and Applied Biochemistry* **9**, 1–11.

JEMMERSON, R. AND STIGBRAND, T. (1984). Monoclonal antibodies block trypsin cleavage site on human placental alkaline phosphatase. *FEBS Letters* **173**, 357–359.

JENNING, M. AND BEACHAM, I. (1990). Analysis of the *E. coli* gene encoding

L-asparaginase II ansB, and its regulation by cyclic AMP receptor and FNR proteins. *Journal of Bacteriology* **172**, 1491–1498.

KOHLER, G. AND MILSTEIN, C. (1975). Continuous cultures of fused cells secreting antibodies of predefined specificity. *Nature* **256**, 495–497.

LAEMMLI, U.K. (1970). Cleavage of structural proteins during the assembly of the head of the bacteriophage T4. *Nature* **227**, 680–685.

MAITA, T., MOROKUMA, K. AND MATSUDA, G. (1974). Amino acid sequence of L-asparaginase from *E. coli*. *Journal of Biochemistry* **76**, 1351–1354.

MATHER, I.H., NACE, C.S., JOHNSON, V.G. AND GODLSBY, L.H. (1980). Preparation of monoclonal antibodies to xanthine oxidase and other protein of bovine milk-fat-globule membrane. *Biochemical Journal* **188**, 925–928.

MCLENDON, G. AND RADANY, E. (1978). Is protein turnover thermo-dynamically controlled? *Journal of Biological Chemistry* **253**, 6335–6337.

MELCHERS, F. AND MESSERS, W. (1970). Enhanced stability against heat denaturation of *E. coli* wild type and mutant β-galactosidase in the presence of specific antibodies. *Biochemical and Biophysical Research Communications* **40**, 570–575.

MESSER, W. AND MELCHERS, F. (1969). The activation of mutant β-galactosidase by specific antibodies. In *The Lactose Operon* (J.R. Beckwith and D. Zipser, Eds), pp. 305–315. Cold Spring Harbor Monograph.

MICHAELI, D., PINTO, J.D., BENJAMINI, E. AND DE BUREN, F.P. (1969). Immunoenzymology of acetylcholinesterase-I. *Immunochemistry* **6**, 101–109.

MOZAHEV, V.V. AND MARTINEK, K. (1984). Structure–stability relationships in proteins: New approaches to stabilizing proteins. *Enzyme and Microbial Technology* **6**, 50–59.

NAKAMURA, S., OHTA, S., ARAI, K., OSHIMA, T. AND KAJIRO, K. (1978). Studies on polypeptide-chain-elongation factors from an extreme thermophile, *Thermus thermophilus* HB8. *European Journal of Biochemistry* **92**, 533–543.

OSHIMA, T. (1982). A pentaamine is present in an extreme thermophile. *Journal of Biological Chemistry* **257**, 9913–9914.

PARK, S.S., PERSSON, A.V., COON, M.J. AND GELBOIN, H.V. (1980). Monoclonal antibodies to rabbit liver cytochrome P-450 LM2. *FEBS Letters* **116**, 231–235.

POZNANSKY, M.J., SHANDLING, M., SALKIE, M.A., ELLIOT, J. AND LAU, E. (1982). Advantages in the use of L-asparaginase–albumin polymer as an antitumor agent. *Cancer Research* **42**, 1020–1025.

PRASAD, A.R. AND MAHESHWARI, R. (1978). Purification and properties of trehalase from thermophilic fungus *Humicola lanuginosa*. *Biochimica Biophysica Acta* **525**, 162–170.

RAMJEESINGH, M., ZYWULKO, M., ROTHSTEIN, A. AND SHAMI, E.Y. (1992). Monoclonal antibodies can protect L-asparaginase against inactivation by trypsin. *Bio/Technology* **10**, 442–445.

REES, A.R., ROBERTS, S., WEBSTER, D. AND CHEETHAM, J.C. (1988). Modifying antibody affinity and specificity. *ICSU Short Reports* **8**, 172–173.

REICHMANN, L., CLARK, M., WALDMAN, H. AND WINTER, G. (1988). Reshaping human antibodies for therapy. *Nature* **332**, 323–327.

ROSENBLUM, M.G., UNGAR, B.W., GUTTERMAM, J.U., HERSH, E.M., DAVID, G.S. AND FRINCKE, J.M. (1985). Modification of human leukocyte interferon pharmacology with a monoclonal antibody. *Cancer Research* **45**, 2421–2424.

ROSS, M.E., REIS, D.J. AND JOH, T.H. (1981). Monoclonal antibodies to tyrosine hydroxylase: Production and characterization. *Brain Research* **208**, 493–498.

ROTMAN, M.B. AND CELADA, F. (1968). Antibody-mediated activation of a defective β-D-galactosidase extracted from *Escherichia coli* mutant. *Proceedings of the National Academy of Sciences, USA* **60**, 660–667.

RÜHLMANN, A., KULKA, D., SCHWAGER, P., BARTELS, K. AND HUBER, K. (1973). Structure of the complex formed by bovine trypsin and bovine pancreatic trypsin inhibitor. *Journal of Molecular Biology* **77**, 417–436.

SHAMI, E.Y., ROTHSTEIN, A. AND RAMJEESINGH, M. (1988). Stabilization of biologically active proteins. *Trends in Biotechnology* **7**, 186–190.

SHAMI, E.Y., ROTHSTEIN, A. AND RAMJEESINGH, M. (1989). Use of antibody/antigen interactions to protect or modulate biological activity. European Patent Office Publication No EP 0 298 654 A2.

SHAMI, E.Y., ROTHSTEIN, A., RAMJEESINGH, A. AND ZYWULKO, M. (1991). Stabilization of enzymes by their specific antibodies. *Enzyme and Microbial Technology* **13**, 424–429.

SHERIFF, S., SILVERTON, E.W., PALDAN, E.A., COHEN, G.H., SMITH-GILL, S.J., FINZEL, B.C. AND DAVIES, D.R. (1987). Three-dimensional structure of an antibody-antigen complex. *Proceedings of the National Academy of Sciences, USA* **84**, 8075–8079.

SOLOMON, B., MOAV, N., PINES, G. AND KATCHALSKI-KATZIR, E. (1984). Interaction of carboxipeptidase-A with monoclonal antibodies. *Molecular Immunology* **21**, 1–11.

STERNBERGER, I.A., HARDY, P.H. JR, CUCULIS, J.J. AND MEYER, H.G. (1970). The unlabeled antibody enzyme method of immunohistochemistry. *Journal of Histochemistry and Cytochemistry* **18**, 315–333.

STOOPS, J.K., AWAD, E.S., ARSLANIAN, M.J., GUNSNBERG, S. AND WAKIL, S.J. (1978). Studies on yeast fatty acid synthetase. *Journal of Biological Chemistry* **253**, 4464–4475.

SWAIN, A.L., JASKOLSKI, M., HOUSSET, D., MOHANA RAO, J.K. AND WLODAWER, A. (1993). Crystal structure of *E. coli* L-asparaginase, an enzyme used in cancer therapy. *Proceedings of the National Academy of Science USA* **90**, 1474–1478.

SWEADNER, K.J. (1991). Trypsin inhibitor paradoxically stabilizes trypsin activity in sodium dodecyl sulfate, facilitating proteolytic fingerprinting. *Analytical Biochemistry* **194**, 130–135.

SWEET, R.M., WRIGHT, H.T., JANIN, J., CHOTHIA, C.H. AND BLOW, D.M. (1974). Crystal structure of the complex of porcine trypsin with soybean trypsin inhibitor (Kunitz) at 2.6-Å resolution. *Biochemistry* **13**, 4212–4228.

TAKAHASHI, K. AND STURTEVANT, J.M. (1981). Thermal denaturation of *Streptomyces* subtilisin inhibitor, subtilisin BPN′, and the inhibitor–subtilisin complex. *Biochemistry* **20**, 6185–6190.

TANFORD, C. (1970). Protein denaturation, Part C. Theoretical models for the mechanism of denaturation. *Advances in Protein Chemistry* **24**, 1–95.

TOMBS, M.P. (1985). Stability of enzymes. *Journal of Applied Biochemistry* **7**, 3–24.

TRIA, E. (1939). Anticatalase. *Journal of Biological Chemistry* **129**, 377–383.

WEBSTER, R.G., HINSHAW, V.S. AND LAVER, W.G. (1982). Selection and analysis of antigenic variants of the neuraminidase of N2 influenza viruses with monoclonal antibodies. *Virology* **117**, 93–104.

WETZEL, R. (1988). Active immunoglobulin fragments synthesized in *E. coli*: From Fab to scantibodies. *Protein Engineering* **2**, 169–170.

ZYK, N. (1973). Modification of L-asparaginase EC-2 by homologous antibodies. *Biochemica Biophysica Acta* **302**, 420–428.

10

Analytical Ultracentrifugation and the Genetic Engineering of Macromolecules

STEPHEN E. HARDING

National Centre for Macromolecular Hydrodynamics, School of Agriculture, University of Nottingham, Sutton Bonington, UK

Introduction

There has developed an interesting 'cause and demand' hierarchy in molecular biology. The great advances in nucleic acid technology produced first of all a vast variety of recombinant and modified 'mutant' macromolecules, either directly (proteins) or indirectly (such as glycoproteins). It was then realized that these needed to be understood in terms of their chemical nature: protein chemists suddenly found they, too, were involved in this molecular biology revolution. Then it was realized that the physical properties needed understanding too. The stage was set for the physical chemist.

Physical chemistry, or 'biophysics', offers a wide variety of techniques for characterizing both the genetically engineered proteins (and derivatives thereof) and also those molecules used as the tools for their production, namely nucleic acids and nucleic acid binding proteins. These techniques can be classed as 'hydrodynamic' (e.g. viscometry, sedimentation velocity and sedimentation equilibrium in the analytical ultracentrifuge, gel permeation chromatography and gel electrophoresis), which give primarily overall size and shape information, 'scattering' (static and dynamic light scattering and X-ray scattering), 'spectroscopic' (absorption, fluorescence, NMR and ESR) and, finally, 'imaging' techniques (electron microscopy – including image reconstruction – optical or X-ray diffraction by fibres and X-ray crystallography), which can give low- and high-resolution structure of macromolecules.

Abbreviations: AMP, adenosine 5′-phospate; ATCase, aspartate transcarbamyolase; ATP, adenosine triphosphate; ATPase, adenosine triphosphatase; CAM, cell adhesion molecule; cDNA, circular DNA; CMC, *N*-cyclohexyl-*n′*-β-(4-methylmorpholinium) ethyl carbodiimide-*p*-toluene sulphonate; CSF, colony stimulating factor; CTP, cytidine triphosphate; EcoSSB, *E. coli* single-strand protein; ESR, electron spin resonance; GC, guanine cytosine; IFN-γ, natural human interferon; mRNA, messenger RNA; NANA, *N*-acetylneuraminic acid; NMR, nuclear magnetic resonance; PDGF, platelet-derived growth factor; RTP, replication terminator protein; SCF, stem cell factor; SDS-PAGE, sodium dodecyl sulphate polyacrylamide gel electrophoresis; TF, tissue factor; T_m, melting temperature (DNA, RNA); TNF, tumour necrosis factor; tRNA, transfer RNA; UV, ultraviolet.

Biotechnology and Genetic Engineering Reviews — Vol. 11, December 1993
0264–8725/93/11/317–356 $20.00 + $0.00 © Intercept Ltd, P.O. Box 716, Andover, Hampshire SP10 1YG, UK

Table 1. Techniques for characterizing the physical properties of macromolecules

Technique	Type of macromolecule examined	Minimum amount of material required[a]	Information obtained
Hydrodynamic			
Viscometry	All	10 mg	Intrinsic viscosity → shape and molecular weight
Sedimentation velocity	All	100 μg–1 mg	1. Homogeneity 2. Sedimentation coefficient → shape and molecular weight[a] 3. Interaction studies
Sedimentation equilibrium	All	100 μg–1 mg	1. Molecular weight 2. Interaction studies
Analytical density gradient sedimentation	All		1. Purity, homogeneity 2. Composition (including G-C fraction for nucleic acids)
Diffusion analysis (in the analytical ultracentrifuge)	Reasonably monodisperse samples		1. Diffusion coefficient 2. Interface transport phenomena
Calibrated gel permeation chromatography	Proteins ($M<10^6$)	10 μg–1 mg	1. Homogeneity 2. Apparent molecular weight
SDS-PAGE	Proteins and nucleic acids	10μg	1. Homogeneity/subunit composition 2. Apparent molecular weight of polypeptide/polynucleotide
Scattering (from solutions)			
Static light scattering	All ($M>50\ 000$)	10 mg–1 g	1. Molecular weight 2. Rg → shape
X-ray scattering	All	100 mg–1 g	Rg → shape
Dynamic light scattering	All ($M>50\ 000$)	10 mg–1 g	Translational diffusion coefficient → shape and molecular weight[a]
Spectroscopic			
Absorption spectroscopy	1. Protein	100 μg	Concentration, ligand-induced conformational changes
	2. Nucleic acid		% G-C content (from T_m); % helical content (if composition known)
Circular dichroism	Proteins and nucleic acids		Secondary structure
Fluorescence	Macromolecules with natural or synthetic chromophore	μg–mg quantities	Concentration, ligand-induced conformational changes and (from depolarization studies) rotational diffusion
Nuclear magnetic resonance	1. Small proteins and nucleic acids 2. tRNA; globular proteins; membranes 3. Polysaccharides	100 mg+(depends on size of macromolecule)	1. Parts of 3-D structure in atomic detail 2. Ligand-induced conformation changes 3. Water relations
Electron spin resonance			Ligand-induced conformational changes
Mass spectrometry	Molecules <150 000 mol. wt.		Precise molecular weight of covalently intact species

Table 1. cont.

Technique	Type of macromolecule examined	Minimum amount of material required[a]	Information obtained
Imaging			
Image reconstruction from electron microscopy	Large macromolecules (e.g. polysaccharides, mucins), macromolecular assemblies, membrane proteins in 2-D array	μg–mg	Clarification of EM images, low-resolution structure, polydispersity
Optical or X-ray diffraction by fibres	Proteins, polysaccharides, DNA	Grams	Distance between repeat units
X-ray crystallography	Crystalline globular proteins and tRNAs	Grams	Electron density map, complete tertiary structure (provided sequence known)

[a] Ballpark figures only. Adapted from Furth and Moore (1986).

Table 1 summarizes the information available and the amount of material required. The interested reader can find a very useful summary of all these techniques in Furth and Moore (1986). Progressively more detailed considerations can be found in van Holde (1985), Cantor and Schimmel (1980) and the now dated, but still regarded by many as the authoritative, book of Tanford (1961).

From *Table 1* it is quite clear that the four analytical ultracentrifuge techniques – sedimentation velocity, sedimentation equilibrium, analytical isopycnic density gradient analysis and diffusion analysis – represent only a small part of the armoury of techniques available. Despite this, there are many features which make them highly useful for characterizing macromolecules, particularly engineered ones and assemblies thereof. Compared with other techniques, they are generally:

- relatively rapid;
- non-invasive;
- do not require vast amounts of material (as is necessary with, for example, NMR and X-ray crystallography);
- give size, structural (including subunit), heterogeneity and interaction information – and in some cases, in surprising detail.

These reasons make it worthy of serious consideration by the molecular biologist. It is no surprise that the demand for techniques to characterize 'gene products' has coincided with a revival of interest in analytical ultracentrifugation. It is worth stressing, however, that as with any 'low-resolution' technique like this, it is at its most powerful when used in conjunction with other complementary techniques, particularly solution X-ray scattering (conformation analysis), dynamic light scattering (size analysis), gel permeation chromatography (size distribution analysis) and gel electrophoresis (subunit composition and homogeneity analysis).

What sort of information about the macromolecules concerned with, and the products of, genetic engineering can we get from analytical ultracentrifugation? It depends on the type of ultracentrifuge technique we apply, which are all possible with the same sort of instrumentation. *Sedimentation velocity* (see, e.g. Harding, 1993a) can provide us with information on the sample homogeneity, shape information and also interaction information (in the case of ligand binding, assaying for what we call co-sedimentation phenomena). At lower rotor speeds, *sedimentation equilibrium* (see, e.g. Harding, 1993b) can provide absolute (i.e. not requiring calibration using standards) molecular weight and subunit composition information, together with association constants for a self-associating system. *Analytical isopynic density gradient equilibrium* (see, e.g. Creeth and Horton, 1977) can provide us with information on the purity and composition of a sample: the classic experiments of Meselson and Stahl (1958), in which DNA replication was shown to be semi-conservative, were performed using this technique. The analytical ultracentrifuge – or at least its optical system – can also be used as a powerful tool for *diffusion analysis* (see, e.g. Harding and Tombs, 1989). Although dynamic light scattering is now the method of choice for measuring translational diffusion coefficients, the optical system on the analytical ultracentrifuge is proving very useful for investigating the diffusion of molecules through matrices and towards and through interfaces between two phase systems (providing a good way of modelling membrane-based processes). A volume in which the state of the art of these various sedimentation techniques can be obtained has recently appeared (Harding, Rowe and Horton, 1992).

The technique is not new – the principles and general shape of the equipment have remained relatively unchanged since the inception of the technique by Svedberg and co-workers in the 1920s (much of this early work is reviewed in Rånby, 1987). Without doubt the great boom period was between 1950–70, when almost every biochemistry department was in possession of one of these instruments, most usually the famous Model E from Beckman Instruments (Palo Alto, CA, USA). These instruments were used mostly for routine sedimentation coefficient and molecular weight measurements. By the end of the 1970s, biochemists and molecular biologists were not content with such information (approximate molecular weights could be obtained anyway by using the much simpler and less expensive techniques of gel permeation chromatography and gel electrophoresis), and crystallography and nuclear magnetic resonance (NMR) had become the main focus of attention. By the mid-1980s, it was clear to many that NMR – at least in high-resolution mode – and crystallography could only be applied to a relatively limited number of biological macromolecules, and that techniques like gel filtration and gel electrophoresis were nowhere near as reliable for getting precise molecular weights, something that was essential for evaluating the subunit composition of assemblies of macromolecules. The need for a relatively rapid method for investigating the sizes, shapes [in terms of simple ellipsoids, with either two *or three* unequal axes (Harding, 1989), or sophisticated bead models (Garcia de la Torre, 1989)], subunit composition and interaction properties became acute with the appearance of this glut of newly engineered molecules. The final *coup de grâce* was the launch in 1991 of a new computer-interfaced analytical ultracentrifuge. A good precis of this history has been given by Schachman (1992). Much of the published work using the analytical ultracentrifuge over the last few years – which this chapter attempts to review – has been on molecules concerned with genetic engineering, and using this and other more 'seasoned' analytical ultracentrifuges. It is appropriate to comment here that the author was stunned to discover so many references: a clear manifestation of the re-emergence of sedimentation methods in modern molecular biology.

Physical principles of analytical ultracentrifugation

An analytical ultracentrifuge is simply an ultracentrifuge with an appropriate optical system for observing and recording solute distributions both during and at the end of a sedimentation process (see, e.g. van Holde, 1985). A typical analytical ultracentrifuge cell contains one or two sector shape channels which can take up to 0.8 ml of solution or reference solvent. The amounts required depend on the type of ultracentrifuge experiment and the information required.

SEDIMENTATION VELOCITY: HOMOGENEITY, CONFORMATION AND INTERACTION ANALYSIS

In a *sedimentation velocity* experiment (see, e.g. Harding, 1993a), the rotor speed is sufficiently high to sediment the macromolecular species. If the macromolecule is a protein or nucleic acid, one can detect the position of the sedimentating boundary using absorption optics at the appropriate wavelength. Concentrations as low as 0.1 mg ml^{-1} can be routinely used, depending on the path length of the cell. Even lower concentrations are possible if a fluorescent chromophore is used (see, e.g. Schmidt and Riesner, 1990, 1992). For these, and macromolecules such as polysaccharides without sufficient chromophore, we can also use one of two types of optical system based on the refractive index: Rayleigh interference optics (this can also be extremely sensitive; see, e.g. Stafford, 1992) and Schlieren (viz. 'refractive index gradient') systems. The latter requires concentrations in excess of 1 mg ml^{-1}, but gives the classical 'Schlieren peak', which most biochemists associate with the analytical ultracentrifuge and, indeed, which has adorned the literature for the last 50 years. The sedimenting boundary is recorded on photographic film and then digitized or recorded via photomultipliers onto chart paper (or as analague output directly into a computer). The presence or not of more than one sedimenting boundary has been one of the classical ways of demonstrating homogeneity or heterogeneity (see, e.g. Morgan *et al.*, 1989, for the Pf1 gene 5 protein system).

The rate of movement of the sedimenting boundary per unit centrifugal field gives the sedimentation coefficient s, which is one of our shape parameters. By correcting this using simple formulae and standard conditions (water as solvent at 20°C) and extrapolating this (or the reciprocal thereof) to zero concentration, one can get from the intercept the corrected sedimentation coefficient, $s^{o}_{20,w}$ (unit = seconds or Svedbergs, S, where 1 S = 10^{-13} s), and from the slope we can get the sedimentation concentration regression coefficient, k_s (unit = ml g^{-1}), and then from *both parameters* our shape information.

For conformation analysis there are three lines of attack. If the macromolecule is fairly rigid, one can combine the sedimentation coefficient with other techniques such as intrinsic viscosity, rotational diffusion (from fluorescence depolarization or electric birefringence measurements) or the radius of gyration (from 'static' light scattering or low angle X-ray scattering) to model the conformation in terms of simple ellipsoids of revolution, general triaxial ellipsoids (see, e.g. Harding, 1989) or sophisticated but beautiful bead models (see, e.g. Garcia de la Torre, 1989), and these approaches have been paticularly successful for the study of proteins, including antibodies and complement (see, e.g. Perkins, 1989). For more extended and more flexible macromolecules like nucleic acids and polysaccharides, one can model the conformation in terms of more general shapes using the 'Wales/van Holde' ratio of k_s to the intrinsic viscosity [η], or the 'Mark-Houwink-Kuhn-Sakurada' b coefficient from the relation between $s^{o}_{20,w}$ and the molecular weight, *M* (similar coefficients exist for the intrinsic viscosity, the diffusion coefficient, and the radius of gyration with *M*; see, e.g. Harding *et al.*, 1991), to permit the modelling of the conformation in terms of general shapes, between the three extremes of compact sphere, rigid rod and random coil. A useful general construction for representing this type of modelling is the 'Haug triangle' (Smidsrød and Andresen, 1979; see also Harding *et al.*, 1991). We can use similar relations of $s^{o}_{20,w}$ etc. *vs M* to model the flexibility of the molecule in terms of the ratio of the contour length (*L*) to the persistence length (*a*) (Freire and Garcia de la Torre, 1992).

Sedimentation velocity is also a powerful probe into interaction phenomena. For self-associating systems, the dependence of the sedimentation coefficient $s^{o}_{20,w}$ on concentration can be used to estimate the stoichiometry and strength of the association (a good example is for tubulin; see Weisenberg, Borisy and Taylor, 1968), although sedimentation equilibrium is a more absolute probe into these phenomena. It also provides a quick and easy-to-use assay for interactions in a mixed solute system (including ligand binding) using the principle of co-sedimentation; a recent example of this, for the interaction of methylmalonyl mutase with its B12 cofactor, is given by Marsh and Harding (1993).

The sedimentation coefficient can also be used to give an absolute measure of molecular weight (by 'absolute' we mean not requiring calibration standards as are necessary with SDS-PAGE – which gives directly only subunit molecular weight of proteins – or gel permeation chromatography), but only when combined with the translational diffusion coefficient, which is now usually measured using dynamic light scattering (see, Pusey, 1989; Bloomfield, 1981). A more direct way is to use a different analytical ultracentrifuge technique known as *sedimentation equilibrium*.

SEDIMENTATION EQUILIBRIUM: MOLECULAR WEIGHT AND SUBUNIT STRUCTURE ANALYSIS

Whereas in a sedimentation velocity experiment at relatively high rotor speeds (for a polysaccharide, say 40 000–50 000 rev min^{-1}) the sedimentation rate and hence sedimentation coefficient depends on the size and shape of the molecule, at much lower speeds (say 20 000 rev min^{-1} or less) in a sedimentation *equilibrium* experiment (see, e.g. Harding, 1993b) the forces of sedimentation and diffusion on the macromolecule become comparable, and instead of getting a sedimenting boundary one obtains a steady-state equilibrium distribution of macromolecule with a low concentration at the meniscus building up to a high concentration at the cell base. This final steady-stage pattern is a function *only* of molecular weight and related parameters (virial coefficients and association constants where appropriate) and not of molecular shape (apart from entering into non-ideality considerations, again when appropriate), since at equilibrium there is no further movement of the macromolecule. Hence frictional effects through shape variation do not come into play and we have an absolute way of getting (weight-average) molecular weights.

Solute concentration distributions at sedimentation equilibrium are recorded most accurately using Rayleigh interference optics. These can be read directly on-line into a computer (Laue, 1992) or 'off-line', i.e. recorded on photographic film and then digitized into a computer (Rowe *et al.*, 1992), using, for example, a laser densitometer of the type found in many biochemistry departments used for scanning electrophoresis gels. Alternatively, the absorption optics can be used; although less precise, these can be much more convenient to use for proteins and nucleic acids. There are various ways of processing the concentration versus distance data.

1. Molecular weight or 'molar mass' (weight average) of all the macromolecular contents in the ultracentrifuge cell: M°_{w}. From routines (such as 'MSTAR': see Harding, Horton and Morgan, 1992) *almost* equivalent to getting the average slope (it is not *quite* as easy as that) of plots of log concentration against distance from the centre of the rotor squared, one can get the apparent weight average molecular weight, $M_{w,app}$, which after an extrapolation (of $M_{w,app}$ or $1/M_{w,app}$) to zero concentration (to remove non-ideality problems) gives us the molecular weight (this will be a weight average for a heterogeneous system). Number and *z*-averages can also be obtained if the data are good enough. For most proteins and nucleic acids at low concentration (0.5 mg ml^{-1} and less), non-ideality corrections can be quite small and a concentration extrapolation is often not necessary. The molecular weight so found will be of the intact molecule (not, like SDS-PAGE or mass spectrometry, of the subunits) (see, e.g. Marsh *et al.*, 1989). For multi-subunit systems where the subunits are of comparable size, by use of appropriate dissociative solvents (such as 6 M GuHCl) it is possible also to obtain the subunit molecular weight, hence the subunit composition. When the subunits are not equal in size, some combination of sedimentation equilibrium, SDS-PAGE (for proteins) and possibly sedimentation velocity is

necessary to give similar information. It is worth mentioning that although the favoured unit of molecular weight or molar mass is g mol^{-1}, since most molecular biologists think in terms of Da or kDa, we will tend to use this unit here.

2. Point average molecular weights (usually weight averages). These can be obtained from local slopes of the log concentration versus distance squared data, and are quite useful for obtaining the stoichiometry and association constants for self-associating systems. A good example has been given by Tang and Adams (1973), who used this procedure to monitor the effect of temperature on the dissociation equilibria of beta-lactoglobulin dimers. Again, point number and point *z*-averages can also be obtained if the data are good enough (see below).

3. Omega function analysis. Association constants can also be obtained by modelling the concentration versus distance squared data directly (see, e.g. Ralston and Morris, 1992).

4. Molecular weight distribution analysis. This is useful for polydisperse macromolecules, like nucleic acids and polysaccharides. There are four principal ways of obtaining distributions (see, e.g. Harding, in press), the most useful being 'Method IV', i.e. a combined approach with gel permeation chromatography.

5. Number and *z*-average molecular weights. These are useful for representing heterogeneous systems, whether they be self-associative, complex forming or polydisperse. If conventional absorption or Rayleigh interference optics are used, the weight average molecular weight is the principal average or 'moment', but it is in principle possible to obtain number average molecular weights [but only for those systems where the so-called 'meniscus depletion' method (Yphantis, 1964) can be employed, a method applicable only to a rather limited number of macromolecular systems) and also *z*-average molecular weights, but only after a noise-sensitive double differentiation of the basic concentration versus distance record. However, if the Schlieren optical system is used, *z*-averages can be obtained directly, and these optics can be as sensitive as the Rayleigh optical system (Rowe *et al.*, 1992).

ANALYTICAL ISOPYCNIC DENSITY GRADIENT ULTRACENTRIFUGATION

This is really a modified sedimentation equilibrium method (see, e.g. Creeth and Horton, 1977), the modification being a heavy salt – usually CsCl (also CsBr, Cs_2SO_4 and LiBr) is used at the appropriate molar concentration as solvent. At sedimentation equilibrium, the salt will redistribute, giving an equilibrium density gradient between the meniscus and cell base. This gradient will be a function of the salt, the rotor speed and the solution column length. A suspended macromolecule will at equilibrium 'band' at the isodensity point, which will either be at the meniscus, the base, or, if the density gradient has been appropriately chosen, somewhere between these two extremes. For example, the buoyant densities of proteins and polysaccharides are ~1.3 g ml^{-1} and ~1.6 mg ml^{-1}, respectively. The buoyant density of nucleic acids is ~ 1.7 g ml^{-1}, depending on the GC content.

This method has been invaluable for assaying the purity of macromolecular preparatons, the extent of glycosylation of proteins and the base content (see, e.g. Schildkraut, Marmur and Doty, 1962) and extent of ligand binding to nucleic acids, and indeed was the classical method used to show that the replication of DNA is semi-conservative (Meselson, Stahl and Vinograd, 1957; Meselson and Stahl, 1958).

DIFFUSION ANALYSIS

Athough still used occasionally, the days of boundary spreading in the ultracentrifuge being used *on a routine basis* for the measurement of translational diffusion coefficients (see, e.g. Tanford, 1961) have long since gone, being replaced by far faster and *usually* more accurate dynamic light-scattering methods (see, e.g. Pusey, 1989). However, the optical systems on the analytical ultracentrifuge, coupled with the stabilizing effect (against convection phenomena) that a small centrifuge field gives (run at ~2000 rev min^{-1}) make it ideal for monitoring the diffusion of small proteins or other molecules through matrices (polysaccharide or other; see, e.g. Comper and Preston, 1992), or for monitoring the diffusion of proteins towards and through interfaces separating two or more incompatible (aqueous or non-aqueous) phase systems (Tombs and Harding, 1988; Harding and Tombs, 1989). The significance of this for monitoring membrane-based and commercial processes has been recognized (involving, for example, lipases), although no work has yet been done on the performance of *engineered* proteins of this type.

THE MACROMOLECULES REVIEWED HERE

Although the main focus of this review is on the engineered macromolecules themselves – largely protein and glycoprotein – it would be instructive to first briefly review the contribution ultracentrifuge methods have made to our understanding of those macromolecules associated with the engineering process, namely nucleic acids and nucleic acid binding proteins. Unfortunately, those arguably of the greatest interest – restriction endonucleases and ligases – responsible for the scissision and re-glueing together of genetic information – are themselves only produced in tiny quantities; they are themselves ideal candidates for gene cloning and overproduction for physical study.

Nucleic acids

The application of the ultracentrifuge to nucleic acid research dates back almost 50 years. For example, Kahler (1948) investigated the concentration dependence of the sedimentation coefficient of DNA. The classic application was the demonstration by Meselson and Stahl (1958; see also Meselson, Stahl and Vinograd, 1957) using analytical density gradient ultracentrifugation, and its ability to separate DNA labelled with ^{15}N from unlabelled ^{14}N DNA, that DNA replication was semi-conservative.

SIMPLE HOMOGENEITY/PURITY AND COMPOSITION CHECKS

Buoyant density measurements using CsCl isopyenic density gradient methods (see, e.g. Weinblum, Geisert and Ostwald, 1990) are still regarded as prime purity assays for a nucleic acid preparation (see, e.g. Durante *et al.*, 1987; Ramachandran and Narayan, 1990). Bouyant density measurements are routinely used to establish the base composition (Schildkraut, Marmur and Doty, 1962; Kirchhoff and Flossdorf, 1987; McIntyre *et al.*, 1987; Ligon *et al.*, 1989) and even estimates for molecular weight from the band shape (see, e.g. Thanaraj and Pandit, 1987). Sedimentation velocity is also routinely used as a criterion for purity (see, e.g. Weller and Hill, 1992), and reproducibility of a sedimentation coefficient is regarded as an indicator of successful reconstitution of rRNA subunits (Bogdanov *et al.*, 1988).

STRUCTURAL ANALYSES

In DNA, sedimentation equilibrium and velocity analyses have been used in conjunction with other techniques such as circular dichroism and gel electrophroresis to investigate the formation of hairpin and cruciform structures (Howard *et al.*, 1991; Ross, Howard and Lewis, 1991; Scaria, Shire and Shafer, 1992), linking number anomalies (Ringquist, Shinn and Hanlon, 1989), B-Z transition phenomena (Chen, Ringquist and Hanlon, 1987), the structure of a short promoter region, in conjunction with neutron and dynamic light scattering (Lederer *et al.*, 1986), and salt-dependent changes in the structure and dynamics of circular DNA, in conjunction with dynamic light scattering (Schaper *et al.*, 1991), and the association properties of DNA nonamers (Braswell *et al.*, 1992). For mRNA, a combined approach (including Mark-Houwink-Kuhn-Sakurada analyses) has been used (Freerksen *et al.*, 1990). Broitman, Im and Fresco (1987) have looked at the triple helical poly (AAU). Steger *et al.* (1986) have also used a combined approach (sedimentation velocity and equilibrium with electron microscopy, high-performance liquid chromatography and spectrophotometry) to investigate the structural properties of the RNA transcripts from cloned oligomeric potato spindle tuber viroid.

LIGAND-BINDING STUDIES

Several recent studies have involved either sedimentation velocity or isopycnic density gradient methods. For example, Bickhardt *et al.* (1991) used sedimentation velocity to look at the interaction of clupeine with DNA. Thomas, Nessler and Katterman (1989) examined the degree of 5-bromodeoxyuridine binding to DNA using isopycnic density gradient analysis in CsCl, and Wang, Sullivan and Libowitz (1989) used the same type of analysis to monitor the extent of binding of CMC (*N*-cyclohexyl-*n*-β-(4-

methylmorpholinium) ethyl carbodiimide-*p*-toluene sulphonate) to supercoiled DNA as a probe for unpaired and mismatched sites in DNA. Richard (1987) had earlier applied scaled particle theory to gels formed by DNA in the ultracentrifuge to obtain estimates for the effective length and radius of a DNA particle, before and after treatment with ethidium bromide (causing extensive lengthening, interpreted in terms of denaturation) or acridine orange (causing modest changes, interpreted in terms of binding). Trohalaki, Frisch and Lerman (1991) have also used scaled particle theory to investigate the effects of LiCl, RbCl, CsCl and $MgCl_2$ on the close packing of persistence length DNA fragments. Westkaemper and Richard (1987) used sedimentation equilibrium analysis to investigate the effects of diadenosine tetraphosphate and ATP on the higher-order structure of DNA.

DNA–RNA INTERACTIONS

White, Wood and Hill (1988) used sedimentation velocity in conjunction with dynamic light scattering and sucrose density gradient centrifugation (on radiolabelled material) to investigate the extent and effects of the interaction between a cDNA 'probe' and the α-sarcin region of 23S ribosomal RNA from *Escherichia coli* while still integral with the 50S ribosomal subunit from which it is derived. This interaction was explored under conditions where the ribosome was in its native state and under 'collapsed state' conditions. Binding of the probe was confirmed while the subunit was in a collapsed state, although the probe itself was shown not to be responsible for the collapse itself, refuting earlier beliefs.

Nucleic acid binding proteins

DNA PACKING PROTEINS

These have been the most widely studied with an extensive literature. Most involve studies on *histones* and their binding to DNA in nucleosomes and chromatin. The application of ultracentrifuge methods ranges from simple purity checks using CsCl isopynic density gradient equilibrium (Amero *et al.*, 1988), right through to specific details of the interaction.

The basic unit of chromatin is the nucleosome: a histone core protein complex about which the DNA is wrapped (see, e.g. van Holde, 1989). Sedimentation analysis combined with gel filtration studies were instrumental in establishing that the histone core complex is an octameric assembly of eight histone protein molecules (two each of histones H2A, H2B, H3 and H4) (Godfrey, Eickbush and Moudrianakis, 1980). Recent studies have focussed on the nature of the interaction between the core proteins themselves and the DNA. Moehs *et al.* (1992) used meniscus depletion sedimentation equilibrium with gel filtration to show that the plant variants of H2A and H2B are responsible for enhanced stability of the octamer compared with its verterate counterpart. The octamer itself is thought to self-associate to give larger structures. Prevelige and Fasman (1987) used similar approaches to investigate the effects of acetylation on this and the conformation, whereas Baxevannis, Godfrey and Moudrianakis (1991) looked at the effect of ionic strength on the association of the $(H3–H4)_2$ tetramer. Other studies have focused on the core proteins themselves (Ausio, 1988; Jutglar *et al.*, 1991) including the effect of H2a specific proteoloysis as a probe into the nature of the assembly process (Eickbush *et al.*, 1988).

In so far as the interactions with DNA are concerned, sedimentation velocity has been used to look at the cooperative inducing effect of Mg^{2+} (Triebel *et al.*, 1988), limitations of the polyglutamic acid reconstitution method for nucleosomes (Pennings, Muyldermans and Wyns, 1986), oligomerization of nucleosomes (Hansen *et al.*, 1989; Hansen, van Holde and Lohr, 1991), the significance of mobile histone tails in nucleosomes (Smith and Hill, 1989), two-sturt double helex formation (Osipora *et al.*, 1990) interactions at low ionic strength (Hirai *et al.*, 1988), the reversibility of the low salt transition of the core particles (Libertini and Small, 1987), the deleterious effect of higher ionic strengths (Yager, McMurray and van Holde, 1989), the influence of chromatin folding on inhibition of transcription initiation and elongation by RNA polymerase III (Hansen and Wolffe, 1992), the effect on the interaction of UV damage to the DNA (Gale and Smerdon, 1988) and B-Z DNA transition phenomena in reconstituted nucleosomes (Ausio, Zhou and van Holde, 1987). Sedimentation velocity distributions have been used to model the distribution of nucleosome positioning (Robert and van Holde, 1992). Sedimentation velocity and sedimentation equilibrium have been used to investigate the role of histone 'tails' in the stabilization of the nucleosome (Ausio, Dong and van Holde, 1989), in H1-depleted oligonucleosome folding (Garcia-Ramirez, Dong and Ausio, 1992) and the role of the $(H3\text{-}H4)_2$ tetramer in positioning the nucleosome (Dong and van Holde, 1991).

Besides histones, other packing-related proteins have been studied. Using sedimentation velocity analysis with low angle X-ray scattering, Herranz *et al.* (1990) have shown that conformational changes caused by proteolysis of the ϕ*29 connector* prevents DNA binding activity.

TOPOISOMERASES

Besides packing-proteins, other non-specific binding proteins have been investigated. Topoisomerases, for example, are enzymes that catalyse the breakage and rejoining of the DNA backbond. Riou *et al.* (1986) used isopycnic density gradient equilibrium to show that topoisomerase preparations from *Plasmodium berghei* were free from leukocytes, and combined the sedimentation coefficient measured in glycerol with the Stokes radius from gel filtration to obtain a molecular weight of 300 kDa, consistent with a dimer, whereas topoisomerase I gave 104 kDa, interpreted as monomer.

OTHER NON-SPECIFIC DNA BINDING PROTEINS

Brenner, Zlotnick and Stafford (1990) have used sedimentation equilibrium to study the self-association properties of RecA (37.8 kDa) from *E. coli*. Monomers were in reversible equilibria with trimers, hexamers and dodecamers to an extent depending on the temperature, the ionic strength, pH and presence or not of ATP. Sarfert *et al.* (1989) have used sedimentation velocity analysis to investigate the DNA binding properties of the transcription affecting *Streptomyces hygroscopicus* DNA binding protein.

GENE5 PROTEIN

The gene5 protein of the filamentous bacteriophage Pf1 is a dimeric protein (30 kDa) which binds to single stranded DNA during phage replication to form a helical nucleoprotein complex. Sedimentation velocity and equilibrium studies (Morgan *et al.*, 1989) have been used to characterize solutions of the protein. Sedimentation diagrams have shown the existence of mostly a 2.6S species – corresponding to the monomer, with a significant proportion of a high molecular weight aggregate (35S). Point weight average molecular weight analysis confirmed that the monomer form was the predominant species, with point average molecular weights of 32 500$\pm$3000 dominating, except near the cell base.

REPLICATION TERMINATOR PROTEIN (RTP)

Specific DNA replication terminators operate by slowing down or arresting replication forks. Using sedimentation, Lewis *et al.* (1990) have shown that *Bacillus subtilis* RTP (14.5 kDa) exists as a dimer (29 kDa) at neutral pH and concentrations above 0.2 mg ml^{-1}.

POLYMERASES

RNA polymerase from *E. coli* (450 kDa) is a dimeric protein under normal solution conditions. Using co-sedimentation experiments, Blazy, Takahashi and Baudras (1980) have shown that RNA polymerase interacts with the cyclic AMP-cyclic AMP receptor protein complex in solution. Further sedimentation velocity experiments by Pinkney and Hoggett (1988) have shown that the cyclic AMP receptor protein itself binds predominantly to the monomeric rather than the dimeric form of the enzyme. Butzow, Oehrl and Eichorn (1991) have used sedimentation diagrams from sedimentation velocity to assay the proportion of free and template-bound *E. coli* RNA polymerase. Suzuki (1990) has shown from a comparison of sedimentation coefficients of synthetic peptides that a heptad repeat in the largest subunit of RNA polymerase II binds to DNA, a result supported by fluorescence quenching measurements. Hansen and Wolffe (1992) investigated the influence of chromatin folding on the inhibition of transcription initiation and elongation by RNA polymerase III. Sedimentation equilibrium studies

(Lewis *et al.*, 1992) have shown that DNA β-polymerase (39 kDa, consisting of 8 kDa and 31 kDa domains), a symmetric monomer, did not self-associate on binding of oligodeoxynucleotide. By investigating the binding as a function of temperature, the enthalpies and entropies of the reaction were evaluated.

POLYMERASE 'ACCESSORY PROTEINS'

The polymerase 'accessory proteins' of phage T4, such as the gene 44, 62 and 45 proteins which show ATPase activity, are able to carry out leading strand DNA synthesis *in vivo*. Jarvis, Paul and von Hippel (1989) used sedimentation equilibrium and velocity, together with dynamic light scattering and gel filtration, to show that the gene 44 and 62 proteins associate to form a tight complex containing four gene 44 protein subunits and one gene 62 subunit, with a total molecular mass of 163.7 kDa, with an axial ratio of 5 : 1 (assuming a prolate ellipsoid model). Sedimentation equilibrium and chemical cross-linking studies suggest the gene 45 protein self-associates to form a trimer, which is also asymmetric.

OTHER GENE CONTROL PROTEINS

Overman, Bujalowski and Lohman (1988) used sedimentation velocity to estimate the average lengths of synthetic homopolynucleotides (based on polyA and polyU calibrations) to investigate cation and anion effects and polynucleotide specificity on the equilibrium binding of *E. coli* single-strand protein (EcoSSB) to single stranded nucleic acids in the $(SSB)_{65}$ binding mode. Using model fitting sedimentation coefficient versus concentration curves, Callaci *et al.* (1990) have shown that the *Xenopus* transcription factor III complexes with 5S RNA in a 1 : 1 ratio, which self-associates to a dimer.

A combination of sedimentation velocity with gel filtration has shown that the Nu1 subunit (21.2 kDa) of bacteriophage γ terminase exists as a $M>500$ kDa aggregate (Paris *et al.*, 1988). The physical properties and potential for self-association as a function of ionic strength of the *E. coli* transcription factors σ and NusA have been examined by Gill, Yager and von Hippel (1991), using sedimentation velocity and equilibrium coupled with chemical cross-linking, dynamic light scattering and low-angle X-ray scattering.

Jaenicke *et al.* (1990) have used a combined sedimentation velocity/equilibrium approach to show that the raf repressor protein from *E. coli* under intracellular conditions exists as a 144 kDa tetramer ('dimer of dimers'), although it undergoes concentration-dependent dissociation at lower concentrations, under meniscus depletion conditions. Sedimentation equilibrium has been used by Richet and Raibaud (1987) to show that the MalT protein, the transcription activator of the *E. coli* maltose regulon, exists as a monomer (100 kDa), while using the same technique Starovasnik *et al.* (1992) have shown the myogenic determination factor MyoD, a sequence-specific DNA binding protein, is best described as a dimer-tetramer, with a dissociation constant of 17.3 μM.

Geiselmann *et al.* (1992) have used the same technique to give a molecular weight of 264 kDa for the *E. coli* transcription termination factor rho I, confirming a hexamer, in agreement with the 'Svedberg equation molecular weight' (sedimentation coefficient combined with the translational diffusion coefficient, the latter from dynamic light-scattering measurements), and to give estimates for the gross conformation consistent with electron microscopy and X-ray scattering that the hexamer is planar. Further studies have investigated the binding of ATP (Geiselmann and von Hippel, 1992) and the binding of DNA to termination factor rho II (Geiselmann, Yager and von Hippel, 1992).

TRANSLATION FACTORS

A similar approach adopted by Sam *et al.* (1990), combining sedimentation velocity and equilibrium measurements with dynamic light scattering, showed that under non-self-associating conditions (as confirmed by sedimentation analysis), the bacterial elongation factor EF-Tu : GDP complex is a near-spherical particle in solution. Bommer *et al.* (1988) have also used sedimentation equilibrium (molecular weight), together with sedimentation velocity and diffusion analysis (molecular weight and frictional ratio), to investigate the size and shape of the eukaryotic initiation factor eIF-2. Sedimentation velocity together with fluorescence quenching and chemical modification studis have been used to investigate the interaction of tryptophanyl-tRNA synthetase with $tRNA^{Trp}$ (Fournier *et al.*, 1987).

INTEGRASE

Retroviral integrase is responsible for the integration of viral DNA into host DNA, with regard to the endonucleolytic processing of viral DNA ends and the cleavage and joining of host DNA to the processed viral DNA termini. In an attempt to understand the nature of the enzyme and the kinetics of the process, short-column sedimentation equilibrium studies were employed by Jones *et al.* (1992) which revealed a substrate-dependent reversible equilibrium among the monomeric (31.7 kDa) dimeric and tetrameric forms of retroviral integrase from Rous sarcoma virus. This information, combined with steady-state kinetic studies, suggests that the minimal functional unit of the enzyme required for both the processing and joining of each viral DNA end is the dimer.

REVERSE TRANSCRIPTASE

Reverse transcriptase from avian myeloblasts is an $\alpha\beta$ type of protein. Lin *et al.* (1991) used sedimentation velocity and equilibrium to investigate the association state under native and glycerol-containing buffer systems. In the absence of glycerol, it was deduced that the enzyme behaved as a heterodimer (152 kDa). In the presence of glycerol, after taking into account solvent effects, the enzyme dimerized to give an $(\alpha\beta)_2$ structure. Recombinant HIV-1 reverse transcriptase, also an αβ heterodimer, has also been the subject of extensive analysis by ultracentrifugation, and will be considered below (p. 341).

RETROTRANSPOSON VIRUS-LIKE PARTICLES

Without wanting this review to drift into the realm of viruses – DNA–protein complexes in the broadest sense – it is worth mentioning that the RNA-based retrotransposon particles, which replicate by a reverse transcriptase step followed by an integration reaction into the host genome, have also been characterized by the 'Svedberg equation route', viz sedimentation coefficients combined with dynamic light-scattering translational diffusion coefficient measurements to size these particles. A value of 142 kDa has been obtained (Burns *et al.*, 1992).

RIBOSOMAL PROTEINS

Sedimentation velocity and equilibrium analysis, together with dynamic light-scattering diffusion coefficient measurements, have shown that the ribosomal protein S4 from *E. coli* was monomeric (23kDa), but asymmetric with an axial ratio of 5 : 1 based on a prolate ellipsoid model (Dodd and Hill, 1987). Georgalis *et al.* (1989) have used sedimentation equilibrium to show that another *E. coli* ribosomal protein complex, L7/L12, is tetrameric (50 kDa), consistent with measurements from other less accurate techniques.

Engineered proteins and glycoproteins

The last few years has seen a plethora of published work on engineered proteins, ranging from small peptides like interleukins and hirudin right through to large multi-enzyme complexes. The sort of questions being addressed include: Does a recombinant protein have the same degree of monodispersity or a tendency to associate as its native or 'wild-type' analogue? (see, e.g., Pingoud *et al.*, 1988). Does site-directed mutagenesis significantly affect the activity of an enzyme by altering its conformation or its relations with a cofactor? (see, e.g., Slodowski *et al.*, 1991). Can we probe the significance of particular amino acids using mutagenesis? Does altering a protein subunit affect the mode of self-assembly into a multi-enzyme complex or alter the shape and size of a viral coat?

MONOCLONAL ANTIBODIES

Because of their flexibility, there is as yet no high-resolution crystallographic structure for an intact, immunologically active antibody molecule, only structures worked out for fragments and mutant antibodies which lack a hinge region. Sedimentation analysis, combined with other solution techniques, such as low-angle solution X-ray scattering, have provided the only real handle on the conformation of the intact molecule in solution.

Byron (1992) used sedimentation equilibrium and velocity to confirm that chimeric monoclonal B72.3 does not self-associate in dilute solution and used hydrodynamic bead modelling of the sedimentation coefficient and radius of gyration to show that the Fab and Fc regions do not form a planar structure. Morgan, Byron and Harding (1992) used the same techniques to show the monodispersity of preparations of Fab and $(Fab)_2$ fragments of this antibody, and used bead models to show that the Fab domain has dimensions approximately 50 × 80 × 50Å, inclusive of hydration.

Wilhelm *et al.* (1987) have looked at size using sedimentation equilibrium analysis and conformation by low-angle X-ray scattering on complexes of monoclonal Fab fragments with the β_2-subunit of *E. coli* tryptophan synthase as the 'antigen', to show (1) that the C-terminal F_2 domains lie at the distal ends of the subunit, and (2) how steric hindrance stops the (dimeric) β_2 from binding more than one Fab 93-6 fragment per dimer. Similar binding studies, involving the whole antibody complexed with ricin (via a galactose-binding domain), have been performed by Colombatti *et al.* (1987). Jentoft and Bolinger (1987) have used sedimentation velocity to investigate the interaction of monoclonal IgG1 with human serum albumin. King *et al.* (1993) have shown by velocity and equilibrium sedimentation analysis that the two Fv domains of the chimaeric B72.3 monoclonal are associated at high concentrations at pH values close to neutral, but dissociate at concentrations lower than ~0.5 mg ml^{-1}.

The behaviour of antibody fragments and constructs thereof ('mini-antibodies') is of topical interest at the moment. Pack and Plückthun (1992) have shown by sedimentation velocity analysis that a mini-antibody (consisting of two single chain Fv domains joined with a flexible hinge region from IgG3 and an amphiphatic helix fused to the C-terminus) dimerizes in solution. In another study using the 'combined approach', Morgan *et al.* (1988) used a combination of sedimentation velocity analysis with monoclonal antibody binding to probe the molecular determinants of haemopexin-mediated haem transport. Haem binding causes domain I of the protein to become more compact and induces an association with domain II.

RECOMBINANT COMPLEMENT

One of the biggest problems of recombinant technology is post-translational modification of the gene product: the problem is particularly acute for glycoproteins, of which complement is an example. Luo *et al.* (1992) measured the sedimentation coefficients of recombinant complement Cl subcomponent lacking in beta-hydroxyasparagine, sialic acid and one of its two carbohydrate chains, and found that it still reassembles with Clq and Clr sub-components to form a functional Cl complex.

RECOMBINANT INTERLEUKINS

Other recombinant proteins from the immune defence system have also been studied. Proudfoot *et al.* (1990) used sedimentation velocity to demonstrate the homogeneity of human interleukin 5 preparations (after gene expression in *E. coli*), and using sedimentation equilibrium showed that the recombinant protein behaves as a homodimer (viz a covalently linked dimer with no evidence for further self-association behaviour), consistent with observations from gel permeation chromatography. Both recombinant interleukin 1α (Wingfield *et al.*, 1987) and 1β (Wingfield *et al.*, 1986) also show no tendency to self-associate in dilute solution, and adopt compact globular conformations.

RECOMBINANT MACROPHAGE FACTORS

Wingfield *et al.* (1988) used a combination of sedimentation velocity and equilibrium ultracentrifugation in conjunction with urea-gradient polyacrylamide gel electrophoresis and several spectroscopic methods to study the conformation and stability of recombinant human and mouse granulocyte-macrophage colony stimulating factors (after its gene had been expressed in *E. coli*). The proteins were demonstrated to be physically homogeneous monomers with compact globular shapes, findings consistent with gel permeation chromatography studies.

RECOMBINANT HUMAN TISSUE FACTOR (TF)

Human tissue factor, a membrane anchored cell-surface protein that initiates coagulation upon tissue damage, is an essential activator for the serine protease Factor VIIa. The gene for the soluble cytoplasmic domain (sTF) of TF has been cloned and expressed in sufficient quantity to enable sTF : VIIa complexes to be assayed and Laue *et al.* (1992b) have used sedimentation velocity analysis to show that sTF, VIIa and the complex are probably highly asymmetric (high frictional ratio) with sTF and VIIa associating end-to-end rather than side-to-side in the complex. This shape could be physiologically significant in terms of perturbation of blood flow (see also Waxman *et al.*, 1992). Another human blood factor, XIIIa, has had its gene expressed in *Sacharomyces cerevisae* in sufficient quantity for a sedimentation equilibrium study (Bishop *et al.*, 1990); see also Rinas *et al.* (1990).

RECOMBINANT HUMAN TUMOUR NECROSIS FACTOR (TNF)

There are two types, TNFα and TNFβ. TNFα is essentially another macrophage factor, since its cellular origin is macrophages, whereas TNFβ is a related cytokine produced by lymphocytes. Arakawa and Yphantis (1987) demonstrated by sedimentation equilibrium analysis using both the short column and long solution column methods that recombinant TNFα (after its gene expression in *E. coli*) exists as trimeric structures in solution with a (z-average) molecular weight of 52 kDa, almost exactly the trimer value and only slightly higher than an earlier relative molecular weight determination by gel filtration, suggesting the trimeric molecule has a compact structure. Schoenfeld *et al.* (1991) obtained a slightly lower result (49 kDa), suggesting the presence of some dimer in their preparations. They also examined TNFβ, again expressed in *E. coli*, and obtained a result (56 kDa) that was also consistent with a trimeric structure.

Recombinant TNF receptor, and its complexes with TNFα and TNFβ have also been studied (see, e.g., Sreekrishna *et al.*, 1989). Using sedimentation equilibrium, Loetscher *et al.* (1991) found a molecular mass of 25 kDa for the receptor (less than half of the value estimated by gel filtration – probably a reflection of a non-compact structure) and 140 kDa for complexes of the receptor with either TNFα or TNFβ, strongly suggesting a stoichiometry of three receptor molecules binding to each TNFα or TNFβ trimer. Values for the sedimentation coefficients and also estimates for the translational diffusion coefficient from dynamic light scattering suggested an extended conformation for both the receptor and its complexes. In an independent study, Pennica *et al.* (1992) obtained similar results for the molecular weights of the receptor determined by sedimentation equilibrium (25 kDa) and gel filtration, although the molecular weight of the complex was somewhat less (114 kDa). The authors concluded that 2–3 receptor molecules bind to each TNFα (see also Shire, Pennica and Goeddel, 1992).

RECOMBINANT HEAT SHOCK PROTEIN HSP25

The heat shock protein hsp25 (23 kDa), which is associated with Ehrlich ascites carcinoma, has had its gene expressed in *E. coli* in sufficient quantities for detailed hydrodynamic analysis. Behlke *et al.* (1991) have used a combination of sedimentation velocity (sedimentation coefficient determination) and boundary spreading (translational diffusion coefficient) measurements in the ultracentrifuge with electron microscopy to observe the formation of high-order association products. A molecular mass of 750 ± 25 kDa was obtained from the Svedberg equation method, corresponding to a 32 subunit structure. From the frictional ratio, a compact hexagonal type of packing was proposed, consistent with electron microscopy. The resistance of the complex to disruption by detergent was also investigated.

RECOMBINANT HUMAN STEM CELL FACTOR (SCF)

Stem cell factor (SCF), also known as 'mast cell growth factor', acts on primitive hematopoietic progenitor cells of the marrow, and can operate synergistically in conjunction with other factors such as the interleukins Il-1, IL-3, IL-6 and IL-7, granulocyte and granulocyte-macrophage colony stimulating factors and erythropoietin (the latter considered below). In a recent study (Arakawa *et al.*, 1991), sufficient quantities of this recombinant glycoprotein (in both unglycosylated form, after gene expression in *E. coli*, and glycosylated form, as expressed in Chinese hamster ovary cells) have been produced to enable reasonably detailed physicochemical investigations. Both the glycosylated and unglycosylated forms were shown to be dimeric by combining the sedimentation equilibrium molecular weights (53 090 and

36 080 Da, respectively) with estimates for the monomer molecular weights from SDS-PAGE (18 500 and 28 000–35 000 Da, respectively) after allowance for anomalous binding of SDS to the glycosylated form: glycosylation does not appear to hold the key to the self-association process. The finding that SCF exists as a dimer is interesting in that other similar factors are also dimeric, such as the macrophage colony stimulating factor (CSF) and the platelet-derived growth factor (PDGF).

RECOMBINANT HUMAN ERYTHROPOIETIN

This factor is related to the stem cell factor and the gene for it has also been expressed – as a glycoprotein – in Chinese hamster ovary cells in sufficient quantities for physicochemical study (Davis *et al.*, 1987). Conventional long column sedimentation equilibrium was used to obtain a molecular weight (z-average) of 30 400 Da with no evidence for any polydispersity or self-association behaviour. This compares with an amino acid sequence molecular weight of 18 399 Da, suggesting the weight fraction of carbohydrate, like that for stem cell factor, is very high (0.395 ± 0.008). Comparison with the estimated hydrodynamic radius of the protein from gel filtration suggests either high expansion through solvation or considerable asymmetry.

RECOMBINANT CELL ADHESION MOLECULES

Cell adhesion molecules (CAMs) play a crucial role in the development and upkeep of adult tissues of multicellular organisms. A synthetic peptide has been produced, reproducing the Ca^{2+} binding motif of the cell adhesion molecule uvomorulin (Ozawa, Engel and Kemier, 1990). After point mutagenesis, where the first Asp was replaced by a Lys, a loss of Ca^{2+} binding and cell adhesion properties was observed. Sedimentation equilibrium analysis has confirmed that the monomeric state of this peptide is still retained after mutagenesis.

RECOMBINANT INTERFERONS

Natural human interferon γ (IFN-γ) is known to be unstable at low pH, although the nature of this instability, in terms of associative behaviour or other mechanisms, is largely unknown. In an attempt to address this issue, Yphantis and Arawaka (1987), and Arakawa and Hsu (1987) used sedimentation equilibrium in conjunction with circular dichroism to look at the effect of lowering the pH on the state of association and conformation of IFN-γ after its gene had been expressed in sufficient quantities in *E. coli*. After due care about the selection of the rotor speed because of the large sizes of the macromolecular species present in some solvent conditions – notably in the presence of 0.1 M NaCl – sedimentation equilibrium (using Rayleigh optics) yielded a largely unknown state of self-association of human interferon γ, expressed in *E. coli*. Acid-induced unfolding was confirmed by circular dichroism. At low pH (3.5) in the absence of salt, almost complete dissociation of the protein into monomers was observed; in the presence of salt, large aggregates were present at this pH. At a higher pH (6.9), the 'natural' dimeric form is reproduced; the presence of salt (0.1 M NaCl) does not appear to produce significant amounts of aggregate as observed at pH 3.5. The association state of human and murine interferon γ has also been investigated by Nagata *et al.* (1987).

The stoichiometry of the interaction between interferon γ and its receptor produced in both *E. coli* (unglycosylated) and baculovirus infected insect Sf9 and CHO cells (the latter giving glycosylated receptor molecules), also cloned in *E. coli*, has been studied by Fountoulakis *et al.* (1992). Single species only were observed in sedimentation diagrams from sedimentation velocity if the molecule and its (glycosylated) receptor were in a 1 : 2 molar ratio mixture. Both sedimentation equilibrium and the sedimentation coefficient (combined with the translational diffusion coefficient from dynamic light scattering) were used to obtain estimates of the molecular weight of the complex, and were in general agreement. The presence of glycosylation appeared to be important, with a binding stoichiometry (ligand : receptor) of 1 : 1 for unglycosylated and 1 : 2 for glycosylated receptors. This is another good demonstration of the importance of post-translational processing on the properties of recombinant proteins, and the perils of ignoring this fact.

RECOMBINANT HIRUDIN

Hirudin, a small peptide hormone (7 kDa), is so named because of the organism that produces it naturally (*Hirudo medicinalis*, the leech), and is responsible for the anti-blood coagulation reputation of this organism. Attempts have been made to clone its gene so that quantities can be made available without destroying the world's supply of leeches.

Recombinant hirudin produced in yeast was studied using sedimentation equilibrium analysis by Rowe *et al.* (1989) using Rayleigh optics with a novel data capture system. A value for the (weight average) molecular weight of 7080 ± 200 Da was obtained, almost in exact agreement with the amino-acid sequence molecular weight of 6964 Da, confirming that the recombinant peptide, like the wild-type peptide (Triebel and Walsmann, 1966), does not self-associate under dilute solution conditions. This observation was confirmed over a range of different solvent conditions by Otto and Seckler (1991) from both sedimentation equilibrium and sedimentation coefficient measurements.

PYRUVATE DEHYDROGENASE POINT MUTANTS

One of the most popular 'metabolic pathway' proteins studied by sedimentation analysis is pyruvate dehydrogenase (see, e.g. Gilbert and Gilbert, 1980). Attention has recently focused on the effects of site-directed mutagenesis on the mode of assembly of this multi-enzyme complex. Using sedimentation velocity analysis combined with gel filtration, Schulze *et al.* (1991) investigated the effect of point mutations in the E2p (dihydrolipoyl transacetylase) component on its binding of the peripheral components E1p and E3, and showed that for example large deletions of the protease-sensitive region of E2p resulted in a total loss of the E1p and E3 binding. In another study (Schulze *et al.*, 1992), *chimeric* E2p components were constructed by exchanging the three domains between E2p from *E. coli* and *Azotobacter vinelandii* at the gene level, and the six chimeric E2p's have been expressed and purified from *E. coli*. Sedimentation coefficient analysis, in conjunction with gel filtration, was again used, this time to study the binding or 'reconstitution' of these chimeric E2p's with E1p and E3. *Escherichia coli* E3 was shown to interact with all the chimeras, whereas *A. vinelandii* only interacted with those chimeras containing its own binding domain. The binding of the E2p's to either the *E. coli* or *A. vinelandii* E1p's also depended strongly on whether or not the catalytic or lipoyl domains were from the same organism or not.

RECOMBINANT NITRITE REDUCTASE

This dimeric enzyme from *Pseudomonas aeruginosa* has been difficult to clone, because of the presence of two different prosthetic groups (haem c and haem d1), which rule out cloning in *E. coli*. The gene for the protein has, however, been successfully cloned in the related *Pseudomonas putida*: the recombinant protein, containing only the chemically bound haem c is water-soluble (unlike the chemically prepared apoprotein), and available in sufficient quantities for characterization (Silverstrini *et al.*, 1992). Based on a comparison of sedimentation coefficients (6.55S native, 6.2S cloned), the authors conclude that this difference is small enough for the cloned enzyme also to be in a dimeric state.

RECOMBINANT PHOSPHOLIPASE

The significance of lipases in membrane-based processes is well recognized. The gene for human phospholipase has been cloned and expressed in sufficient quantities for sedimentation equilibrium analysis (Levin *et al.*, 1992).

RECOMBINANT 4-OXALOCROTONATE TAUTOMERASE

4-Oxalocrotonate tautomerase is a bacterial enzyme and is part of a set of enzymes that oxidatively catabolize toluene, *m*- and *p*-xylene, 3-ethyltoluene and 1,2,4-trimethylybenzene to intermediates in the Krebs cycle. The entire pathway is encoded by the TOL plasmid pWW0 and enables strains of soil bacteria carrying this plasmid to utilize simple aromatic hydrocarbons such as these as their sole sources of carbon and energy.

Chen *et al.* (1992) obtained a molecular mass of 32 000 Da from the sedimentation coefficient and diffusion coefficient for the recombinant enzyme of *Pseudomonas putida* expressed in *E. coli*. Because of the lack of tyrosine and tryptophan, far-uv absorption optics had to be employed to record solute distributions. This value compares with a value of 37 000 Da obtained by gel filtration for both the cloned and the wild-type enzyme. After comparison with the sequence molecular mass of 6811 Da, a pentameric structure for both the wild-type and cloned enzyme was proposed.

RECOMBINANT AMINOTRANSFERASE

This enzyme enables one pathway of amino acid catabolism, leading to the synthesis of intermediate metabolites that can enter the Krebs cycle. Although the molecular weight of the mammalian enzyme is known (50 kDa) and that it exists as a 'homodimer' (two covalently linked subunits), not much else is known because of insufficient purified material. It has been over-expressed in sufficient quantities in both *Saccharomyces* and *E. coli*. Dietrich, Lorber and Kern (1991) have used a combination of sedimentation velocity (sedimentation coefficient) and boundary spreading (translational diffusion coefficient) in the ultracentrifuge with other techniques (notably gel filtration, SDS and native PAGE) to assay the physical characteristics of the recombinant protein. The results for the molecular weight (110 ± 10 kDa) and frictional ratio (1.8 ± 0.2) of the protein are consistent only with a dimeric protein that is either asymmetric and/or significantly hydrated.

ASPARTATE TRANSCARBAMOYLASE POINT MUTANTS

Aspartate transcarbamoylase (ATCase) is one of the most extensively studied multi-subunit enzymes in terms of ligand binding, conformational change and its allosteric activity (see, e.g. Schachman *et al.*, 1984), and has been the subject of several studies on the effects of point mutagenesis. This 'new tool' has been invaluable, since previously quantitative information on the energetics of the various interchain interactions had been difficult to obtain because of the strength and multiplicity of the subunit contacts stabilizing the quaternary structure of the wild-type enzyme (see, e.g. Eisenstein, Markby and Schachman, 1989). ATCase catalyses the first committed reaction in pyrimidine biosynthesis in *E. coli*, exhibiting classical sigmoidal kinetics and subject to feedback inhibition by CTP and activation by ATP. It consists of 12 subunits, two catalytic trimers and three regulatory dimers linked non-covalently. The complex exists in either of two states: a low-affinity 'R' ('relaxed' or swollen) state or a high-affinity 'T' ('taut') state, affected by CTP and ATP binding. Boundary spreading sedimentation velocity experiments on the wild-type enzyme compared with a missense 'T-state' mutant ATCase554 (in which serine at position 52 in the catalytic chains was replaced by phenylalanine) indicated no significant intermediate conformation states (Werner and Schachman, 1989; see also Werner *et al.*, 1987) in the reaction.

Eisenstein, Markby and Schachman (1989) reported the effects of a number of point mutations in a limited region of the zinc-binding domain of the regulatory subunits. They compared changes in activity with changes in conformation, as manifested by a fractional change in the sedimentation coefficient of the enzyme, in the absence and in the presence of the ATP and

CTP ligands. For example, replacement of an asparagine residue at the interface between a regulatory and a catalytic chain in the opposing catalytic trimer causes a complete loss of the homotropic and heterotropic effects characteristic of wild-type ATCase. This corresponded with observations from the fractional change in sedimentation coefficient upon binding a bisubstrate ligand 'PALA' (viz 'difference sedimentation coefficient' experiments), that this mutant enzyme exists in the R conformation in the absence of active site ligands due to preferential destabilization of the T conformation relative to the R state.

To assist in the detection of any possible global conformation change caused by the binding of the inhibitor CTP, Eisenstein, Markby and Schachman (1990) replaced Lys143 in the regulatory chain Ala, thereby perturbing interactions at the regulatory–catalytic interface and destabilizing the T-state. Difference sedimentation velocity experiments involving measurements of the changes caused by the binding of a bisubstrate analogue demonstrated that the sedimentation coefficient of the mutant enzyme was intermediate between that observed for the T and R states of the wild-type ATCase. Together with equilibrium binding data, these results provide convincing evidence that heterotropic effectors cause an alteration in the gross conformation of ATCase and thereby regulate the activity of the enzyme by perturbing the $T \rightleftharpoons R$ equilibrium, refuting suggestions that only local changes occur on binding of CTP or ATP.

In point mutation studies on the catalytic chain (Newell and Schachman, 1990), a Lys164-Glu, Glu239-Lys or both showed that the mutationally altered enzymes were devoid of both homotropic and heterotropic effects, supporting the view that the allosteric properties of the wild-type enzyme are linked to a ligand-induced alteration in quaternary structure.

Advantage has been taken of crystallographic studies which have pinpointed the nucleotide (i.e. CTP or ATP) binding region on the regulatory chains. Wente and Schachman (1991) have shown that different point mutations at the same positions (Lys-60 and Lys-94) cause 'bewildering' changes in allosteric properties, again using the fractional change in the secimentation coefficient upon binding of the bisubstrate ligand as the conformational probe. Similar studies have explored the effects of replacement of the active site residues Gln231 (Peterson, Burman and Schachman, 1992) and Gln288 in the near-C-terminus α-helical region (Peterson and Schachman, 1992) in the catalytic chains.

RECOMBINANT AND POINT MUTANT NUCLEIC ACID-BINDING PROTEINS

The tools of genetic engineering have themselves been the subject of recombinant technology and point mutagenesis. The HIV-1 transactivating protein HIV-1 Rev (13 kDa) has been produced in *E. coli*. This regulates gene expression by binding to specific regions of viral mRNA. Wingfield *et al.* (1991) used both sedimentation velocity and sedimentation equilibrium to investigate the solution properties of the recombinant protein. Upward curved log concentration versus distance squared plots were observed from sedimentation equilibrium, consistent with a significant self-association [whole cell weight average (M_w°) 56 kDa, with point average molecular weights ranging from 47 000 (cell meniscus) to 74 000 (cell base) at the rotor speed used]. This evidence for a self-association was confirmed by sedimentation velocity with a classical increase in the sedimentation coefficient with concentration.

Chakerian *et al.* (1991) used sedimentation velocity measurements in conjunction with other techniques to look at evidence for a zipper motif in the Lac repressor protein, and in a related study (Chakerian and Mathews, 1991) investigated the effects of point mutations at positions 281 or 282 on the oligomerization domain of the protein. The wild-type protein is a tetramer, with an $s°_{20,w}$ of 7.5S. Depending on the substition, up to a three-fold decrease in $s°_{20,w}$ was observed, indicating this region is essential for oligomerization.

Blondel and Bedouelle (1990) investigated the hydrodynamic properties (sedimentation velocity and gel filtration) of the gene5 protein of phage M13 hybridized with the maltose binding protein of *E. coli* and expressed in the latter. This hybridization route is considered a possible way of producing and purifying proteins like gene5 protein. Both sedimentation velocity and gel filtration are consistent with a monomer–dimer equilibrium for this hybrid.

Wallis *et al.* (1992) have over-produced colicin E9 immunity protein, a bacterial anti-suicide DNase protein, in sufficient quantity for sedimentation velocity and equilibrium analysis. The molecular weight (9.4 kDa) came out close to the sequence value, confirming a monomer. Some concern was expressed about the very high concentration dependence of the (apparent) molecular weight with concentration for a molecule so small. The sedimentation coefficient of 1.14S combined with the translational diffusion coefficient from dynamic light scattering yields a very similar molecular mass.

HIV-1 reverse transcriptase, as considered above, is an αβ heterodimer (M_α=66 kDa, M_β=51kDa). The α and β subunits, known as 'p66' and 'p51', respectively (Becerra *et al.*, 1991), have been expressed individually in *E. coli*, and have been the subject of extensive study by sedimentation equilibrium analysis. Studies in a solvent of moderate ionic strength (0.25 M NaCl) revealed that p66 was in a monomer–dimer equilibrium (K_A 5.1 × 10^4 M^{-1}), whereas p51 was monomeric. Mixing the p66 and p51 resulted in reconstitution of the 1 : 1 heterodimer with K_A 4.9 × 10^5 M^{-1}. This binding was resistant to a high salt concentration (1 M NaCl), suggesting a strong hydrophobic interaction, whereas a 20 kDa C-terminal truncation of p51 eliminated all complex formation with p66 and a peptide corresponding to the C-terminal 15 kDa of p66 was not able to bind to p66, suggesting that the central region of RT may contain sequences required for the dimerization process.

Point mutagenesis has been used to explore the tetramerization and function of the single-stranded DNA binding protein, EcoSSB of *E. coli* (Curth *et al*, 1991). In the native state, EcoSSB exists as a tetramer. Amino acid 55 (His) in the polypeptide chain was thought to play a key role: mutants were produced with tyr, glu, lys, phe and ile in this position. A combination of sedimentation velocity (concentration dependence of the sedimentation coefficient), NMR, gel filtration and fluorimetric detection methods revealed that whereas the phe and ile substitutions did not change the properties of the protein appreciably, the others caused dissociation and loss of affinity for poly(dT).

The class II restriction endonuclease EcoRI has also been the subject of point mutagenesis studies (Geiger *et al.*, 1989). Sedimentation velocity, coupled with gel filtration, isoelectric focusing and circular dichroism studies have shown an altered monomer/dimer equilibrium for the Gln44Lys145 and Gln144Lys145Lys200 mutants.

RECOMBINANT POST-TRANSLATIONAL PROTEASES

Post-translational processing has been the thorn in the flesh for recombinant technology, until this was at least partially circumvented by the use of eukaryotic hosts, such as Chinese hamster ovary cells. One such post-translational processing protein is the MAS gene-encoded processing protease of yeast mitochondria, which cleaves amino-terminal presequences of proteins imported from the cytoplasm across the mitochondrial inner membrane. This has two subunits, the MAS1 gene encoded subunit and the MAS2 gene encoded subunit. These have themselves been the subject of over-production experiments and the MAS2 gene encoded subunit has been produced in sufficient quantities for sedimentation analysis (Geli *et al.*, 1990). Sedimentation velocity and sedimentation equilibrium experiments have shown that the isolated MAS2 subunit is a monomer, with a molecular weight (52 kDa) close to that predicted by the gene sequence, but with the presence (20–30%) of some larger species of molecular weights 90 kDa and 254 kDa.

Glycosylation is one of the most crucial forms of post-translational processing; among many other things, carbohydrate moieties are the key elements of molecular recognition phenomena. One of the key sugars is *N*-acetylneuraminic acid (NANA). Zapata *et al.* (1989) have cloned and expressed the K1 CMO *N*-acetylneuraminic acid synthetase (49 kDa) gene from *E. coli* and shown by sedimentation equilibrium analysis and gel filtration that the enzyme is active as a monomer, although it may form aggregates.

RECOMBINANT ANTENNAPEDIA HOMEODOMAIN

Homeotic genes and several other genes controlling development share a characteristic DNA segment, the 'homeobox'. Müller *et al.* (1988) have expressed the Antennapedia gene of *Drosophila* in *E. coli* and analysed the properties of the protein using sedimentation equilibrium and velocity analyses. Under reducing conditions, the protein behaves as a monomer, with a sedimentation equilibrium weight average molecular weight (9.04 ± 0.34 kDa) close to the sequence value (8.545 kDa). By combining this with the sedimentation coefficient (1.0 ± 0.1S), a globular structure was inferred, from the axial ratio (1.4–1.8) of the equivalent prolate ellipsoid, assuming 'typical' hydration. In the absence of reducing conditions, the protein showed a tendency to dimerize.

RECOMBINANT LAMIN

Chicken lamin B2, a structural filament protein related to myosin, has been expressed in *E. coli*. The structure and assembly properties of the recombinant protein were examined by a combination of electron microscopy and sedimentation velocity and equilibrium ultracentrifugation (Heitlinger *et al.*, 1991). This protein was shown to form myosin-like 3.1S dimers, which associate longitudinally to form polar head-to-tail polymers.

RECOMBINANT N-TERMINAL GLOBULE OF COLLAGEN VI

The size and shape of a recombinant collagen fragment, N9–N2, has been studied using sedimentation equilibrium and velocity analyses by Specks *et al.* (1992). A molar mass from sedimentation equilibrium of 180 ± 5 kDa was inferred, consistent with an octameric structure. From the frictional ratio, a high degree of asymmetry was inferred (axial ratio 9, assuming a prolate ellipsoid model and 'typical' hydration), consistent with electron microscopy.

RECOMBINANT ADENOVIRUS PROTEIN IIIa

The gene for this has been cloned and expressed in *E. coli*, in both wild-type and ts112 mutant forms (Cuillel *et al.*, 1989). Both proteins have been shown by sedimentation velocity analysis (in sucrose density gradients) to be monomeric. Both the hydrodynamic and electron microscopic data were consistent with a cylindrical conformation of 20 nm length and 2.8 nm diameter.

VIRAL COAT PROTEIN MUTANTS

Shire *et al.* (1990) have looked at the self-assembly properties of recombinant DNA derived tobacco mosaic virus protein, expressed in *E. coli*. The recombinant protein lacks an acetyl group on the amino terminus of the polypeptide chain, so by comparison with earlier results of self-assembly of wild-type protein, important conclusions about the role of amino terminal acetylation in the self-assembly process could be inferred. Sedimentation velocity and equilibrium analysis was used in conjunction with electron microscopy and circular dichroism to demonstrate a reversible endothermic self-association, although the detail of the assembly process was different from the wild-type. For example, in contrast to the wild-type assembly process, the recombinant assemblies contain some unpolymerized disk-like structures, which suggest either two structures for the 28S aggregates (not seen in the wild-type), two parallel polymerization pathways or both. The effect of the extra charge through acetylation was thus shown to have important consequences for quaternary structure, even though this end of the protein chain is on the outside of the protein in the assembly and in the virus.

Molina-Garcia *et al.* (1992) used sedimentation velocity and equilibrium analyses to investigate the effect of coat protein point mutations on the self-assembly of bacteriophage fd. Significant increases in the molecular weight of the virus (from 15 000 to 20 000 kDa) were observed by the substitution of the lysine residue in position 48 with either glutamine or alanine. Combining with the measured sedimentation coefficients, these results imply an increase in length of the virus from 900 to 1400 nm, consistent with observations from electron microscopy.

Engineered polysaccharides

Genetic engineering – recombinant technology or mutagenesis – produces directly engineered protcins, and also engineered glycoproteins so long as an appropriate host organism is used. The technology is also having an impact on the synthesis and properties of polysaccharides, although the only example I could find where ultracentrifuge analyses are involved was one involving ourselves at Nottingham.

Pectins (polyuronides), with molar masses from a few thousand to several million, are an important constituent of the cell walls of fruits, and are used extensively by the food industry as thickening and gelling agents. In tomatoes, the softening of the tomato during the green–red–rotten ripening process is associated with a progressive degradation of the cell wall pectins: a comparative sedimentation equilibrium and gel filtration study on green and red tomato pectins has confirmed this (Seymour and Harding, 1986). One of the key enzymes responsible for the degradation of cell wall pectins is polygalacturonase. Smith *et al.* (1990) have produced novel mutant tomato lines in which expression of the polygalacturonase gene has been downregulated by antisense RNA: this antisense gene was inherited stably after two generations. Molar masses (apparent values at a single low concentration) of pectins from red tomatoes – which had either 20% (GR95 progeny) or 1% (GR105 progeny) residual polygalacturonase activity – were measured by sedimentation equilibrium and compared with values for green and wild-type red tomato pectins. Samples from green tomatoes of antisense and wild-type plants had a similar apparent M_w of 158 kDa. As ripening progressed, pectin breakdown occurred, resulting in an increase in the amount but decrease in the size of soluble polyuronide fragments. In normal fruit, M_w had dropped to 80 kDa by 14 days, whereas it was 95 kDa for GR95 and as high as 135 kDa for GR105. These results demonstrate that while *in vivo* solubilization of pectin was not affected by substantially reduced polygalacturonase levels, depolymerization of the pectin was largely prevented.

The engineering of polysaccharides is clearly an untapped gold mine – and not just for the ultracentrifuge.

References

AMERO, S.A., OGLE, R.C., KEATING, J.L., MONTOYA, V.L., MURDOCH, W.L. AND GRAINGER, R.M. (1988). The purification of ribosomal RNA gene chromatin from *Physarum polycephalum*. *Journal of Biological Chemistry* **263**, 10 725–10 733.

ARAKAWA, T. AND HSU, Y.-R. (1987). Acid unfolding and self-association of recombinant *Escherichia coli* derived human interferon γ. *Biochemistry* **36**, 5428–5432.

ARAKAWA, T. AND YPHANTIS, D.A. (1987). Molecular weight of recombinant human tumor necrosis factor-α. *Journal of Biological Chemistry* **262**, 7484–7485.

ARAKAWA, T., YPHANTIS, D.A., LARY, J.W., NARHI, L.O., LU, H.S., PRESTRELSKI, S.J., CLOGSTON, C.L., ZSEBO, K.M., MENDIAZ, E.A., WYPYCH, J. AND LANGLEY, K.E. (1991). Glycosylated and unglycosylated recombinant-derived human stem cell factors are dimeric and have extensive regular secondary structure. *Journal of Biological Chemistry* **266**, 18 942–18 948.

AUSIO, J. (1988). An unusual cystein-containing histone H1-like protein and two protamine-like proteins are the major nuclear proteins of the sperm of the bivalve mollusc *Macoma nasuta*. *Journal of Biological Chemistry* **263**, 10 141–10 150.

AUSIO, J., DONG, F. AND VAN HOLDE, K.D. (1989). Use of selectively trypsinized nucleosome core particle to analyze the role of the histone 'tails' in the stabilization of the nucleosome. *Journal of Molecular Biology* **206**, 451–463.

AUSIO, J., ZHOU, G. AND VAN HOLDE, K. (1987). A reexamination of the reported B-Z DNA transition in nucleosomes reconstituted with poly(dG-m^{5}dC)-poly(dG-m^{5}dC). *Biochemistry* **26**, 5595–5599.

BAXEVANIS, A.D., GODFREY, J.E. AND MOUDRIANAKIS, E.N. (1991). Associative behaviour of the histone $(H3\text{-}H4)_2$ tetramer: Dependence on ionic environment. *Biochemistry* **30**, 8817–8823.

BECERRA, S.P., KUMAR, A., LEWIS, M.S., WIDEN, S.G., ABBOTTS, J., KARAWYA, E.M., HUGHES, S.H., SHILOACH, J. AND WILSON, S.H. (1991). Protein–protein interactions of HIV-I reverse transcriptase: Implication of central and C-terminal regions in subunit binding. *Biochemistry* **30**, 11 707–11 719.

BEHLKE, J., LUTSCH, G., GAESTEL, M. AND BIELKA, H. (1991). Supramolecular structure of the recombinant murine small heat shock protein hsp25. *FEBS Letters* **288**, 119–122.

BICKHARDT, S., EBERT, G., NISHI, N. AND HAGIWARA, K. (1991). Sedimentation velocity analysis of the DNA–protamine interaction. *Progress in Colloid and Polymer Science* **86**, 36–40.

BISHOP, P.D., TELLER, D.C., SMITH, R.A., LASSER, G.W., GILBERT, T. AND SEARLE, R.L. (1990). Expression, purification, and characterization of human factor XIII in *Saccharomyces cerevisiae*. *Biochemistry* **29**, 1861–1869.

BLAZY, B., TAKAHASHI, M. AND BAUDRAS, A. (1980). *Molecular Biology Reports* **6**, 39–43.

BLONDEL, A. AND BEDOUELLE, H. (1990). Export and purification of a cytoplasmic dimeric protein by fusion to the maltose-binding protein of *Escherichia coli*. *European Journal of Biochemistry* **193**, 325–330.

BLOOMFIELD, V.A. (1981). Quasi-elastic light scattering applications in biochemistry and biology. *Annual Review of Biophysics and Bioengineering* **10**, 421–450.

BOGDANOV, A.A., CHICHKOVA, N.V., KOPYLOV, A.M., MANKIN, A.S. AND SKRIPKIN, E.A. (1988). Surface topography of ribosomal RNA. In *Methods in Enzymology* (H.F. Noller and K. Moldave, Eds), Vol. 164, pp. 440–456. Academic Press, San Diego, CA.

BOMMER, U.-A., LUTSCH, G., BEHLKE, J., STAHL, J., NESYTOVA, N., HENSKE, A. AND BIELKA, H. (1988). Shape and location of eukaryotic initiation factor eIF-2 on the 40S ribosomal subunit of rat liver: Immuno-electron-microscopic and hydrodynamic investigations. *European Journal of Biochemistry* **172**, 653–662.

BRASWELL, E.H., ZHANG, C., ZHU, D. AND COLE, R. (1992). Sedimentation equilibrium studies of several DNA nonamers. *Biophysical Journal* **61**, A361.

BRENNER, S.L., ZLOTNICK, A. AND STAFFORD, W.F., III (1990). RecA protein self-assembly. II. Analytical equilibrium ultracentrifugation studies of the entropy-driven self-association of RecA. *Journal of Molecular Biology* **216**, 949–964.

BROITMAN, S.L., IM, D.D. AND FRESCO, J.R. (1987). Formation of the triple-stranded polynucleotide helix poly (AAU). *Proceedings of the National Academy of Sciences, USA* **84**, 5120–5124.

BURNS, N.R., SAIBIL, H.R., WHITE, N.S., PARDON, J.F., TIMMINS, P.A., RICHARDSON, S.M.H., RICHARDS, B.M., ADAMS, S.E., KINGSMAN, S.M. AND KINGSMAN, A.J. (1992). Symmetry, flexibility and permeability in the structure of yeast retrotransposon virus-like particles. *EMBO Journal* **11**, 1155–1164.

BUTZOW, J.J., OEHRL, L.L. AND EICHORN, G.L. (1991). Selectivity of *Escherichia coli* RNA polymerase for template conformation. *Biochemistry* **30**, 6454–6464.

BYRON, O. (1992). Solution studies on the conformation and assembly of the monoclonal antibody B72.3. PhD dissertation, University of Nottingham, UK.

CALLACI, T.P., CAI, G.-Z., LEE, J.C., DALY, T.J. AND WU, C. (1990). Assembly of *Xenopus* transcription factor III A–5S RNA complex. *Biochemistry* **29**, 1653–4659.

CANTOR, C.R. AND SCHIMMEL, P.R. (1980). *Biophysical Chemistry. Part II: Techniques for the Study of Biological Structure and Function*. W.H. Freeman, San Francisco.

CHAKERIAN, A.E. AND MATHEWS, K.S. (1991). Characterization of mutations in oligomerization domain of lac repressor protein. *Journal of Biological Chemistry* **266**, 22 206–22 214.

CHAKERIAN, A.E., TESMER, V.M., MANLY, S.P., BRACKETT, J.K., LUNCH, M.J., HOH, J.T. AND MATTHEWS, K.S. (1991). Evidence of leucine zipper motif in lactose repressor protein. *Journal of Biological Chemistry* **266**, 1371–1374.

CHEN, C., RINGQUIST, S. AND HANLON, S. (1987). Covalent attachment of an alkylamine prevents the B to Z transition in poly (dG-dC). *Biochemistry* **26**, 8213–8221.

CHEN, L.H., KENYON, G.L., CURTIN, F., HARAYAMA, S., BEMBENEK, M.E., HAJIPOUR, G. AND WHITEMAN, C.P. (1992). 4-Oxalocrotonate tautomerase, an enzyme composed of 62 amino acid residues per monomer. *Journal of Biological Chemistry* **267**, 11 716–11 721.

COLOMBATTI, M., JOHNSON, V.G., SKOPICKI, H.A., FENDLEY, B., LEWIS, M.S. AND YOULE, R.J. (1987). Identification and characterization of the monoclonal antibody recognizing a galactose-binding domain of the toxin ricin. *Journal of Immunology* **138**, 3339–3344.

COMPER, W.D. AND PRESTON, B.N. (1992). The analytical ultracentrifuge as a tool for diffusion measurements: Cross-diffusion effects in ternary polymer : polymer-solvent systems. In *Analytical Ultracentrifugation in Biochemistry and Polymer Science* (S.E. Harding, A.J. Rowe and J.C. Horton, Eds), pp. 428–442. Royal Society of Chemistry, Cambridge.

CREETH, J.M. AND HORTON, J.R. (1977). Macromolecular distribution near the limits of density-gradient columns. Some applications to the separation and fractionation of glycoproteins. *Biochemical Journal* **161**, 449.

CUILLEL, M., CORTOLEZZIS, B., CHROBOCZEK, J., LANGOWSKI, J., RUIGROK, R.W.H. AND JACROT, B. (1989). Purification and characterization of wild-type and ts112 mutant protein IIIa of human edenovirus 2 expressed in *Escherichia coli*. *Virology* **175**, 222–231.

CURTH, U., BAYER, I., GREIPEL, J., MAYER, F., URBANKE, C. AND MAASS, G. (1991). Amino acid 55 plays a central role in tetramerization and function of *Escherichia coli* single-stranded DNA binding protein. *European Journal of Biochemistry* **196**, 87–93.

DAVIS, J.M., ARAKAWA, R., STRICKLAND, T.W. AND YPHANTIS, D.A. (1987).

Characterization of recombinant human erythropoietin produced in Chinese hamster ovary cells. *Biochemistry* **26**, 2633–2638.

DIETRICH, J.B., LORBER, B. AND KERN, D. (1991). Expression of mammalian tyrosine aminotransferase in *Saccharomyces cereivisiae* and *Escherichia coli. European Journal of Biochemistry* **201**, 399–407.

DODD, J. AND HILL, W.E. (1987). Physical characteristics of ribosomal protein S4 from *Escherichia coli. Journal of Biological Chemistry* **262**, 2478–2484.

DONG, F. AND VAN HOLDE, K.E. (1991). Nucleosome positioning is determined by the $(H3\text{-}H4)_2$ tetramer. *Proceedings of the National Academy of Sciences, USA* **88**, 10 596–10 600.

DURANTE, M., CREMONINI, R., TAGLIASACHHI, A.M., FORINO, L.M.C. AND CIOINA, P.G. (1987). Characterization and chromosomal localization of fast renaturing and satellite DNA sequences in *Phaseolus coccineus. Protoplasma* **137**, 100–108.

EICKBUSH, R.H., GODFREY, J.E., ELIA, M.C. AND MOUDRIANAKIS, E.N. (1988). H2a-specific proteolysis as a unique probe in the analysis of the histone octamer. *Journal of Biological Chemistry* **263**, 18 972–18 978.

EISENSTEIN, E., MARKBY, D.W. AND SCHACHMAN, H.K. (1989). Changes in stability and allosteric properties of dispartate transcarbomoylase resulting from amino acid substitutions in the zinc-binding domain of the regulatory chains. *Proceedings of the National Academy of Sciences, USA* **86**, 3094–3098.

EISENSTEIN, E., MARKBY, D.W. AND SCHACHMAN, H.K. (1990). Heterotropic effectors promote a global conformational change in aspartate transcarbamoylase. *Biochemistry* **29**, 3724–4394.

FOUNTOULAKIS, M., ZULAUF, M., LUSTIG, A. AND GAROTTA, G. (1992). Stoichiometry of interaction between interferon γ and its receptor. *European Journal of Biochemistry* **208**, 781–787.

FOURNIER, M., PLANTARD, C., LABOUESSE, B. AND LABOUESSE, J. (1987). Binding stoichiometry of $tRNA^{Trp}$ and tryptophanyl-tRNA synthetase from bovine pancreas under pH conditions of maximum activity: Analysis by ultracentrifugation, fluorescence quenching and chemical modification. *Biochimica Biophysica Acta* **916**, 350–357.

FREERKSEN, D.L., SHIH, P.C.-F., VASTA-RUSSELL, J.F., HORLICK, R.A. AND YAU, W.W. (1990). Single stranded RNA molecular weight and shape determination by differential pressure capillary viscometry, sedimentation velocity and gel electrophoresis. *Analytical Biochemistry* **189**, 163–168.

FREIRE, J.J. AND GARCIA DE LA TORRE, J. (1992). Sedimentation coefficients of flexible chain polymers. In *Analytical Ultracentrifugation in Biochemistry and Polymer Science* (S.E. Harding, A.J. Rowe and J.C. Horton, Eds), pp. 346–358. Royal Society of Chemistry, Cambridge.

FURTH, A. AND MOORE, R. (1986). *Self-Assembly of Macromolecules*, pp. 50–99. Open University Press, Milton Keynes.

GALE, J.M. AND SMERDON, M.J. (1988). UV-induced pyrimidine dimers and trimethylsporalen cross-links do not alter chromatin folding *in vitro. Biochemistry* **27**, 7197–7205.

GARCIA-RAMIREZ, M., DONG, F. AND AUSIO, J. (1992). Role of the histone 'tails' in the folding of oligonucleosomes depleted of histone H1. *Journal of Biological Chemistry* **267**, 19 587–19 595.

GARCIA DE LA TORRE, J. (1989). Hydrodynamic properties of macromolecular assemblies. In *Dynamic Properties of Biomolecular Assemblies* (S.E. Harding and A.J. Rowe, Eds), pp. 3–31. Royal Society of Chemistry, Cambridge.

GEIGER, R., RÜTER, T., ALVES, J., FLIESS, A., WOLFES, H., PINGOUD, V., URBANKE, C., MAASS, G., PINGOUD, A., DÜSTERHÖFT, A. AND KRÖGER, M. (1989). Genetic engineering of EcoRI mutants with altered amino acid residues in the DNA binding site: Physiochemical investigations give evidence for an latered monomer/dimer equilibrium for the Gln44Lys145 and Gln144Lys145Lys200 mutants. *Biochemistry* **28**, 2667–2677.

GEISELMANN, J. AND VON HIPPEL, P.H. (1992). Functional interactions of ligand cofactors with *Escherichia coli* transcription factor rho I: Binding of ATP. *Protein Science* **2**, 850–860.

GEISELMANN, J., YAGER, T.D. AND VON HIPPEL, P.H. (1992). Functional interactions of ligand cofactors with *Escherichia coli* transcription termination factor rho II: Binding of RNA. *Protein Science* **1**, 861–873.

GEISELMANN, J., YAGER, R.D., GILL, S.C., CALMETTES, P. AND VON HIPPEL, P.H. (1992). Physical properties of the *Escherichia coli* transcription termination factor rho I: Association states and geometry of the rho hexamer. *Biochemistry* **31**, 111–121.

GELI, V., YANG, M., SUDA, K., LUSTIG, A. AND SCHATZ, G. (1990). The MAS-encoded processing protease of yeast mitochondria: Overproduction and characterization of its two nonidentical subunits. *Journal of Biological Chemistry* **265**, 19 216–19 222.

GEORGALIS, Y., DIJK, J., LABISCHINSKI, H. AND WILLS, P.R. (1989). The tetrameric form of ribosomal protein L7/L12 from *Escherichia coli. Journal of Biological Chemistry* **264**, 9210–9214.

GILBERT, G.A. AND GILBERT, L.M. (1980). Determination in the ultracentrifuge of protein heterogeneity by computer modelling, illustrated by pyruvate dehydrogenase multi-enzyme complex. *Journal of Molecular Biology* **144**, 405–408.

GILL, S.C., YAGER, T.D. AND VON HIPPEL, P.H. (1991). Escherichia coli σ^{70} and NusA proteins, II. Physical properties and self-association states. *Journal of Molecular Biology* **220**, 325–333.

GODFREY, J.E., EICKBUSH, T.H. AND MOUDRIANAKIS, E.N. (1980). Reversible association of calf thymus histones to form the symmetrical octanes $(H2AH2BH3H4)_2$: A case of a mixed-associating system. *Biochemistry* **19**, 1339–1346.

HANSEN, J.C. AND WOLFFE, A.P. (1992). Influence of chromatin folding on transcription initation and elongation by RNA polymerase III. *Biochemistry* **31**, 7977–7988.

HANSEN, J.C., VAN HOLDE, K.E. AND LOHR, D. (1991). The mechanism of nucleosome assembly onto oligomers of the sea urchin 5S DNA positioning sequence. *Journal of Biological Chemistry* **266**, 4276–4282.

HANSEN, J.C., AUSIO, J., STANIK, V.H. AND VAN HOLDE, K.E. (1989). Homogeneous reconstituted oligonucleosomes: Evidence for salt-dependent folding in the absence of histone H1. *Biochemistry* **28**, 9129–9136.

HARDING, S.E. (1989). Modelling the gross conformation of assemblies using hydrodynamics: The whole body approach. In *Dynamic Properties of Biomolecular Assemblies* (S.E. Harding and A.J. Rowe, Eds), pp. 32–56. Royal Society of Chemistry, Cambridge.

HARDING, S.E. (1993a). Determination of macromolecular homogeneity, shape and interactions using sedimentation velocity analytical ultracentrifugation. In *Methods in Molecular Biology, Vol. 22: Optical Spectroscopy and Macroscopic Techniques* (C. Jones, B. Mulloy and A.H. Thomas, Eds), pp. 61–72. Humana Press, Totowa, New Jersey, USA.

HARDING, S.E. (1993b). Determination of absolute molecular weights using sedimentation equilibrium analytical ultracentrifugation. In *Methods in Molecular Biology, Vol. 22: Optical Spectroscopy and Macroscopic Techniques* (C. Jones, B. Mulloy and A.H. Thomas, Eds), pp. 72–84. Humana Press, Totowa, New Jersey, USA.

HARDING, S.E. (in press). Sedimentation equilibrium analysis of glycopolymers. *Colloid and Polymer Science*.

HARDING, S.E. AND TOMBS, M.P. (1989). Protein diffusion through interfaces. In *Interactions in Protein Systems* (K.D. Schwenke and B. Raab, Eds), pp. 65–72.

Abhandlungen der Akademie der Wissenschaften der DDR, Akademie-Verlag, Berlin.

HARDING, S.E., HORTON, J.C. AND MORGAN, P.J. (1992). MSTARA: A FORTRAN program for the model independent molecular weight analysis of macromolecules using low speed or high speed sedimentation equilibrium. In *Analytical Ultracentrifugation in Biochemistry and Polymer Science* (S.E. Harding, A.J. Rowe and J.C. Horton, Eds), pp. 275–294. Royal Society of Chemistry, Cambridge.

HARDING, S.E., ROWE, A.J. AND HORTON, J.C. (Eds) (1992). *Analytical Ultracentrifugation in Biochemistry and Polymer Science*. Royal Society of Chemistry, Cambridge.

HARDING, S.E., VÅRUM, K.-M., STØKKE, B.-T. AND SMIDSRØD, O. (1991). Molecular weight determination of polysaccharides. In *Advances in Carbohydrate Analysis* (C. White, Ed.), Vol. 1, pp. 63–144. JAI Press, New York.

HEITLINGER, E., PETER, M., HÄNER, M., LUSTIG, A., AEBI, U. AND NIGG, E.A. (1991). Expression of chicken lamin B_2 in *Escherichia coli*: Characterization of its structure, assembly, and molecular interactions. *Journal of Cell Biology* **113**, 485–495.

HERRANZ, L., BORDAS, J., TOWNS-ANDREWS, E., MENDEZ, E., USOBIAGA, P. AND CARRASCOSA, J.L. (1990). Conformational changes in bacteriophage ϕ29 connector prevents DNA-binding activity. *Journal of Molecular Biology* **213**, 263–273.

HIRAI, M.K., NIIMURA, N., MITA, K., ICHIMURA, S., TOKUNAGA, F. AND ISHIKAWA, Y. (1988). Interparticle interactions and structural changes of nucleosome core particles in low-salt solution. *Biochemistry* **27**, 7924–7931.

HOWARD, F.B., CHEN, C., ROSS, P.D. AND MILES, H.T. (1991). Hairpin formation in the self-complementary dodecamer *d*-GGTACGCGTACC and its derivatives containing GA and IA mispairs. *Biochemistry* **30**, 779–782.

JAENICKE, R., MUIZNIEKS, I., ASLANIDIS, C. AND SCHMITT, R. (1990). Ultracentrifugal analysis of the quaternary structure of the raf repressor from *Escherichia coli*. *FEBS Letters* **260**, 233–235.

JARVIS, T.C., PAUL, L.S. AND VON HIPPEL, P.H. (1989). Structural and enzymatic studies of the T4 DNA replication system. I. Physical characterization of the polymerase accessory protein complex. *Journal of Biological Chemistry* **264**, 12 709–12 716.

JENTOFT, J.E. AND BOLINGER, L. (1987). Forms of a self-associating autoantibody complex between monoclonal human IgG1 and human serum albumin. *Molecular Immunology* **24**, 163–169.

JONES, K.S., COLEMAN, J., MERKEL, G.W., LAUE, T.M. AND SKALKA, A.M. (1992). Retroviral integrase functions as a multimer and can turn over catalytically. *Journal of Biological Chemistry* **267**, 16 037–16 040.

JUTGLER, L., BORRELL, J.I. AND AUSIO, J. (1991). Primary, secondary and tertiary structure of the core of a histone H1-like protein from the sperm of *Mytilus*. *Journal of Biological Chemistry* **266**, 8184–8191.

KAHLER, H. (1948). *Journal of Physical and Colloid Chemistry* **52**, 676.

KING, D.J., BYRON, O.D., MOUNTAIN, A., WEIR, N., HARVEY, A., LAWSON, A.D.G., PROUDFOOT, K.A., BALDOCK, D., HARDING, S.E., YARRANTON, G.T. AND OWENS, R.J. (1993). Expression, purification and characterization of B72-3 Fv fragments. *Biochemical Journal* **290**, 723–729.

KIRCHHOFF, H. AND FLOSSDORF, J. (1987). Determination of the guanine-plus-cytosine content of *Mollicutes* by isopycnic gradient centrifugation. *International Journal of Systematic Bacteriology* **37**, 454–455.

KITAGISHI, K., OBATTA, H. AND KONDO, T. (1987). Effect of zinc deficiency on 80S ribosome content of meristematic tissues of rice plant. *Soil Science and Plant Nutrition* **33**, 423–429.

LAUE, T.M. (1992). On-line data acquisition and analysis from the Rayleigh interferometer. In *Analytical Ultracentrifugation in Biochemistry and Polymer Science*

(S.E. Harding, A.J. Rowe and J.C. Horton, Eds), pp. 63–89. Royal Society of Chemistry, Cambridge.

LAUE, T.M., KLEVIT, R.E., STAROVASNIK, M.A. AND WEINTRAUB, H. (1992a). The reversible micellization of myocyte differentiation factor (MyoD). *Biophysical Journal* **61**, A476.

LAUE, T.M., WAXMAN, E., KONIGSBERG, W.K., NEMERSON, Y. AND ROSS, J.B.A. (1992b). Testing a working model for human tissue factor (TF): Hydrodynamics of TF : VIIa complex by centrifugation. *Biophysical Journal* **61**, A65.

LEDERER, H., MAY, R.P., KJEMS, J.K., BAER, G. AND HEUMANN, H. (1986). Solution structure of a short DNA fragment studied by neutron scattering. *European Journal of Biochemistry* **161**, 191–196.

LEVIN, W., DANIEL, R.F., STONER, C.R., STOLLER, T.J., WARDWELL-SWANSON, J.A., ANGELILLO, Y.M., FAMILLETI, P.C. AND CROW, R.M. (1992). Purification of recombinant human secretory phospholipase A_2 (Group II) produced in long-term immobilized cell culture. *Protein Expression and Purification* **3**, 27–35.

LEWIS, M.S., KIM, S.-J., KUMAR, A. AND WILSON, S.H. (1992). Thermodynamic parameters of oligodeoxynucleotide binding to DNA-β-polymerase revealed by equilibrium ultracentrifugation. *Biophysical Journal* **61**, A489.

LEWIS, P.J., RALSTON, G.B., CHRISTOPHERSON, R.I. AND WAKE, R.G. (1990). Identification of the replication terminator protein binding sites in the terminus region of the *Bacillus subtilis* chromosome and stoichiometry of the binding. *Journal of Molecular Biology* **214**, 73–84.

LIBERTINI, L.J. AND SMALL, E.W. (1987). Reversibility of the low-salt transition of chromatin core particles. *Nucleic Acids Research* **15**, 6655–6664.

LIGON, J.M., BOLEN, P.L., HILL, D.S., BOTHAST, R.J. AND KURTZMAN, C.P. (1989). Physical and biological characterization of linear DNA plasmids of the yeast *Pichia inositovora. Plasmid* **21**, 185–194.

LIN, T.-H., QUINN, T., WALSH, M., GRANDGENETT, D. AND LEE, J.C. (1991). Avian myeloblastosis reverse transcriptase: Effects of glycerol on its hydrodynamic properties. *Journal of Biological Chemistry* **266**, 1635–1640.

LOETSCHER, H., GENTZ, R., ZULAUF, M., LUSTIG, A., TABUCHI, H., SCHLAEGER, E.-J., BROCKHAUS, M., GALLATI, H., MANNBERG, M. AND LESSLAUER, W. (1991). Recombinant 55-kDa tumor necrosis factor (TNF) receptor: Stoichiometry of binding to TNFα and TNFβ and inhibition of TNG activity. *Journal of Biological Chemistry* **266**, 18 324–18 329.

LUO, C., THIELENS, N.M., GAGNON, J., GAL, P., SARVARI, M., TSENG, Y., TOSI, M., ZAVODSKY, P., ARLSUD, G.J. AND SCHUMAKER, V.N. (1992). Recombinant human complement subcomponent C2s lacking β-hydrodyasparagine, sialic acid, and one of its two carbohydrate chains still reassembles with C1q and C1r to form a functional C1 complex. *Biochemistry* **31**, 4254–4262.

MARSH, E.N. AND HARDING, S.E. (1993). Methylmalonyl-CoA mutase from *Propionibacterium shermanii*: Characterization of the cobalamin-inhibited form and subunit-cofactor interactions studied by analytical ultracentrifugation. *Biochemical Journal* **290**, 551–555.

MARSH, E.N., HARDING, S.E. AND LEADLEY, P.F. (1989). Subunit interactions in *Propionibacterium shermanii* methylmalonyl-CoA mutase studied by analytical ultracentrifugation. *Biochemical Journal* **260**, 353–358.

MCINTYRE, M., ROSENBAUM, V., RAPPOLD, W., DESSELBERGER, M., WOOD, D. AND DESSELBERGER, U. (1987). Biophysical characterization of rotavirus particles containing rearranged genomes. *Journal of General Virology* **68**, 2961–2966.

MESELSON, M. AND STAHL, F.W. (1958). *Proceedings of the National Academy of Sciences, USA* **44**, 671.

MESELSON, M., STAHL, F.W. AND VINOGRAD, J. (1957). *Proceedings of the National Academy of Sciences, USA* **43**, 581.

MOEHS, C.P., BAXEVANIS, A.D., MOUDRIANAKIS, E.N. AND SPIKER, S. (1992).

Enhanced stability of histone octamers from plant nucleosomes: Role of H2A and H2B histones. *Biochemistry* **31**, 10 844–10 851.

MOLINA GARCIA, A.D., HARDING, S.E., DIAZ, F.G., GARCIA DE LA TORRE, J., ROWITCH, D. AND PERHAM, R.N. (1992). Effect of coat protein mutations in bacteriophage fd studied by sedimentation analysis. *Biophysical Journal* **63**, 1293–1298.

MORGAN, P.J., BYRON, O.D. AND HARDING, S.E. (1992). The solution conformation of novel antibody fragments studied using the Optima™ XL-A Analytical Ultracentrifuge. Technical Information DS-834. Spinco Business Unit, Palo Alto, CA.

MORGAN, P.J., HARDING, S.E., PLYTE, S.E. AND KNEALE, G.G. (1989). Sedimentation analysis of the Pf1 gene 5 protein. *Biochemical Society Transactions* **17**, 234–235.

MORGAN, W.T., MUSTER, P., TATUM, F.M., MCCONNELL, J., CONWAY, T.P., HENSLEYM, P. AND SMITH, A. (1988). Use of hemopexin domains and monoclonal antibodies to hemopexin to probe the molecular determinants of hemopexin-mediated heme transport. *Journal of Biological Chemistry* **163**, 8220–8225.

MÜLLER, M., AFFOLTER, M., LEUPIN, W., OTTING, G., WUTHRICH, K. AND GEHRING, W.J. (1988). Isolation and sequence-specific DNA binding of the Antennapedia homeodomain. *EMBO Journal* **7**, 4299–4304.

NAGATA, K., IZUMI, T., KITAGAWA, T. AND TOSHIDA, N. (1987). Oligomeric structure of recombinant human and murine immune interferons by means of sedimentation equilibrium. *Journal of Interferon Research* **7**, 313–320.

NEWELL, J.O. AND SCHACHMAN, H.K. (1990). Amino acid substitutions which stabilize aspartate transcarbamyolase in the R state disrupt both homotropic and heterotropic effects. *Biophysical Chemistry* **37**, 183–196.

OSIPOVA, T.N., KARPOVA, W.V. AND VOROB'EV, V.I. (1990). Chromatin higher-order structure: Two-starch double superhelix formed by zig-zag shaped nucleosome chain with folded linker DNA. *Journal of Biomolecular Structure and Dynamics* **8**, 11–22.

OTTO, A. AND SECKLER, R. (1991). Characterization, stability and refolding of recombinant hirudin. *European Journal of Biochemistry* **202**, 67–73.

OVERMAN, O.B., BUJALOWSKI, W. AND LOHMAN, T.M. (1988). Equilibrium binding of *Escherichia coli* single-strand binding protein to single-stranded nucleic acids in the $(SSB)_{65}$ binding mode: Cation and anion effects and polynucleotide specificity. *Biochemistry* **27**, 456–471.

OZAWA, M., ENGEL, J. AND KEMIER, R. (1990). Single amino acid substitutions in one Ca^{2+} binding site of uvomorulin abolish the adhesive function. *Cell* **63**, 1033–1038.

PACK, P. AND PLÜCKTHUN, A. (1992). Miniantibodies: Use of amphipathic helices to produce functional, flexibly linked dimeric F_v fragments with high avidity in *Escherichia coli*. *Biochemistry* **31**, 1579–1584.

PARIS, W., DAVIDSON, A., KEELER, C.L. JR AND GOLD, M. (1988). The Nu1 subunit of bacteriophage γ terminase. *Journal of Biological Chemistry* **263**, 8413–8419.

PENNICA, D., KOHR, W.J., FENDLY, B.M., SHIRE, S.J., RAAB, H.E., BORCHARDT, P.E., LEWIS, M. AND GOEDDEL, D.V. (1992). Characterization of a recombinant extracellular domain of the type 1 tumor necrosis factor receptor: Evidence for tumor necrosis factor-α induced receptor aggregation. *Biochemistry* **31**, 1134–1141.

PENNINGS, S., MUYLDERMANS, S. AND WYNS, L. (1986). Limitations of the poly-(glutamic acid) reconstitution method in the reassembly of mono- and dinucleosomes. *Biochemistry* **25**, 5043–5051.

PERKINS, S.J. (1989). Hydrodynamic modelling of complement. In *Dynamic Properties of Biomolecular Assemblies* (S.E. Harding and A.J. Rowe, Eds), pp. 226–245. Royal Society of Chemistry, Cambridge.

PETERSON, C.B. AND SCHACHMAN, H.K. (1992). Long range effects of amino acid substitutions in the catalytic chain of aspartate transcarbamyolase: Localized replacements in the carboxyl-terminal α-helix cause marked alterations in allosteric properties and inter-subunit interactions. *Journal of Biological Chemistry* **267**, 2443–2450.

PETERSON, C.B., BURMAN, D.L. AND SCHACHMAN, H.K. (1992). Effects of replacement of active site residue glutamine 231 on activity and allosteric properties of aspartate transcarbamoylase. *Biochemistry* **31**, 8508–8515.

PINGOUD, A., WEHRMANN, M., PIEPER, U., GAST, F.-U., URBANKE, C., ALVES, J., FEUERSTEIN, J. AND WITTINGHOFER, A. (1988). Spectroscopic and hydrodynamic studies reveal structural differences in normal and transforming H-ras gene products. *Biochemistry* **27**, 4735–4740.

PINKNEY, M. AND HOGGETT, J.G. (1988). Binding of the cyclic AMP receptor protein of *Escherichia coli* to RNA polymerase. *Biochemical Journal* **250**, 897–902.

PREVELIGE, P.E. JR AND FASMAN, G.D. (1987). Structural studies of acetylated and control inner core histones. *Biochemistry* **26**, 2944–2955.

PROUDFOOT, A.E.I., FARRAH, D., KAWASHIMA, E.H., BERNARD, A. AND WINGFIELD, P.T. (1990). Preparation and characterization of human interleukin-5 expressed in recombiant *Escherichia coli. Biochemical Journal* **270**, 357–361.

PUSEY, P.N. (1989). Dynamic light scattering by partially fluctuating media. In *Dynamic Properties of Biomolecular Assemblies* (S.E. Harding and A.J. Rowe, Eds), pp. 90–102. Royal Society of Chemistry, Cambridge.

RALSTON, H.B. AND MORRIS, M.B. (1992). The use of the omega function for sedimentation equilibrium analysis. In *Analytical Ultracentrifugation in Biochemistry and Polymer Science* (S.E. Harding, A.J. Rowe and J.C. Horton, Eds), pp. 253–274. Royal Society of Chemistry, Cambridge.

RAMACHANDRAN, C. AND NARAYAN, R.K.J. (1990). Satellite DNA specific to knob heterochromatin in *Cucumis metuliferus* (Cucurbitaceae). *Genetica* **80**, 129–138.

RÅNBY, B. (1987). *Physical Chemistry of Colloids and Macromolecules*. Blackwell Scientific, Oxford.

RICHARD, A.J. (1987). A comparison by ultracentrifugation of the effects on DNA of ethidium bromide and of acridine orange at low ionic strength. *Biophysical Chemistry* **27**, 191–195.

RICHET, E. AND RAIBAUD, O. (1987). Purification and properties of the MalT protein, the transcription activator of the *Escherichia coli* maltose regulon. *Journal of Biological Chemistry* **262**, 12 647–12 653.

RINAS, U., RISSE, B., JAENICKE, R., BROKER, M., KARGES, H.E., KEPPER, H.A. AND ZETTLMEISSL, G. (1990). Characterization of recombinant factor XIIIa produced in *Saccharomyces cerevisiae. Bio/Technology* **8**, 543–546.

RINGQUIST, S., SHINN, R. AND HANLON, S. (1989). Linking number anomalies in DNA under conditions close to condensation. *Biochemistry* **28**, 1076–1085.

RIOU, J.-F., GABILLOT, M., PHILIPPE, M., SCHREVEL, J. AND RIOU, G. (1986). Purification and characterization of *Plasmodium berghei* DNA topoisomerases I and II: Drug action, inhibition of decatenation and relaxation, and stimulation of DNA cleavage. *Biochemistry* **25**, 1471–1479.

ROBERT, C.H. AND VAN HOLDE, K.E. (1992). Prediction of sedimentation distributions for positioned nucleosomes on DNA templates at arbitrary saturation. *FASEB Journal* **6**, A66.

ROSS, P.D., HOWARD, F.B. AND LEWIS, M.S. (1991). Thermodynamics of antiparallel hairpin-double helix equilibria in DNA oligonucleotides from equilibrium ultracentrifugation. *Biochemistry* **30**, 6269–6275.

ROWE, A.J., WYNNE JONES, S., THOMAS, D.G. AND HARDING, S.E. (1989). Analysis of solute concentration and concentration derivative distribution by means of frameshift Fourier and other algorithms applied to Rayleigh interferometric and Fresnel fringe patters. SPIE Vol. 1163, *Fringe Pattern Analysis*, pp. 138–148.

SPIE-International Society for Optical Engineering, Bellingham, Washington, USA.

ROWE, A.J., WYNNE JONES, S., THOMAS, D.G. AND HARDING, S.E. (1992). Methods for off-line analysis of sedimentation velocity and sedimentation equilibrium pattern. In *Analytical Ultracentrifugation in Biochemistry and Polymer Science* (S.E. Harding, A.J. Rowe and J.C. Horton, Eds), pp. 49–62. Royal Society of Chemistry, Cambridge.

SAM, T., PLEY, C. AND MANDEL, M. (1990). A hydrodynamic study with quasielastic light scattering and sedimentation of bacterial elongation factor EF-Tu-guanosine-5′-diphosphate complex under nonassociating conditions. *Biopolymers* **30**, 299–3080.

SARFERT, E., SEDOVA, V., TRIEBEL, H., BAR, H. AND ZIMMER, C. (1989). DNA binding protein from *Streptomyces hygroscopicus*: Detection of binding by gel retardation, sedimentation and effects on the transcriptional activity *in vitro*. *Biomedica Biochimica Acta* **48**, 633–643.

SCARIA, P.V., SHIRE, S.J. AND SHAFER, R.H. (1992). Quadruplex structure of $d(G_3T_4G_3)$ stabilized by K^+ or Na^+ is an asymmetric hairpin dimer. *Proceedings of the National Academy of Sciences, USA* **89**, 10 336–10 340.

SCHACHMAN, H.K. (1992). Is there a future for the ultracentrifuge? In *Analytical Ultracentrifugation in Biochemistry and Polymer Science* (S.E. Harding, A.J. Rowe and J.C. Horton, Eds), pp. 3–15. Royal Society of Chemistry, Cambridge.

SCHACHMAN, H.K., PAUZA, C.D., NAVRE, M., KARELA, M.J., WU, L. AND YANG, Y.R. (1984). Location of amino acid alterations in mutants of aspartate transcarbamoylase. Structural aspects of interallelic complementation. *Proceedings of the National Academy of Sciences, USA* **81**, 115–119.

SCHAPER, A., URBANKE, C., KOHRING, C.-W. AND MAASS, G. (1991). Structure and dynamics of the complex of single stranded DNA binding protein of *Escherichia coli* with circular single stranded DNA deduced from boundary sedimentation. *Journal of Biomolecular Structure and Dynamics* **3**, 887–898.

SCHILDKRAUT, C.L., MARMUR, J. AND DOTY, P. (1962). Determination of the base composition of deoxyribonucleic acid from its buoyant density in CsCl. *Journal of Molecular Biology* **4**, 430–433.

SCHMIDT, B. AND RIESNER, B. (1992). A fluorescence detection system for the analytical ultracentrifuge and its application to proteins, nucleic acids, viroids and viruses. In *Analytical Ultracentrifugation in Biochemistry and Polymer Science* (S.E. Harding, A.J. Rowe and J.C. Horton, Eds), pp. 176–207. Royal Society of Chemistry, Cambridge.

SCHMIDT, B., RAPPOLD, W., ROSENBAUM, V., FISCHER, R. AND RIESNER, D. (1990). A fluorescence detection system for the analytical ultracentrifuge and its application to proteins, nucleic acids and viruses. *Colloid and Polymer Science* **268**, 45–54.

SCHOENFELD, J.-J., POESCHL, B., FREY, J.R., LOETSCHER, H., HUNZIKER, W., LUSTIG, A. AND ZULAUF, M. (1991). Efficient purification of recombinant human tumor necrosis factor-β from *Escherichia coli* yields biologically active protein with trimeric structure that binds to both tumor necrosis factor receptors. *Journal of Biological Chemistry* **266**, 3863–3869.

SCHULZE, E., WESTPHAL, A.H., BOUMANS, H. AND DE KOK, A. (1991). Site-directed mutagenesis of the dihydrolipoyl transacetylase component (E2p) of the pyruvate dehydrogenase complex from *Azotobacter vinelandii*: Binding of the peripheral components E1p and E3. *European Journal of Biochemistry* **202**, 841–848.

SCHULZE, E., WESTPHAL, A.H., VEEGER, C. AND DE KOK, A. (1992). Reconstitution of pyruvate dehydrogenase multienzyme complexes based on chimeric core structures of *Azotobacter vinelandii* and *Escherichia coli*. *European Journal of Biochemistry* **206**, 427–435.

SHIRE, S.J., PENNICA, D. AND GOEDDEL, D.V. (1992). TNF-α induced receptor aggregation as assessed by analytical ultracentrifugation. *Biophysical Journal* **61**, A95.

SHIRE, S.J., MCKAY, P., LEUNG, D.W., CACHIANES, G.J., JACKSON, E. AND WOOD, W.I. (1990). Preparation and properties of recombinant DNA derived tobacco mosaic virus coat protein. *Biochemistry* **29**, 5119–5126.

SILVERSTRINI, M.C., CUTRUZZOLA, F., D'ALESSANDRO, R., BRUNORI, M., FOCHESATO, N. AND ZENNARO, E. (1992). Expression of *Pseudomonas putida* nitrite reductase and characterization of the recombinant protein. *Biochemical Journal* **285**, 661–666.

SLODOWSKI, O., BOHM, J., SCHONE, B. AND OTTO, B. (1991). carboxy-terminal truncated rhuIFN-γ with a substitution of Gln133 or Ser132 to leucine leads to higher biological activity than in the wild type. *European Journal of Biochemistry* **202**, 1133–1140.

SMIDSRØD, O. AND ANDRESEN, I.-L. (1979). *Biopolymerkjemi*. Tapir Press, Trondheim.

SMITH, C.J., WATSON, C.F., MORRIS, P.C., BIRD, C.R., SEYMOUR, G.B., GRAY, J.E., ARNOLD, C., TUCKER, G.A., SCHUCH, W., HARDING, S. AND GRIERSON, D. (1990). Inheritance and effect on ripening of antisense polygalacturonase in transgenic tomatoes. *Plant Molecular Biology* **14**, 369–379.

SMITH, R.M. AND RILL, R.L. (1989). Mobile histone tails in nucleosomes: Assignments of mobile segments and investigations of their role in chromatin folding. *Journal of Biological Chemistry* **264**, 10 574–10 581.

SPECKS, U., MAYER, U., NISCHT, R., SPISSINGER, R., MANN, K., TIMPL, R., ENGEL, J. AND CHU, M.-L. (1992). Structure of recombinant N-terminal globule of type VI collagen α3 chain and its binding to heparin and hyaluronan. *EMBO Journal* **11**, 4281–4290.

SREEKRISHNA, K., NELLES, L., POTENZ, R., CRUZE, J., MAZZAFERRO, P., FISH, W., FUKE, M., HOLDEN, K., PHELPS, D., WOOD, P. AND PARKER, K. (1989). High-level expression, purification, and characterization of recombinant human tumor necrosis factor synthesized in the methylotrophic yeast *Pichia pastoria*. *Biochemistry* **28**, 4117–4125.

STAFFORD, W.F. III (1992). Methods for obtaining sedimentation coefficient distributions. In *Analytical Ultracentrifugation in Biochemistry and Polymer Science* (S.E. Harding, A.J. Rowe and J.C. Horton, Eds), pp. 359–393. Royal Society of Chemistry, Cambridge.

STAROVASNIK, M.A., BLACKWELL, T.K., LAUE, T.M., WEINTRAUB, H. AND KLEVIT, R.E. (1992). Folding topology of the disulfide-bonded dimeric DNA-binding domain of the myogenic determination factor MyoD. *Biochemistry* **31**, 9891–9903.

STEGER, G., BRÜGGERMANN, W., COLPAN, M., KLOTZ, G., SÄNGER, H.-L. AND RIESNER, D. (1986). Structure of viroid replicative intermediates: Physicochemical studies on SP6 transcripts of cloned oligomeric potato spindle tuber viroid. *Nucleic Acids Research* **14**, 9613–9630.

SUZUKI, M. (1990). The heptad repeat in the largest subunit of RNA polymerase II binds by intercalating into DNA. *Nature* **344**, 562–565.

TANFORD, C. (1961). *Physical Chemistry of Macromolecules*. John Wiley, New York.

TANG, L.H. AND ADAMS, E.T. JR (1973). *Archives of Biochemistry and Biophysics* **157**, 520.

THANARAJ, T.A. AND PANDIT, M.W. (1987). Buoyant density and molecular weight calculations on an Apple II microcomputer from equilibrium banding experiments on analytical ultracentrifuge. *Computer Methods and Programs in Biomedicine* **24**, 189–193.

THOMAS, J.C., NESSLER, C. AND KATTERMAN, F. (1989). Interruption of somatic embryogenesis in *Daucus carota* L. by 5-bromodeoxyuridine. *Plant Physiology* **90**, 921–927.

TOMBS, M.P. AND HARDING, S.E. (1988). Protein transport processes in the water–

water interface in incompatible two-phase systems. In *Advances in Separations Using Aqueous Phase Systems in Cell Biology and Biotechnology* (D. Fisher and I. Sutherland, Eds), pp. 229–232. Plenum Press, New York.

TRIEBEL, H. AND WALSMANN, P. (1966). Die Bestimmung des Molekulargewichtes von Hirudin in der Ultrazentrifuge. *Biochimica Biophysica Acta* **120**, 137–140.

TRIEBEL, H., VON MICKWITZ, C.-U., BÄR, H. AND BURCKHARDT, G. (1988). Mg^{2+}-induced transition to strongly cooperative binding of histone H1 to linear DNA. *International Journal of Biological Macromolecules* **10**, 322–328.

TROHALAKI, S., FRISCH, H.L. AND LERMAN, L.S. (1991). The effects of lithium, rubidium, cesium and magnesium ions on the close packing of persistence-length DNA fragments. *Biophysical Chemistry* **40**, 197–205.

VAN HOLDE, K.E. (1985). *Physical Biochemistry*, 2nd edn. Prentice-Hall, Englewood Cliffs, NJ.

VAN HOLDE, K.E. (1989). *Chromatin*. Springer, New York.

WALLIS, R., REILLY, A., ROWE, A., MOORE, G.R., JAMES, R. AND KLEANTHOUS, C. (1992). *In vivo* and *in vitro* characterization of overproduced colicin E9 immunity protein. *European Journal of Biochemistry* **207**, 687–695.

WANG, A.A., SULLIVAN, J.K. AND LEBOWITZ, J. (1989). Immunoassays for carbodiimide modified DNA-detection of unpaired transitions in super coiled ColE1 DNA. *Nucleic Acids Research* **17**, 9957–9977.

WAXMAN, E., ROSS, J.B.A., LAUE, T.M., GUHA, A., THIRUVIKRAMAN, S.V., LIN, T.C., KONIGSBERG, W.H. AND NEMERSON, Y. (1992). Tissue factor and its extracellular soluble domain: The relationship between intermolecular association with factor VIIa and enzymatic activity of the complex. *Biochemistry* **31**, 3998–4003.

WEINBLUM, D., GEISERT, M. AND OSTWALD, E. (1990). Numerical analysis of density gradient centrifugation profiles from eukaryotic DNA. *Colloid and Polymer Science* **268**, 55–59.

WEISENBERG, R.C., BORISY, G.G. AND TAYLOR, E.W. (1968). The colchicine-binding protein of mammalian brain and its relation to microtubules. *Biochemistry* **7**, 4466.

WELLER, J.W. AND HILL, W.E. (1992). Probing dynamic changes in tRNA conformation in the 30S subunit of the *Escherichia coli* ribosome. *Biochemistry* **31**, 2748–2757.

WENTE, S.R. AND SCHACHMAN, H.K. (1991). Different amino acid substitutions at the same position in the nucleotide-binding of aspartate transcarbamoylase have diverse effects on the allosteric properties of the enzyme. *Journal of Biological Chemistry* **266**, 20 833–20 839.

WERNER, W.E. AND SCHACHMAN, H.K. (1989). Analysis of the ligand-promoted global conformational change in aspartate transcarbamoylase: Evidence for a two-state transition from boundary spreading sedimentation velocity experiments. *Journal of Molecular Biology* **206**, 221–230.

WERNER, W.E., CANN, J.R. AND SCHACHMAN, H.K. (1989). Boundary spreading in sedimentation velocity experiments on partially liganded aspartate transcarbamoylase: A ligand-mediated isomerization. *Journal of Molecular Biology* **206**, 231–237.

WESTKAEMPER, R.B. AND RICHARD, A.J. (1987). Effects of diadenosine 5′, 5″-P1,P4-tetraphosphate and adenosine 5′-triphosphate on the higher order structure of calf thymus DNA. *Biochemistry and Biophysics Research Communications* **143**, 617–623.

WHITE, G.A., WOOD, T. AND HILL, W.E. (1988). Probing the α-sarcin region of *Escherichia coli* 23S rRNA with a cDNA oligomer. *Nucleic Acids Research* **16**, 10 817–10 831.

WILHELM, P., FRIGUET, B., DJAVADI-OHANIANCE, L., PILZ, I. AND GOLDBERG, M.E. (1987). Epitope localization in antigen–monoclonal antibody complexes by small angle X-ray scattering: An approach to domain organization in the β_2

subunit of *Escherichia cli* tryptophan synthase. *European Journal of Biochemistry* **164**, 103–109.

WINGFIELD, P., PAYTON, M., TAVERNIER, J., BARNES, J., SHAW, A., ROSE, K., SIMONA, M.G., DEMCZUK, S., WILLIAMSON, K. AND DAYER, J.-M. (1986). Purification and characterization of human interleukin-1β expressed in recombinant *Escherichia coli*. *European Journal of Biochemistry* **160**, 491–497.

WINGFIELD, P., PAYTON, M., GRABER, P., ROSE, K., DAYER, J.-M., SHAW, A.R. AND SCHMEISSNER, J. (1987). Purification and characterization of human interleukin-1α produced in *Escherichia coli*. *European Journal of Biochemistry* **165**, 537–541.

WINGFIELD, P., GRABER, P., MOONEN, P., CRAIG, S. AND PAIN, R.H. (1988). The conformation and stability of recombinant-derived granulocyte-macrophage colony stimulating factors. *European Journal of Biochemistry* **173**, 65–72.

WINGFIELD, P.T., STAHL, S.J., PAYTON, M.A., VENKATESAN, S., MISRA, M. AND STEVEN, A.C. (1991). HIV-1 Rev expressed in recombinant *Escherichia coli*: Purification, polymerization and conformational properties. *Biochemistry* **30**, 7527–7534.

YAGER, T.D., MCMURRAY, C.T. AND VAN HOLDE, K.E. (1989). Salt-induced release of DNA from nucleosome core particles. *Biochemistry* **28**, 2271–2281.

YPHANTIS, D.A. (1964). Equilibrium ultracentrifugation of dilute solutions. *Biochemistry* **3**, 297–317.

YPHANTIS, D.A. (1987). Sedimentation equilibrium measurements of recombinant DNA derived human interferon γ. *Biochemistry* **26**, 5422–5427.

ZAPATA, G., VANN, W.F., AARONSON, W., LEWIS, M.S. AND MOOS, M. (1989). Sequence of the cloned *Escherichia coli* K1 CMP-*N*-acetylneuraminic acid synthetase gene. *Journal of Biological Chemistry* **264**, 14 769–14 774.

11
DNA Mismatch Repair Systems: Mechanisms and Applications in Biotechnology

MIROSLAV RADMAN AND ROBERT WAGNER

Institute Jacques Monod, Paris, France

Mismatch repair systems: Mechanisms and specificities

The detection of mispaired or unpaired bases in DNA is a central element of mutation avoidance (Radman *et al.*, 1979) genetic recombination (Jones, Wagner and Radman, 1987a) and speciation (Rayssiguier, Thaler and Radman, 1989). Mispaired or unpaired bases may arise as a result of replication errors, strand exchange between homologous but non-identical sequences or deamination of 5-methyl-cytosine (5-meC) to thymine. Both replication errors and deaminations of 5-meC are changes in DNA sequences which, if not rectified, will give rise to heritable mutations.

Mismatch repair systems can be divided in two broad classes on the basis of their mechanisms of action (for reviews, see Radman and Wagner, 1986; Claverys and Lacks, 1986; Modrich, 1991). The long patch mismatch repair (LPMR) systems act by removing a number of nucleotides (often 1000–3000) from one strand of a DNA double helix in the region of a mispaired or unpaired base (Wildenberg and Meselson, 1975; Wagner and Meselson, 1976; Mejean and Claverys, 1984; Brooks *et al.*, 1989). The resulting gap is then filled by DNA synthesis (Lahue, Au and Modrich, 1989; Brooks *et al.*, 1989). Short patch mismatch repair (SPMR) systems act by removing only the mismatched nucleotide or the mismatched nucleotide and a few surrounding nucleotides (Lieb, 1983; Jones, Wagner and Radman, 1987a,c; Radicella, Clark and Fox, 1988; Au *et al.*, 1988, 1989; Hennecke *et al.*, 1992).

Long patch mismatch repair systems are often referred to as 'generalized' mismatch repair systems because they act to repair many of the different mismatches formed as replication errors, i.e. mismatches formed as a result of errors in the polymerase nucleotide selection process which escape detection by the polymerase proofreading system (Radman and Wagner, 1986; Jones, Wagner and Radman, 1987b). Long patch mismatch repair systems also

---

Abbreviations: dGTP-ase, deoxyguanosine triphosphate; LPMR, long patch mismatch repair; 5-meC, 5-methyl-cytosine; SPMR, short patch mismatch repair.

Biotechnology and Genetic Engineering Reviews — Vol. 11, December 1993
0264–8725/93/11/357–365 $20.00 + $0.00 © Intercept Ltd, P.O. Box 716, Andover, Hampshire SP10 1YG, UK

recognize and repair small (1–4 base) addition and deletion errors (Dohet, Wagner and Radman, 1986; Gasc, Sicard and Claverys, 1989; Parker and Marinus, 1992). Except for those systems requiring recombinational interactions between homologous DNA duplexes, any mismatch repair system whose mission is the correction of replication errors must be able to discriminate newly synthesized strands from parental strands and act specifically to preserve the sequence of the latter. Such strand discrimination is accomplished in *Escherichia coli* and some other bacteria by restricting LPMR to regions of DNA where one strand is undermethylated (Wagner and Meselson, 1976) with respect to adenine methylation of GATC sequences (Radman *et al.*, 1980; Pukkila *et al.*, 1983). The adenines in GATC sequences in non-replicating *E. coli* DNA are normally methylated (Marinus and Morris, 1973). This methylation inhibits the *E. coli* LPMR system (Radman *et al.*, 1980; Pukkila *et al.*, 1983). However, adenine methylation lags somewhat behind replication such that, immediately behind the replication fork, newly synthesized strands are transiently undermethylated (Lyons and Schendel, 1984). It is in this region that LPMR can act to remove replication errors. Another basis of strand discrimination in bacteria and in other organisms may be the discontinuities in newly synthesized DNA strands (Claverys and Lacks, 1986; Laengle-Rouault, Maenhaut-Michel and Radman, 1987; Lahue, Su and Modrich, 1987; Holmes, Clark and Modrich, 1990; Thomas, Roberts and Kunkel, 1991). In any event, it appears that LPMR systems generally act in the vicinity of the replication fork, although it has been shown that they do not require replication for their action (Radman and Wagner, 1986).

Short patch mismatch repair systems, which are considered to be 'specialized' mismatch repair systems, can act on non-replicating DNA to repair specific mismatches in a directed fashion (for a review, see Modrich, 1991). Two such systems have been identified in *E. coli*. One repairs only those G:T mismatches which result, or which could have resulted, from the deamination of 5-meC (Lieb, 1983; Jones, Wagner and Radman, 1987a,c; Zell and Fritz, 1987). Thus, this SPMR system acts only on a specific base (T) of a specific mismatch (G:T) at a specific position (the cytosine methylation position) in a specific sequence (a tetranucleotide sequence derived from the pentanucleotide cytosine methylase recognition sequence, CCWGG) (Lieb, Allen and Reid, 1986).

The other *E. coli* SPMR system acts on G:A mismatches to repair them to G:C (Au *et al.*, 1988; Radicella, Clark and Fox, 1988; Tsai-Wu, Radicella and Lu, 1991). This system apparently evolved because many G:A mismatches are not well recognized by the LPMR system (Dohet, Wagner and Radman, 1985; Jones, Wagner and Radman, 1987b), even though G:A mismatches are among the most common mistakes of DNA polymerases (Fersht and Knill-Jones, 1983). Because SPMR systems are not strand-directed, only those G:A mismatches in which A is on the newly synthesized strand will be repaired in such a way as to remove the replication error. In those cases where A is on the template strand, the action of this system would fix the mutation. However, the dGTPase activity of the polymerase-associated MutT protein (Akiyama *et al.*, 1989; Bhatnagar and Bessman, 1988) may function to prevent the formation of G:template-A mismatches during replication and, thus, eliminate them as potential substrates for the SPMR system. Moreover, the MutT protein is likely to be involved in prevention of mutagenic incorporation of oxidatively damaged substrates for DNA synthesis such as the 8-oxoguanine triphosphate (Maki and Sekiguchi, 1992), whereas the MutY protein may prevent adenine misincorporation opposite 8-oxoG in the template (Michaels *et al.*, 1992).

The existence of mismatch repair was originally postulated to account for genetic marker effects in the studies of genetic recombination (Holliday, 1964; Ephrussi-Taylor and Gray, 1966) and it has been found that elements of both LPMR and SPMR do have a significant impact on the outcome of genetic recombination events (Jones, Wagner and Radman, 1987a). A key intermediate in the process of genetic recombination is a hybrid or heteroduplex overlap – a region of DNA in which one strand of the double helix is derived from each of the two recombining molecules (see *Figure 1*). It is this hybrid overlap that aligns recombining molecules and assures that recombination will occur only between homologous sequences. Mismatches will be formed whenever sequences that are not completely identical are included within the hybrid overlap (*Figure 1*). Whenever mismatches within hybrid overlaps are substrates for SPMR, they will be repaired according to the rules of the particular system involved (see *Figure 1*). It is this type of repair that is responsible for the elevated recombination of close markers (high negative interference) (Lieb, 1983; Jones, Wagner and Radman, 1987c; Meselson, 1988; Raposa and Fox, 1987; Dzidic and Radman, 1989) and will generally result in diversification of related, but non-identical, sequences (*Figure 1*).

Initially, it was not at all clear what, if any, role LPMR plays in genetic recombination. In fact, it was suggested that LPMR does not operate at all on recombinational intermediates (Lieb, 1981, 1983; Gussin, Rosen and Wulff, 1981). Such might be the case if the mismatch repair process was completely inhibited by methylation of GATC sequences. However, it has been demonstrated that, at least *in vitro*, LPMR proceeds sequentially and only breaking of the DNA strand containing the incorrect base, one of the later steps, requires unmethylated GATC sequences (Laengle-Rouault, Maenhaut-Michel and Radman, 1987; Au, Welsh and Modrich, 1992). Thus, the early steps of LPMR, including mismatch recognition, will occur even in fully methylated DNA. Given the structure of recombinational intermediates, it may be that the strand discontinuities in the donor DNA preceding and following the donor strand exchange serve to direct LPMR to the incoming strand and result in the strand being removed whenever it creates mismatches with the recipient strand (*Figure 1*). Thus, LPMR will prevent recombination between non-homologous sequences by 'aborting' the recombination process or by destroying the heteroduplex intermediate (Jones, Wagner and Radman, 1987c; Rayssiguier, Thaler and Radman, 1989; Shen and Huang, 1989; see *Figure 1*).

The 'editing' of recombinational intermediates and suppression of recombination by LPMR can account for the stability of eukaryotic genomes in which there are hundreds of thousands of dispersed repeated genetic elements.

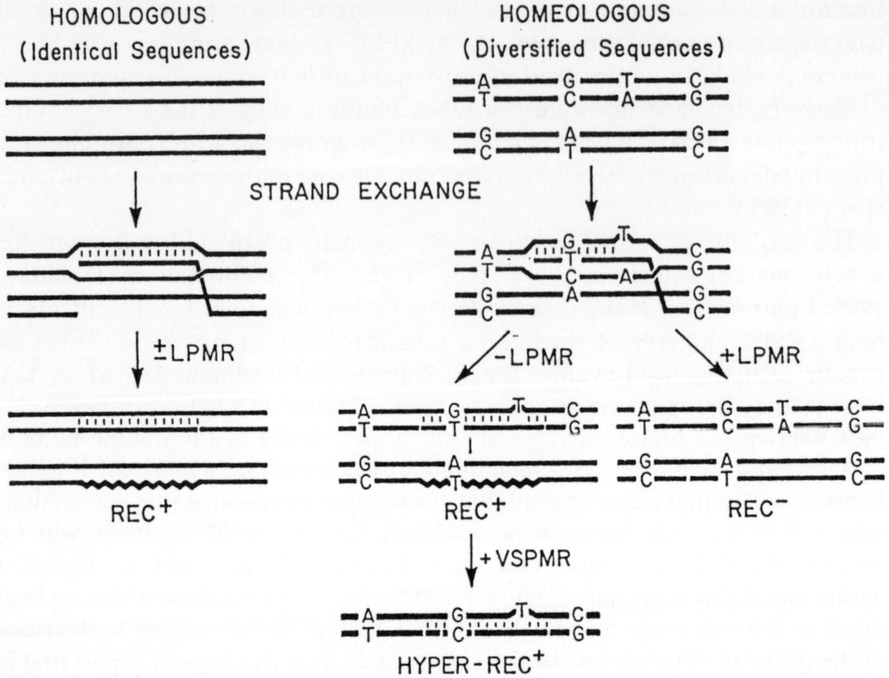

Figure 1. Specific prevention of recombination between partially homologous (homeologous) DNAs by the LPMR systems and hyper-recombination of specific allelic point mutations by the SPMR systems. LPMR proteins act by aborting the mismatched intermediates (hybrid DNA joints) generated by recombination proteins, thus creating the 'REC-' phenotype. Under LPMR-deficient conditions, the barrier to homeologous recombination is broken, allowing for inter-repeat and inter-species recombination (see text). An active SPMR (or VSPMR) can further process the mismatched intermediate to generate fine structure diversity at the nucleotide sequence level (HYPER-REC phenotype).

However, the repeated elements are sufficiently diverged (2–15%) that hybrid overlaps formed between them will contain mismatches and therefore, be substrates for action by a LPMR system. Long patch mismatch repair can prevent homologous interactions between repeated DNA sequences (Petit *et al.*, 1991) and between genomes of closely related species (Rayssiguier, Thaler and Radman, 1989), thereby protecting an organism against deleterious chromosomal rearrangements and acting as an effective genetic barrier between related species (Radman, 1991). *Escherichia coli* mutants deficient in mismatch detection proteins (MutS and MutL) show an increased rate of duplication and deletion formation as a result of recombination between copies of repeated, diverged sequences (Petit *et al.*, 1991). In addition, mutations which suppress LPMR allow interspecies *E. coli*/*Salmonella typhimurium* recombination at about 10% of intraspecies levels (*mut*+ bacteria allow only about 0.001% of intraspecies levels) (Rayssiguier, Thaler and Radman, 1989; Rayssiguier, Dohet and Radman, 1991). This high efficiency of hybrid DNA formation involving non-identical sequences provides substrates for SPMR and thus can result in high recombination frequencies (Jones, Wagner and Radman, 1987c).

Implications and applications of mismatch repair systems in genetic engineering and biotechnology

The ability to control the fidelity of DNA replication and recombination processes by controlling LPMR and SPMR opens new doors in biotechnology, genetic engineering and medicine. Not only is it now possible to use powerful mutator mutants (i.e. mutants deficient in LPMR and SPMR) to alter DNA sequences rapidly but randomly, but it is also possible to generate new genes and even new species by allowing heretofore forbidden recombination between related genes and species. Recombination between diverged DNA seuqences coding for proteins of similar function can be facilitated under LPMR-deficient conditions to allow 'protein engineering' analogous to the strategies employed in nature to generate diversity in the immune (Reynaud *et al.*, 1987), histocompatibility (Jaulin *et al.*, 1985) and parasite surface antigen (Thon, Baltz and Eisen, 1989) systems (*Figure 2*). In these systems, partially homologous DNA sequences act as donor sequences to a single expressed 'mosaic' gene. In addition, in yeast plasmid transformation experiments, recombination between diverged genes produced mosaic genes which coded for enzymes with altered specificities (Mezard, Pompon and Nicolas, 1992; Goguel, Delahodde and Jacq, 1992). By selecting appropriate donor sequences, it should be possible to set up similar systems of directed evolution in LPMR-deficient bacteria to create 'new' genes coding for new or improved proteins. Because the initial donor sequences can be selected by virtue of their ability to code for functional proteins, most recombinants will presumably be functional, an enormous advantage over current systems of random mutagenesis. Thus, it will be necessary only to select or screen for the desired alteration (e.g. improved stability, increased activity or altered specificity) from among a population of bacterial (or phage) clones coding for functional proteins. Recombinational repair between broken diverged (homeologous) genes is frequent in yeast plasmid transformation experiments, in which it was demonstrated that the resulting mosaic genes encode for enzymes with altered specificities (Mezard, Pompon and Nicolas, 1992; Goguel, Delahodde and Jacq, 1992).

The process of creating new genes can perhaps be made even more efficient by performing some of the steps *in vitro*. For example, heteroduplex DNA can be formed *in vitro* between two diverged genes (Abastado *et al.*, 1984) and be used to transfect LPMR-deficient *E. coli*. The heteroduplexes will then be substrates for SPMR, which will act to generate 'patchwork' sequences, which will represent increased repertoires of recombination products between the two genes.

Inhibition of LPMR will allow cloning of partially homologous genes from diverse species. By preparing a 'probe' sequence from a previously cloned gene of identical function to the desired gene (and, therefore, presumably partially homologous to it), it will be possible, by means of recombination between partially homologous sequences, to extract efficiently the desired gene from a library of genes of a given species. Obviously, such a procedure is not restricted to coding sequences, but may be used to obtain any partially

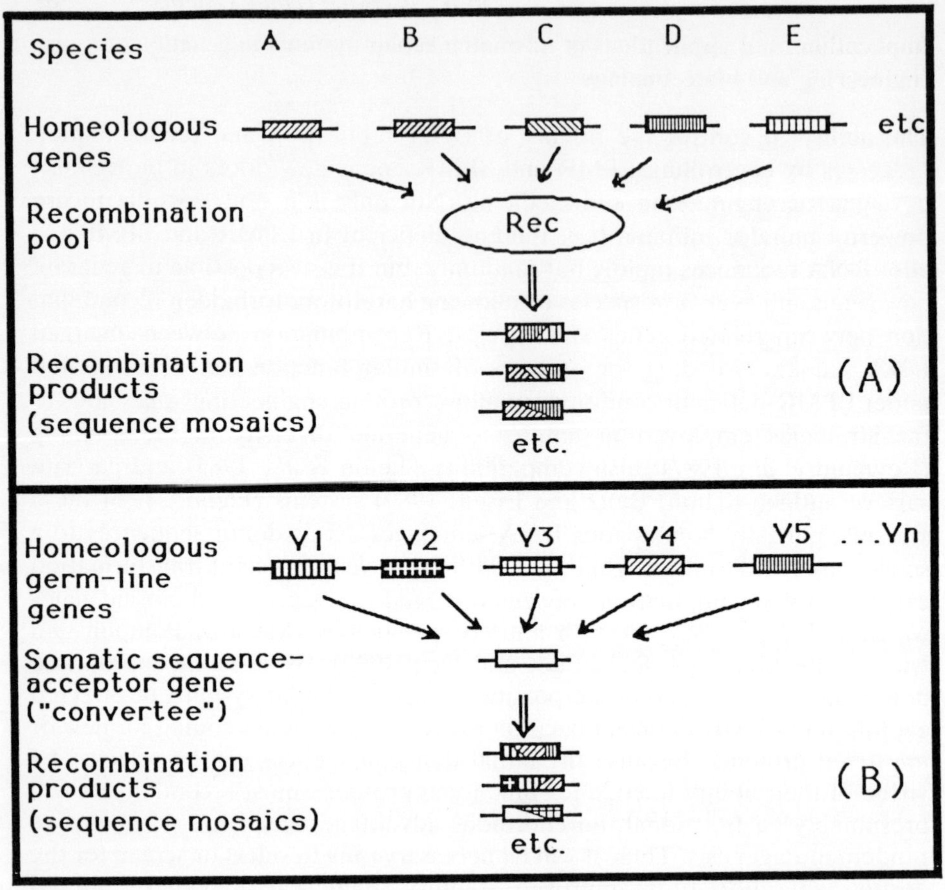

Figure 2. 'Instant evolution' of genes and the encoded proteins by maximized homeologous recombination in the absence of LPMR and with active SPMR systems (see *Figure 1*). Box B presents an oversimplified version of natural systems generating protein diversity through gene conversion processes (see text). Box A illustrates how one can mimic such diversification processes by homeologous recombination of cloned genes in bacteria that are LPMR–, SPMR+.

homologous sequences from any species. A similar method was previously developed, but did not employ LPMR-deficient bacteria and, therefore, was inefficient due to the high fidelity of genetic recombination in wild-type *E. coli* (Seed, 1983).

The same principle can serve as a basis for gene inactivation. An inactive but partially homologous sequence to the gene in question can be constructed and, under LPMR-deficient conditions, allowed to recombine with the target gene to inactivate it. The ability to inactivate genes selectively can be useful in gene mapping and may eventually have diagnostic or therapeutic applications, particularly as it becomes possible to inhibit, perhaps transiently, LPMR in higher organisms. If, as suggested by Radman (1988, 1989, 1991), mitotic recombination is prevented by the differential sequence polymorphisms carried by the two homologous chromosomes, then gene replacement, gene inactivation or 'gene therapy' by homologous recombination with exogenous DNA may also be prevented by the sequence polymorphism carried by the target plant, animal or human cells. Using cloned 'isogenic'

DNA sequences (from the same genetic background as the recipient cells) (Te Riele, Maandag and Berns, 1992) or inactivating mismatch detection and/or repair proteins may be necessary for efficient gene 'targeting' or 'therapy'.

The ability to remove genetic barriers between species by allowing them to recombine under LPMR-deficient conditions, as has been demonstrated in *Escherichia–Salmonella* crosses (Rayssiguier, Thaler and Radman, 1989), means that it is now possible to create new species of micro-organisms. The potential applications appear almost limitless. Attenuated live vaccines can be created by selectively recombining pathogenic bacteria, such as *Salmonella* and *Shigella*, with *E. coli* to create non-pathogenic strains that retain the immunogenicity of the pathogens. It may also be possible to create universal salmonella-based vectors for any antigen presentation which could be used as oral vaccines illiciting mucosal immunity (McGhee *et al.*, 1992).

Finally, it has been demonstrated that the mismatch-binding proteins, such as the bacterial MutS protein, can be used to detect substitution and small addition/deletion mutations. Such applications have wide ranging potentials for diagnosing human hereditary and somatic genetic diseases and in the study of polymorphism in human and other genomes (R. Wagner and M. Radman, unpublished).

Acknowledgements

M.R. was supported by grants from INSERM and the Ministry of Science and Technology of France (MRT, Section Biotechnologies).

References

ABASTADO, J.-P., CAMI, B., HUYNH DINH, T., IGOLEN, J. AND KOURILSKY, P. (1984). Processing of complex heteroduplexes in *Escherichia coli* and *Cos-1* monkey cells. *Proceedings of the National Academy of Sciences, USA* **81**, 5792–5796.

AKIYAMA, M., MAKI, H., SEKIGUCHI, M. AND HORIUCHI, T. (1989). A specific role of MutT protein; to prevent dG.dA mispairing in DNA replication. *Proceedings of the National Academy of Sciences, USA* **86**, 3949–3952.

AU, K.G., WELSH, K. AND MODRICH, P. (1992). Initiation of methyl-directed mismatch repair. *Journal of Biological Chemistry* **267**, 12 142–12 148.

AU, K.G., CABRERA, M., MILLER, J.H. AND MODRICK. P. (1988). *Escherichia coli mutY* gene product is required for specific A-G to C-G mismatch correction. *Proceedings of the National Academy of Sciences, USA* **85**, 9163–9166.

AU, K.G., CLARK, S., MILLER, J.H. AND MODRICH, P. (1989). The *Escherichia coli* mutY gene encodes an adenine glycosylase active on G-A mispairs. *Proceedings of the National Academy of Sciences, USA* **86**, 8877–8881.

BHATANAGAR, S.K. AND BESSMAN, M.J. (1988). Studies on the mutator gene, *mutT* of *Escherichia coli*: molecular cloning of the gene, purification of the gene product, and identification of a novel nucleoside triphosphatase. *Journal of Biological Chemistry* **263**, 8953–8957.

BROOKS, P., DOHET, C., ALMOUZNI, G., MÉCHALI, M. AND RADMAN, M. (1989). Mismatch repair involving localised DNA synthesis in Xenopus egg extracts. *Proceedings of the National Academy of Sciences, USA* **86**, 4425–4429.

CLAVERYS, J.-P. AND LACKS, S.L. (1986). Heteroduplex deoxyribonucleic acid base mismatch repair in bacteria. *Annual Review of Microbiology* **50**, 133–165.

DOHET, C., WAGNER, R. AND RADMAN, M. (1985). Repair of defined single base pair mismatches in *E. coli*. *Proceedings of the National Academy of Sciences, USA* **82**, 503–505.

DOHET, C., WAGNER, R. AND RADMAN, M. (1986). Methyl-directed repair of frameshift mutations in heteroduplex DNA. *Proceedings of the National Academy of Sciences, USA* **83**, 3395–3397.

DZIDIC, S. AND RADMAN, M. (1989). Genetic requirements for hyperrecombination by very short patch mismatch repair: involvement of *E. coli* DNA polymerase I. *Molecular and General Genetics* **217**, 254–256.

EPHRUSSI-TAYLOR, H. AND GRAY, T.C. (1966). Genetic studies of recombining DNA in pneumococcal transformation. *Journal of General Physiology* **49**, 211–231.

FERSHT, A.R. AND KNILL-JONES, J.W. (1983). Contribution of 3′→5′ exonuclease activity of DNA polymerase III holoenzyme from *Escherichia coli* to specificity. *Journal of Molecular Biology* **165**, 669–682.

GASC, A.M., SICARD, M. AND CLAVERYS, J.-P. (1989). Repair of single- and multiple-substitution mismatches during recombination in *Streptococcus pneumoniae*. *Genetics* **121**, 29–36.

GOGUEL, V., DELAHODDE, A. AND JACQ, C. (1992). Connections between RNA splicing and DNA intron mobility in yeast mitochondria: RNA maturase and DNA endonuclease switching experiments. *Molecular and Cellular Biology* **12**, 696–705.

GUSSIN, G.N., ROSEN, E.D. AND WULFF, D.L. (1981). Mapability of very close markers of bacteriophage lambda. *Genetics* **96**, 1–24.

HENNECKE, F., KOLMAR, H., BRUENDEL, K. AND FRITZ, H.-J. (1992). The vsr gene product of *E. coli* K-12 is a strand- and sequence-specific DNA mismatch endonuclease. *Nature* **353**, 776–778.

HOLLIDAY, R. (1964). A mechanism for gene conversion in fungi. *Genet Res* **5**, 282–304.

HOLMES, J.J., CLARK, S. AND MODRICH, P. (1990) Strand-specific mismatch correction in nuclear extracts of human and *Drosophila melanogaster* cell lines. *Proceedings of the National Academy of Sciences, USA* **87**, 5837–5841.

JAULIN, C., PERRIN, A., ABASTADO, J.-P., DUMAS, B., PAPAMETHEAKIS, J. AND KOURILSKY, P. (1985). Polymorphism in mouse and human class I *H-2* and *HLA* genes is not the result of random independent point mutations. *Immunogenetics* **22**, 453–470.

JONES, M., WAGNER, R. AND RADMAN, M. (1987a). Mismatch repair of deaminated 5-methyl-cytosine. *Journal of Molecular Biology* **194**, 155–159.

JONES, M., WAGNER, R. AND RADMAN, M. (1987b). Repair of a mismatch is influenced by the base composition of the surrounding nucleotide sequence. *Genetics* **115**, 605–610.

JONES, M., WAGNER, R. AND RADMAN, M. (1987c). Mismatch repair and recombination in *E. coli*. *Cell* **50**, 621–626.

LAENGLE-ROUAULT, F., MAENHAUT-MICHEL, G. AND RADMAN, M. (1987). GATC sequences, DNA nicks and the mutH function in *Escherichia coli* mismatch repair. *EMBO Journal* **6**, 1121–1127.

LAHUE, R.S., AU, K.G. AND MODRICH, P. (1989). DNA mismatch correction in a defined system. *Science* **245**, 160–164.

LAHUE, R.S., SU, S.-S. AND MODRICH, P. (1987). Requirement for d(GATC) sequences in *Escherichia coli mutHLS* mismatch correction. *Proceedings of the National Academy of Sciences, USA* **84**, 1482–1486.

LIEB, M. (1981). A fine structure map of spontaneous and induced mutations in the lambda repressor gene, including insertions of IS elements. *Molecular and General Genetics* **184**, 364–371.

LIEB, M. (1983). Specific mismatch correction in bacteriophage lambda crosses by very short patch repair. *Molecular and General Genetics* **191**, 118–125.

LIEB, M., ALLEN, E. AND REID, D. (1986). Very short patch mismatch repair in phage lambda: repair sites and length of repair tracts. *Genetics* **114**, 1041–1060.

LYONS, S. AND SCHENDEL, P.F. (1984). Kinetics of methylation in *Escherichia coli* K-12. *Journal of Bacteriology* **159**, 421–423.

MAKI, H. AND SEKIGUCHI, M. (1992). MutT protein specifically hydrolyses a potent mutagenic substrate for DNA synthesis. *Nature* **355**, 273–275.

MARINUS, M.G. AND MORRIS, N.R. (1973). Isolation of deoxyribonucleic acid methylase mutants of *Escherichia coli K-12*. *Journal of Bacteriology* **114**, 1143–1150.

MCGHEE, J.R., MESTECKY, J., DERTZBAUGH, M.T., ELDRIDGE, J.H., HIRASAVA, M. AND KIYONO, H. (1992). The mucosal immune system: from fundamental concepts to vaccine development. *Vaccine* **10**, 75–88.

MEJEAN, V. AND CLAVERYS, J.-P. (1984). Effect of mismatched base pairs on the fate of donor DNA in transformation of *Streptococcus pneumoniae*. *Molecular and General Genetics* **197**, 467–471.

MESELSON, M. (1988). Methyl-directed repair of DNA mismatches. In *Recombination of the Genetic Material* (K.B. Low, Ed.), pp. 91–113. Academic Press, San Diego, CA.

MEZARD, C., POMPON, D. AND NICOLAS, A. (1992). Recombination between similar but not identical DNA sequences during yeast transformation occurs within short stretches of identity. *Cell* **70**, 659–670.

MICHAELS, M.L., CRUZ, C., GROLLMAN, A.P. AND MILLER, J.H. (1992). Evidence that MutY and MutM combine to prevent mutations by an oxidatively damaged form of guanine in DNA. *Proceedings of the National Academy of Sciences, USA* **89**, 7022–7025.

MODRICH, P. (1991). Mechanisms and biological effects of mismatch repair. *Annual Review of Genetics* **25**, 229–253.

PARKER, B.O. AND MARINUS, M.G. (1992). Repair of DNA heteroduplexes containing small heterologous sequences in *Escherichia coli*. *Proccedings of the National Academy of Sciences, USA* **89**, 1730–1734.

PETIT, M.-A., DIMPFL, J., RADMAN, M. AND ECHOLS, H. (1991). Control of chromosomal rearrangements in *E. coli* by the mismatch repair system. *Genetics* **129**, 327–332.

PUKKILA, P.J., PETERSON, J., HERMAN, G., MODRICH, P. AND MESELSON, M. (1983). Effects of high levels of DNA adenine methylation on methyl-directed mismatch repair in *Escherichia coli*. *Genetics* **104**, 571–582.

RADICELLA, J.P., CLARK, E.A. AND FOX, M.S. (1988). Some mismatch repair activities in *Escherichia coli*. *Proceedings of the National Academy of Sciences, USA* **85**, 9674–9678.

RADMAN, M. (1988). Mismatch repair and genetic recombination. In *Genetic Recombination* (R. Kutcherlapati and G.R. Smith, Eds), pp. 169–192. American Society for Microbiology, Washington, DC.

RADMAN, M. (1989). Mismatch repair and the fidelity of genetic recombination. *Genome* **31**, 68–73.

RADMAN, M. (1991). Avoidance of inter-repeat recombination by sequence divergence and a mechanism of neutral evolution. *Biochimie* **73**, 357–361.

RADMAN, M. AND WAGNER, R. (1986). Mismatch repair in *Escherichia coli*. *Annual Review of Genetics* **20**, 523–538.

RADMAN, M., VILLANI, G., BOITEUX, S., KINSELLA, A.R., GLICKMAN, B.W. AND SPADARI, S. (1979). Replication fidelity: mechanisms of mutation evoidance and mutation fixation. *Cold Spring Harbor Symposia on Quantum Biology* **43**, 937–946.

RADMAN, M., WAGNER, R.E., GLICKMAN, B.W. AND MESELSON, M. (1980). DNA methylation, mismatch correction and genetic stability. In *Progress in Environmental Mutagenesis* (M. Alacevic, Ed.), pp. 121–130. Elsevier, Amsterdam.

RAPOSA, S. AND FOX, M.S. (1987). Some features of base pair mismatch and heterology repair in *Escherichia coli. Genetics* **117**, 381–390.

RAYSSIGUIER, C., DOHET, C. AND RADMAN, M. (1991). Interspecific recombination between *Escherichia coli* and *Salmonella typhimurium* occurs by the RecBCD pathway. *Biochimie* **73**, 371–374.

RAYSSIGUIER, C., THALER, D.S. AND RADMAN, M. (1989). The barrier to recombination between *Escherichia coli* and *Salmonella typhimurium* is disrupted in mismatch-repair mutants. *Nature* **342**, 396–401.

REYNAUD, C.-A., ANQUEZ, V., GRIMAL, H. AND WEILL, J.-C. (1987). A hyperconversion mechanism generates the chicken light chain preimmune repertoire. *Cell* **48**, 379–387.

SEED, B. (1983). Purification of genomic sequences from bacteriophage libraries by recombination and selection in vivo. *Nucleic Acids Research* **11**, 2427–2445.

SHEN, P. AND HUANG, H.V. (1989). Effect of base pair mismatches on recombination via the RecBCD pathway. *Molecular and General Genetics* **218**, 358–360.

TE RIELE, H., MAANDAG, E.R. AND BERNS, A. (1992). Highly efficient gene targeting in embryonic stem cells through homologous recombination with isogenic DNA constructs. *Proceedings of the National Academy of Sciences, USA* **89**, 5128–5132.

THOMAS, D.C., ROBERTS, J.D. AND KUNKEL, T.A. (1991). Heteroduplex repair in extracts of human HeLa cells. *Journal of Biological Chemistry* **266**, 3744–51.

THON, G., BALTZ, T. AND EISEN, H. (1989). Antigenic diversity by the recombination of pseudogenes. *Genes Dev* **3**, 1247–1254.

TSAI-WU, J.-J., RADICELLA, J.P. AND LU, A.-L. (1991). Nucleotide sequence of the *Escherichia coli micA* gene required for A/G-specific mismatch repair: identity of MicA and MutY. *Journal of Bacteriology* **173**, 1902–1910.

WAGNER, R. AND MESELSON, M. (1976).Repair tracts in mismatched DNA heteroduplexes. *Proceedings of the National Academy of Sciences, USA* **73**, 4135–4139.

WILDENBERG, J. AND MESELSON, M. (1975). Mismatch repair in heteroduplex DNA. *Proceedings of the National Academy of Sciences, USA* **72**, 2202–2206.

ZELL, R. AND FRITZ, H.-J. (1987). DNA mismatch-repair in *Escherichia coli* counteracting the hydrolytic deamination of 5-methylcytosine residues. *EMBO Journal* **6**, 1809–1815.

Index

Biotechnology & Genetic Engineering Reviews VOLUME 4 (September 1986); ISBN 0 946707 07 3; ISSN 0264–8725; vii + 436 pp; 234 × 156 mm; illustrated; comprehensive index; hardback.

CONTENTS

Biotechnology & Genetic Engineering Reviews VOLUME 5 (September 1987); ISBN 0 946707 10 3; ISSN 0264–8725; x + 419 pp; 234 × 156 mm; illustrated; comprehensive index; hardback. Retail Price £80.00.

CONTENTS

Biotechnology & Genetic Engineering Reviews VOLUME 6 (September 1988); ISBN 0 946707 12 X; ISSN 0264–8725; x + 457 pp; 234 × 156 mm; illustrated; comprehensive index; hardback.

CONTENTS

Biotechnology & Genetic Engineering Reviews VOLUME 7 (December 1989); ISBN 0 946707 22 7; ISSN 0264–8725; xii + 324 pp; 234 × 156 mm; illustrated; comprehensive index; hardback.

CONTENTS

Biotechnology & Genetic Engineering Reviews VOLUME 8 (November 1990); ISBN 0 946707 32 4; ISSN 0264–8725; xiii + 410 pp; 234 × 156 mm; illustrated; comprehensive index; hardback.

CONTENTS

Biotechnology & Genetic Engineering Reviews VOLUME 9 (December 1991); ISBN 0 946707 39 1; ISSN 0264–8725; x + 425 pp; 234 × 156 mm; illustrated; comprehensive index; hardback.

CONTENTS

Biotechnology & Genetic Engineering Reviews VOLUME 10 (December 1992); ISBN 0 946707 39 1; ISSN 0264–8725; xi + 510 pp; 234 × 156 mm; illustrated; comprehensive index; hardback.

CONTENTS